Plant-Pest Interactions: From Molecular Mechanisms to Chemical Ecology

Indrakant K. Singh • Archana Singh
Editors

Plant-Pest Interactions: From Molecular Mechanisms to Chemical Ecology

Springer

Editors
Indrakant K. Singh
Molecular Biology Research Lab
Deshbandhu College, University of Delhi
New Delhi, India

Archana Singh
Department of Botany
Hans Raj College, University of Delhi
New Delhi, India

ISBN 978-981-15-2466-0 ISBN 978-981-15-2467-7 (eBook)
https://doi.org/10.1007/978-981-15-2467-7

This Springer imprint is published by the registered company Springer Nature Singapore Pte Ltd.
The registered company address is: 152 Beach Road, #21-01/04 Gateway East, Singapore 189721, Singapore

Foreword

Technical reference books on entomology and botany are readily available, but the integration of these two disciplines into the field of insect–plant interactions is relatively recent, with far fewer comprehensive volumes devoted to the subject. It is, therefore, a pleasure to discover a new book exploring this rapidly growing area of research. The editors of *Plant Interactions with Insects* have compiled a diverse array of contributions ranging from behavior and ecology to the molecular aspects of plant defense and signaling cascades, insect adaptations, and tri-trophic interactions.

From the initial historical overview through chapters on proteomics, transcriptomics, miRNA, phytohormones, herbivore-associated elicitors, volatiles, mitogen-activated protein kinases, proteinase inhibitors, microbial influences, and evolution, this volume provides an up-to-date compendium of references coupled with clear explanations of new molecular approaches. Older, classic books on plant–animal relationships focus primarily at the organismal level to explore behavior and co-evolutionary interactions. In contrast, *Plant Interactions with Insects* emphasizes biochemical and genetic aspects of the ecological relationships between plants and insects and the potential of these research techniques for agriculture. The need for new approaches to insect pest management is acute; reliance on traditional insecticides is unsustainable and insect resistance to modern crops genetically modified to express Bt endotoxins is becoming a serious concern. This book will be valuable in spurring further research into the practical application of molecular techniques to the chemical ecology of insect–crop plant interactions.

Several contributors are former members of the Insect Behavior Lab directed by Professor Ashok Kumar Singh at the University of Delhi. I was fortunate to have the opportunity to conduct research in this lab, collaborating with Professor Singh both as a Fulbright Scholar in India (2001/2) and a University of Delhi Fellow (2005/6). This volume provides evidence of Professor Singh's lasting influence on the study of phytophagous insect behavior and on the current generation of researchers in this field, and I am honored to be asked to write this Foreword. I congratulate the editors and all the authors for their successful efforts in producing an excellent addition to the literature.

Professor Emerita of Biology, Winthrop University Paula Levin Mitchell
Rock Hill, SC, USA
24 November 2019

Preface

Plants being the producers of food have been constantly under attack by insects. During evolution, plants have not just developed different mechanisms to defend themselves but have also refined the ways of implementing them to maintain the ecological balance. During insect attack, first of all effector molecules such as herbivore-associated molecular patterns (HAMPs) are perceived by plant recognition receptors, after which the receptors interact with additional trans-membrane proteins that act as signaling adapters or amplifiers to achieve defense response. Phytohormones, such as Jasmonic acid (JA), Ethylene (ET), Abscisic Acid (ABA), and Salicylic acid (SA), also take part in the signaling cascade to strengthen the signal and transport it systemically. Further, cross talk among signaling molecules, Ca^+ ions, enzymes, and genetic factors allows the plants to undergo transcriptional reprogramming and proteomic alterations, which induces the synthesis of proteins responsive against the pest and prevents further damage at the wounded site. Plants also develop biochemical mechanisms such as synthesis of insecticidal proteins (proteinase inhibitors, Cry, lectins) and synthesis of secondary metabolites (phenols, terpenoids, alkaloid) that are toxic and interfere with the digestive system of the infesting insects. The production of volatile compounds is known to be another approach towards defense. These compounds are produced to attract the predators to repel the herbivores and to communicate with the neighboring plants to trigger their defenses against the pathogens. This book is an attempt to portray various strategies employed by plants to ensure their continued survival and success against insects. Since only few books are available on this theme, this proposed book will bridge the gap about the latest information and will make it available to a wide range of audience. Overall, this volume will deliver an overview of our current knowledge on insect–plant interactions and it is mandatory to understand this process for the genetic improvement of crops. This book is conceived for researchers and professionals in the field of agriculture, plant pathology, entomology, cell biology, molecular biology, and genetics.

We present this book with an objective to provide most recent information on basic as well as the most advanced insights on molecular mechanism of plant–insect Interactions and chemical ecology. The book has 18 chapters to broadcast the most updated information and detailed overviews on insect–plant interactions. The contents of the book offer a complete pictorial illustration on the subject, starting from the perception of the insects to the molecular and biochemical alterations that occur in plants during insect infestation. This book is an essential reading for researchers and professionals in Molecular Ecology, Genomics, Plant Pathology, and Entomology. Overall this volume will convey an overview of insect–plant interactions, which will be helpful in understanding this process for the genetic improvement of crops.

We are grateful to the authors of various chapters of this book for writing their chapters methodically and with great responsibility. Our heartful acknowledgment goes to Prof. Ashok K. Singh and Dr. Praveen K Verma, our mentors, for educating us about this amazing world of insect–plant interactions. We are extremely thankful to Dr. Rama, Principal, Hans Raj College, University of Delhi, and Dr. Rajiv Aggrawal, Principal, Deshbandhu College, University of Delhi, for providing overall support for our research and academic pursuits. We convey our gratitude to Dr. Paula Mitchell for useful comments and unconditional support in our academic endeavors. We would like to extend our sincere thanks to Dr. V. K. Kawatra, Dr. P K Singh, Dr. Shalendra S Chauhan, and Mr. Jagmohan Kaushik for their encouragement and support. We appreciate the beautiful ambiance created by our little angels Saumya and Kimaya, which allowed us to work tirelessly and gave us all emotional support. We are grateful to our parents for their constant support and blessings. The editors acknowledge Science and Engineering Research Board (SERB), DST, Ministry of Science and technology, Govt. of India, New Delhi, India, for the financial supports given to IKS and AS [SB/YS/LS-59/2013 (IKS); ECR/2017/002478 (AS)]. Last but not the least, our sincere thanks to handling editors and publisher.

We are optimistic that this book will be effective in conveying the latest knowledge on insect–plant interactions.

New Delhi, India

Indrakant K. Singh
Archana Singh

Contents

Editors and Contributors

About the Editors

Indrakant K. Singh received his Ph.D. in Zoology from the University of Delhi (DU) and is currently working as an Assistant Professor in the Department of Zoology, Deshbandhu College, University of Delhi. Dr. Indrakant has received Young Scientist Scheme Award from the SERB, Ministry of Science and Technology, Govt. of India; Raman Post-Doctoral Fellowship from the UGC, MHRD, Govt. of India; Max Planck Fellowship from the Max Planck Society, Germany; and **ICMR-DHR International Fellowships For Young Indian Biomedical Scientist** from MoHFW, Govt. of India . Dr. Indrakant has published his research in prominent peer-reviewed journals such as *JXB* (Oxford), *Scientific Reports*, *IJOBioMAc*, and many more . His research interests include the molecular basis of plant–insect interactions and human health.

Archana Singh obtained her Ph.D. in PMB from the NIPGR, Delhi, India, and is currently working as an Assistant Professor in the Department of Botany, Hans Raj College, University of Delhi. She has received Young Scientist Award and Early Career Research Award from the SERB, Ministry of Science and Technology, Govt. of India; Raman Post-Doctoral Fellowship from the MHRD, Govt. of India; and an EMBO fellowship from EMBO, Heidelberg, Germany. She has published her research in prominent peer-reviewed journals. Over the years, she has expanded her research interests from PMB to genomics and proteomics.

Contributors

Qasim Ali Faculty of Forestry and Wood Sciences, EXTEMIT-K, Czech University of Life Sciences, Prague, Czech Republic

Rahul Arora Division of Biosciences, University College London, London, UK

Muhammad Binyameen Department of Entomology, Faculty of Agricultural Sciences and Technology, Bahauddin Zakariya University, Multan, Pakistan
Faculty of Forestry and Wood Sciences, EXTEMIT-K, Czech University of Life Sciences, Prague, Czech Republic

Amrita Chakraborty Faculty of Forestry and Wood Sciences, EVA 4.0, Czech University of Life Sciences, Prague, Czech Republic
Department of Plant Protection Biology, Swedish University of Agricultutral Sciences, Alnarp, Sweden

Ritu Chaturvedi St. John's College, D.B.R.A University, Agra, UP, India

Vibha Gulyani Checker Department of Botany, Kirori Mal College, University of Delhi, Delhi, New Delhi, India

Samira Chugh Department of Botany, Gargi College, Siri Fort, New Delhi, India

Arvind Gupta Department of Biotechnology, Hemvati Nandan Bahuguna Garhwal (Central) University, Srinagar Garhwal, Uttarakhand, India

Sumanti Gupta Department of Botany, Rabindra Mahavidyalaya, Champadanga, WB, India

Sunila Hooda Ram Lal Anand College, University of Delhi, Delhi, New Delhi, India

Jinu Jacob Biotechnology Division, ICAR-Indian Institute of Millets Research (IIMR), Rajendranagar, Telangana, India

Sengodan Karthi Division of Biopesticides and Environmental Toxicology, Manonmaniam Sundaranar University, Alwarkurichi, Tamil Nadu, India

Manjeet Kaur Department of Plant Sciences, School of Life Sciences, University of Hyderabad, Hyderabad, Telangana, India

Deepak Kumar Department of Botany, Central University of Jammu, Jammu, Jammu & Kashmi, India

Mohit Kumar Department of Zoology, University of Delhi, Delhi, New Delhi, India

Rakesh Kumar Lab 15/18, Insect-Plant Interactions Group. Department of Botany, University of Delhi, Delhi, New Delhi, India

Pragati Kumari Scientist Hostel-S-02, Srinagar Garhwal, Uttarakhand, India

Punam Kumari Department of Biosciences and Biotechnology, Fakir Mohan University, Odisha, India

P. Madhu ICRISAT, Genomics and Trait Discovery Theme-Genetic Gains Programme, Hyderabad, India

Garima Malik R.G. (PG) College, C.C.S University, Meerut, Uttar Pradesh, India

Vachaspati Mishra Bangalore Bioinnovation Centre, Bangalore, India

Anjana Singha Naorem Department of Zoology, Cotton University, Guwahati, Assam, India

R. Poorniammal Horticultural College and Research Institute, Periyakulam, Theni, India

Amit Roy Faculty of Forestry and Wood Sciences, EXTEMIT-K, Czech University of Life Sciences, Prague, Suchdol, Czech Republic

Anjana Rustagi Department of Botany, Gargi College, Siri Fort, New Delhi, India

Mahaswetta Saikia Lab 15/18, Insect-Plant Interactions Group. Department of Botany, University of Delhi, Delhi, New Delhi, India

A. R. Sakthi Tamil Nadu Agricultural University, Coimbatore, India

Fredrik Schlyter Faculty of Forestry and Wood Sciences, EXTEMIT-K, Czech University of Life Sciences, Prague, Czech Republic
Division of Chemical Ecology, Department of Plant Protection Biology, Swedish University of Agricultural Sciences, Alnarp, Sweden

C. Selvi Tamil Nadu Agricultural University, Coimbatore, India

Anupam Varshney Sharma Department of Zoology, Hindu College, University of Delhi, Delhi, New Delhi, India

Meenakshi Sharma Department of Botany, Daulat Ram College, University of Delhi, Delhi, New Delhi, India

Shweta Sharma Department of Botany, Maitreyi College, Chanakyapuri, New Delhi, India

Pallee Shree Department of Zoology, University of Delhi, Delhi, New Delhi, India

Archana Singh Department of Botany, Hans Raj College, University of Delhi, Delhi, New Delhi, India

Dileep K. Singh Department of Zoology, University of Delhi, Delhi, New Delhi, India

Indrakant K. Singh Molecular Biology Research Lab. Department of Zoology, Deshbandhu College, Kalkaji, University of Delhi, Delhi, New Delhi, India

Sujata Singh Molecular Biology Research Lab. Department of Zoology, Deshbandhu College, Kalkaji, University of Delhi, Delhi, New Delhi, India

Anupama Tiku Department of Botany, Ramjas College, University of Delhi, Delhi, New Delhi, India

R. Vinodh Biotechnology Division, ICAR-Indian Institute of Millets Research (IIMR), Rajendranagar, Telangana, India

Saurabh Yadav Department of Biotechnology, Hemvati Nandan Bahuguna Garhwal (Central) University, Srinagar Garhwal, Uttarakhand, India

Sunita Yadav Department of Botany, Sri Venkateswara College, University of Delhi, Delhi, New Delhi, India

Abbreviations

2D-DIGE	Two-dimensional difference gel electrophoresis
2D-PAGE	Two Dimensional Polyacrylamide Gel Electrophoresis
ABA	Abscisic acid
ABA	*Agaricus bisporus* agglutinin
ACE	Acetyl choline esterase
ACS	1-Aminocyclopropane-1-carboxylate synthase
ALP	Alkaline phosphatase
ALs	Antennal lobes
AMF	Arbuscular mychorrizal fungi
AmiRNAs	Artificial miRNAs
AMTL	*Amorphophallus paeonifolius* tuber agglutinin.
AOS	Allene oxide synthase
AP	Action potential
APN	Aminopeptidase
APX	Ascorbate peroxidase
ASAL	*Allium sativum* leaf agglutinin
AtSr1	Arabidopsis SIGNAL RESPONSIVE1
BBMV	Brush border membrane villi
BC	Bovine Chymotrypsin
BPH	Brown plant hopper
BR	Brassinosteroids
Bt	Bacillus *thuringenesis*
BXDs	Benzoxazinoids
CaM	Calmodulin (CaM)
CBD	Carbohydrate binding protein
CBL	Calcineurin B-like
CDPK	Calcium-dependent protein kinase
CDPKs	Calcium-dependent protein kinases
CEA	*Colocasia esculenta* tuber agglutinin
CEBiP	Chitin elicitor binding protein
CIPK	Calcium interacting protein kinases
CITRX	Cf-9-interacting TRX
CK	Cytokinins

CML	Calmodulin like (CML) proteins
cMOL	*Moringa oleifera* lectin
COEs	Carboxylesterases
Con A	Concanavalin A
COR	COR: Coronatine
CPK	Calcium dependent protein kinases
CRISPR	Clustered regularly interspersed short palindromic repeats
Cry	Crystal
CV-N	Cyanovirin- N
DAHP	3-deoxy-d-arabinoheptulosonate-7 phosphate
DAMPs	Damage-associated molecular patterns
DCL1	Dicer Like-1
DMAPP	Dimethylallyl pyrophosphate
Dpi	Days post infection
dsRNA	Double stranded RNA
E4P	Erythrose 4-phosphate
E-64	L-trans-epoxy succinylleucylamide[4-guanido] butane
EDTA	Ethylene diamine tetraacetic acid
EEA	Euonymus europaeus agglutinin
ENS	Empty neuron system
eQTL	Expression quantitative trait loci
ERK	Extracellular signal-regulated kinase
ESMI	Epithiospecific Modifier Protein
ESP	Epithiospecific Protein
ESTs	Expressed sequence tags
ET	Ethylene
ETI	Effector-triggered immunity
ETS	Effector-triggered susceptibility
FACs	Fatty acid-amino acid conjugates
FITC	Fluorescein isothiocyanate
FPP	Farnesyl pyrophosphate
FPPS	Farnesyl pyrophosphate synthase
GA	Gibberellins
GGPP	Geranylgeranyl pyrophosphate
GGPPS	Geranylgeranyl pyrophosphate synthase
GLVs	Green leaf volatiles
GM	Genetically modified
GNA	Galanthus *nivalis* agglutinin
GOX	Glucose oxidase
GPI	Glycosyl phophatidyl inositol
GPP	Geranyl pyrophosphate
GPPS	Geranyl pyrophosphate synthase
GRF	Growth regulating Factor
GSH	Glutathione
GSNO	S-nitrosoglutathione

GSNOR	S-nitrosoglutathione reductase
GSSG	Oxidized glutathione disulfide
GST	Glutathione-S-transferases
GUS	beta-glucuronidase
H2O2	Hydrogen peroxide
HAEs	Herbivore-associated elicitors
HAMPs	Herbivore-associated molecular patterns
HCN	Hydrogen Cyanide
HEK-cell	Human embryonic kidney cell
HGT	Horizontal Gene Transfer
HIPVs	Herbivore-induced plant volatiles
HIVOCs	Herbivore-induced plant VOCs
HPR	Host plant resistance
HTI	Herbivore-triggered immunity
IAA	Indole acetic acid
ICAT	Isotope-Coded Affinity Tags
ICP	Insecticidal crystal protein
IPM	Integrated pest management
IPP	Isopentenyl pyrophosphate
ISR	Induced Systemic Resistance
ITRAQ	Isobarric Tagged for Relative and Absolute Quantitation
IUBMB	International Union of Biochemistry and Molecular Biology
JA	Jasmonic acid
JA-Ile	Jasmonoyl-L-isoleucine
JAZ	Jasmonic acid ZIM-domain
JBTX	Jaburetox
JH	Juvenile Hormone
LH	Lateral horn
LNs	Local interneurons
LOXs	Lipoxygenases
LRR-RLK	Leucine-rich repeat receptor-like protein kinase
LsyM	Lysin domain
LT	Long-trichoid
MAMPs	Microbial-associated molecular patterns
MAPK	Mitogen activated protein kinase
MAPKK	Mitogen activated protein kinase kinase
MAPKKK	Mitogen activated protein kinase kinase kinase
MBs	Mushroom bodies
MCAs	Microbial control agents
MeJA	Methyl jasmonate
MEP	Methyl-erythritol-phosphate
MGC	Macro glomerulus complex
MIR	Mycorrhiza-induced resistance
miRNA	Micro-RNA

MTI	MAMP-triggered immunity
MMBL	Monocot mannose binding lectins
MPI	Maize protease inhibitor
mRNAs	messenger RNAs
MTI	Mustard trypsin Inhibitor
MudPIT	Multidimensional Protein Identification Technology
MVA	Mevalonic acid
mVOCs	Microbial volatile organic compounds
NADPH	Nicotinamide adenine dinucleotide phosphate hydrogen
nat-siRNAs	natural antisense small interfering RNAs
ncRNAs	non-protein coding RNAs
NGS	Next-generation sequencing
NHVs	Non-host plant volatiles
NICTABA	*Nicotiana tabacum* agglutinin
NIF	Nitrogen-fixing gene
NLA	Nitrogen Limitation Adaptation
NMR	Nuclear Magnetic Resonance
NO	Nitric oxide
NOD	Rhizobial nodule forming gene
NOS	Nitric oxide synthase
NPP	Neryl pyrophosphate
NPR1	Non-expressor of Pathogenesis-Related gene1
NPV	Nucleopolyhedrovirus
NTR	NADP-thioredoxin reductase
OBPs	Olfactory binding proteins
ODE	Odor degrading enzymes
OIPVs	Oviposition induced plant volatiles
OPDA	12-oxo-phytodienoic acid
ORs	Olfactory receptors
OS	Oral Secretion
OSNs	Olfactory sensory neurons
PA	Pyrrolizidine alkaloids
PAL	Phenylalanine ammonia lyase
PAMP	Pathogen associated molecular pattern
PAMPs	Pathogen-associated molecular patterns
PBPs	Pheromone binding proteins
PCI	Potato polypeptide chymotrypsin Inhibitors
PCR	Polymerase Chain Reaction
PCWDe	Plant Cell Wall-Degrading enzymes
PDLPs	Plasmodesmata located proteins
PEP	Phosphoenolpyruvate
PGPF	Plant growth promoting fungi
PGPR	Plant growth promoting rhizobacteria
PHA	*Phaseolus vulgaris* agglutinin
Phe	Phenylalanine

PI	Proteinase inhibitor
PIs	Proteinase inhibitors
PM	Peritrophic membrane
PMM	Perimicrovillar membrane
PNs	Projection neurons
POD	Peroxidase
PODs	Peroxidases
PPI	Plant Proteinase Inhibitors
PPO	Polyphenol oxidase
PPOs	Polyphenol oxidases
PR	Pathogenesis-related
PRR	Pattern recognition receptors
PSA	*Pisum sativum* agglutinin
PTI	Pattern-triggered immunity
PUFA	Polyunsaturated fatty acid
qRT-PCR	quantitative Reverse Transcriptase-Polymerase Chain Reaction
R proteins	Resistant proteins
RBOH	Respiratory burst oxidase homologs
RCB	Red Cotton Bug
RIPs	Ribosome-inactivating proteins
RISC	RNA-induced silencing complex
RNAi	RNA Interference
RNAseq	RNA sequencing
RNS	Reactive Nitrogen species
RobpsCRA	Chitinase-related agglutinin from *Robinia pseudoacacia*
ROS	Reactive Oxygen Species
SA	Salicylic acid
SAMK	Stress-activated MAPK
SCN	Soybean Cyst Nematode
SERCA	Sarcoplasmic reticulum type Ca++ ATPase
SG-HM	Slow growth high mortality
SIMK	Salt stress inducible MAPK
SIPK	Stress-induced MAPK
siRNAs	small interfering RNAs
SNA	*Sambucus nigra* agglutinin I
SNMPs	Sensory neuron membrane proteins
SOD	Superoxide dismutase
SP	Serine Proteinases
SPL	Squamosa promoter binding protein like
SSR	Single sensillum recording
SteLL	S.*terebinthifolius* leaf lectin
STI	Soybean Trypsin Inhibitor(Kunitz)
tasi-RNAs	transacting small interfering RNAs
TLP	Thioredoxin-like superfamily protein
Trx	Thioredoxin

UDP-GlcNAcP	UDP-N-acetylglucosamine pyrophosphorylase
UGTs	UDP-glucosyl transferase
UTR	Untranslated Region
VIGS	Virus induced gene silencing
VIP	Vegetative insecticidal protein
VOC	Volatile organic compound
VOCs	Volatile organic compounds
VP	Variation potential
WGA	Wheat germ agglutinin
WIPK	Wound-induced MAPK
WIR	Wound-induced resistance

Role of Herbivore-Associated Molecular Patterns (HAMPs) in Modulating Plant Defenses

Garima Malik, Ritu Chaturvedi, and Sunila Hooda

Abstract

Being sessile organisms, plants have evolved a vast range of resistance mechanism to offset biotic stress caused by insect herbivores. The coevolution of plants and insect herbivores has not only generated advanced defense strategies in plants but also led to development of feeding strategies and counter-adaptive mechanisms in insects. Several plant species can differentiate insect attack from mechanical damage by the perception of a suite of chemical signals or herbivore-associated elicitors (HAEs), also known as herbivore-associated molecular patterns (HAMPs), produced by the insect. HAMPs could arise from insect oral secretions (OS), saliva, digestive waste products, and ovipositional fluids. Apart from elicitors, OS from some insect herbivores also contain effectors that suppress plant antiherbivore defenses. HAEs are dissimilar in their origin and structure, ranging from FACs (fatty acid-amino acid conjugates) such as volicitin, chemically related oxylipins, sulfur-containing fatty acids (caeliferins), peptides (systemins and inceptins) to high-molecular-weight enzymes (glucose oxidase and glucosidase). The perception of HAEs leads to the commencement of specific physiological processes in plants in order to defend themselves from insect attack. These responses can vary from changes in plant's metabolic activity and gene expression pattern to changes in their overall growth and development. Some HAEs are also known to counteract the defense response of plants. However, relatively less is known about the molecular components used by plants to perceive and recognize HAEs and the downstream signaling pathways leading

G. Malik
R.G. (PG) College, C.C.S University, Meerut, U.P., India

R. Chaturvedi
St. John's College, D.B.R.A University, Agra, U.P., India

S. Hooda (✉)
Ram Lal Anand College, University of Delhi, New Delhi, Delhi, India

I. K. Singh, A. Singh (eds.), *Plant-Pest Interactions: From Molecular Mechanisms to Chemical Ecology*, https://doi.org/10.1007/978-981-15-2467-7_1

to the initiation of plant response. In this chapter, we will focus on the recent developments made in the field of insect HAEs and their role in modulating plant defenses which will provide novel insights into our understanding of the interaction between plant and insects.

Keywords

Herbivory · Elicitors · HAMPs · FACs · HAEs · Volicitin · Systemin

1 Introduction

Plant-herbivore interaction is a multifaceted dynamic set of reactions, spanning across multiple habitats (terrestrial and aquatic) and employing a diverse range of taxa and organismal size. Interactions between producers ranging from algae to trees and consumers ranging from arthropods to mammals are vital since they generate animal biomass merely from sunlight. These pivotal reactions are channeled by a range of biomolecules generated by both the herbivore and plant. These biomolecules including proteins, sugars, and lipopolysaccharides are decisive factors in determining the fate of interaction. While herbivore-derived molecules determine the success of herbivory, plant-derived molecules participate in the recognition of invader attack and are involved in initiating the defense response. Though the relationship seems to be parasitic, but it benefits both the landscape vegetation and the herbivore itself by virtue of nutrient recycling phenomenon (Coleman and Sollenberger 2007). Both positive and negative impacts are attributed on plants through insect herbivory. On one hand, flowering plants are particularly benefitted by insect herbivores by exploiting them as pollinators; insect herbivores also serve to be potential vectors of plant diseases besides causing physical damage and potential loss of crop productivity.

Herbivores are particularly significant ecologically since they are known to affect the fitness of plants adversely and also occupy key positions in the food web. Since herbivores are deterrent to plant survival, plants have evolved defense mechanisms for their sustenance. These defense mechanisms either alone or in synchronization make use of mechanical, chemical, or mimicry strategies (Ruttan and Lortie 2015). Mechanical defense strategies are attributed to morphological adaptations of plants such as spinescence (presence of thorns and spines), pubescence (trichomes), sclerophylly (toughened or hardened leaves), etc. which reduce the palatability of tissues and thereby provide first line of defense against herbivore attack. Chemical resistance involves constitutive or induced production of secondary metabolites to impart defense. The constitutive defense metabolites that are stored in inactive forms, known as phytoanticipins, get activated by β-glucosidase during herbivore attack, which further mediates the release of various aglycone metabolites that are biocidal in nature (Morant et al. 2008). Phytoanticipins include glucosinolates (hydrolyzed by myrosinases upon tissue disruption) and benzoxazinoids (widely distributed in members of family Poaceae) (War et al. 2012). Metabolites triggered in response to herbivory are known as phytoalexins and include isoflavonoids, terpenoids, alkaloids, etc., which are known to affect the development and survival of herbivores (Walling 2000). Besides, plant phenolics, flavonoids, tannins, lectins, proteinase inhibitors (PIs), peroxidases (PODs), polyphenol oxidases (PPOs), lipoxygenases (LOXs), and phytohormones such as jasmonic acid (JA), salicylic acid (SA), and ethylene (ET) also play salient role in imparting host-plant resistance (HPR) (War et al. 2012). In areas where insects are principal herbivores, sap secretion from injured areas of plants serves as a way to trap and prevent insects from further damage. Although not well-documented, plants also mimic toxic plants or insects (e.g., orchids) in sight, smell, and taste in order to avoid herbivores (Lev-Yadun and Ne'eman 2012). Production of leaves with yellow spots in *Passiflora* species serves as an excellent example of mimicry strategy adopted by plants to avoid insect herbivory whereby yellow spots on leaves mimic the eggs of heliconid butterflies. Since butterflies avoid ovipositing on plants already containing eggs in order to minimize unnecessary competition for resources, the plant defends itself from larvae that are its specialized herbivores (Pannell and Farmer 2016).

The defense mechanisms involved against herbivores are quite different from that of pathogens. In general, plants can respond to pathogen attack by apoptosis and resist the infection from spreading. But insect herbivores are not prone to this isolate-and-kill strategy due to their mobility and migration ability. Strategies such as go-away-or-die and slow-them-down are effective against insect herbivores, and both share common physiological characteristics. In both these approaches, plants channelize a series of events that will hamper the growth and development of insect herbivore on one hand and reallocation of resources on the other (Kant et al. 2015). In order to affect the physiology of herbivore, antiherbivore defense proteins such as PIs and PPOs are secreted by plants that interfere with the digestion process in the gut of insects (Zhu-Salzman et al. 2008). At the same time, reallocation of resources, particularly carbon and nitrogen containing compounds, to either reproductive or storage organs, will secure them for future use and also deprive the feeding insects.

Plants regulate these defense mechanisms at the expense of their growth and reproduction due to the high-energy cost involved (Puentes and Ågren 2013). Plants not only recognize herbivory from mechanical damage caused to host tissues but also from secretions (insect saliva, regurgitant, frass) deposited on wounded tissues (Ray et al. 2015). Molecules from these depositions that induce defense responses in plants are known as elicitors or herbivore-associated molecular patterns (HAMPs) and are analogous to pathogen-associated molecular patterns (PAMPs) or microbe-associated molecular patterns (MAMPs) (Schmelz 2015). In order to counteract HAMP-mediated defense responses of plants, insects' secretions do contain effector molecules that determine the success of herbivory (Felton and Tumlinson 2008). The defense mechanisms are regulated and specific not only at plants' level; recent studies indicate that soil microorganisms may also confer broad-spectrum resistance against insect herbivores (Rashid and Chung 2017). Beneficial rhizospheric microbes are known to prime the entire plant and increase defense against insect herbivores by induced systemic resistance (ISR) mechanism (Pieterse et al. 2014). In general, ISR involves protection through physical or chemical barriers of plants and can be activated by biological or abiotic factors against pathogens or herbivores. Though plant-microbe and plant-insect interactions are interconnected via molecular pathways, little is known about the tri-trophic level interactions involving plant, microbes, and insects (Pineda et al. 2010).

The study of plant-herbivore interactions is multidisciplinary and involves thrust areas like coevolution, chemical ecology, nutritional ecology, and ecological stoichiometry (Burkepile and Parker 2017). Technological advances in the field of genetics, phylogenetics, DNA barcoding, GPS, and remote sensing have led to recent developments in this field. The interaction between plant and herbivore is being discoursed at molecular level, and genes involved in imparting defense are being identified. *Resistance* (*R*) genes that are present across plant genomes and confer resistance by producing R proteins are being identified and studied (Kourelis and van der Hoorn 2018). Products of *R* genes are not only involved in conveying resistance against pathogens but are now believed to provide resistance against phloem-feeding insects also, as suggested by genetic evidences of crop plants (Smith and Boyko 2007). Specific plant NBS-LRR (nucleotide-binding site-leucine-rich repeat) proteins have been identified in tomato and melon that confer resistance to some aphids (Rossi et al. 1998; Dogimont et al. 2009). The transfer of such resistance-coding genes to economically important plants can be of immense commercial application.

Ecologically, it is an enthralling area and has profound scope for utilization in plant protection. Agricultural losses (~10–20%) because of insect herbivores are a crucial factor in limiting food production. In order to design strengthened crop plants against herbivores, induced responses can be engineered genetically, resulting in the constitutive production of defense compounds. The commercial introduction of genetically modified crops expressing genes encoding the entomocidal δ-endotoxin from *Bacillus thuringiensis* (Bt) has been a milestone in the field of crop protection. Further, the potential of metabolites produced during induced resistance and HAMPs can be harnessed for pest management, which will in turn reduce the harms and amount of insecticides employed in pest control. Thus, knowledge of plant-herbivore interactions and various HAMPs is an indispensable component of integrated pest management and for sustainable crop production.

2 Herbivore Behavior and Feeding Styles

Among herbivores, insects are phenomenally abundant and diverse. Around half of the million known insect species rely on green plant tissues for their survival (Wu and Baldwin 2010). Insect herbivores employ diverse feeding strategies on plants to acquire nutrients for growth, development, and reproduction. Some insect herbivores are food generalists (polyphagous), i.e., feed on a wide variety of plant families, while others are specialists and possess restricted options by limiting themselves on members of a particular family, particular species, or even specific plant parts or tissues such as leaves or phloem (i.e., monophagous and oligophagous). Around two-thirds of insect herbivores comprising of members of order

Table 1 Insects and their feeding behavior

Order	Members	Feeding mechanism	References
Coleoptera	Beetles	Chewing, snipping	Schoonhoven et al. (1998)
Diptera	Mosquitoes and flies	Sucking	Yoshinaga et al. (2007)
Orthoptera (largest order of herbivore insects)	Grasshoppers, crickets	Chewing, snipping	Yoshinaga et al. (2007)
Hemiptera	Aphids, plant hoppers, leaf hoppers and mealybugs	Sucking	Howe and Jander (2008)
Lepidoptera	Butterflies and moths	Chewing, tearing (larvae), siphoning (sucking without piercing) (adults)	Schoonhoven et al. (1998)
Thysanoptera	Thirps	Puncturing and sucking	Kindt et al. (2003)

Coleoptera (beetles) or Lepidoptera (caterpillars) injure and feed on leaves using mouthparts evolved for chewing, tearing, or snipping (Schoonhoven et al. 1998). Members of order Thysanoptera (thirps) and Trombidiformes (spider mites) feed on plants by puncturing and sucking the contents from lacerated cells using tube-like structures. Many Hemipterans (aphids and whiteflies) possess specialized stylets and insert them between cells to penetrate into phloem (Howe and Jander 2008). A summary of various insect orders and their feeding mechanisms is presented in Table 1. This diversity of insects and their feeding patterns provides an insight on wide evolution that led to the food partitioning and development of specialized feeding styles such as external chewing, spore feeding, phloem feeding, sap sucking, gall making, etc. The high diversification in the feeding apparatus has not only restricted the diet breadth and feeding behavior of insects, but, at the same time, adaptations in chewing functions has led to convergence or strikingly similar morphology of mouthparts in insects that share similar diets (Bernays 1998).

Though insect herbivores employ advanced feeding mechanisms to extract nutrients from above and below ground plant parts', the success of insect herbivory is confounding since the meager protein content of nonreproductive tissues in most plants makes them an inadequate food source. Although herbivorous insects have small bodies, rapid growth and are poikilotherms, still they have a higher relative need for proteins than carbohydrates. In order to fulfil the greater demand of proteins than most of herbivores, insects possess a substantial gut size and brisk throughput of ingested food. Besides, the feeding behavior also gets altered according to the nutrient needs of insects. Insects are less likely to feed if the concentration of nutrients is high within, whereas if the concentration is low, the insects opt for compensatory feeding. Some herbivorous insects also undergo supplemental feeding on nonplant sources in order to gain nutrients (Bernays 1998). Feeding pattern of insects may also depend on the stages in their life cycles. In holometabolous insects (i.e., insects that undergo complete metamorphosis), the food requirements of larvae are completely different than adults. On the other hand, the food preferences during larval and adult stages are identical in hemimetabolous insects (i.e., insects that undergo incomplete metamorphosis) indicating their capability of retaining memory and experiences of feeding (Matsumoto and Mizunami 2002). Also herbivore foraging behavior can be altered due to risk of predation, which subsequently affects the distribution, abundance, and diversity of plants (Burkepile and Parker 2017) as shown in a mesocosm study where reduction of grasshopper (herbivore) population due to active hunting spiders (predator) resulted in increased plant species diversity in grasslands (Schmitz 2008).

3 Plant Immune System and Defense Mechanisms

Unlike the adaptive forms of immunity in vertebrates, plants' immunity comprises their innate ability to recognize and respond defensively to herbivore attack. Each cell is capable of perceiving danger signals and transmitting the information systemically followed by induction of direct and indirect defense responses (Howe and Jander 2008). Plants are believed to harness multiple surveillance systems for recognition of insects showing diverse lifestyles and feeding behavior. Herbivore recognition requires prompt signaling cascades at the plant-herbivore interface much before changes at host's genetic and metabolic level (Maffei et al. 2007). Plants have well-developed mechanisms to distinguish herbivory from mechanical shearing in order to avoid waste of defense responses. Since plants are sessile organisms, a range of direct, indirect, and constitutive defense mechanisms are present that act cumulatively to protect plants from herbivores. The coevolutionary struggle between herbivore – to consume and plants – to not being consumed, has led to an advancement of metabolism and genetic diversity of plants (Jander and Howe 2008). The first line of defense is imparted by physical barriers such as thick leaves, hair, trichomes, barbs, thorns, and spines. Also the prevention of attack by herbivores relies largely on secondary metabolites (chemical compounds) that provide direct defense (Howe and Jander 2008). Plants react to herbivory by synthesizing defense metabolites (phytoanticipins and phytoalexins) and proteins (PIs, PPO, LOXs, POD) that impose toxic, repellant, or antinutritive effects on herbivores, rather than behaving as silent sufferers in these interactions (Zhu-Salzman et al. 2008). Constitutive defenses such as accumulation of defense metabolites in tissues are imparted by preformed barriers irrespective to the threat of herbivory. Beneath the direct defense layer is present a layer of indirect defense given by herbivore-induced plant volatiles (Kessler and Baldwin 2002). Herbivore-provoked plants do emit volatile compounds such as terpenes, green leaf volatiles (GLVs), and volatile aromatic compounds, in order to attract insect predators and protect them against herbivory (Holopainen and Blande 2013). Secondary metabolites secreted by plants play a large role in deciphering whether the herbivore will feed upon a host or not. These specialized compounds acting as chemical deterrents and attractants are a result of coevolutionary struggle between herbivores and plants (Becerra 2007). Coevolution between plants and herbivores has served to reduce the effectiveness of these

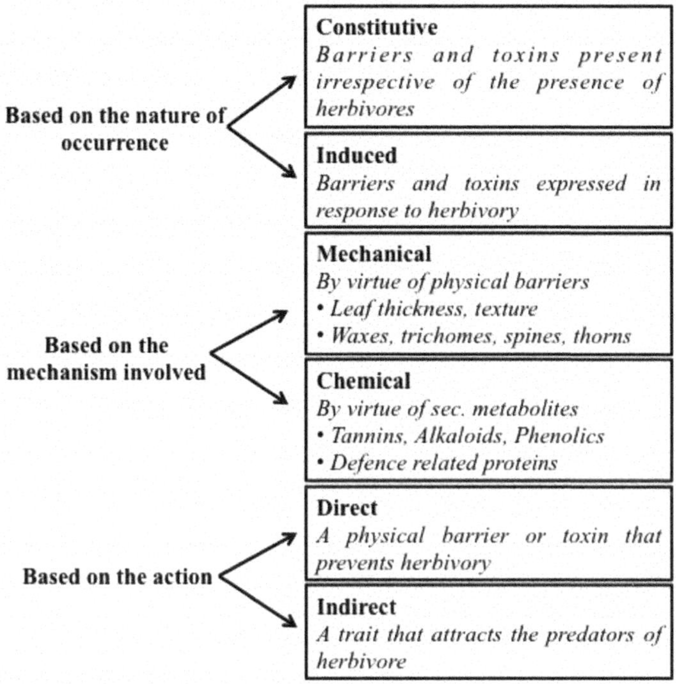

Fig. 1 Mechanisms of plant defense against insect herbivores

defenses via developing tolerance or counter-defense mechanism by the herbivores (Mithöfer and Boland 2012). The coevolution of specialist herbivores with their preferred plants has been to an extent that they may employ the secondary metabolites as nutrient sources (Bennett and Wallsgrove 1994). A summary of defense mechanisms adopted by plants has been illustrated in Fig. 1.

Plants are capable of recognizing compounds present in OS of insect herbivores (HAMPs) as shown by induction of defense responses (release of terpenes, GLVs, and ET) upon application of insect-derived factors on artificial wounds. Herbivory or application of OS to wounded tissues generates a disparate or acute response than mechanical injury alone (Schmelz et al. 2006). The recognition of herbivore OS and signal cascades from injured plant tissues initiates an enterprising form of immunity comprising of phosphorylation chain reactions, calcium ion fluxes, and biochemical pathways leading to synthesis of JA, SA, and ET which orchestrate defense induction (Stotz et al. 2000). Oxidative stress is also generated upon insect feeding and itself serves as a defense mechanism. Production of reactive oxygen species (ROS) and apoptosis are key strategies adopted by plants to protect themselves from insects (Lei et al. 2014). Enhanced production of H_2O_2 and other ROS can destroy the intestines of insects and kill them directly (Rashid and Chung 2017). Besides, plants also possess mechanisms to estimate the quality and quantity of tissue damage. Temporal and spatial patterns of mechanical injury are also critical determinants of host defense response. Induction of defense responses has been seen to occur when mechanical devices were used repeatedly to mimic herbivore injury (Mithöfer et al. 2005). Also salivary secretions have been shown to affect the defense response qualitatively in experiments where plants were challenged with larvae whose labial salivary glands have been removed (Musser et al. 2006).

Besides employing their morphological and biochemical characteristics, surrounding plants are also known to make associational effects on focal plants to provide defense against herbivores (Huang et al. 2016). In order to cut the energy costs of defense mechanisms and allocate resources for growth and reproduction, some plants do not harness individual-level defense mechanisms and instead gets benefitted from anti-herbivory mechanisms of adjacent plants (Barbosa et al. 2009). These associational effects are justified by at least two main hypotheses, i.e., attractant-decoy hypothesis and repellant-plant hypothesis, which propose that herbivory can be avoided by associating with either higher or lower palatable plants (Ruttan and Lortie 2015). The attractant-decoy hypothesis makes use of association of susceptible plants with more preferred species in a high-quality patch, in order to divert the herbivore to favorable neighbors, while the repellant-plant hypothesis holds the reverse concept of being located in a low-quality patch to avoid herbivory. This is because herbivores avoid visiting such patches in the advent of using their efforts on beneficial patches (Huang et al. 2016). Both these hypotheses are of particular significance and dependent whether the herbivores make decisions on plant species scale or patch scale, respectively. In practical, herbivores employ both these scales simultaneously to make decisions, and the net foraging decision depends upon spatial scale at which the enormity of selection by the herbivore is higher (Huang et al. 2016).

4 Elicitors: Molecules that Elicit Plant Defense Against Predators

Plants are constantly exposed to wide array of attacking biotrophs, but due to absence of an adaptive immune system, plant defense depends on the innate immune system of each cell to recognize and respond to predators and on systemic cues generated from site of attack or infection (Henry et al. 2012). When a plant comes in contact with predators, a number of cellular responses are initiated in the host plant, such as changes in ion flux, rapid perturbations of the plasma membrane potential [V (m)], mitogen-activated protein kinase (MAPK) activation, formation of ROS and oxylipins, as well as expression of typical defense-linked genes (Boller 1995; Nürnberger et al. 2004; Maffei et al. 2007; Nühse et al. 2000; Asai et al. 2002; Apel and Hirt 2004; Ramonell and Somerville 2002; Kessler and Baldwin 2002). To successfully combat predators, plants rely on a highly receptive and explicit recognition system with the skill to detect "danger" and respond promptly to induce a locally or systemically expressed resistance in order to ward off potentially hazardous microbes and insect herbivores. To identify their invaders, plants required the evolution of cell surface localized receptor proteins that might identify and bind predator's signatures or distinctive "patterns" that act as a "molecular identity card," whose structure are generally conserved among various categories of predators and thus perceived as non-self/foreign molecules by plants (Albert 2013; Choi and Klessig 2016).

In plant biology, the term elicitors is widely used for molecular compounds that are perceived by plants and can stimulate induced defense responses against pathogen/insects in the host plant after they are exposed to it. These elicitors could be herbivore-derived elicitors, plant-derived or endogenous elicitors, conjugates of plants and herbivores-derived elicitors, and artificial or synthetic elicitors (Kant et al. 2015). Elicitors are mainly divided in two categories: general or nonspecific elicitors and specific elicitors. General elicitors are present either constitutively in the pathogens or are produced during pathogen attack by receptors present on cell surface (Jones and Dangl 2006). They are involved in general resistance and do not considerably vary in their impact on diverse cultivars within a plant species (Montesano et al. 2003). On the other hand, specific elicitors are produced by specialized pathogen strains and work only in those plant cultivars that have the related disease resistance gene (Montesano et al. 2003). The high level of specificity in the case of specific elicitors indicates coevolution of the antagonists, host plant, and pathogen, respectively (Mithöfer and Boland 2008). The mode of action of elicitors may differ, but increasingly some elicitors are reported to act as ligands that bind plant receptors (likely membrane-bound extracellular leucine-rich repeat receptor kinases) initiating an intricate cascade of signal transduction leading to cytosolic Ca^{+2} fluctuations, MAPK activation, and stimulation of defense-related phytohormones (JA, SA, ET), which serve as signals for subsequent direct and/or indirect defense responses (Alborn and Schmelz 2008) (Fig. 2).

PAMPs or MAMPs are pathogen/microbial molecules or general elicitors that alert plant about the intruding pathogens and act as inducers of defense response (Newman et al. 2013). Well-known examples for such MAMPs develop from characteristic microbial structures such as fungal chitin, bacterial peptidoglycan, or flagellin (Felix et al. 1993, 1999; Gust et al. 2007). Additionally, various pathogens employ cleaving and degrading enzymes during plant invasion that injure plant cells and produce typical degradation products that might act as endogenous elicitors or "damage-associated molecular patterns" (DAMPs), for instance, cutin monomers, plant cell wall fragments (oligogalacturonides or cellulose fragments), or peptides derived from cleaved and ruined proteins (Lotze et al. 2007; Kauss et al. 1999; Ortmann et al. 2006; Nühse 2012; Yamaguchi and Huffaker 2011).

Similarly, the term HAMPs or herbivore-associated elicitors (HAEs) is used for the herbivore-derived elicitors present in the oral secretions, saliva, regurgitate, digestive waste products, oviposition fluids, or eggs of herbivores that indicate the presence of herbivores to plants and induce defense response (Felton and Tumlinson 2008; Wu and Baldwin 2010; Zipfel 2014). A wide range of structurally diverse HAMPs has been identified, including enzymes, modified lipids, sulfur-containing fatty acids, cell wall fragments, and peptides (Fig. 3). However, unlike PAMPs, most of these HAMPs do not function as general elicitors of antiherbivore responses in every plant species but serve as specific elicitors, typically limited to specific plant-

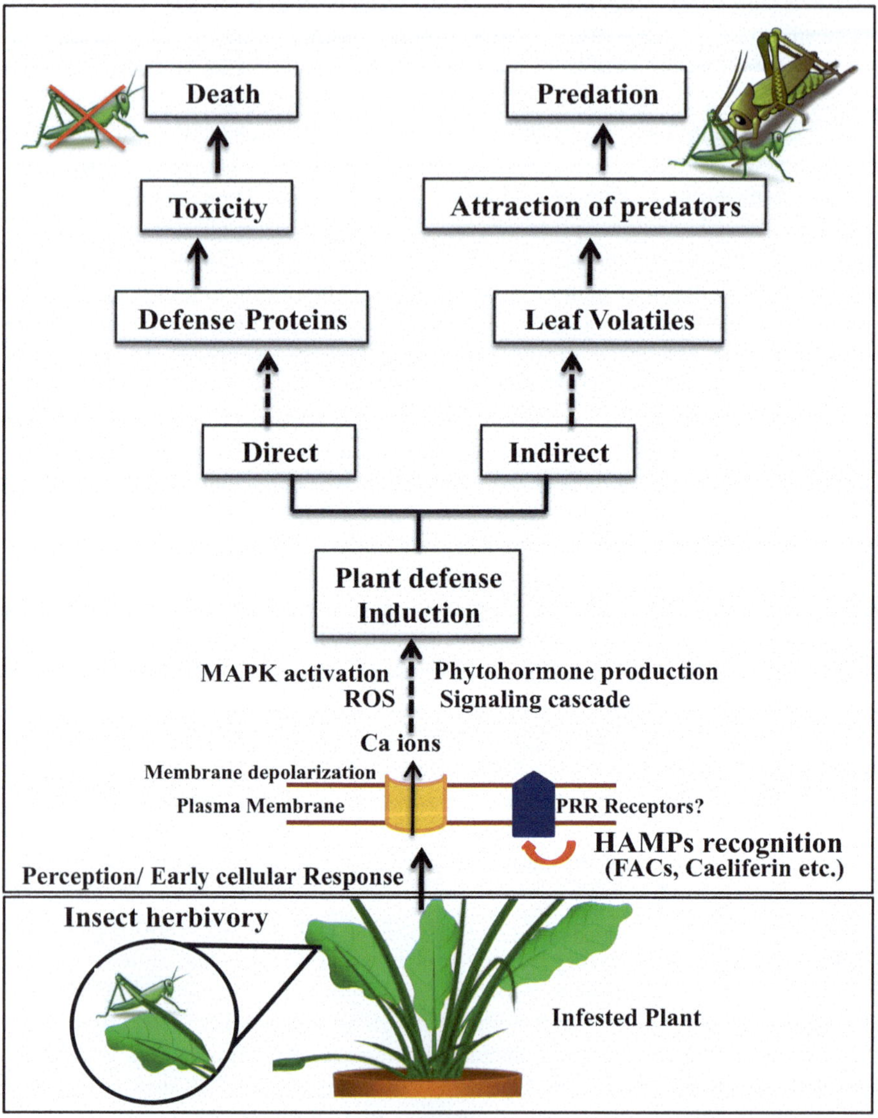

Fig. 2 Role of HAMPs in plant defense initiation

insect associations (Bonaventure et al. 2011) (Table 2). HAMPs are apparently recognized by pattern recognition receptors (PRR) that evolved to identify conserved pathogen and herbivore-derived molecules or motifs (Erb et al. 2012).

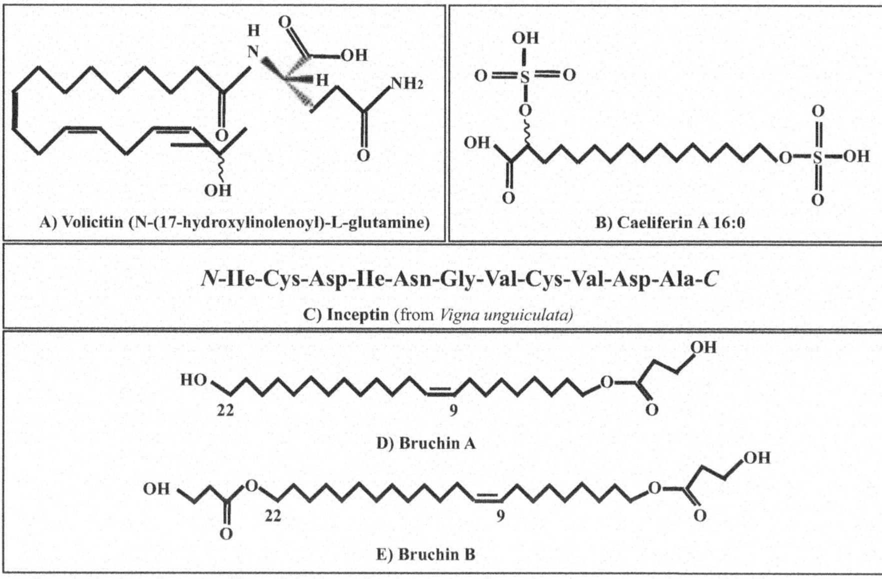

Fig. 3 Structure of some known HAMPs

5 HAMPs: Herbivore-Associated Molecular Patterns

Several HAMPs have been isolated and characterized from the OS of herbivores that initiate plant defense against herbivore attack. Many HAMPs comprise plant-derived compounds that have been modified by herbivore enzymes during feeding such as pectin, oligogalacturonide fragments, oligosaccharides, or chloroplastic ATP-synthase fragments (Doares et al. 1995; Creelman and Mullet 1997; Bergey et al. 1999; Schmelz et al. 2006). Some of the HAMPs are discussed below:

5.1 Fatty Acid-Amino Acid Conjugates (FACs)

As the name suggests, FACs are conjugates of two moieties: one is plant-derived fatty acid, i.e., linolenic acid (18:3) or linoleic acid (18:2), released from membranes by lipases activity or wounding and herbivore-derived amino acid, either glutamate (Glu) or glutamine (Gln) (Alborn et al. 1997; Paré et al. 1998; Halitschke et al. 2001; Yoshinaga et al. 2008). Both compounds are conjugated in the gut of herbivore probably to boost its nitrogen assimilation efficiency (Paré et al. 1998; Yoshinaga et al. 2008). Many insect species produce FACs elicitor including most caterpillar larvae of moths and butterflies (Lepidoptera), crickets *Teleogryllus taiwanemma* and *Teleogryllus emma* (Orthoptera), and the fruit fly *Drosophila melanogaster* (Diptera) (Yoshinaga et al. 2007; Bonaventure et al. 2011). Volicitin (N-(17-hydroxylinolenoyl)-

Table 2 List of some HAMPs that induce specific responses in plants during insect herbivory

HAMPs	Insect species	Plant species	References
N-Acyl-amino acids (FACs)	*Spodoptera exigua*, *Manduca sexta*, *Teleogryllus taiwanemma*, *Drosophila melanogaster* and several *Lepidoptera*	*Zea mays* (maize), *Glycine max* (soybean), *Solanum melongena* (eggplant), *Nicotiana attenuata* (coyote tobacco), *Solanum nigrum*	Alborn et al. (1997), Takabatake et al. (2006), Hu et al. (2008), Bede et al. (2006), Eichenseer et al. (2010) Schmelz et al. (2009), Yoshinaga et al. (2007, 2010), Halitschke et al. (2001) and Pohnert et al. (1999)
Caeliferins	*Schistocerca americana*	*Z. mays*, *Arabidopsis thaliana*	Alborn et al. (2007) and Schmelz et al. (2009)
Inceptin	*Spodoptera frugiperda*	*Vigna unguiculata* (cowpea), some *Fabaceae*	Schmelz et al. (2006) and Schmelz et al. (2009)
Glucose oxidase (GOX)	*Helicoverpa zea*, *S. exigua*, *Helicoverpa armigera*, other *Lepidoptera* and *Hymenoptera*	*Nicotiana tabacum* (tobacco), *N. attenuata*, *Medicago truncatula*, *Solanum lycopersicum* (tomato)	Musser et al. (2002), Diezel et al. (2009), Hu et al. (2008), Bede et al. (2006), Eichenseer et al. (2010) and Musser et al. (2005)
b-Glucosidase	*Pieris brassicae*	Brassica oleracea (cabbage), *Phaseolus lunatus* (lima beans), *Z. mays*	Mattiacci et al. (1995) and Hopke et al. (1994)

Source: Bonaventure et al. (2011)

L-glutamine) was the first fully characterized FAC, identified in the OS of beet armyworm (*Spodoptera exigua*) caterpillars (Alborn et al. 1997) (Fig. 3a). When applied to damaged leaves, volicitin induces the *Zea mays* seedlings to discharge a combination of volatile compounds (terpenoids and indole) which in turn attract female parasitic wasps, usual enemies of the *Spodoptera exigua* larvae (Alborn et al. 1997). The existence of FAC receptors or plasma membrane binding proteins in plants has been confirmed by discern binding of volicitin to enriched plasma membrane fractions isolated from *Zea mays* leaves (Truitt et al. 2004). The main components in OS from *Manduca sexta* and several other lepidopteran insects accountable for the differential activation of *Nicotiana attenuata* genes, involved in primary and secondary metabolism, have been identified as FACs (Giri et al. 2006). The function of FACs in stimulating the activity of the salicylic acid-induced protein kinase and the wound-induced protein kinase in *N. attenuata* when their leaves were wounded by caterpillars or treated with OS was also reported (Wu et al. 2007).

2-Hydroxyoctadecatrienoic acid (2-HOT), a newly identified HAMP from OS of *M. sexta*, participates in defense activation of *Nicotiana attenuata* against insect feeding. It is derived from linolenic acid through the action of the α-dioxygenase of tobacco plant in the midgut of *M. sexta*. 2-HOT helps tobacco to sense the development of the catterpillar's attack and to continue its production of JA that stimulate antiherbivore defenses (Gaquerel et al. 2009, 2012).

5.2 Caeliferins

Caeliferins elicitors are saturated and monounsaturated sulfated α-hydroxy fatty acids (C15-C20) first isolated from regurgitate of the American bird grasshopper species *Schistocerca americana* (Alborn et al. 2007) (Fig. 3b). Like FACs, caeliferins are able to induce discharge of parasitoid-attracting volatile when applied to injured seedlings of corn. However, till now caeliferins are found to be restricted to only within the suborder Caelifera (e.g., grasshoppers) and are not widespread as FACs (Alborn et al. 2007). In *Arabidopsis thaliana*, treatment with synthetic caeliferin A 16:0 did not stimulate volatile compound emission but robustly induced both ET and JA production within 2 hours and a modest increase in SA after 4 hours of application to injured leaves suggesting central role of caeliferins in grasshopper oral secretion elicited responses (Schmelz et al. 2009). These preliminary research indicates that more in-depth studies about biological mode of action of the caeliferins' and discovery of their specific receptors in plants may enhance our understanding to utilize caeliferins as natural plant defense products.

5.3 Inceptins

Inceptins are plant-derived small disulfide-bonded peptides (11–13 amino acids) elicitors produced by proteolytic degradation of ATP-synthase gamma-subunit regulatory regions in the insect gut (Fig. 3c). They induce and amplify local and systemic defense responses (Yamaguchi and Huffaker 2011). Inceptin recognition indicates that plants have not only evolved the method to sense insects directly by their secretions or movements but also indirectly by observing the appearance of catabolic products suggestive of an insect that effectively feeds and digests (Schmelz et al. 2009). Inceptins were first isolated from the OS of *Spodoptera frugiperda* (fall armyworm) larva feeding on cowpea (*Vigna unguiculata*) and maize (Schmelz et al. 2006, 2007). Later, inceptins were reported to be present in the OS of larvae of many Lepidoptera species (Schmelz et al. 2012). The mechanism of inceptin perception is not yet identified; however, two prospective mechanisms have been put forward for inceptin perception: R gene guard protein mediated and receptor mediated (Schmelz et al. 2012; Maischak et al. 2007). As a result of insect gut proteolysis, the amino acid sequences of inceptins can be somewhat dissimilar. As compared to their N terminus, the C terminus is more vital for inceptins bioactivity as they are highly responsive to selected modifications near the C terminus but is tolerant to substitutions at the N terminus (Schmelz et al. 2007). When the wounded tissue was treated with inceptin in cowpea plants, they induce swift production of phytohormones, viz., ET, SA, and JA, and subsequently stimulate various direct and indirect defense reactions including emission of volatiles such as methyl salicylate (MeSA) and homoterpene (E)-4,8-dimethyl-1,3,7-nonatriene (DMNT) (James 2005; Schmelz et al. 2007). Inceptin-like sequences are ubiquitous in all plant chloroplastic ATP-synthase γ-subunits, but the elicitor activity of inceptins seems to be limited to the genera *Phaseolus* and *Vigna* of Fabaceae (Schmelz et al. 2007). One of the recent study revealed that the inceptin-related peptide having a C-terminal truncation, present in the OS of the legume-specializing velvetbean caterpillar (*Anticarsia gemmatalis*), does not stimulate but rather antagonizes defenses (Schmelz et al. 2012).

5.4 Bruchins

Bruchins are mono- and bis-(3-hydroxypropanoate) esters of C22 and C24 α,ω-diols (Oliver et al. 2002) (Fig. 3d, e). They were first isolated from ovipositional fluid of adult insects of two species of Bruchidae, viz., pea weevils (*Bruchus pisorum* L.) and cowpea weevils (*Callosobruchus maculatus* F.) (Doss et al. 2000). During oviposition, when female *Bruchus pisorum* interacts with their host plant, pea (*Pisum sativum*), bruchins stimulate cell division and callus development (neoplasms) on pods at the site of egg attachment. Because of neoplasms growth, the eggs are displaced from the oviposition site which in turn impedes larval entry into the pod tissue and thus exposing them to predators and desiccation (Doss et al. 2000). The bruchins are the first natural products found with the capacity to stimulate neoplasm development when applied to pods of peas. This distinctive form of induced resistance is conditioned by the dominant allele, neoplastic pod (*Np*), present in the host-plant genotype (Dodds and Matthews 1966). At present, the receptors have not been identified for these oviposition-associated elicitors. Moreover, application of bruchins to pea pods with *Np* leads to upregulation of various defense metabolic pathways genes. The level of a putative isoflavone synthase gene, CYP93C18, was reported to be amplified within 8 hours of bruchin treatment along with an increase in the expression of the isoflavonoid phytoalexin pisatin, indicating the role of bruchin B in activating other plant defense responses (Cooper et al. 2005).

5.5 Enzymes

Beta-glucosidase present in the regurgitant of white cabbage butterfly (*Pieris brassicae*) larvae induces the discharge of volatiles from cabbage (*Brassica capitata*) leaves that attracts parasitic wasps (*Cotesia glomerata*) to attack the herbivores (Mattiacci et al. 1995). However, enzyme glucose oxidase (GOX) secreted by labial glands of corn earworm (*Helicoverpa zea*) has been shown to repress injury-inducible herbivore defenses of host plant, tobacco (*Nicotiana tabacum*), and another solanaceous plant, tomato (*Lycopersicon esculentum*) (Eichenseer et al. 1999; Musser et al. 2005). Similarly, alkaline phosphatase identified in the salivary gland of adult silverleaf whitefly (*B. argentifolii*) has been shown to aid the species feeding on different plant species (Funk 2001; Yan et al. 2011).

In contrast to chewing insects, phloem-feeding Hemiptera such as aphids, whiteflies, etc. cause comparatively little tissue damage while feeding from phloem sieve elements. Hemiptera insects-derived elicitors are not very well characterized; however, plants exposed to hemipteran attack are able to escalate unique metabolic and transcriptional responses (De Vos et al. 2005; Kempema et al. 2007; Mewis et al. 2006; Voelckel et al. 2004). Aphid salivary enzymes such as peroxidase, pectinase, and GOX may act as elicitors of plant direct and indirect defense responses (Miles 1999; Harmel et al. 2008). Polyphenol oxidases present in the saliva of the grain aphid *Sitobion avenae* and *S. graminum* are reported to elicit JA and terpene signaling pathways in wheat plants (Ma et al. 2010). Similarly, when pectinases in the saliva of *S. avenae* are applied to wheat, they trigger indirect plant defense response by activating the discharge of volatiles that draw the parasitoid wasp *Aphidius avenae* (Liu et al. 2009). *M. persicae* salivary protein, a heat-sensitive peptide between 3 and 10 kDa, can induce defense against the aphid in *A. thaliana* plants and reduce aphid performance (De Vos and Jander 2009). Several other salivary HAMPs of *M. persicae* have been shown to be harmful to aphids and decrease their fecundity on *A. thaliana* and *N. tabacum* by initiating defense reactions (Bos et al. 2010; Elzinga et al. 2014).

6 Functional Genomics Approach for Identification of HAMPs

Plant defenses in response to the insect attack involve a complex and coordinated set of events leading to diverse changes at the genomic, transcriptomic, proteomic, and metabolomic levels which also involve a signaling cascade. The insect-derived compounds HAMPs and a large number of salivary effectors can be produced by a single herbivore species, making the interaction more complicated and dynamic. There are lots of studies available on identification and functional characterization of biochemical, signaling, and molecular effectors in the plant-herbivore interactions (Barah and Bones 2015; Foyer et al. 2015; War et al. 2012; Oates et al. 2016; Basu et al. 2018). Yet, the crucial role of HAMP's in inducing host plant response is not properly understood. While it is important to explore the mechanisms in plants, the role of insect effectors leading to wide-scale reprogramming in plants also needs to be probed. The current functional genomics tools may provide supporting clues in understanding the plant-insect interactions using this bidirectional approach. It is known that wide-scale changes in genomes occur in plants in response to herbivory involving the upregulation or downregulation of more than thousands of genes (Barah and Bones 2015; Foyer et al. 2015; Oates et al. 2016). The information on the vast array of proteomic and metabolomic changes is also available (Jansen et al. 2009; Tian et al. 2012; Acevedo et al. 2017; Tzin et al. 2015). It is clear that the use of "omics" approach has significantly improved our understanding and has provided important clues about the mechanisms involved in interaction in host plants and insect effectors. Yet, the area remains relatively underexplored on a genome/transcriptome/proteome/metabolome scale, and we are far from system level understanding. System biology along with the recent advances in genomics, transcriptomics, and proteomics techniques along with next-generation sequencing (NGS) technologies and RNA sequencing may help us in providing some insights into the dynamic and continuously evolving plant-insect interaction mechanisms. Also, eQTL mapping (expression quantitative trait loci) has come up as a promising technique as it helps in analyzing thousands of traits simultaneously. The technique has been successfully used in many plants especially *Arabidopsis* and rice.

In the last few years, transcriptomics and proteomics approaches have increasingly been applied (Chen et al. 2009; Zhang et al. 2013a, b; War et al. 2012; Tian et al. 2012; Acevedo et al. 2017; Tzin et al. 2015) in many plants. Though all these studies have not been done on a single plant-insect interaction system, yet the combined analysis of the work done in different plant systems would provide important clues. Also, genomics, transcriptomics, proteomics, and metabolomic studies using HAMPs as an elicitor of plant response in this complex interaction have not been undertaken so far.

6.1 Transcriptomics

The availability of whole-genome microarrays and gene expression arrays has led to the development of transcriptomics in the field of plant and herbivorous insect interaction. Microarray analysis of host defense in *Arabidopsis* against *Pieris rapae* and *Brassica oleracea* var. *capitata* L. and *Brassica nigra* L. or the aphid *Brevicoryne brassicae* L. has been investigated (Broekgaarden et al. 2007; Reymond et al. 2004). A number of transcriptomic studies have been done to investigate the plant response against insect attack (Thompson and Goggin 2006). Many of these studies were done in the plants infested with insects to understand the repertoire of genes involved in plant-insect interaction. Also recently, many studies were done in the saliva of insects to understand the effectors or elicitors of the response.

In a transcriptomic study, use of aphid salivary gland expressed sequence tags (ESTs) identified a total of 48 effector candidates from *Myzus persicae* (green peach aphid) on the basis of similarity to plant-pathogen effectors (Bos et al. 2010). The recent advances in the genomics of plant adaptation in two major insect orders, namely, Hemipterans and Lepidopterans, infesting a wide variety of crops have been recently reviewed (Simon et al. 2015).

An integrated study involving comparative assessment of global gene expression profiles in *Arabidopsis* plants against pathogens (*Pseudomonas syringae* pv. *tomato*, *Alternaria brassicicola*) and insects (*Pieris rapae*, *Frankliniella occidentalis*, *Myzus persicae*) with a very different feeding mechanism showed considerable overlap, and stress-related transcripts were highly represented in all (De Vos et al. 2005).

Transcriptomic analysis of two developmental stages (nymphs and adults) of green peach aphid, *Myzus persicae*, revealed a total of 2244 differentially expressed genes in two developmental stages. The genes primarily belonged to the processes involved in metabolism, hormone production, cuticle formation, food digestion, and absorption detoxification (Ji et al. 2016). Transcriptomic analyses of transcripts encoding secreted proteins from the salivary glands (SSGPs) of wheat midge larvae (*Sitodiplosis mosellana*) resulted in the identification of 97 groups of transcripts encoding SSGPs. This study may lead to identification and characterization of molecular mechanisms of wheat midge-wheat interactions (Al-jbory et al. 2018).

Thompson and Goggin (2006) suggested that in order to understand the basis of plant defense, changes in gene expression profiles should be simultaneously investigated within the insect as well as in host plant as studied in the brown planthopper-rice interaction (Zhang et al. 2004; Yang et al. 2005; Yuan et al. 2005).

It is well-known that the response of a plant to insects attack is highly dynamic and complex involving an array of wide-scale changes in the transcriptome. The differentially regulated defense transcriptome varies with time and space. In order to understand the coordinated regulation of plant defense transcriptome, analysis at different time points over a period of 48–96 hours is necessary. This approach has been used to study *Arabidopsis* defense against *Botrytis cinerea* (Windram et al. 2012). More recently, whole-genome transcriptomic studies were reported in *Solanum dulcamara* (Lortzing et al. 2017), cotton, and other cereals (Chen et al. 2017).

With the advent of NGS technology and the availability of insect genome sequences, genome-wide prediction of insect effectors at a wide scale and comparative genomics studies are facilitated. Recently, avirulence gene Avr5 of the tomato leaf mold pathogen, *Cladosporium fulvum* f. sp. *melonis* (Fom), was identified by comparative genomic approaches (Mesarich et al. 2014).

6.2 Proteomics

Whole proteome analysis is expected to yield information not only on the repertoire of proteins expressed or regulated in response to insect attack in plants but can also be used to reveal the proteome of insect saliva providing critical insights into the mechanism of plant-insect interaction. In this chapter we will present some of proteomic studies which have played a key role in understanding the role of HAMP in insect-plant interactions.

A comparative proteome analysis of *A. thaliana Ril* (*P. xylostella*-resistant Rils and pooled *P. xylostella*-susceptible Rils) by 2D-PAGE coupled with MS/MS identified 29 differentially expressed proteins (Collins et al. 2010). A total of 500 leaf proteins were monitored by 2-DE in a comparative potato proteome analysis using *Solanum tuberosum* L. infested with the Colorado potato beetle (*Leptinotarsa decemlineata)* and potato aphid (*Macrosiphum euphorbiae*) (Duceppe et al. 2012). Differential protein expression was reported in *Oryza sativa* L. mutants infested by brown planthopper, *Nilaparvata lugens* (Sangha et al. 2013), and in *Zea mays* infested with a chewing insect (*Spodoptera littoralis*) and a boring insect (*Busseola fusca*) (George et al. 2011).

Salivary proteins from three aphids, *Acyrthosiphon pisum*, *Megoura viciae* and *Myzus persicae*, were comparatively analyzed on a whole proteome level. A significant variability among the salivary proteomes of three aphids was revealed. Majority of the proteins belonged to three categories: DNA-binding proteins, GTP-binding proteins, and proteins with oxidoreductase activity with 22%, 19%, and 19%, respectively (Vandermoten et al. 2014). In another proteomic study, Miao et al. (2018) identified a total of 161 proteins out of which 21 were secretory proteins in the saliva of *S. furcifera using* shotgun liquid chromatography-tandem mass spectrometry (LC-MS/MS). It may provide important clues to understand the mechanism of interaction with host rice plants. In another study, saliva of three different aphid species, *M. persicae*, *M. cerasi*, and *R. padi*, were investigated through combined transcriptomic and proteomic analysis in order to identify the effector proteins (Thorpe et al. 2016). Many other proteomic studies were reported for identification and characterization of effector molecules in insect salivary proteins (Carolan et al. 2009, 2011; Rao et al. 2013; Harmel et al. 2008; Atamian et al. 2013).

Proteomic analysis of *Helicoverpa zea*-secreted salivary proteins identified 33 proteins with GOX as the most abundant protein, suggesting its primary role in induction in plant defense response in tomato (Tian et al. 2012). Another proteomics study was attempted to identify proteins present in saliva of the fall armyworm (FAW), *Spodoptera frugiperda*, by comparing the salivary proteome of insects of two different host plants – corn and rice. The study identified 13 differentially identified proteins in each strain (Acevedo et al. 2017). More recently, the changes in saliva proteome in response to different diets –cabbage, tomato, and an artificial pinto bean in *Trichoplusia ni* using iTRAQ labeling, and LC-MS/MS, were reported (Rivera-Vega et al. 2018). 63 proteins were found to be differentially regulated from a total of 254 proteins identified.

6.3 Metabolomics

The plant-insect interactions involve a large and complex array of metabolites and other effector molecules. A global metabolic profiling would enable us to gain insights into the plant as well as insect metabolome. The significance of global metabolomic analyses as discussed above in plant defense involves constitutive or induced production of secondary metabolites as well as primary metabolites. While it is difficult to analyze the enormous diversity of metabolites by traditional approaches, the plant metabolomics has received a great thrust with the recent developments in high-throughput profiling of metabolites and more sensitive imaging methods with improved spatial and temporal resolution (Maag et al. 2015; Sumner et al. 2018).

The targeted as well as nontargeted approaches have been used in many plants such as maize, *Chrysanthemum*, tomato leaves, while some studies focused only on high-throughput metabolite analyses, integrated studies involving transcriptomics and proteomics are also providing new insights into the plant-insect interactions either using the whole plant or specific plant parts (Leiss et al. 2009a, b; Gómez et al. 2012; Jansen et al. 2009; Tzin et al. 2015; Zhang et al. 2016). A comparative analysis of metabolites of five resistant and five susceptible *Chrysanthemum* varieties using

NMR revealed the presence of chlorogenic acid (CGA) and feruloylquinic acid (FQA), suggesting that these metabolites play a significant role in providing resistance to *Chrysanthemum* to thrips (Leiss et al. 2009b).

Simultaneous analysis of metabolome in host plant *Brassica oleracea* and insect *Pieris rapae* analyzed using ultra-performance liquid chromatography/time-of-flight mass spectrometry (UPLCT/MS) showed induction and higher levels of coumaroylquinic acids (Jansen et al. 2009). Similar nontargeted approach revealed primary metabolites involvement in insect-plant interaction such as increased levels of glucose-6-phosphate and maltose and other monomeric sugars in tomato leaves (Gómez et al. 2012; Steinbrenner et al. 2011).

The altered levels of specific benzoxazinoids and terpenes in maize-corn leaf aphid interactions were reported by high-throughput RNA sequencing (RNA-seq) and metabolite profiling using untargeted liquid chromatography-time-of-flight mass spectrometry (LC-TOF-MS) in a time course study from 0 to 96 hours (Tzin et al. 2015).

Metabolomics approach has been excellently used to understand the dynamic changes in phytohormones in the plant host as a result of insect attack. Yamaguchi et al. 2012 demonstrated for the first time that the insect gall-inducing sawfly (*Pontania* spp.) synthesize indole acetic acid (IAA) during interaction with host plant *Salix japonica* using liquid chromatography coupled with tandem mass spectrometry (LC-MS-MS) approach. An increase in the concentrations of cytokinin, JA, and SA and a decrease of abscisic acid were reported during the interaction between apple (*Malus domestica*) and *Phyllonorycter blancardella* using integrated approaches involving LC-MS-MS and microarray expression profiling (Zhang et al. 2016).

Initially in order to characterize herbivory-associated cues/molecular patterns, the genome-wide gene expression profiles or proteome or metabolome profiles of the response of plant to insect attack were generated along with studies in effectors in insects saliva, as discussed above. Since any transcriptional reprogramming is manifested as changes in the metabolic components, transcription factors, and signaling molecules, so the need for an integrated omics approach was percieved. Also, the need to analyze the insect effectors found in the saliva attacking insects led to the multi-omic approach to identify the effective repertoires of regulators/proteins/metabolites in the plant pathogens (Table 3). Taken together they may be then used to characterize the complex and cross-linked molecular mechanisms in plant-insect interaction.

6.4 Other Functional Genomics Techniques

The potential of RNAi for understanding of plant-insect interactions and an effective pest control mechanism was also investigated recently in two insects, *Myzus persicae* (the green peach aphid) and *Bactericera cockerelli* (the potato/tomato psyllid), using osmoregulatory genes as targets (Tzin et al. 2015). Similarly, RNAi-mediated studies in chewing insects have given excellent results (Baum et al. 2007; Bolognesi et al. 2012; Hu et al. 2016; Mao and Zeng 2014; Zhu et al. 2011). This approach will help to identify additional insect HAMPs and effectors and will

Table 3 Plant-insect interaction studies using multi-omics approach

Insect	Plant	Technique	References
Tomato fruitworm, *Helicoverpa zea*	*S. lycopersicum*	Shotgun proteomic analysis	Tian et al. (2012)
Grain aphid, *Sitobion avenae*	Wheat and other cereals	Transcriptomics and RNA-seq	Zhang et al. (2013a)
Sesamia nonagrioides	Maize	Transcriptomic, RNA-seq and metabolomics	Rodríguez et al. (2012)
Mustard leaf beetle, *Phaedon cochleariae*	Mustard	2-DE, LC-MS/MS profiling	Kirsch et al. (2012)
European corn borer, *Ostrinia nubilalis*	Maize	Microarray and iTRAQ protein separation	Dafoe et al. (2013)
Spodoptera littoralis	Maize	Ultra-high-pressure liquid chromatography time-of-flight mass spectrometry (UHPLC-TOF-MS)-based metabolomics	Marti et al. (2013)
Cotton bollworm, *Helicoverpa armigera*	Cereals	2-DE, MALDI-TOF/TOF mass spectrometry, GC-MS analysis	Zhang et al. (2013b)
Brown planthopper, *Nilaparvata lugens*	Rice	2-DE comparative proteomic and RNA-seq transcriptomic	Zhai et al. (2013)
Macrosyphum euphorbiae	*Solanum lycopersicum*	Transcriptome and 2-DE, MALDI-TOF-MS	Coppola et al. (2013)
Manduca sexta	*Nicotiana attenuata*	Microarray and metabolomics	Gulati et al. (2014)
Corn leaf aphid, *Rhopalosiphum maidis*	Maize	High-throughput RNA sequencing (RNA-seq) with metabolomics	Tzin et al. (2015)
Brown planthopper, *Nilaparvata lugens* (Stål)	Rice	Comparative transcriptomics with Illumina short-read sequencing technology	Ji et al. (2013)
Tarnished plant bug, *Lygus lineolaris*	Cotton, alfalfa, fruits and vegetable crop	Transcriptomics using Illumina sequencing	Showmaker et al. (2016)
Wheat midge (*Sitodiplosis mosellana*) (Géhin)	Wheat	Transcriptomics with sanger sequencing	Al-jbory et al. (2018)

also unravel the mechanisms of insect HAMPs perception by plants. It will also help in the understanding of complex interactions between plants and insect herbivores.

Genome-editing tools such as CRISPR-Cas9 can, potentially, be used to understand the function of insect HAMPs and effectors in modulating plant defenses.

Gene-editing approach has been elegantly used in *Helicoverpa armigera* to disrupt the cadherin gene (HaCad) (Wang et al. 2016).

System biology is another approach which can be used for a systems-level understanding of the complex and dynamic insect-plant interactions (Oates et al. 2016). The interactome consists of complex networks of biochemical, molecular, and metabolic processes occurring within and between plant and herbivorous insects. Though the interactive analysis is at the nascent stage and most of plant-insect interactions are currently being analyzed using multi-omics technologies only, yet the multilayered approach of system biology enabling information from molecular to organismal level has the potential of suggesting key targets for the plant-insect interaction. More recently, with the setting up of high-throughput phenotyping, using advanced phenomics facilities at many places will further contribute to elucidate plant-insect interactions (Goggin et al. 2015).

7 Conclusion and Future Outlook

Plants being exposed to various biotic as well as abiotic attacks have a plastic phenotype which is modulated by epigenetic factors in response to these biotic and abiotic stresses. The plants perceive a range of HAEs or HAMPs from insect OS, saliva, digestive waste products, etc. It is well-known that herbivore-induced plant defense involves a range of dynamic and complex interactions that have coevolved over a long period of time. These can be well explained by the use of genome-wide transcriptomic, proteomic tools, and functional genomics tool that are increasingly being used and started giving some insights into the spatiotemporal factors influencing the plant-insect interactions. Together these "omics" technologies along with RNAi, QTL mapping, and system level approach would provide definitive clues to the understanding of the complex interactome and repertoire of molecules involved in plant-insect interaction. It may further give us cues on the genetic and molecular mechanisms involved in plant-insect interactions leading to the evolutionary success. Along with wider insights, short-term and long-term adaptations will open new opportunities for large-scale sustainable pest management. Further, there is a need to analyze plant response through multi-omics approach using HAMPs as an elicitor due to the lack of such studies in this complex interaction.

References

Acevedo FE, Stanley BA, Stanley A, Peiffer M, Luthe DS, Felton GW (2017) Quantitative proteomic analysis of the fall armyworm saliva. Insect Biochem Mol Biol 86:81–92

Albert M (2013) Peptides as triggers of plant defence. J Exp Bot 64(17):5269–5279

Alborn HT, Schmelz EA (2008) Arthropod-associated plant effectors (AAPEs): elicitors and suppressors of crop defense. In: Capinera JL (ed) Encyclopedia of entomology. Springer, Dordrecht

Alborn HT, Turlings TCJ, Jones TH et al (1997) An elicitor of plant volatiles from beet armyworm oral secretion. Science 276:945–949

Alborn HT, Hansen TV, Jones TH et al (2007) Disulfooxy fatty acids from the American bird grasshopper *Schistocerca* americana, elicitors of plant volatiles. Proc Natl Acad Sci U S A 104:12976–12981

Al-jbory Z, Anderson KM, Harris MO, Mittapalli O, Whitworth RJ, Chen MS (2018) Transcriptomic analyses of secreted proteins from the salivary glands of wheat midge larvae. J Insect Sci 18(1):17

Apel K, Hirt H (2004) Reactive oxygen species: metabolism, oxidative stress, and signal transduction. Annu Rev Plant Biol 55:373–399

Asai T, Tena G, Plotnikova J et al (2002) MAP kinase signalling cascade in Arabidopsis innate immunity. Nature 415:977–983

Atamian HS, Chaudhary R, Cin VD, Bao E, Girke T, Kaloshian I (2013) In planta expression or delivery of potato aphid *Macrosiphum euphorbiae* effectors Me10 and Me23 enhances aphid fecundity. Mol Plant-Microbe Interact 26:67–74

Barah P, Bones AM (2015) Multidimensional approaches for studying plant defence against insects: from ecology to omics and synthetic biology. J Exp Bot 66:479–493

Barbosa P, Hines J, Kaplan I et al (2009) Associational resistance and associational susceptibility: having right or wrong neighbors. Annu Rev Ecol Evol Syst 40:1–20

Basu S, Varsani S, Louis J (2018) Altering plant defenses: Herbivore-associated molecular patterns and effector arsenal of chewing herbivores. Mol Plant-Microbe Interact 31(1):13–21. https://doi.org/10.1094/MPMI-07-17-0183-FI

Baum JA, Bogaert T, Clinton W, Heck GR, Feldmann P, Ilagan O et al (2007) Control of coleopteran insect pests through RNA interference. Nat Biotechnol 25:1322–1326

Becerra JX (2007) The impact of herbivore-plant coevolution on plant community structure. Proc Natl Acad Sci U S A 104:7483–7488

Bede JC, Musser RO, Felton GW, Korth KL (2006) Caterpillar herbivory and salivary enzymes decrease transcript levels of *Medicago truncatula* genes encoding early enzymes in terpenoid biosynthesis. Plant Mol Biol 60:519–531

Bennett RN, Wallsgrove RM (1994) Secondary metabolites in plant defence mechanisms. New Phytol 127:617–633

Bergey DR, Orozco-Cardenas M, de Moura DS et al (1999) A wound- and systemin-inducible polygalacturonase in tomato leaves. Proc Natl Acad Sci U S A 96(4):1756–1760

Bernays EA (1998) Evolution of Feeding Behavior in Insect Herbivores: success seen as different ways to eat without being eaten. Bioscience 48(1):35–44

Boller T (1995) Chemoperception of microbial signals in plant cells. Annu Rev Plant Physiol Plant Mol Biol 46:189–214

Bolognesi R, Ramaseshadri P, Anderson J, Bachman P, Clinton W, Flannagan R et al (2012) Characterizing the mechanism of action of double-stranded RNA activity against western corn rootworm (*Diabrotica virgifera virgifera* LeConte). PLoS One 7:e47534

Bonaventure G, Van Doorn A, Baldwin IT (2011) Herbivore-associated elicitors: FAC signaling and metabolism. Trends Plant Sci 16:294–299

Bos JIB, Prince D, Pitino M, Maffei ME, Win J, Hogenhout SA (2010) A functional genomics approach identifies candidate effectors from the aphid species *Myzus persicae* (Green Peach Aphid). PLoS Genet 6(11):e1001216

Broekgaarden C, Poelman EH, Steenhuis G, Voorrips RE, Dicke M, Vosman B (2007) Genotypic variation in genome-wide transcription profiles induced by insect feeding: *Brassica oleracea--Pieris rapae* interactions. BMC Genomics 8:239

Burkepile DE, Parker JD (2017) Recent advances in plant-herbivore interactions. F1000Research 6:119. https://doi.org/10.12688/f1000research.10313.1

Carolan JC, Fitzroy CI, Ashton PD, Douglas AE, Wilkinson TL (2009) The secreted salivary proteome of the pea aphid *Acyrthosiphon pisum* characterised by mass spectrometry. Proteomics 9(9):2457–2467

Carolan JC, Caragea D, Reardon KY et al (2011) Predicted effector molecules in the salivary secretome of the pea aphid (*Acyrthosiphon pisum*): a dual transcriptomic/proteomic approach. J Proteome Res 10:1505–1518

Chen Z, Zheng Z, Huang J, Lai Z, Fan B (2009) Biosynthesis of salicylic acid in plants. Plant Signal Behav 4:493–496

Chen D, Chen F, Chen C, Chen X, Mao Y (2017) Transcriptome analysis of three cotton pests reveals features of gene expressions in the mesophyll feeder *Apolygus lucorum*. Life Sci 60(8)

Choi HW, Klessig DF (2016) DAMPs, MAMPs, and NAMPs in plant innate immunity. BMC Plant Biol 16:232. https://doi.org/10.1186/s12870-016-0921-2

Coleman SW, Sollenberger LE (2007) Plant-herbivore interactions. In: Barnes RF et al (eds) Forages – The science of grassland agriculture. Blackwell Publishing, Ames, pp 123–136

Collins RM, Afzal M, Ward DA, Prescott MC, Sait SM, Rees HH et al (2010) Differential proteomic analysis of *Arabidopsis thaliana* genotypes exhibiting resistance or susceptibility to the insect herbivore, *Plutella xylostella*. PLoS One 5(4):e10103

Cooper LD, Doss RP, Price R et al (2005) Application of Bruchin B to pea pods results in the up-regulation of CYP93C18, a putative isoflavone synthase gene, and an increase in the level of pisatin, an isoflavone phytoalexin. J Exp Bot 56(414):1229–1237

Coppola V, Coppola M, Rocco M et al (2013) Transcriptomic and proteomic analysis of a compatible tomato–aphid interaction reveals a predominant salicylic acid-dependent plant response. BMC Genomics 14:515

Creelman RA, Mullet JE (1997) Oligosaccharins, brassinolides, and jasmonates: nontraditional regulators of plant growth, development, and gene expression. Plant Cell 9(7):1211–1223

Dafoe NJ, Thomas JD, Shirk PD, Legaspi ME, Vaughan MM, Huffaker A et al (2013) European corn borer (*Ostrinia nubilalis*) induced responses enhance susceptibility in maize. PLoS One 8: e73394

De Vos M, Jander G (2009) *Myzus persicae* (green peach aphid) salivary components induce defence responses in *Arabidopsis thaliana*. Plant Cell Environ 32:1548–1560

De Vos M, Van Oosten VR, Van Poecke RM, Van Pelt JA, Pozo MJ, Mueller MJ, Buchala AJ, Métraux JP, Van Loon LC, Dicke M et al (2005) Signal signature and transcriptome changes of *Arabidopsis* during pathogen and insect attack. Mol Plant-Microbe Interact 18:923–937

Diezel C, von Dahl CC, Gaquerel E, Baldwin IT et al (2009) Different lepidopteran elicitors account for crosstalk in herbivory-induced phytohormone signaling. Plant Physiol 150:1576–1586

Doares SH, Syrovets T, Weiler EW, Ryan CA (1995) Oligogalacturonides and chitosan activate plant defensive genes through the octadecanoid pathway. Proc Natl Acad Sci U S A 92 (10):4095–4098

Dodds KS, Matthews P (1966) Neoplastic pod in pea. J Hered 57:83–85

Dogimont C, Bendahmane A, Pitrat M et al (2009) Gene resistant to Aphis gossypii United States of America patent no 0070016977

Doss RP, Oliver JE, Proebsting WM et al (2000) Bruchins: Insect-derived plant regulators that stimulate neoplasm formation. PNAS 97(11):6218–6223

Duceppe M-O, Cloutier C, Michaud D (2012) Wounding, insect chewing and phloem sap feeding differentially alter the leaf proteome of potato, *Solanum tuberosum L.* Proteome Sci 10:73

Eichenseer H, Mathews MC, Bi JL et al (1999) Salivary glucose oxidase: Multifunctional roles for *Helicoverpa zea*? Arch Insect Biochem Physiol 42:99–109

Eichenseer H, Mathews MC, Powell JS, Felton GW (2010) Survey of a salivary effector in caterpillars:glucose oxidase variation and correlation with host range. J Chem Ecol 36:885–897

Elzinga DA, De Vos M, Jander G (2014) Suppression of plant defenses by a *Myzus persicae* (green peach aphid) salivary effector protein. Mol Plant-Microbe Interact 27:747–756

Erb M, Meldau S, Howe GA (2012) Role of phytohormones in insect-specific plant reactions. Trends Plant Sci 17:250–259

Felix G, Regenass M, Boller T (1993) Specific perception of subnanomolar concentrations of chitin fragments by tomato cells. Induction of extracellular alkalinization, changes in protein phosphorylation, and establishment of a refractory state. Plant J 4:307–316

Felix G, Duran JD, Volko S et al (1999) Plants have a sensitive perception system for the most conserved domain of bacterial flagellin. Plant J 18:265–276

Felton GW, Tumlinson JH (2008) Plant-insect dialogs: complex interactions at the plant-insect interface. Curr Opin Plant Biol 11:457–463

Foyer CH, Verrall SR, Hancock RD (2015) Systematic analysis of phloem-feeding insect-induced transcriptional reprogramming in Arabidopsis highlights common features and reveals distinct responses to specialist and generalist insects. J Exp Bot 66:495–512

Funk CJ (2001) Alkaline phosphatase activity in whitefly salivary glands and saliva. Arch Insect Biochem Physiol 46:165–174

Gaquerel E, Weinhold A, Baldwin IT (2009) Molecular interactions between the specialist herbivore *Manduca sexta* (Lepidoptera, Sphingidae) and its natural host *Nicotiana attenuata*. VIII. An unbiased GCxGC-ToFMS analysis of the plant's elicited volatile emissions. Plant Physiol 149:1408–1423

Gaquerel E, Steppuhn A, Baldwin IT (2012) Nicotiana attenuata α-DIOXYGENASE1 through its production of 2-hydroxylinolenic acid is required for intact plant defense expression against attack from Manduca sexta larvae. New Phytol 196:574–585

George D, Babalola OO, Gatehouse AMR (2011) Differential protein expression in maize (*Zea mays*) in response to insect attack. Afr J Biotechnol 10(39):7700–7709

Giri AP, Wünsche H, Zavala JA et al (2006) Molecular interactions between the specialist herbivore *Manduca sexta* (Lepidoptera, Sphingidae) and its natural host *Nicotiana attenuata* VII Changes in the plant's proteome. Plant Physiol 142:1621–1641

Goggin LF, Lorence A, Topp CN (2015) Applying high-throughput phenotyping to plant–insect interactions: picturing more resistant crops. Curr Opin Insect Sci 9:69–76

Gómez S, Adam S, Sonia S (2012) From shoots to roots: transport and metabolic changes in tomato after simulated feeding by a specialist Lepidopteran. Entomol Exp Appl 144:101–111

Gulati J, Baldwin IT, Gaquerel E (2014) The roots of plant defenses: integrative multivariate analyses uncover dynamic behaviors of gene and metabolic networks of roots elicited by leaf herbivory. Plant J 77(6):880–892

Gust AA, Biswas R, Lenz HD et al (2007) Bacteria-derived peptidoglycans constitute pathogen-associated molecular patterns triggering innate immunity in *Arabidopsis*. J Biol Chem 282:32338–32348

Halitschke R, Schittko U, Pohnert G et al (2001) Molecular interactions between the specialist herbivore *Manduca sexta* (Lepidoptera, Sphingidae) and its natural host *Nicotiana attenuata*. III. Fatty acid-amino acid conjugates in herbivore oral secretions are necessary and sufficient for herbivore specific plant responses. Plant Physiol 125:711–717

Harmel N, Létocart E, Cherqui A, Giordanengo P, Mazzucchelli G, Guillonneau F et al (2008) Identification of aphid salivary proteins: a proteomic investigation of *Myzus persicae*. Insect Mol Biol 17(2):165–174

Henry G, Thonart P, Ongena M (2012) PAMPs, MAMPs, DAMPs and others: an update on the diversity of plant immunity elicitors. Biotechnol Agron Soc Environ 16(2):257–268

Holopainen JK, Blande JD (2013) Where do herbivore-induced plant volatiles go? Front Plant Sci 4 (185):1–13

Hopke J, Donath J, Blechert S, Boland W (1994) Herbivore-induced volatiles–the emission of acyclic homoterpenes from leaves of *Phaseolus Lunatus* and *Zea Mays* can be triggered by a b-glucosidase and jasmonic acid. FEBS Lett 352:146–150

Howe GA, Jander G (2008) Plant immunity to insect herbivores. Annu Rev Plant Biol 59:41–66

Hu YH, Leung DW, Kang L, Wang CZ (2008) Diet factors responsible for the change of the glucose oxidase activity in labial salivary glands of *Helicoverpa armigera*. Arch Insect Biochem Physiol 68:113–121

Hu X, Richtman NM, Zhao J-Z, Duncan KE, Niu X, Procyk LA, Oneal MA, Kernodle BM, Steimel JP, Crane VC, Sandahl G, Ritland JL, Howard RJ, Presnail JK, Lu AL, Wu G (2016) Discovery of midgut genes for the RNA interference control of corn rootworm. Sci Rep 6:30542

Huang Y, Wang L, Wang D et al (2016) How does the foraging behavior of large herbivores cause different associational plant defenses? Sci Rep 6:20561
James DG (2005) Further field evaluation of synthetic herbivore-induced plant volatiles as attractants for beneficial insects. J Chem Ecol 1:481–495
Jander G, Howe G (2008) Plant interactions with arthropod herbivores: state of the field. Plant Physiol 146:801–803
Jansen JJ, Allwood JW, Marsden-Edwards E, van der Putten WH, Goodacre R, van Dam NM (2009) Metabolomic analysis of the interaction between plants and herbivores. Metabolomics 5 (1):150
Ji R, Yu H, Fu Q, Chen H, Ye W, Li S, Lou Y (2013) Comparative transcriptome analysis of salivary glands of two populations of rice brown planthopper, N*ilaparvata lugens*, that differ in virulence. PLoS One 8(11):e79612
Ji R, Wang Y, Cheng Y, Zhang M, Zhang H-B, Zhu L, Fang J, Zhu-Salzman K (2016) Transcriptome analysis of Green Peach Aphid (*Myzus persicae*): insight into developmental regulation and inter-species divergence. Front Plant Sci 7:1562
Jones JDG, Dangl JL (2006) The plant immune system. Nature 444:323–329
Kant MR, Jonckheere W, Knegt B et al (2015) Mechanisms and ecological consequences of plant defence induction and suppression in herbivore communities. Ann Bot 115:1015–1051. pmid:26019168
Kauss H, Fauth M, Merten A et al (1999) Cucumber hypocotyls respond to cutin monomers via both an inducible and a constitutive H_2O_2-generating system. Plant Physiol 120:1175–1182
Kempema LA, Cui X, Holzer FM, Walling LL (2007) *Arabidopsis* transcriptome changes in response to phloem-feeding silverleaf whitefly nymphs. Similarities and distinctions in responses to aphids. Plant Physiol 143:849–865
Kessler A, Baldwin IT (2002) Plant responses to insect herbivory: the emerging molecular analysis. Annu Rev Plant Biol 53:299–328
Kindt F, Joosten NN, Peters D et al (2003) Characterisation of the feeding behavior of western flower thrips in terms of electrical penetration graph (EPG) waveforms. J Insect Physiol 49:183–191
Kirsch R, Wielsch N, Vogel H, Svatos A, Heckel DG, Pauchet Y (2012) Combining proteomics and transcriptome sequencing to identify active plant-cell-wall-degrading enzymes in a leaf beetle. BMC Genomics 13:587
Kourelis J, van der Hoorn RAL (2018) Defended to the nines: 25 years of resistance gene cloning identifies nine mechanisms for R protein function. Plant Cell 30(2):285–299. https://doi.org/10.1105/tpc.17.00579
Lei J, Finlayson SA, Salzman RA, Shan L, Zhu-Salzman K (2014) BOTRYTIS-INDUCED KINASE1 modulates *Arabidopsis* resistance to green peach aphids via PHYTOALEXIN DEFICIENT4. Plant Physiol 165:1657–1670. https://doi.org/10.1104/pp.114.242206
Leiss KA, Choi YH, Abdel-Farid IB, Verpoorte R, Klinkhamer PG (2009a) NMR metabolomics of thrips (*Frankliniella occidentalis*) resistance in *Senecio* hybrids. J Chem Ecol 35:219–229
Leiss KA, Maltese F, Choi YH, Verpoorte R, Klinkhamer PGL (2009b) Identification of chlorogenic acid as a resistance factor for thrips in *Chrysanthemum*. Plant Physiol 150:1567–1575
Lev-Yadun S, Ne'eman G (2012) Does bee or wasp mimicry by orchid flowers also deter herbivores? Arthropod Plant Interact 6:327–332
Liu Y, Wang WL, Guo GX, Ji XL (2009) Volatile emission in wheat and parasitism by *Aphidius avenae* after exogenous application of salivary enzymes of *Sitobion avenae*. Entomol Exp Appl 130:215–221
Lortzing T, Firtzlaff V, Nguyen D et al (2017) Transcriptomic responses of *Solanum dulcamara* to natural and simulated herbivory. Mol Ecol Resour 17:e196–e211
Lotze MT, Zeh HJ, Rubartelli A et al (2007) The grateful dead: damage-associated molecular pattern molecules and reduction/oxidation regulate immunity. Immunol Rev 220:60–81
Ma R, Chen JL, Cheng DF, Sun JR (2010) Activation of defense mechanism in wheat by polyphenol oxidase from aphid saliva. J Agric Food Chem 58:2410–2418

Maag D, Erb M, Glauser G (2015) Metabolomics in plant-herbivore interactions: challenges and applications. Entomol Exp Appl 157(1):18–29

Maffei ME, Mithöfer A, Boland W (2007) Before gene expression: early events in plant-insect interaction. Trends Plant Sci 12(7):310–316

Maischak H, Grigoriev PA, Vogel H et al (2007) Oral secretions from herbivorous lepidopteran larvae exhibit ion channel forming activities. FEBS Lett 581:898–904

Mao J, Zeng F (2014) Plant-mediated RNAi of a gap gene-enhanced tobacco tolerance against the *Myzus persicae*. Transgenic Res 23:145–152

Marti G, Erb M, Boccard J, Glauser G, Doyen GR, Villard N et al (2013) Metabolomics reveals herbivore-induced metabolites of resistance and susceptibility in maize leaves and roots. Plant Cell Environ 36(3):621–639

Matsumoto Y, Mizunami M (2002) Lifetime olfactory memory in the cricket *Gryllus bimaculatus*. J Comp Physiol A Neuroethol Sens Neural Behav Physiol 188:295–299

Mattiacci L, Dicke M, Posthumus MA (1995) Beta-glucosidase-an elicitor of herbivore-induced plant odor that attracts host-searching parasitic wasps. Proc Natl Acad Sci U S A 92:2036–2040

Mesarich CH, Griffiths SA, van der Burgt A, Okmen B, Beenen HG, Etalo DW, Joosten MHAJ, de Wit PJGM (2014) Transcriptome sequencing uncovers the Avr5 avirulence gene of the tomato leaf mold pathogen *Cladosporium fulvum*. Mol Plant-Microbe Interact 27:846–857

Mewis I, Tokuhisa JG, Schultz JC, Appel HM, Ulrichs C, Gershenzon J (2006) Gene expression and glucosinolate accumulation in *Arabidopsis thaliana* in response to generalist and specialist herbivores of different feeding guilds and the role of defense signaling pathways. Phytochemistry 67:2450–2462

Miao YT, Deng Y, Jia HK, Liu YD, Hou ML (2018) Proteomic analysis of watery saliva secreted by white-backed planthopper, *Sogatella furcifera*. PLoS One 13(5):e0193831

Miles PW (1999) Aphid saliva. Biol Rev 74:41–85

Mithöfer A, Boland W (2008) Recognition of herbivory-associated molecular patterns. Plant Physiol 146:825–831

Mithöfer A, Boland W (2012) Plant defense against herbivores: chemical aspects. Annu Rev Plant Biol 63:431–450

Mithöfer A, Wanner G, Boland W (2005) Effects of feeding Spodoptera littoralis on lima bean leaves. II Continuous mechanical wounding resembling insect feeding is sufficient to elicit herbivory-related volatile emission. Plant Physiol 137:1160–1168

Montesano M, Brader G, Palva ET (2003) Pathogen derived elicitors: searching for receptors in plants. Mol Plant Pathol 4:73–79

Morant AV, Jørgensen K, Jørgensen C, Paquette SM, Sánchez-Pérez R, Møller BL et al (2008) beta-Glucosidases as detonators of plant chemical defense. Phytochemistry 69:1795–1813. https://doi.org/10.1016/j.phytochem.2008.03.006

Musser RO, Hum-Musser SM, Eichenseer H, Peiffer M, Ervin G, Murphy JB, Felton GW (2002) Herbivory: caterpillar saliva beats plant defences. Nature 416:599–600

Musser RO, Cipollini DF, Hum-Musser SM et al (2005) Evidence that the caterpillar salivary enzyme glucose oxidase provides herbivore offense in solanaceous plants. Arch Insect Biochem Physiol 58(2):128–137

Musser RO, Cipollini DF, Hum-Musser SM et al (2005) Evidence that the caterpillar salivary enzyme glucose oxidase provides herbivore offense in solanaceous plants. Arch Insect Biochem Physiol 58(2):128–137

Musser RO, Farmer E, Peiffer M et al (2006) Ablation of caterpillar labial salivary glands: technique for determining the role of saliva in insect-plant interactions. J Chem Ecol 32:981–992

Newman MA, Sundelin T, Nielsen JT et al (2013) MAMP (microbe-associated molecular pattern) triggered immunity in plants. Front Plant Sci 4:139. https://doi.org/10.3389/fpls.2013.00139

Nühse TS (2012) Cell wall integrity signaling and innate immunity in plants. Front Plant Sci 3:280

Nühse TS, Peck SC, Hirt H et al (2000) Microbial elicitors induce activation and dual phosphorylation of the *Arabidopsis thaliana* MAPK 6. J Biol Chem 275:7521–7526

Nürnberger T, Brunner F, Kemmerling B et al (2004) Innate immunity in plants and animals: striking similarities and obvious differences. Immunol Rev 198:249–266

Oates CN, Denby KJ, Myburg AA, Slippers B, Naidoo S (2016) Insect gallers and their plant hosts: from omics data to systems biology. Int J Mol Sci 17(11):1891

Oliver JE, Doss RP, Marquez B et al (2002) Bruchins, plant mitogens from weevils: structural requirements for activity. J Chem Ecol 28:2503–2513

Ortmann I, Conrath U, Moerschbacher BM (2006) Exopolysaccharides of *Pantoea agglomerans* have different priming and eliciting activities in suspension-cultured cells of monocots and dicots. FEBS Lett 580:4491–4494

Pannell JR, Farmer EE (2016) Mimicry in plants. Curr Biol 26(17):R784–R785

Paré P, Alborn H, Tumlinson J (1998) Concerted biosynthesis of an insect elicitor of plant volatiles. Proc Natl Acad Sci 95:13971–13975

Pieterse CM, Zamioudis C, Berendsen RL, Weller DM, Van Wees SC, Bakker PA (2014) Induced systemic resistance by beneficial microbes. Annu Rev Phytopathol 52(1):347–375

Pineda A, Zheng SJ, VanLoon JJA, Pieterse CMJ, Dicke M (2010) Helping plants to deal with insects: the role of beneficial soil-borne microbes. Trends Plant Sci 15:507–514. https://doi.org/10.1016/j.tplants.2010.05.007

Pohnert G, Jung V, Haukioja E, Lempa K, Boland W (1999) New fatty acid amides from regurgitant of lepidopteran (Noctuidae, Geometridae) caterpillars. Tetrahedron 55:11275–11280

Puentes A, Ågren J (2013) No trade-off between trichome production and tolerance to leaf and inflorescence damage in a natural population of Arabidopsis lyrata. J Plant Ecol 7:373–383

Ramonell KM, Somerville S (2002) The genomics parade of defense responses: to infinity and beyond. Curr Opin Plant Biol 5:291–294

Rao SAK, Carolan JC, Wilkinson TL (2013) Proteomic profiling of cereal aphid saliva reveals both ubiquitous and adaptive secreted proteins. PLoS One 8:e57413

Rashid MH-O, Chung YR (2017) Induction of systemic resistance against insect herbivores in plants by beneficial soil microbes. Front Plant Sci 8:1816. https://doi.org/10.3389/fpls.2017.01816

Ray S, Gaffor I, Acevedo FE et al (2015) Maize plants recognize herbivore-associated cues from caterpillar frass. J Chem Ecol 41:781–792

Reymond P, Bodenhausen N, Van Poecke RMP, Krishnamurthy V, Dicke M, Farmer EE (2004) A conserved transcript pattern in response to a specialist and a generalist herbivore. Plant Cell 16:3132–3147

Rivera-Vega LJ, Stanley BA, Stanley A, Felton GW (2018) Proteomic analysis of labial saliva of the generalist cabbage looper (*Trichoplusia ni*) and its role in interactions with host plants. J Insect Physiol 107:97–103

Rodríguez VM, Santiago R, Malvar RA, Butrón A (2012) Inducible maize defense mechanisms against the corn borer *Sesamia nonagrioides*: a transcriptome and biochemical approach. Mol Plant-Microbe Interact 25:61–68

Rossi M, Goggin FL, Milligan SB et al (1998) The nematode resistance gene Mi of tomato confers resistance against the potato aphid. Proc Natl Acad Sci U S A 95:9750–9754

Ruttan A, Lortie J (2015) A systematic review of the attractant-decoy and repellent plant hypotheses: do plants with heterospecific neighbours escape herbivory? J Plant Ecol 8(4):337–346

Sangha JS, Chen YH, Kaur J et al (2013) Proteome analysis of rice (*Oryza sativa* L.) mutants reveals differentially induced proteins during brown planthopper (*Nilaparvata lugens*) infestation. Int J Mol Sci 14:3921–3945

Schmelz EA (2015) Impacts of insect oral secretions on defoliation-induced plant defense. Curr Opin Insect Sci 9:7–15

Schmelz EA, Carroll MJ, LeClere S et al (2006) Fragments of ATP synthase mediate plant perception of insect attack. Proc Natl Acad Sci U S A 103(23):8894–8899

Schmelz EA, LeClere S, Carroll MJ et al (2007) Cowpea chloroplastic ATP synthase is the source of multiple plant defense elicitors during insect herbivory. Plant Physiol 144:793–805

Schmelz EA, Engelberth J, Alborn HT et al (2009) Phytohormone-based activity mapping of insect herbivore-produced elicitors. Proc Natl Acad Sci U S A 106(2):653–657

Schmelz EA, Huffaker A, Carroll MJ et al (2012) An amino acid substitution inhibits specialist herbivore production of an antagonist effector and recovers insect-induced plant defenses. Plant Physiol 160:1468–1478

Schmitz OJ (2008) Effects of predator hunting mode on grassland ecosystem function. Science 319 (5865):952–954

Schoonhoven LM, Jermey T, van Loon JJA (1998) Insect-plant biology: from physiology to evolution. Chapman & Hall, London

Showmaker KC, Bednářová A, Gresham C, Hsu CY, Peterson DG, Krishnan N (2016) Insight into the salivary gland transcriptome of *Lygus lineolaris* (Palisot de Beauvois). PLoS One 11(1): e0147197

Simon J-C, d'Alencon E, Guy E et al (2015) Genomics of adaptation to host-plant in herbivorous insects. Brief Funct Genomics 2015:1–11

Smith CM, Boyko EV (2007) The molecular bases of plant resistance and defense responses to aphid feeding: current status. Entomol Exp Appl 122:1–16

Steinbrenner AD, Gomez S, Osorio S, Fernie AR, Orians CM (2011) Herbivore-induced changes in tomato (*Solanum lycopersicum*) primary metabolism: a whole plant perspective. J Chem Ecol 37:1294–1303

Stotz HU, Pittendrigh BR, Kroymann J et al (2000) Induced plant defense responses against chewing insects. ethylene signaling reduces resistance of arabidopsis against egyptian cotton worm but not diamondback moth. Plant Physiol 124(3):1007–1018

Sumner LW, Yang DS, Bench BJ, Watson BS, Li C, Jones DA (2018) Spatially resolved plant metabolomics. Annual plant reviews book series. vol 43. Biology of plant metabolomics. https://doi.org/10.1002/9781119312994

Takabatake R, Seo S, Ito N, Gotoh Y, Mitsuhara I, Ohashi Y (2006) Involvement of wound-induced receptorlike protein kinase in wound signal transduction in tobacco plants. Plant J 47:249–257

Thompson GA, Goggin FL (2006) Transcriptomics and functional genomics of plant defence induction by phloem-feeding insects. J Exp Bot 57:755–766

Thorpe P, Cock PJA, Bos J (2016) Comparative transcriptomics and proteomics of three different aphid species identifies core and diverse effector sets. BMC Genomics 17:1

Tian D, Peiffer M, Shoemaker E, Tooker J, Haubruge E et al (2012) Salivary glucose oxidase from caterpillars mediates the induction of rapid and delayed-induced defenses in the tomato plant. PLoS One 7(4):e36168

Truitt CL, Wei HX, Paré PW (2004) A plasma membrane protein from *Zea mays* binds with the herbivore elicitor volicitin. Plant Cell 16:523–532

Tzin V, Fernandez-Pozo N, Richter A, Schmelz EA, Schoettner M, Schäfer M et al (2015) Dynamic maize responses to aphid feeding are revealed by a time series of transcriptomic and metabolomic assays. Plant Physiol 169:1727–1743

Vandermoten S, Harmel N, Mazzucchelli G, De Pauw E, Haubruge E, Francis F (2014) Comparative analyses of salivary proteins from three aphid species. Insect Mol Biol 23:67–77

Voelckel C, Weisser WW, Baldwin IT (2004) An analysis of plant–aphid interactions by different microarray hybridization strategies. Mol Ecol 13:3187–3195

Walling LL (2000) The myriad plant responses to herbivores. J Plant Growth Regul 19:195–216

Wang J, Zhang H, Wang H, Shan Z, Yayun Z, Yihua Y, Wu Y (2016) Functional validation of cadherin as a receptor of Bt toxin Cry1Ac in Helicoverpa armigera utilizing the CRISPR/Cas9 system. Insect Biochem Mol Biol 76:11–17

War AR, Paulraj MG, Ahmad T, Buhroo AA, Hussain B, Ignacimuthu S, Sharma HC (2012) Mechanisms of plant defense against insect herbivores. Plant Signal Behav 7(10):1306–1320

Windram O, Madhou P, McHattie S, Hill C, Hickman V, Cooke E et al (2012) *Arabidopsis* defense against *Botrytis cinerea*: chronology and regulation deciphered by high-resolution temporal transcriptomic analysis. Plant Cell Online 24:3530–3557

Wu JQ, Baldwin IT (2010) New insights into plant responses to the attack from insect herbivores. Annu Rev Genet 44:1–24. https://doi.org/10.1146/annurev-genet-102209-163500

Wu J, Hettenhausen C, Meldau S et al (2007) Herbivory rapidly activates MAPK signaling in attacked and unattacked leaf regions but not between leaves of *Nicotiana attenuata*. Plant Cell 19:1096–1122

Yamaguchi Y, Huffaker A (2011) Endogenous peptide elicitors in higher plants. Curr Opin Plant Biol 14:351–357

Yamaguchi H, Tanaka H, Hasegawa M, Tokuda M, Asami T, Suzuki Y (2012) Phytohormones and willow gall induction by a gall-inducing sawfly. New Phytol 196:586–595

Yan Y, Peng L, Liu WX et al (2011) Host plant effects on alkaline phosphatase activity in the whiteflies, *Bemisia tabaci* Biotype B and *Trialeurodes vaporariorum*. J Insect Sci 11(1):9

Yang Z, Zhang F, He Q, He G (2005) Molecular dynamics of detoxification and toxin-tolerance genes in brown planthopper (*Nilaparvata lugens* Stål., Homoptera: Delphacidae) feeding on resistant rice plants. Arch Insect Biochem Physiol 59:59–66

Yoshinaga N, Aboshi T, Ishikawa C et al (2007) Fatty acid amides, previously identified in caterpillars, found in the cricket *Teleogryllus taiwanemma* and fruit fly *Drosophila melanogaster* larvae. J Chem Ecol 33:1376–1381

Yoshinaga N, Aboshi T, Abe H et al (2008) Active role of fatty acid amino acid conjugates in nitrogen metabolism in *Spodoptera litura* larvae. Proc Natl Acad Sci 105:18058–18063

Yoshinaga N, Alborn HT, Nakanishi T, Suckling DM, Nishida R, Tumlinson JH, Mori N (2010) Fatty acid-amino acid conjugates diversification in lepidopteran caterpillars. J Chem Ecol 36 (3):319–325

Yuan HY, Chen XP, Zhu LL, He GC (2005) Identification of genes responsive to brown planthopper *Nilaparvata lugens* Stal (Homoptera, Delphacidae) feeding in rice. Planta 221:105–112

Zhai Y, Zhang J, Sun Z, Dong X, He Y, Kang K, Liu Z, Zhang W (2013) Proteomic and transcriptomic analyses of fecundity in the brown planthopper *Nilaparvata lugens* (Stål). J Proteome Res 12(11):5199–5212

Zhang F, Zhu L, He G (2004) Differential gene expression in response to brown planthopper feeding in rice. J Plant Physiol 161:53–62

Zhang M, Zhou Y, Wang H et al (2013a) Identifying potential RNAi targets in grain aphid (*Sitobion avenae* F.) based on transcriptome profiling of its alimentary canal after feeding on wheat plants. BMC Genomics 14:560–575

Zhang Q, Lu YX, Xu WH (2013b) Proteomic and metabolomic profiles of larval hemolymph associated with diapause in the cotton bollworm, *Helicoverpa armigera*. BMC Genomics 14:751

Zhang H, de Bernonville TD, Body M, Glevarec G, Reichelt M, Unsicker S, Brunea M, Renou JP, Huguet E, Dubreuil G et al (2016) Leaf-mining by *Phyllonorycter blancardella* reprograms the host-leaf transcriptome to modulate phytohormones associated with nutrient mobilization and plant defense. J Insect Physiol 84:114–127

Zhu F, Xu J, Palli R, Ferguson J, Palli SR (2011) Ingested RNA interference for managing the populations of the Colorado potato beetle, *Leptinotarsa decemlineata*. Pest Manag Sci 67:175–182

Zhu-Salzman K, Luthe DS, Felton GW (2008) Arthropod-inducible proteins: broad spectrum defenses against multiple herbivores. Plant Physiol 146:852–858

Zipfel C (2014) Plant pattern-recognition receptors. Trends Immunol 35:345–351. https://doi.org/10.1016/j.it.2014.05.004

Transcriptomics Studies Revealing Enigma of Insect-Plant Interaction

Sujata Singh, Archana Singh, and Indrakant K. Singh

Abstract

The study of insect-plant interaction is a complex and dynamic process that employs multidimensional and multidisciplinary approaches. A comprehensive understanding of plant defense mechanisms against insect plunderer is fundamental, to build up a profitable and effective pest management strategy. In the last few decades, the technological evolution of various high-throughput omics technologies (i.e., genomics, transcriptomics, proteomics, metabolomics) enabled a qualitative as well as the quantitative record of several biological molecules. Transcriptomics involves genome-wide analysis studies that delineate gene expression pattern of cells and tissues. The extensive use of transcriptomics in the field of agriculture offers an excellent route of genomic research beyond traditional "model" organisms. It easily provides accessible and affordable data for almost any organism, both model and non-model plants, which can be exploited in developing pest-resistant crops. This chapter aims to summarize contemporary research conducted using transcriptomics techniques to decipher the mystery of insect-plant interaction along with a brief analysis of its limitations, technological expansion, and prospects of omics in solving the enigma of plant-insect interaction.

Keywords

Plant · Insect · Herbivory · Defense · Transcriptome · RNA-seq · Insect adaptation · Insect pest management

S. Singh · I. K. Singh (✉)
Molecular Biology Research Lab. Department of Zoology, Deshbandhu College, Kalkaji, University of Delhi, New Delhi, India
e-mail: iksingh@db.du.ac.in

A. Singh (✉)
Department of Botany, Hans Raj College, University of Delhi, New Delhi, India
e-mail: archanasingh@hrc.du.ac.in

I. K. Singh, A. Singh (eds.), *Plant-Pest Interactions: From Molecular Mechanisms to Chemical Ecology*, https://doi.org/10.1007/978-981-15-2467-7_2

1 Introduction

Insects and plants share an inextricable bond from millions of years. This coevolutionary struggle is influenced by numerous biochemical reactions occurring at plant-insect interface. Plants utilize a multitude of toxic compounds and a web of interconnected signaling cascades to tolerate herbivory (Fig. 1) (Ferry et al. 2004; Howe and Jander 2008; Wu and Baldwin 2010). In contrast, insects also undergo some transcriptional adjustments to procure essential nutrients from plant tissue armed with gut-active toxins, required for its growth and development (Howe and Herde 2015). This entire process has been evolutionarily molded over a course of billions of years. In order to elucidate the complexity of insect-plant interaction, knowledge of plants anatomical features, metabolites, and signaling pathways involved in limiting insect infestation must be accompanied by understanding of insect counter-defense and trade-off switches. It will assist in the development of better pest control strategies. A major task in solving this mystery is that most of the molecular studies are performed on the individual insect-plant model, exploring the host allelochemicals induced in response to effector molecules secreted by insects belonging to different feeding guild, i.e., chewing vs. piercing-sucking herbivores, followed by its behavioral and physiological response on the herbivore. Researchers failed to draw some general conclusions on insect-plant behavior (Heidel-Fischer et al. 2018). Such fundamental information underpins different functional studies performed at molecular, biochemical, and cellular levels. Later, it should be

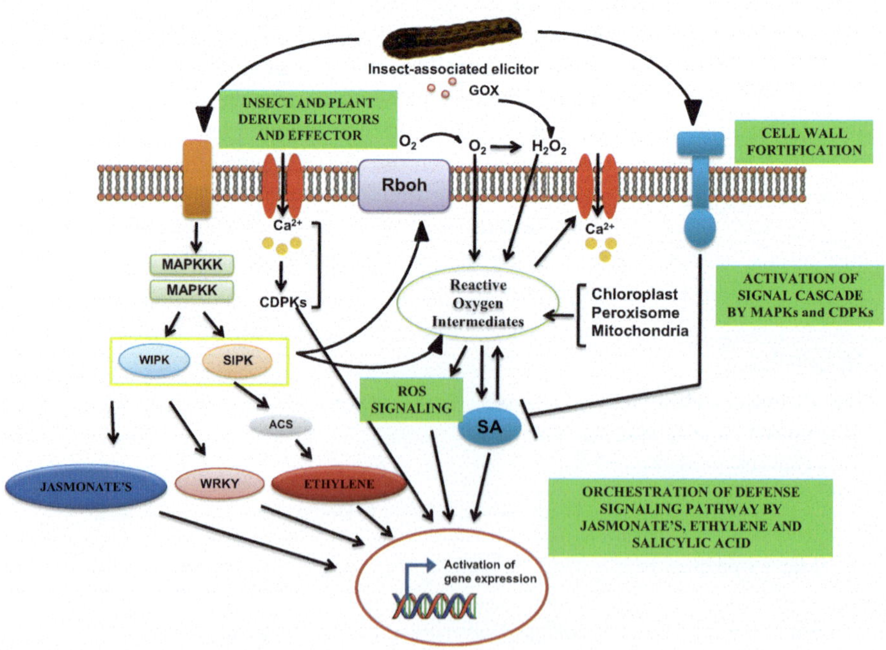

Fig. 1 Plant defense cascade deciphered by global gene expression profiling tools

integrated and evaluated on ecological and evolutionary scale to make it more valuable for crop improvement. But it is turning gruesome by emerging plant diversity, behavioral heterogeneity among the herbivore species, variability in the results among different research groups, and variation in the approach and systems used for insect-plant studies (Heidel-Fischer et al. 2018). The field of omics has emerged as a miraculous tool in probing these critical mind-boggling factors. Transcript profiling (transcriptomics) has been one of the globally measurable approaches of phenotypic response in both plants and insects in response to herbivory for decades (Heidel-Fischer et al. 2018; Vogel et al. 2018). It allows simultaneous analysis of thousands of genes in cells during different states (Wang et al. 2009). It began with a low-throughput Northern blot involving single-gene detection and progressed to a high-throughput multiple-gene expression profiling microarray technology, followed by next-generation sequencing that revolutionized transcriptomics by multidimensional assessment of cellular transcriptomics with single-base resolution. NGS is ruling the regime of genomics by generating massive amount of data overnight from both model and non-model organism. Since the success of first microarray analysis of 45 Arabidopsis gene using two-color fluorescent hybridization (Schena et al. 1995) microarray became the most popular genome-wide expression profiling tool. It expanded our basics about the genomic expression and genetic alteration of both the interacting allies which will enlighten the idea of pest-resistant crop. It is a superior approach of insect pest control compared to indiscriminate pesticide usage, which is undesirable for our ecosystem. It has embarked a new phase in designing crop resistant to insect technology.

2 Transcriptomics: A Miraculous Prop

An intriguing enigma of phenotypic diversity among different insect-plant interaction has been of prime interest to researchers for decades. Transcriptomics emerged as a panacea in all the facets of biological studies including insect herbivory and pest management (Fig. 2). It has become an exciting field of biological inquiry in the post-genomic era by solving enigma behind the correlation of cellular fate and function with gene expression patterns, in addition to supporting proteomics research. It decodes transcriptional status of the entire genome by cataloging change in the expression of all form of RNA (mRNA, non-coding RNAs, and small RNA) in a cell during various stress and developmental conditions (Wang et al. 2009). It advanced from candidate gene-based detection of RNAs, i.e., Northern blot to high-throughput expression profiling, i.e., microarrays. With the evolution of DNA sequencing technologies, i.e., next-generation sequencing, transcriptomics field was further revolutionized in terms of high-throughput expression data and cost-effectivity (Morozova et al. 2009). Initial transcriptomics analysis, i.e., Northern blots or microarray studies, highly relied on hybridization-based technologies and hybridization intensity, which is an indirect measure of the abundance of a transcript. But, the introduction of sequencing-based approaches to transcriptomics studies provided a best alternative to microarray-based analysis of gene expression due to

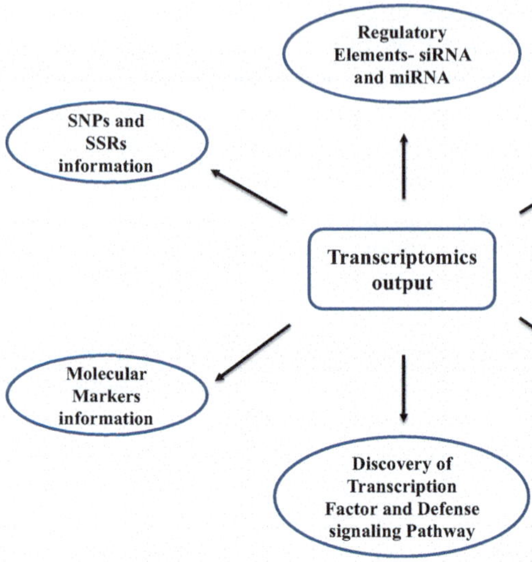

Fig. 2 Transcriptomics output generated by gene expression tools

its ability to directly determine the identity and the abundance of a transcript. The advent of high-throughput and cost-effective next-generation DNA sequencing (NGS) technologies tremendously transformed transcriptomics by allowing RNA inquiry at massive scale through cDNA sequencing approach termed as RNA-sequencing (Ozsolak and Milos 2011). Microarray-based gene expression profiling turned out as a routine in laboratories dealing with insect-plant interaction studies around the world. Besides being an indirect method relying on hybridization intensity, it also requires previous knowledge of a set of relevant transcripts. Other techniques utilized in probing differential gene regulation of previously unknown transcripts among different treatments include real-time PCR (qPCR), cDNA amplified length polymorphisms (cDNA-AFLP), and suppression subtractive hybridization (SSH). In the past few years, RNA-seq sculptured the transcriptomics landscape by providing more precise information of differentially expressed transcripts and their isoforms using deep-sequencing technologies. Transcriptional analysis of non-model insects and plant species is just an overnight task. It is important to highlight that these new transcriptomics approaches are not designed to replace conventional techniques. Standard techniques like Northern blot and qPCR are still utilized for selecting false-positive results and detail follow-ups of selected genes.

In this chapter, we will focus on various transcriptional studies conducted in both insect and plant upon insect herbivory, especially the expression data gained via global transcriptome analysis tools.

3 Transcriptomics Deciphering Plant's Enigmatic Response to Insect Herbivory

The plant genome is under constant alert to varied abiotic and biotic stresses. Insect herbivores occupy approximately one-quarter of all known eukaryotic species, emanate as one of the major stresses confronted by plants. As an aftermath of persistent selection pressure imposed by the insects, plant genome has evolved multitude of defense mechanisms to recognize and deploy appropriate defense responses against multiple insect species (Reymond et al. 2004; Vogel et al. 2007). The heterogeneity of plant defense response, its specificity and sensitivity with time, space, age, developmental stage, and organ, exhibits its long coevolutionary struggle with a diverse group of insects in community dynamics. This relationship is also influenced by changing global climatic conditions, i.e., increased atmospheric temperature and CO_2 concentration. Recent advances in transcript profiling technologies have profoundly decoded the complexity of insect-plants interaction in some model plants like *Nicotiana attenuata* (Hermsmeier et al. 2001; Hui et al. 2003; Halitschke et al. 2003; Roda et al. 2004; Giri et al. 2006) and *Arabidopsis thaliana* (Reymond et al. 2000, 2004; Cheong et al. 2002; Kempema et al. 2007; Consales et al. 2011; Appel et al. 2014) and economically significant crops like tomato (Kant et al. 2004; Musser et al. 2005, 2006, 2012; Scranton et al. 2013), cotton (Firmino et al. 2013; Dubey et al. 2013; Huang et al. 2015), maize (Tamayo et al. 2000; Johnson et al. 2011), chickpea (Singh et al. 2008), citrus (Mozoruk et al. 2006), poplar (Ralph et al. 2006a, b; Major and Constabel 2006; Philippe et al. 2010), and soybean (Wang et al. 2014, 2015) (Table 1). Initially, significant research was focused on pairwise interaction, elucidating plant defense mechanisms in response to single insect invader at a time which is far from natural ecological standards. In an ecological scenario, plants are exposed to multiple herbivores belonging to different feeding guilds simultaneously, involving a unique and complex web of agonist, antagonistic, and synergistic interactions among the plant metabolites which are missing in pairwise interaction studies. Hanna and other co-authors (Heidel-Fischer et al. 2018) reviewed most of the plant transcriptomics studies performed during the past two decades of transcriptomics revolution. It covered various studies undertaken to understand complicacy of plant defense orchestration in response to mechanical wounding vs. insect herbivory, differential feeding by chewing vs. piercing/sucking insect pest, and generalist vs. specialist insect feeders and insect herbivory indistinguishable in their feeding pattern.

3.1 Recognition of Insect Attackers Based on Feeding Guilds, HAMPs, and Herbivore-Specific Chemical Cues

Besides constitutive defense, plants also utilize sophisticated induced defense mechanism and deploy an exclusive set of molecules for recognizing insect attackers. It also avoids wasteful and unspecific defense responses. Plants can well discriminate

Table 1 Solving enigma of insect-plant interaction using transcriptomics techniques

Plant/ insecticide	Insect	Transcriptome profiling tool	Enigma resolved	References
Nicotiana tabacum	*Bemisia tabaci*	Suppression subtractive hybridization (SSH) and cDNA microarray	Interaction between *B. tabaci* and secondary defense metabolites	Alon et al. (2012)
Tomato	*Helicoverpa zea*	Microarray	Caterpillar labial saliva is an important component of herbivory that can alter plant gene expression	Musser et al. (2012)
Pyrethroid insecticide	*Helicoverpa armigera*	Microarray	Generalist insect's adaptation to major host plant secondary metabolites enhances defense against other toxic chemicals such as insecticide	Tao et al. (2012)
Potato	*Leptinotarsa decemlineata*	"Suppression subtractive hybridization" library using cDNA	Impact of protease inhibitors on CPB larvae	Petek et al. (2012)
Pirimicarb	*Myzus persicae*	Microarray	Transcriptional plasticity responsible for adaptation to different insecticides	Silva et al. (2012)
Tomato	*Macrosiphum euphorbiae*	Microarray	Tomato-aphid interaction, stress-responsive SA-dependent genes	Coppola et al. (2013)
Zea mays	*Spodoptera frugiperda*	RNA-Seq	Mir1-CP from maize was found to degrade insect intestinal mucin. It led to compensatory upregulation of genes in the MG that encode proteins involved in PM production and food digestion	Fescemyer et al. (2013)
Solanum tuberosum	*Leptinotarsa decemlineata*	Pyrosequencing	Generation of dataset of insect gut transcriptome and its analysis for the presence of RNAi-related genes	Swevers et al. (2013)
Apple and snowberry	*Rhagoletis* flies	Microarray	Mechanism involved in host race formation in plant-feeding insects	Ragland et al. (2015)

(continued)

Table 1 (continued)

Plant/ insecticide	Insect	Transcriptome profiling tool	Enigma resolved	References
Eucalyptus grandis	*Leptocybe invasa*	RNA-Seq	Transcriptional landscape and terpene profile of a resistant and susceptible *Eucalyptus*	Oates et al. (2015)
Solanum tuberosum, Barbarea verna, Hordeum vulgare	*Myzus persicae, Myzus cerasi, Rhopalosiphum padi*	RNA-Seq	Identified diverse putative effector sets, important for specific plant-aphid interactions. It aids in determining aphid host range	Thorpe et al. (2016)
Malus domestica	*Phyllonorycter blancardella*	Microarray	Reprogramming of plant phytohormonal balance for increased nutrient mobilization and mitigation of plant defense	Zhang et al. (2016)
Camellia sinensis	*Ectropis oblique*	RNA-Seq	Molecular mechanism of response of tea to *Ectropis oblique* attack	Wang et al. (2016)
Solanum lycopersicum	*Agriotes lineatus (root herbivore) Myzus persicae*	Time-course gene-expression analysis (real time PCR)	Deciphered interaction between below- and aboveground pests with different feeding guilds	Coppola et al. (2017)
Arabidopsis thaliana	*Pieris brassicae* (specialist) *Heliothis virescens* (generalist)	RNA-Seq	Glucosinolates had profound and contrasting effect on the performance and transcriptome signature of specialist and generalist	Schweizer et al. (2017)
Cabbage and tobacco	*Spodoptera litura*	RT-PCR	Role of chemosensory proteins in adaptation and specialization to different ecosystems	Yi et al. (2017)
Solanum dulcamara	*Spodoptera exigua*	Microarray	Wounding and application of oral secretion that mimics herbivory	Lortzing et al. (2017)
			This entire row needs to be shifted below Oates et al. (2015)	

insect feeding from mechanical wounding (Mithöfer et al. 2005). Researchers found a distinct transcriptomics signature in response to insect damage, compared to wounding (Reymond et al. 2000; Roda et al. 2004; Ralph et al. 2006b; Lawrence et al. 2008; Consales et al. 2011). The pioneering microarray study of *Arabidopsis-Pieris rapae* interaction (Reymond et al. 2000) was performed using a “boutique” chip, customized of 150 wound-inducible genes. It revealed differential dynamics of *A. thaliana* defense induction in response to mechanical wounding, compared to insect feeding. A temporal analysis upon wounding recognized a potential correlation between wound-induced transcript signature and oxylipin signature. Genes induced by wounding stimulus showed similar behavior as that of genes that are involved in JA synthesis and metabolism of 12-oxo-phytodienoic acid (OPDA), dinoroxophytodienoic acid (dinor OPDA), and jasmonic acid (JA). But the expression analysis of coi1-1 mutant identified expression of some wound-inducible genes in JA-independent manner. However, a distinct transcriptome profile was recognized in Pieris-wounded plants as compared to mechanically wounded plants, especially in the expression of water stress-regulated genes. A reduction in the expression of insect-wounded plant was observed. The failure to mimic insect-derived wounding in mechanical wounding samples could be the reason for variable response (Reymond et al. 2000). Later availability of the entire *A. thaliana* genome-based microarray opened new perspectives for unbiased transcriptome analysis studies. Problem of inadequate stimulus produced by manual clipping or punching holes was compensated with MecWorm (a mechanical caterpillar) that mimics insect damage by exerting continuous wounding. Profiling of plants' volatile emission upon MecWorm-induced damage was similar to that of insect damage-induced volatile bouquet (Mithöfer et al. 2005). An unpublished study by Kroymann, Mithöefer, Boland, and Vogel addressed by Heidel-Fischer and group (Heidel-Fischer et al. 2018) reported 70% overlap of transcript profile between *Plutella xylostella* and MecWorm-damaged plants. This transcript signature was specific to mechanical stimulus, rather chemical cues present in insect-derived secretions. In comparison to *P. xylostella* herbivory-induced response, MecWorm was found to strongly induce expression of heat shock-responsive genes. Mimicking of insect-induced mechanical damage by MecWorm suggested the role of insect-derived chemical secretion in modulating plant gene expression profiling upon herbivory. A sialome study revealed presence of a wide range of crucial factors in caterpillar's labial salivary transcriptome that could potentially reprogram plant's transcriptomics responses upon insect attack (de la Paz Celorio-Mancera et al. 2011). Application of volicitin (a compound present in oral secretion) from *Spodoptera exigua* caterpillar on mechanically wounded leaves of corn released different volatile blends, compared to mechanically wounded leaves without application (Alborn et al. 1997). Transcripts responsible for volatile production were upregulated by volicitin and *Spodoptera exigua* infestation (Lawrence and Novak 2004). From evolutionary point of view, induction of plant defense upon insect herbivory and its suppression by phytophagous insects is a key point. Besides tissue maceration, oral secretion (OS) comprised by labial and mandibular gland salivary secretion along with regurgitant also customizes plants induced defense. Pioneer research published by Musser and group reported glucose oxidase (a salivary enzyme) as the suppressor of wound-inducible defense. According to a previous report, glucose oxidase from *Helicoverpa zea* labial gland secretion was found to counteract nicotine accumulation in tobacco and suppressed it (Musser et al. 2002). Role of glucose oxidase in suppressing plant defense response was also shown in other solanaceous plants, like tomato (*Lycopersicon esculentum*). It has been found to lower trypsin inhibitor level in tomato (Musser et al. 2005). Similar reports of OS-mediated attenuation of wound-inducible responses were observed in *Arabidopsis thaliana* transcriptome. Oral secretion of *Pieris brassicae* and *Spodoptera littoralis* was found to suppress plant defensive responses for minimum 48 h, especially ERF/AP2 transcription factor and protease inhibitors. This suppression was not derived from any of the known OS elicitors, and it was independent of JA or SA pathway (Consales et al. 2011). In contrary, some studies demonstrated the induction of plant defensive response by herbivore-derived OS. Global gene expression profiling of wounded *Solanum tuberosum* leaves upon treatment with Colorado potato beetle (CPB) regurgitant identified wound-responsive and regurgitant-responsive genes. Regurgitant-derived elicitors were found to induce or repress gene expression involved in CPB-potato interaction. Application of CPB regurgitant on wounded potato leaves induced genes involved in general metabolism, secondary metabolism, transcriptional regulation, pathogen response, stress response, and protein expression. It repressed photosynthesis and stress-related genes (Lawrence et al. 2008). A cDNA microarray-based transcript profiling in tomato, pointed caterpillar saliva as the crucial component of *H. zea* herbivory in reorganizing tomato gene expression. Components derived from insect oral secretion facilitates plant in recognizing and reprogramming gene expression upon different types of herbivory, beyond physical damage. Some of the transcripts such as protease inhibitors, arginase, dehydrin, acidic endochitinase, acid phosphatase, arginase, polyphenol oxidase, and threonine deaminase were strongly upregulated in mock ablated *H. zea* larvae, compared to ablated. It was a robust study that identified candidate genes responsive to *H. zea* labial saliva (Musser et al. 2012). Chemical cues (saliva and regurgitant) released by fall armyworm (FAW) during feeding bouts have been found to trigger maize defensive response upon *Spodoptera frugiperda* herbivory. During each feeding bout very less amount of regurgitant has been deposited in wounded leaves by FAW, compared to saliva. Due to variation in the wax composition of cuticle across different developmental stages, maize (B73) defense induction (direct and indirect defense) was found less responsive to the application of regurgitant. However, saliva (potential HAMP) was found to significantly alter the induction of several maize defensive genes (Chuang et al. 2014). Earliest studies on simulated herbivory (application of OS on mechanically wounded sites) measured plant defense dynamics in a time frame of few hours to few days. However, studies characterizing early wound-inducible genes are sporadic due to practical problems. Recently, a transcriptome profiling of early wound-responsive genes within minutes to simulated herbivory was performed in chickpea, using OS derived from *Helicoverpa armigera, which showed* substantial modulation in hormonal networks and growth-associated genes. Within 20 min of wounding, upregulation of genes involved in jasmonic acid and ethylene pathways and downregulation of growth-promoting genes were recorded (Pandey et al. 2017). Besides insect-derived oral secretion, insect eggs are also found to remodel gene expression profile of plants. Various fascinating transcriptomics studies discerned effect of insect's inert stage, i.e., eggs on plant defense induction. A comparison of gene expression profile of Arabidopsis foliage upon egg-laying by *Pieris brassicae* and larval feeding revealed drastic differences. Surprisingly, insect oviposition induced SA-dependent hormonal pathway (Little et al. 2007). Arabidopsis leaves treated with egg extract derived from *P. brassicae* and *S. littoralis* induced SA pathway; it concomitantly suppressed JA-dependent induction of plant defensive metabolites which is responsive against caterpillars (Bruessow et al. 2010). These gene expression profiling studies revealed the intriguing advantage availed by insect from the molecular cross talk between SA and JA pathways, consequently suppress plant defense induction and promote larval growth. Oviposition by *P. brassicae* triggered accumulation of SA in local and systemic leaves. This egg-induced response was similar to the response induced upon pathogen-associated molecular patterns (PAMPs) recognition response. It activated systemic acquired resistance (SAR) that inhibited *Pseudomonas syringae* growth. Bacterial growth had a detrimental effect on *P. brassicae* larval growth. Altogether, *P. brassicae* triggered SA pathway and activated SAR response. It was aimed to reduce the detrimental effect of *Pseudomonas syringae* growth on feeding *P. brassicae* caterpillar (Hilfiker et al. 2014). A recent study explored impact of moth oviposition in plant defense priming. It evaluated *Nicotiana attenuata* transcriptomics response and measured phytohormone level upon *Spodoptera exigua* (generalist) and *Manduca sexta* oviposition. In this study generalist larvae demonstrated stronger impact in regulating gene expression and phytohormone level compared to specialist larvae. It observed a remarkable species-specific customization of plant defense response, specific to ovipositing species.

3.2 Deciphering of Herbivory-Associated Molecular Cross Talk and Phytohormonal Signaling

Plant surveillance system could specifically perceive herbivore-associated cues, allowing recognition and cost-effective response. This specificity was achieved by spatiotemporal modulation of JA-dependent and JA-independent processes. It is an integrative system which includes some of the key JA-modulating hormones such as salicylic acid (SA), ethylene (ET) abscisic acid (ABA), brassinosteroids (BR), gibberellins (GB), auxin, and cytokinins (CK). As compared to pathogens, insects are far more complex multicellular organisms. Plants could recognize insect cues emanating from their lifestyle and behavioral patterns. Even mechano-stimulation was found to induce JA accumulation (Tretner et al. 2008). Jasmonates (JAs) include elite group of secondary metabolites that are highly investigated in past years, especially reflected in several genome-wide gene expression profiling studies in Arabidopsis. JAs could induce production of multiple branches of several classes of secondary metabolites that include the terpenoids, phenylpropanoids, and alkaloids (Zhao et al. 2005). Transcriptome profiling suggested extensive reprogramming of genes involved in cellular metabolism and cell cycle by jasmonates (Goossens et al. 2003; Pauwels et al. 2008). Transcriptional cascade triggered by JA further induces several other waves of pathways, reflected in terms of positive and negative gene regulation. This subsequent wave induced by JA includes signaling activator MYC2, jasmonate ZIM-domain (JAZ) repressor family, and AP2/ERF family (Dombrecht et al. 2007; Shoji et al. 2008). Tissue disruption inflicted by insects triggers production of JA-Ile, a precursor of defense hormone, that mount well-organized signal cascade. But the extent of involvement of herbivory-associated signals (HAMPs, "non-self") and plant-associated signals (DAMPs, "self") in JA-Ile production is still unresolved (Heil 2009). However mechanical wounding could sufficiently induce JA-Ile production which indicated least involvement of herbivory cues in activating response (Koo and Howe 2009; Chung et al. 2008). JA pathway emerged as key signaling machinery that gets activated to both specific and non-specific herbivory attack. The fine-tuning of defense machinery is beyond JA regulation. It could be achieved either by regulating expression of JA modulators in a spatial-temporal pattern or via utilizing parallel pathways in JA-independent manner. Resistance to aphid attack was found to be regulated by Mi-1 gene, a putative receptor involved in inducing SA signaling, independent of JA. It suggests the involvement of JA-independent pathways in inducing resistance to phloem feeders (Li et al. 2006). Most of the insect herbivores inflict more damage compared to phloem feeders and activates resistance via jasmonic acid signaling. In such cases, the herbivore-specific response is achieved by molecular cross talk with other phytohormones. Genomic reprogramming induced by JA is mainly altered by SA and ET. SA acts antagonistically to resistance induced by JA, while ET can alter JA-induced resistance in both positive and negative manner (Erb and Glauser 2010; Zhu et al. 2011). Molecular cross talk mediated by SA-JA-ET phytohormones regulate plant defense to insect herbivores. Microarray-based gene expression analysis of plant tissue mined by leaf miner *Phyllonorycter blancardella* demonstrated accumulation of cytokinin hormone, despite weak expression of cytokinin-related genes in mined plant tissues. This study suggested manipulation of plant physiology by leaf miner to make a favorable nutritional environment (Zhang et al. 2016). Mitogen-activated protein kinases (MAPKs) have been found to get activated upon insect attack and regulates plant defense induction against herbivory. Besides regulating phytohormonal dynamics, MAPKs also induce transcription of defense-related genes and production of defense metabolites (Hettenhausen et al. 2015). Various transcriptome studies performed on Arabidopsis on exogenous application of JA reflected switch over from growth to defense. It invoked transcript expression of pathogenesis-related proteins, oxidative stress signaling, anthocyanin, and cell wall remodeling. Transcriptome analysis provided a comprehensive view of molecular cross talk and the importance of JA homeostasis upon herbivory (Pauwels et al. 2009).

3.3 Plants Transcriptional Signature upon Single and Multiple Insect Herbivory

Plants' transcriptional reprogramming dictated by phytohormonal cross talk in response to insect herbivory shows remarkable specificity. Gene expression data generated upon single vs. sequential herbivory shared common phytohormonal response, suggesting it as a common modulator of plant defense signaling against various stresses. Transcriptional signature produced upon prior stress was found to affect subsequent stress signaling. In Arabidopsis, prior exposure to *Pieris rapae* strongly alters its resistance to fungus *Botrytis cinerea* (Coolen et al. 2016). Such experiments are close to an ecological scenario where plants are exposed concurrently to various stresses. Drought and flooding was also found to vary plants response to insect herbivory. *Spodoptera exigua* showed better performance on flooded *Solanum dulcamara* plant compared to drought-stressed plant. RNA-seq profiling of flooded tissue indicated a lack of accumulation of JA in flooded tissue (Nguyen et al. 2016). Drought was found to strengthen plant response to insect herbivory.

3.4 Deciphering Enigma of Plant Defense Priming Using Transcriptomics

In order to exhibit a quick and efficient response to environmental challenges plants utilize "priming memory." Recent advancement in omics techniques has also elucidated priming phenomenon in plants (Balmer et al. 2015). Priming has no negative impact on plant energy status because it does not utilize such defense repertoire constitutively. The priming component is only utilized upon stress. Both direct defense and primed plant provide enhanced protection, but direct defense incurs high physiological cost on plants compared to priming. It suggests greater importance of priming-based crop protection. Priming has three major stages – (a) priming phase, (b) post-challenge primed state, and (c) transgenerational primed state. As insect oviposition is followed by herbivory, and plants are found to strengthen their defense by inducing direct and indirect defense, to fend off future herbivory. Oviposition by *Helicoverpa zea* on tomato leaves led to stronger induction of pin2 (protease inhibitor2) and JA, compared to the one without oviposition. Therefore, oviposition primed wounding-induced defense in tomato (Kim et al. 2012). Abscisic acid (ABA) was found to be a core regulator of defense priming against insect herbivores in Arabidopsis (Vos et al. 2013). ABA was shown to prime JA-dependent signaling. β-Aminobutyric acid (BABA) is another key inducer of defense priming against insects. BABA application on soybean primed ROS scavenging and SA signaling (Zhong et al. 2014). Beneficial organisms that live in association with plants are also found to prime antiherbivore defense in tomato via JA pathway (Song et al. 2013). Volatile organic compounds (VOCs) that get induced upon insect attack could also prime defense response in conspecific neighboring plants. It enhances indirect defense strategy of the receiving plants (Kost and Heil 2006).

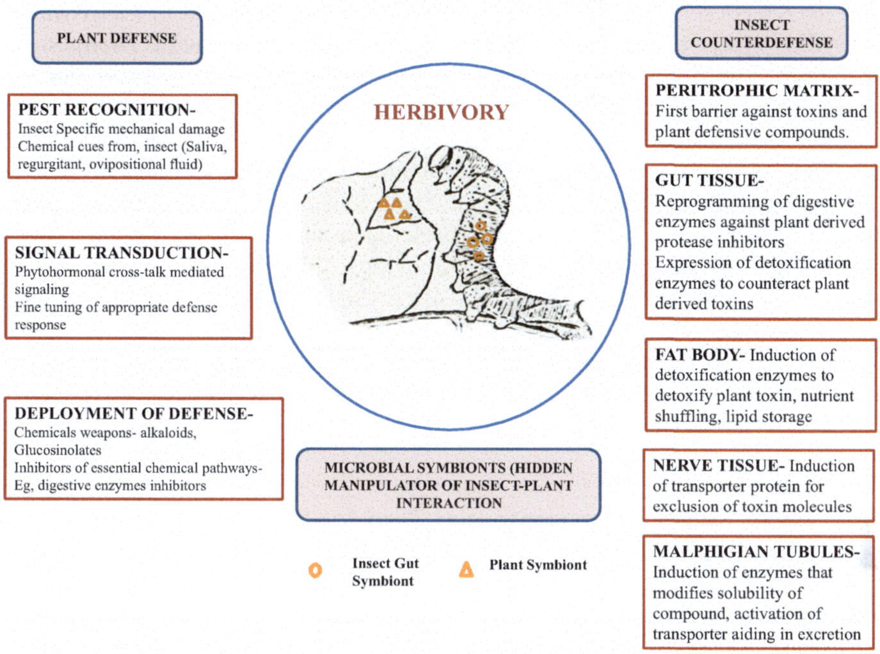

Fig. 3 Enigma of plant-insect interaction elucidated by transcriptome profiling tool

3.5 Deciphered Presence of a Hidden Player in Insect-Plant Interaction, Microbes: A Manipulator of Insect-Plant Interaction

Phytophagous insects could manipulate host plant physiology to get benefits (Pieterse and Dicke 2007). Microbial symbiotic partners are involved in synthesizing bioactive molecules that participate in other interactions (Spiteller et al. 2000). The growing evidence from genomic data and omics techniques has highlighted the importance of microbial symbiosis in influencing plant-insect interaction (Fig. 3). These microbial mutualists benefit insects by manipulating plant physiology and alter host plant range. Besides providing essential nutrients, influencing host behavior and physiology, it is apparent that these insect-associated symbionts can interfere with plant defense cascade induced upon insect attack (Frago et al. 2012; Zhu et al. 2014). Recent insight using transcriptome profiling demonstrated microorganism-induced tilt in plant defense signaling in favor of insects. Microarray-based gene expression profiling of maize fed by antibiotic-treated and untreated *Diabrotica virgifera virgifera* larvae showed higher induction of maize defense-related genes in *Wolbachia*-cured *Diabrotica* beetle. It was the first study that found potential link of insect microbial symbiont in silencing induced defense response of insect host plant (Barr et al. 2010). Similar suppression of plant defense signaling was observed in tomato by endosymbiont "*Candidatus Liberibacter psyllaurous*" (Lps) of

Bactericerca cockerelli. This microarray-based study revealed a selective advantage for psyllids by Lps (Casteel et al. 2012). In another study performed in *Leptinotarsa decemlineata*, bacteria in larval oral secretion suppressed JA-responsive defense response of tomato (Chung et al. 2013). A facultative symbiont Hamiltonella defense in whiteflies was found to suppress JA-mediated defense responses in tomato. This microbial symbiont-associated downregulation of defensive pathway led to enhanced fecundity and survival (Su et al. 2015). Salivary secretion of potato aphids was identified with a chaperonin GroEL (bacterial protein), which is derived from its endosymbiont *Buchnera aphidicola*. In Arabidopsis, GroEL was recognized as an inducer of oxidative burst and BRASSINOSTEROID INSENSITIVE 1-ASSOCIATED RECEPTOR KINASE 1-mediated expression of pattern-triggered immunity (PTI) (Chaudhary et al. 2014). The green island phenotype of leaf miners induced in plants by manipulating plants cytokinin profile is also derived from symbiotic bacteria (Giron and Glevarec 2014).

3.6 RNA-Seq: A Transcriptomics Tool for Non-model Plant Species

In the past decade advancement of computational tools and advent of contemporary next-generation sequencing technology has complemented RNA-seq (RNA sequencing) technology. It has elevated both the quality and quantity of transcriptomics data. RNA-seq is preferred over some of the global gene expression profiling tools such as serial analysis of gene expression (SAGE) and microarray due to its accuracy, reproducibility, and statistical significance. Compared to the microarray, RNA-seq also gives an insight of small RNA, alternative splicing, and allele-specific expression. Transcriptome analysis using RNA-seq is independent of genome annotation, and it circumvents bias that arises in hybridization-based technology. In this technology simultaneous analysis of the repertoire of insect and plant traits in non-model species has become easier, with no requirement of the reference genome and biological information of the target organism. RNA-seq facilitated comparative analysis of phylogenetically unrelated species or across different developmental stages and diverse experimental conditions. In particular, tracking of gene expression using RNA-seq not only explores the transcriptomics landscape of non-model species and serves basic research, but also proved beneficial for crop improvement and supports applied research. Besides deciphering plants enigmatic defense response to insect feeding, it also reveals insect resistance to plant defensive toxins and evolution of polyphagy (Strickler et al. 2012; Oppenheim et al. 2015).

With the latest RNA-seq technology, researchers get explicit knowledge of plant defense fine-tuning against specific phytophagous insects (Lv et al. 2014; Oates et al. 2015; Yang et al. 2015; Wang et al. 2016). It allows comparison of herbivore-induced defense response in the model and non-model plants and facilitates better elucidation of plant defensive responses against insect herbivores. RNA-seq revealed resistance gene BHP15 and its molecular mechanism of resistance against brown planthopper. BHP15 is utilized in rice breeding program for effectively controlling *Nilaparvata lugens* (brown planthopper), a destructive agricultural pest (Lv et al. 2014). The improved knowledge of *Eucalyptus grandis* transcript profile was obtained using RNA-seq, upon *Leptocybe invasa* infestation. It illuminated key defensive genes of *Eucalyptus* that could be manipulated for designing pest-tolerant plants (Oates et al. 2015). Analysis of comparative transcriptome sequencing profile of corn upon *Ostrinia furnacalis* feeding and methyl jasmonate application revealed jasmonate hormone signaling as the crucial pathway in corn defensive response, upon insect herbivory (Yang et al. 2015). The elucidation of *Camellia sinensis* (tea plant)-induced defense response to *Ectropis obliqua*, a serious pest of tea plantation, found comprehensive reprogramming of plants secondary metabolites and signaling pathways. It implied eco-friendly and pesticide-free tea farming (Wang et al. 2016). Application of advanced mapping techniques and RNA-seq in exploring plant defense to herbivores will enable the implying of the molecular logic of insect-plant interaction. Studies on insect transcriptomics tracked various aspects of insect behavior, physiology, and development.

3.7 Transcriptomics Empowering Concept of Sustainable Agriculture

The elaborate elucidation of this intricate and dynamic plant-insect relationship will improve agricultural production globally, without deteriorating environmental quality. Current research on plants' biotic stress has demonstrated great interest in the analysis of gene expression profiling during insect-plant interaction upon herbivory (Gimenez et al. 2018). Efforts of the scientific community in unraveling plant defense dynamics and insect strategies to breach plant defense profile will revitalize pest management strategies and improve sustainable agriculture. Designing of insect-tolerant, transgenic plants are the cornerstone of sustainable agriculture. The modern agriculture has primarily utilized genetic transformation tool in targeting insect pests and enhancing insect-tolerant trait in crops. In the 1980s, gene transfer into a plant genome using *Agrobacterium tumefaciens* (soil bacterium) resident plasmid brought major agricultural revolution. It was embarked with creation of first transgenic tobacco plants (Barton et al. 1983; Herrera-Estrella et al. 1983). Insect-tolerant plant turned out to be a good alternative to insecticide. For designing such plants, identification and characterization of plants' natural defense repository among wild varieties is essential. A combination of omics techniques, molecular genetics tools, and studies involving genome-wide association dissects variation in plants resistance to insect herbivores. It facilitates analysis of defensive genes and molecular markers which assists in the breeding of resistant traits and production of economically important insect-resistant crops (Broekgaarden et al. 2011). Transgenic technology helps in functional annotation of gene and elucidates its role in vivo. In the near future, crop plants will be genetically engineered for enhanced production of defensive compounds. Knowledge of chemical elicitors of plant-induced resistance could be sprayed on crops to confer resistance (Bruce 2010).

4 Unravelling Enigma of Insect Resistance, Adaptation to Xenobiotics and Herbivory-Induced Chemical Pressure in Ecological Context

The spectrum of phytophagous insects-plant interaction is vast, ranging from specialist herbivores munching on related plants or devouring specific plant families to generalist herbivores adapted to distantly related plant families. With time global gene expression profiling tools have overshadowed studies exploring single or few genes expression (Li et al. 2007; Wybouw et al. 2015; Gompert et al. 2015; Ragland et al. 2015; Eyres et al. 2016; Snoeck et al. 2017; Pearce et al. 2017; Schweizer et al. 2017). Over past decades, substantial progress has been made in elucidating transcriptional paradigm of phytophagous insects' response to plant models, economically important crops, genetically modified crops, and synthetic insecticides (xenobiotics). However, recently researchers headed to unravel the genetic basis of molecular and evolutionary aspect of insect adaptation to strong chemical pressure in ecological context. Researchers have critically examined the evidence facilitated by transcriptomics to decode some of the pivotal question pertaining to better insect pest control. It involves:

- How insects circumvent and adapt to diverse chemicals defenses posed by the host plant?
- What is the cost and benefit of host tolerance response and its link to the evolution of insecticide resistance in the ecological landscape? How insects decide and evolve its host plant breadth?
- To what extent adaptation of generalist insect herbivore to its host plants reprograms both herbivore and host transcriptome?
- What is the molecular basis of host race formation and transcriptomics profiling of phenotype responsible for speciation, among different host races?
- How specialists and generalists differ in adaptation to host plant and what is its genetic basis?
- What are the genomic concept and transcriptional plasticity underlying polyphagy?

In this part we have addressed all the above pivotal questions related to insect physiology and adaptation upon plant feeding, focused exclusively on insights gained through transcriptomics profiling techniques.

During the last decade, noteworthy research has been conducted in elucidating molecular aspects of insecticide resistance in various insect pests. Global transcriptional response of insect herbivore has been recorded upon feeding host plant(s), revealing different regulatory mechanisms, detoxification, and compensatory response-related genes. Transcriptional profiling using NGS is an inexpensive technique that can sequence vast transcript data overnight. It has also been proved beneficial in comprehensive gene expression profiling of non-model species of insects, especially belonging to Coleoptera, Lepidoptera, Hemiptera, and Diptera (Vogel et al. 2018). Phytophagous insects face various challenges during host plant selection. Insects surpass plants' inducible defense by using a combination of different counter-defensive strategies that encompass avoidance and detoxification in generalist insects, sequestration, and effective excretion in specialist insect feeders. Gene expression profiling comprehended molecular and evolutionary basis of insect behavior, physiology, and development. The impact of plant hormones (JA, SA, and ET) on insect physiology was delineated by evaluating transcriptional response of *Helicoverpa zea* larvae CYP450 genes using RT-PCR. Dietary supplementation of *H. zea* larvae with jasmonic acid and salicylic acid reported over transcription of detoxification-related genes (CYP450 genes) upon consumption. This "eavesdrop" ability of *H. zea* on plant defensive signals, i.e., phytohormones have been found to protect insects from plant defensive toxins (Li et al. 2002). The shortcoming of Li and colleagues' study on "eavesdropping hypothesis" which only tracked insect detoxification-related genes was overcome by utilizing large-scale transcriptomics tool. This study identified global gene expression reprogramming in *Helicoverpa armigera* upon feeding JA, SA, and ET mixed diet. SA and ET have been found to differentially regulate genes involved in insect defense, immunity, and cell rescue, while 12-oxophytodienoic acid (OPDA), a JA precursor, was toxic *to H. armigera* (Vogel et al. 2018). Global transcriptome studies were conducted to figure out insects' transcriptome-wide response to plant-derived protease inhibitors. Protease inhibitors are "default component of plants' natural defense system against phytophagous insects," and it has been extensively explored for their antiherbivore activity and putative pest control agent (Oppert et al. 2010; Zhu-Salzman and Zeng 2015; Singh et al. 2018). Genomic and transcriptomics analysis of highly polyphagous Noctuidae pest *Spodoptera litura*, *Helicoverpa armigera*, and *H. zea* observed great expansion of detoxification and digestion genes that include CYP450 family, (CYP3 and CYP4), GST family, and gustatory receptor (Pearce et al. 2017; Cheng et al. 2017). Specialist herbivores better adapted to plant-induced defense compared to generalist herbivores (Govind et al. 2010). Specialists can even manipulate the defensive chemistry of their host plant. *Manduca sexta*, a specialist of solanaceous plant, served a crucial model in understanding insect molecular biology. RNAseq analysis of *Manduca sexta* larvae reared on Solanaceous and Brassicaceous (non-host) plant identified specific transcriptional signature and digestive plasticity on both host and non-host plants. In contrary to our expectations, it demonstrated efficient utilization of non-host plant (Koenig et al. 2015), as generalists are adapted to survive on a broad range of host plant. This enigmatic adaptation of polyphagous herbivores involves great extent of transcriptional reprogramming of both herbivore and associated host plants. Generalists arthropod pest *Tetranychus urticae* adapted to tomato demonstrated higher detoxification potential with enriched gene profile of detoxification enzymes and xenobiotic transporters. Besides, it also reduced the production of defensive compounds in hosts plant (Wybouw et al. 2015). A study by Dermauwa and group has given unprecedented view of the transcriptional correlation between polyphagy and evolution of pesticide resistance in insect pests (Dermauw et al. 2013). Helicoverpa larvae that was fed with gossypol demonstrated enhanced expression of CYP450 in the midgut. These cotton bollworm larvae exhibited higher tolerance to deltamethrin (Tao et al. 2012). Analysis of the transcriptional activity of *Bemisia tabaci*, a hemipteran polyphagous pest, fed on transgenic tobacco that is constitutively expressing secondary metabolites identified enhanced expression of transcript encoding for oxidative stress, detoxification, immunity, and other stress-related genes. Its ability to maintain its reproductive performance on such transgenics indicates its high adaptability to those secondary metabolites. This experimental study utilized cDNA microarray and suppression subtractive hybridization (SSH) approaches (Alon et al. 2012). Dietary adaptation of Spodoptera species involved substantial transcript alteration of peritrophic membrane, digestive, transporters, detoxifying enzymes, and immunity-associated genes in the diet-specific manner (Roy et al. 2016). Gene expression analysis by RNA-seq has elucidated both insect adaptation and speciation. Races that arise due to genetic drift and reproductive isolation show variation in their transcriptomic response to the plant on which they are maintained (Eyres et al. 2016). Transcriptome-driven analysis indicated that evolutionary biology of host plant adaptation and insecticide resistance are the same.

The convenience and reliability with which the whole transcriptome can be explored from both model and non-model species even contradicted the 60-year-old "pre-adaptation hypothesis." According to this hypothesis, polyphagous insects are genetically predisposed and more susceptible to develop resistance to insecticides. However, recent studies hinted the other way around. Despite expanded defensome of generalist insects, resistance development is also governed by other factors such as genetic, biological, and operational factors (Dermauw et al. 2018).

5 Conclusion

Transcriptomics brings value to the available gene sequences and allows thorough data mining of the target gene. With global transcriptome analysis tool such as microarray and advanced latest profiling techniques like RNA-seq immense gene expression data could be mined in a very short duration. It allowed well documentation of various host plant defensome, specific plant metabolite, insecticides and xenobiotic-induced genetic reprogramming in insect herbivore pests. It tracked differential molecular response of generalists and specialists insect feeders in response to host and non-host plant feeding. It critically examined reorganization of the genetic architecture of plant defensive response and insect counter-defensive response upon insect herbivory. It also discerned the potential link between polyphagous lifestyle and insecticide resistance. This growing knowledge could be efficiently utilized in designing insect-tolerant plants.

References

Alborn HT, Turlings TCJ, Jones TH, Stenhagen G, Loughrin JH, Tumlinson JH (1997) An elicitor of plant volatiles from beet armyworm oral secretion. Science 276(5314):945–949

Alon M, Elbaz M, Ben-Zvi MM, Feldmesser E, Vainstein A, Morin S (2012) Insights into the transcriptomics of polyphagy: *Bemisia tabaci* adaptability to phenylpropanoids involves coordinated expression of defense and metabolic genes. Insect Biochem Mol Biol 42(4):251–263

Appel HM, Fescemyer H, Ehlting J, Weston D, Rehrig E, Joshi T, Xu D, Bohlmann J, Schultz J (2014) Transcriptional responses of *Arabidopsis thaliana* to chewing and sucking insect herbivores. Front Plant Sci 5:565

Balmer A, Pastor V, Gamir J, Flors V, Mauch-Mani B (2015) The 'prime-ome': towards a holistic approach to priming. Trends Plant Sci 20(7):443–452

Barr KL, Hearne LB, Briesacher S, Clark TL, Davis GE (2010) Microbial symbionts in insects influence down-regulation of defense genes in maize. PLoS One 5(6):e11339

Barton KA, Binns AN, Matzke AJ, Chilton MD (1983) Regeneration of intact tobacco plants containing full length copies of genetically engineered T-DNA, and transmission of T-DNA to R1 progeny. Cell 32(4):1033–1043

Broekgaarden C, Snoeren TA, Dicke M, Vosman B (2011) Exploiting natural variation to identify insect-resistance genes. Plant Biotechnol J 9(8):819–825

Bruce TJ (2010) Tackling the threat to food security caused by crop pests in the new millennium. Food Secur 2(2):133–141

Bruessow F, Gouhier-Darimont C, Buchala A, Metraux JP, Reymond P (2010) Insect eggs suppress plant defence against chewing herbivores. Plant J 62(5):876–885

Casteel CL, Hansen AK, Walling LL, Paine TD (2012) Manipulation of plant defense responses by the tomato psyllid (*Bactericerca cockerelli*) and its associated endosymbiont *Candidatus Liberibacter psyllaurous*. PLoS One 7(4):e35191

Chaudhary R, Atamian HS, Shen Z, Briggs SP, Kaloshian I (2014) GroEL from the endosymbiont *Buchnera aphidicola* betrays the aphid by triggering plant defense. Proc Natl Acad Sci 111 (24):8919–8924

Cheng T, Wu J, Wu Y, Chilukuri RV, Huang L, Yamamoto K, Feng L, Li W, Chen Z, Guo H, Liu J (2017) Genomic adaptation to polyphagy and insecticides in a major East Asian noctuid pest. Nat Ecol Evol 1(11):1747

Cheong YH, Chang HS, Gupta R, Wang X, Zhu T, Luan S (2002) Transcriptional profiling reveals novel interactions between wounding, pathogen, abiotic stress, and hormonal responses in Arabidopsis. Plant Physiol 129(2):661–677

Chuang WP, Ray S, Acevedo FE, Peiffer M, Felton GW, Luthe DS (2014) Herbivore cues from the fall armyworm (*Spodoptera frugiperda)* larvae trigger direct defenses in maize. Mol Plant-Microbe Interact 27(5):461–470

Chung HS, Koo AJ, Gao X, Jayanty S, Thines B, Jones AD, Howe GA (2008) Regulation and function of Arabidopsis JASMONATE ZIM-domain genes in response to wounding and herbivory. Plant Physiol 146(3):952–964

Chung SH, Rosa C, Scully ED, Peiffer M, Tooker JF, Hoover K, Luthe DS, Felton GW (2013) Herbivore exploits orally secreted bacteria to suppress plant defenses. Proc Natl Acad Sci 110 (39):15728–15733

Consales F, Schweizer F, Erb M, Gouhier-Darimont C, Bodenhausen N, Bruessow F, Sobhy I, Reymond P (2011) Insect oral secretions suppress wound-induced responses in Arabidopsis. J Exp Bot 63(2):727–737

Coolen S, Proietti S, Hickman R, Davila Olivas NH, Huang PP, Van Verk MC, Van Pelt JA, Wittenberg AH, De Vos M, Prins M, Van Loon JJ (2016) Transcriptome dynamics of Arabidopsis during sequential biotic and abiotic stresses. Plant J 86(3):249–267

Coppola V, Coppola M, Rocco M, Digilio MC, D'Ambrosio C, Renzone G, Martinelli R, Scaloni A, Pennacchio F, Rao R, Corrado G (2013) Transcriptomic and proteomic analysis of a compatible tomato-aphid interaction reveals a predominant salicylic acid-dependent plant response. BMC Genomics 14(1):515

Coppola V, Soler R, Rao R, Corrado G (2017) Tomato-mediated interactions between root herbivores and aphids: insights into plant defence signalling. Entomol Exp Appl 163 (2):170–176 de la Paz Celorio-Mancera M, Courtiade J, Muck A, Heckel DG, Musser RO, Vogel H (2011) Sialome of a generalist lepidopteran herbivore: identification of transcripts and proteins from *Helicoverpa armigera* labial salivary glands. PLoS One 6(10):e26676

Dermauw W, Wybouw N, Rombauts S, Menten B, Vontas J, Grbić M, Clark RM, Feyereisen R, Van Leeuwen T (2013) A link between host plant adaptation and pesticide resistance in the polyphagous spider mite *Tetranychus urticae*. Proc Natl Acad Sci 110(2):E113–E122

Dermauw W, Pym A, Bass C, Van Leeuwen T, Feyereisen R (2018) Does host plant adaptation lead to pesticide resistance in generalist herbivores? Curr Opin Insect Sci 26:25–33

Dombrecht B, Xue GP, Sprague SJ, Kirkegaard JA, Ross JJ, Reid JB, Fitt GP, Sewelam N, Schenk PM, Manners JM, Kazan K (2007) MYC2 differentially modulates diverse jasmonate-dependent functions in Arabidopsis. Plant Cell 19(7):2225–2245

Dubey NK, Goel R, Ranjan A, Idris A, Singh SK, Bag SK, Chandrashekar K, Pandey KD, Singh PK, Sawant SV (2013) Comparative transcriptome analysis of *Gossypium hirsutum* L. in response to sap sucking insects: aphid and whitefly. BMC Genomics 14(1):241

Erb M, Glauser G (2010) Family business: multiple members of major phytohormone classes orchestrate plant stress responses. Chem Eur J 16(34):10280–10289

Eyres I, Jaquiéry J, Sugio A, Duvaux L, Gharbi K, Zhou JJ, Legeai F, Nelson M, Simon JC, Smadja CM, Butlin R (2016) Differential gene expression according to race and host plant in the pea aphid. Mol Ecol 25(17):4197–4215

Ferry N, Edwards MG, Gatehouse JA, Gatehouse AM (2004) Plant–insect interactions: molecular approaches to insect resistance. Curr Opin Biotechnol 15(2):155–161

Fescemyer HW, Sandoya GV, Gill TA, Ozkan S, Marden JH, Luthe DS (2013) Maize toxin degrades peritrophic matrix proteins and stimulates compensatory transcriptome responses in fall armyworm midgut. Insect Biochem Mol Biol 43(3):280–291

Firmino AAP, de Assis Fonseca FC, de Macedo LLP, Coelho RR, de Souza Jr JDA, Togawa RC, Silva-Junior OB, Pappas-Jr GJ, da Silva MCM, Engler G, Grossi-de-Sa MF (2013) Transcriptome analysis in cotton boll weevil (*Anthonomus grandis*) and RNA interference in insect pests. PLoS One 8(12):e85079

Frago E, Dicke M, Godfray HCJ (2012) Insect symbionts as hidden players in insect–plant interactions. Trends Ecol Evol 27(12):705–711

Gimenez E, Salinas M, Manzano-Agugliaro F (2018) Worldwide research on plant defense against biotic stresses as improvement for sustainable agriculture. Sustainability 10(2):391

Giri AP, Wünsche H, Mitra S, Zavala JA, Muck A, Svatoš A, Baldwin IT (2006) Molecular interactions between the specialist herbivore *Manduca sexta* (Lepidoptera, Sphingidae) and its natural host *Nicotiana attenuata*. VII Changes in the plant's proteome. Plant Physiol 142 (4):1621–1641

Giron D, Glevarec G (2014) Cytokinin-induced phenotypes in plant-insect interactions: learning from the bacterial world. J Chem Ecol 40(7):826–835

Gompert Z, Jahner JP, Scholl CF, Wilson JS, Lucas LK, Soria-Carrasco V, Fordyce JA, Nice CC, Buerkle CA, Forister ML (2015) The evolution of novel host use is unlikely to be constrained by trade-offs or a lack of genetic variation. Mol Ecol 24(11):2777–2793

Goossens A, Häkkinen ST, Laakso I, Seppänen-Laakso T, Biondi S, De Sutter V, Lammertyn F, Nuutila AM, Söderlund H, Zabeau M, Inzé D (2003) A functional genomics approach toward the understanding of secondary metabolism in plant cells. Proc Natl Acad Sci 100 (14):8595–8600

Govind G, Mittapalli O, Griebel T, Allmann S, Böcker S, Baldwin IT (2010) Unbiased transcriptional comparisons of generalist and specialist herbivores feeding on progressively defenseless *Nicotiana attenuata* plants. PLoS One 5(1):e8735

Halitschke R, Gase K, Hui D, Schmidt DD, Baldwin IT (2003) Molecular interactions between the specialist herbivore *Manduca sexta* (Lepidoptera, Sphingidae) and its natural host *Nicotiana attenuata*. VI Microarray analysis reveals that most herbivore-specific transcriptional changes are mediated by fatty acid-amino acid conjugates. Plant Physiol 131(4):1894–1902

Heidel-Fischer HM, Musser RO, Vogel H (2018) Plant transcriptomic responses to herbivory. Annu Plant Rev online:155–196

Heil M (2009) Damaged-self recognition in plant herbivore defence. Trends Plant Sci 14 (7):356–363

Hermsmeier D, Schittko U, Baldwin IT (2001) Molecular interactions between the specialist herbivore *Manduca sexta* (Lepidoptera, Sphingidae) and its natural host *Nicotiana attenuata* I Large-scale changes in the accumulation of growth-and defense-related plant mRNAs. Plant Physiol 125(2):683–700

Herrera-Estrella L, Depicker A, Van Montagu M, Schell J (1983) Expression of chimaeric genes transferred into plant cells using a Ti-plasmid-derived vector. Nature 303(5914):209

Hettenhausen C, Schuman MC, Wu J (2015) MAPK signaling: a key element in plant defense response to insects. Insect Sci 22(2):157–164

Hilfiker O, Groux R, Bruessow F, Kiefer K, Zeier J, Reymond P (2014) Insect eggs induce a systemic acquired resistance in Arabidopsis. Plant J 80(6):1085–1094

Howe GA, Herde M (2015) Interaction of plant defense compounds with the insect gut: new insights from genomic and molecular analyses. Curr Opin Insect Sci 9:62–68

Howe GA, Jander G (2008) Plant immunity to insect herbivores. Annu Rev Plant Biol 59:41–66

Huang XZ, Chen JY, Xiao HJ, Xiao YT, Wu J, Wu JX, Zhou JJ, Zhang YJ, Guo YY (2015) Dynamic transcriptome analysis and volatile profiling of *Gossypium hirsutum* in response to the cotton bollworm *Helicoverpa armigera*. Sci Rep 5:11867

Hui D, Iqbal J, Lehmann K, Gase K, Saluz HP, Baldwin IT (2003) Molecular interactions between the specialist herbivore *Manduca sexta* (Lepidoptera, Sphingidae) and its natural host *Nicotiana attenuata*: V. Microarray analysis and further characterization of large-scale changes in herbivore-induced mRNAs. Plant Physiol 131(4):1877–1893

Johnson ET, Dowd PF, Liu ZL, Musser RO (2011) Comparative transcription profiling analyses of maize reveals candidate defensive genes for seedling resistance against corn earworm. Mol Gen Genomics 285(6):517–525

Kant MR, Ament K, Sabelis MW, Haring MA, Schuurink RC (2004) Differential timing of spider mite-induced direct and indirect defenses in tomato plants. Plant Physiol 135(1):483–495

Kempema LA, Cui X, Holzer FM, Walling LL (2007) Arabidopsis transcriptome changes in response to phloem-feeding silverleaf whitefly nymphs. Similarities and distinctions in responses to aphids. Plant Physiol 143(2):849–865

Kim J, Tooker JF, Luthe DS, De Moraes CM, Felton GW (2012) Insect eggs can enhance wound response in plants: a study system of tomato *Solanum lycopersicum* L. and *Helicoverpa zea* Boddie. PLoS One 7(5):e37420

Koenig C, Bretschneider A, Heckel DG, Grosse-Wilde E, Hansson BS, Vogel H (2015) The plastic response of *Manduca sexta* to host and non-host plants. Insect Biochem Mol Biol 63:72–85

Koo AJ, Howe GA (2009) The wound hormone jasmonate. Phytochemistry 70(13–14):1571–1580

Kost C, Heil M (2006) Herbivore-induced plant volatiles induce an indirect defence in neighbouring plants. J Ecol 94(3):619–628

Lawrence SD, Novak NG (2004) Maize genes induced by herbivory and volicitin. J Chem Ecol 30 (12):2543–2557

Lawrence SD, Novak NG, Ju CJT, Cooke JE (2008) Potato, *Solanum tuberosum*, defense against Colorado potato beetle, *Leptinotarsa decemlineata* (Say): microarray gene expression profiling of potato by Colorado potato beetle regurgitant treatment of wounded leaves. J Chem Ecol 34 (8):1013–1025

Li X, Schuler MA, Berenbaum MR (2002) Jasmonate and salicylate induce expression of herbivore cytochrome P450 genes. Nature 419(6908):712

Li Q, Xie QG, Smith-Becker J, Navarre DA, Kaloshian I (2006) Mi-1-mediated aphid resistance involves salicylic acid and mitogen-activated protein kinase signaling cascades. Mol Plant-Microbe Interact 19(6):655–664

Li X, Schuler MA, Berenbaum MR (2007) Molecular mechanisms of metabolic resistance to synthetic and natural xenobiotics. Annu Rev Entomol 52:231–253

Little D, Gouhier-Darimont C, Bruessow F, Reymond P (2007) Oviposition by pierid butterflies triggers defense responses in Arabidopsis. Plant Physiol 143(2):784–800

Lortzing T, Firtzlaff V, Nguyen D, Rieu I, Stelzer S, Schad M, Kallarackal J, Steppuhn A (2017) Transcriptomic responses of *Solanum dulcamara* to natural and simulated herbivory. Mol Ecol Resour 17(6):e196–e211

Lv W, Du B, Shangguan X, Zhao Y, Pan Y, Zhu L, He Y, He G (2014) BAC and RNA sequencing reveal the brown planthopper resistance gene BPH15 in a recombination cold spot that mediates a unique defense mechanism. BMC Genomics 15(1):674

Major IT, Constabel CP (2006) Molecular analysis of poplar defense against herbivory: comparison of wound-and insect elicitor-induced gene expression. New Phytol 172(4):617–635

Mithöfer A, Wanner G, Boland W (2005) Effects of feeding *Spodoptera littoralis* on lima bean leaves. II Continuous mechanical wounding resembling insect feeding is sufficient to elicit herbivory-related volatile emission. Plant Physiol 137(3):1160–1168

Morozova O, Hirst M, Marra MA (2009) Applications of new sequencing technologies for transcriptome analysis. Annu Rev Genomics Hum Genet 10:135–151

Mozoruk J, Hunnicutt LE, Cave RD, Hunter WB, Bausher MG (2006) Profiling transcriptional changes in *Citrus sinensis* (L.) Osbeck challenged by herbivory from the xylem-feeding leafhopper *Homalodisca coagulata* (Say) by cDNA macroarray analysis. Plant Sci 170 (6):1068–1080

Musser RO, Hum-Musser SM, Eichenseer H, Peiffer M, Ervin G, Murphy JB, Felton GW (2002) Herbivory: caterpillar saliva beats plant defences. Nature 416(6881):599

Musser RO, Cipollini DF, Hum-Musser SM, Williams SA, Brown JK, Felton GW (2005) Evidence that the caterpillar salivary enzyme glucose oxidase provides herbivore offense in solanaceous plants. Arch Insect Biochem Physiol 58(2):128–137

Musser RO, Farmer E, Peiffer M, Williams SA, Felton GW (2006) Ablation of caterpillar labial salivary glands: technique for determining the role of saliva in insect–plant interactions. J Chem Ecol 32(5):981–992

Musser RO, Hum-Musser SM, Lee HK, DesRochers BL, Williams SA, Vogel H (2012) Caterpillar labial saliva alters tomato plant gene expression. J Chem Ecol 38(11):1387–1401

Nguyen D, D'Agostino N, Tytgat TO, Sun P, Lortzing T, Visser EJ, Cristescu SM, Steppuhn A, Mariani C, van Dam NM, Rieu I (2016) Drought and flooding have distinct effects on herbivore-induced responses and resistance in *Solanum dulcamara*. Plant Cell Environ 39(7):1485–1499

Oates CN, Külheim C, Myburg AA, Slippers B, Naidoo S (2015) The transcriptome and terpene profile of *Eucalyptus grandis* reveals mechanisms of defense against the insect pest, *Leptocybe invasa*. Plant Cell Physiol 56(7):1418–1428

Oppenheim SJ, Baker RH, Simon S, DeSalle R (2015) We can't all be supermodels: the value of comparative transcriptomics to the study of non-model insects. Insect Mol Biol 24(2):139–154

Oppert B, Elpidina EN, Toutges M, Mazumdar-Leighton S (2010) Microarray analysis reveals strategies of *Tribolium castaneum* larvae to compensate for cysteine and serine protease inhibitors. Comp Biochem Physiol D Genomics and Proteomics 5(4):280–287

Ozsolak F, Milos PM (2011) RNA sequencing: advances, challenges and opportunities. Nat Rev Genet 12(2):87

Pandey SP, Srivastava S, Goel R, Lakhwani D, Singh P, Asif MH, Sane AP (2017) Simulated herbivory in chickpea causes rapid changes in defense pathways and hormonal transcription networks of JA/ethylene/GA/auxin within minutes of wounding. Sci Rep 7:44729

Pauwels L, Morreel K, De Witte E, Lammertyn F, Van Montagu M, Boerjan W, Inzé D, Goossens A (2008) Mapping methyl jasmonate-mediated transcriptional reprogramming of metabolism and cell cycle progression in cultured Arabidopsis cells. Proc Natl Acad Sci 105(4):1380–1385

Pauwels L, Inzé D, Goossens A (2009) Jasmonate-inducible gene: what does it mean? Trends Plant Sci 14(2):87–91

Pearce SL, Clarke DF, East PD, Elfekih S, Gordon KHJ, Jermiin LS, McGaughran A, Oakeshott JG, Papanikolaou A, Perera OP, Rane RV (2017) Genomic innovations, transcriptional plasticity and gene loss underlying the evolution and divergence of two highly polyphagous and invasive *Helicoverpa* pest species. BMC Biol 15(1):63

Petek M, Turnšek N, Gašparič MB, Novak MP, Gruden K, Slapar N, Popovič T, Štrukelj B, Gruden K, Štrukelj B, Jongsma MA (2012) A complex of genes involved in adaptation of *Leptinotarsa decemlineata* larvae to induced potato defense. Arch Insect Biochem Physiol 79 (3):153–181

Philippe RN, Ralph SG, Mansfield SD, Bohlmann J (2010) Transcriptome profiles of hybrid poplar (*Populus trichocarpa*× deltoides) reveal rapid changes in undamaged, systemic sink leaves after simulated feeding by forest tent caterpillar (*Malacosoma disstria*). New Phytol 188 (3):787–802

Pieterse CM, Dicke M (2007) Plant interactions with microbes and insects: from molecular mechanisms to ecology. Trends Plant Sci 12(12):564–569

Ragland GJ, Almskaar K, Vertacnik KL, Gough HM, Feder JL, Hahn DA, Schwarz D (2015) Differences in performance and transcriptome-wide gene expression associated with *R.hagoletis* (Diptera: Tephritidae) larvae feeding in alternate host fruit environments. Mol Ecol 24 (11):2759–2776

Ralph S, Oddy C, Cooper D, Yueh H, Jancsik S, Kolosova N, Philippe RN, Aeschliman D, White R, Huber D, Ritland CE (2006a) Genomics of hybrid poplar (*Populus trichocarpa*× deltoides) interacting with forest tent caterpillars (*Malacosoma disstria*): normalized and full-length cDNA libraries, expressed sequence tags, and a cDNA microarray for the study of insect-induced defences in poplar. Mol Ecol 15(5):1275–1297

Ralph SG, Yueh H, Friedmann M, Aeschliman D, Zeznik JA, Nelson CC, Butterfield YS, Kirkpatrick R, Liu J, Jones SJ, Marra MA (2006b) Conifer defence against insects: microarray gene expression profiling of Sitka spruce (*Picea sitchensis*) induced by mechanical wounding or feeding by spruce budworms (*Choristoneura occidentalis*) or white pine weevils (*Pissodes strobi*) reveals large-scale changes of the host transcriptome. Plant Cell Environ 29 (8):1545–1570

Reymond P, Weber H, Damond M, Farmer EE (2000) Differential gene expression in response to mechanical wounding and insect feeding in Arabidopsis. Plant Cell 12(5):707–719

Reymond P, Bodenhausen N, Van Poecke RM, Krishnamurthy V, Dicke M, Farmer EE (2004) A conserved transcript pattern in response to a specialist and a generalist herbivore. Plant Cell 16 (11):3132–3147

Roda AMY, Halitschke R, Steppuhn A, Baldwin IT (2004) Individual variability in herbivore-specific elicitors from the plant's perspective. Mol Ecol 13(8):2421–2433

Roy A, Walker III WB, Vogel H, Chattington S, Larsson MC, Anderson P, Heckel DG, Schlyter F (2016) Diet dependent metabolic responses in three generalist insect herbivores Spodoptera spp. Insect Biochem Mol Biol 71:91–105

Schena M, Shalon D, Davis RW, Brown PO (1995) Quantitative monitoring of gene expression patterns with a complementary DNA microarray. Science 270(5235):467–470

Schweizer F, Heidel-Fischer H, Vogel H, Reymond P (2017) Arabidopsis glucosinolates trigger a contrasting transcriptomic response in a generalist and a specialist herbivore. Insect Biochem Mol Biol 85:21–31

Scranton MA, Fowler JH, Girke T, Walling LL (2013) Microarray analysis of tomato's early and late wound response reveals new regulatory targets for leucine aminopeptidase A. PLoS One 8 (10):e77889

Shoji T, Ogawa T, Hashimoto T (2008) Jasmonate-induced nicotine formation in tobacco is mediated by tobacco COI1 and JAZ genes. Plant Cell Physiol 49(7):003–1012

Silva AX, Jander G, Samaniego H, Ramsey JS, Figueroa CC (2012) Insecticide resistance mechanisms in the green peach aphid *Myzus persicae* (Hemiptera: Aphididae) I: a transcriptomic survey. PLoS One 7(6):e36366

Singh A, Singh IK, Verma PK (2008) Differential transcript accumulation in *Cicer arietinum* L. in response to a chewing insect *Helicoverpa armigera* and defence regulators correlate with reduced insect performance. J Exp Bot 59(9):2379–2392

Singh, S., Singh, A., Kumar, S., Mittal, P. & Singh, I.K. (2018). Protease inhibitors: recent advancement in its usage as a potential biocontrol agent for insect pest management. Insect Sci

Snoeck S, Greenhalgh R, Tirry L, Clark RM, Van Leeuwen T, Dermauw W (2017) The effect of insecticide synergist treatment on genome-wide gene expression in a polyphagous pest. Sci Rep 7(1):13440

Song YY, Ye M, Li CY, Wang RL, Wei XC, Luo SM, Zeng RS (2013) Priming of anti-herbivore defense in tomato by arbuscular mycorrhizal fungus and involvement of the jasmonate pathway. J Chem Ecol 39(7):1036–1044

Spiteller D, Dettner K, Boland W (2000) Gut bacteria may be involved in interactions between plants, herbivores and their predators: microbial biosynthesis of N-acylglutamine surfactants as elicitors of plant volatiles. Biol Chem 381(8):755–762

Strickler SR, Bombarely A, Mueller LA (2012) Designing a transcriptome next-generation sequencing project for a nonmodel plant species 1. Am J Bot 99(2):257–266

Su Q, Oliver KM, Xie W, Wu Q, Wang S, Zhang Y (2015) The whitefly-associated facultative symbiont *Hamiltonella defensa* suppresses induced plant defences in tomato. Funct Ecol 29 (8):1007–1018

Swevers L, Huvenne H, Menschaert G, Kontogiannatos D, Kourti A, Pauchet Y, ffrench-Constant-R, Smagghe G (2013) Colorado potato beetle (Coleoptera) gut transcriptome analysis: expression of RNA interference-related genes. Insect Mol Biol 22(6):668–684

Tamayo MC, Rufat M, Bravo JM, San Segundo B (2000) Accumulation of a maize proteinase inhibitor in response to wounding and insect feeding, and characterization of its activity toward digestive proteinases of *Spodoptera littoralis* larvae. Planta 211(1):62–71

Tao XY, Xue XY, Huang YP, Chen XY, Mao YB (2012) Gossypol-enhanced P450 gene pool contributes to cotton bollworm tolerance to a pyrethroid insecticide. Mol Ecol 21 (17):4371–4385

Thorpe P, Cock PJ, Bos J (2016) Comparative transcriptomics and proteomics of three different aphid species identifies core and diverse effector sets. BMC Genomics 17(1):172

Tretner C, Huth U, Hause B (2008) Mechanostimulation of *Medicago truncatula* leads to enhanced levels of jasmonic acid. J Exp Bot 59(10):2847–2856

Tzin V, Fernandez-Pozo N, Richter A, Schmelz EA, Schoettner M, Schäfer M, Ahern KR, Meihls LN, Kaur H, Huffaker A, Mori N (2015) Dynamic maize responses to aphid feeding are revealed by a time series of transcriptomic and metabolomic assays. Plant Physiol 169(3):1727–1743

Vogel H, Kroymann J, Mitchell-Olds T (2007) Different transcript patterns in response to specialist and generalist herbivores in the wild Arabidopsis relative *Boechera divaricarpa*. PLoS One 2 (10):e1081

Vogel H, Musser RO, de la Paz Celorio-Mancera M (2018) Transcriptome responses in herbivorous insects towards host plant and toxin feeding. Annu Plant Rev online:197–233

Vos IA, Verhage A, Schuurink RC, Watt LG, Pieterse CM, Van Wees S (2013) Onset of herbivore-induced resistance in systemic tissue primed for jasmonate-dependent defenses is activated by abscisic acid. Front Plant Sci 4:539

Wang Z, Gerstein M, Snyder M (2009) RNA-Seq: a revolutionary tool for transcriptomics. Nat Rev Genet 10(1):57

Wang Y, Wang H, Fan R, Yang Q, Yu D (2014) Transcriptome analysis of soybean lines reveals transcript diversity and genes involved in the response to common cutworm (*Spodoptera litura* F abricius) feeding. Plant Cell Environ 37(9):2086–2101

Wang Y, Wang H, Ma Y, Yang W, Yang Q, Yu D (2015) Identification of soybean herbivory-regulated genes and a transgenic investigation of their potential in insect resistance. Plant Cell Tiss Organ Cult 123(2):321–340

Wang YN, Tang L, Hou Y, Wang P, Yang H, Wei CL (2016) Differential transcriptome analysis of leaves of tea plant (*Camellia sinensis*) provides comprehensive insights into the defense responses to *Ectropis oblique* attack using RNA-Seq. Funct Integr Genomics 16(4):383–398

Wu J, Baldwin IT (2010) New insights into plant responses to the attack from insect herbivores. Annu Rev Genet 44:1–24

Wybouw N, Zhurov V, Martel C, Bruinsma KA, Hendrickx F, Grbić V, Van Leeuwen T (2015) Adaptation of a polyphagous herbivore to a novel host plant extensively shapes the transcriptome of herbivore and host. Mol Ecol 24(18):4647–4663

Yang F, Zhang Y, Huang Q, Yin G, Pennerman KK, Yu J, Liu Z, Li D, Guo A (2015) Analysis of key genes of jasmonic acid mediated signal pathway for defense against insect damages by comparative transcriptome sequencing. Sci Rep 5:16500

Yi X, Qi J, Zhou X, Hu MY, Zhong GH (2017) Differential expression of chemosensory-protein genes in midguts in response to diet of *Spodoptera litura*. Sci Rep 7(1):296

Zhang H, de Bernonville TD, Body M, Glevarec G, Reichelt M, Unsicker S, Bruneau M, Renou JP, Huguet E, Dubreuil G, Giron D (2016) Leaf-mining by *Phyllonorycter blancardella* reprograms the host-leaf transcriptome to modulate phytohormones associated with nutrient mobilization and plant defense. J Insect Physiol 84:114–127

Zhao J, Davis LC, Verpoorte R (2005) Elicitor signal transduction leading to production of plant secondary metabolites. Biotechnol Adv 23(4):283–333

Zhong Y, Wang B, Yan J, Cheng L, Yao L, Xiao L, Wu T (2014) DL-β-aminobutyric acid-induced resistance in soybean against *Aphis glycines Matsumura* (Hemiptera: Aphididae). PLoS One 9 (1):e85142

Zhu Z, An F, Feng Y, Li P, Xue L, Mu A, Jiang Z, Kim JM, To TK, Li W, Zhang X (2011) Derepression of ethylene-stabilized transcription factors (EIN3/EIL1) mediates jasmonate and ethylene signaling synergy in Arabidopsis. Proc Natl Acad Sci 108(30):12539–12544

Zhu F, Poelman EH, Dicke M (2014) Insect herbivore-associated organisms affect plant responses to herbivory. New Phytol 204(2):315–321

Zhu-Salzman K, Zeng R (2015) Insect response to plant defensive protease inhibitors. Annu Rev Entomol 60:233–252

Plant–Insect Interaction: A Proteomic Approach in Defence Mechanism

Anjana Rustagi, Samira Chugh, Shweta Sharma, Punam Kumari, and Deepak Kumar

Abstract

Plants being immovable are under continuous attack by various pests, predators and pathogens. In order to defend themselves from these varied invasions, the plants trigger a range of molecular defence mechanisms, thus inducing profound alterations in host gene/protein expression. Plants counteract insect attacks both directly and indirectly. Many of these defences are regulated by signalling pathways where phytohormones have major roles. Simultaneously, insects also develop strategies to overcome plant barriers and thereby lead to an intrigued co-evolution of plant–insect relationship. The study of this complex relationship is highly solicited to reduce crop losses due to insect infestations. Proteomics has emerged as a promising tool of biotechnological workflow in the past few decades, and different proteomic tools have efficiently been utilized to decipher the complex enigma of plant–insect interactions. Here, we report the recent studies that have tapped the potential of different proteomic tools to study plant defence against insects.

A. Rustagi (✉) · S. Chugh
Department of Botany, Gargi College, New Delhi, India
e-mail: anjana.rustagi@gargi.du.ac.in

S. Sharma
Department of Botany, Maitreyi College, New Delhi, India

P. Kumari
Department of Biosciences and Biotechnology, Fakir Mohan University, Balasore, Odisha, India

D. Kumar
Department of Botany, Central University of Jammu, Jammu, Jammu & Kashmir, India

I. K. Singh, A. Singh (eds.), *Plant-Pest Interactions: From Molecular Mechanisms to Chemical Ecology*, https://doi.org/10.1007/978-981-15-2467-7_3

1 Introduction

The interaction of insects with plants can be both beneficial and extremely deleterious to plants (Barah and Bones 2015). On the one hand, plants and insects have a mutualistic relationship where insects serve as pollinators and are indispensible to the life cycle of plants, whereas on the other hand, almost half of the total six million insect species are herbivorous and have evolved a variety of feeding mechanisms to acquire nutrients from their host plants. Phytophagous insects cause significant yield losses in agriculture across all major agricultural areas of the world even after considerable investments in control measures (Kerchev et al. 2012).

Plants produce numerous secondary metabolites to defend themselves against herbivores including insects (Berenbaum and Zangerl 2008). Insects also evolve several strategies to overcome plant defence barriers, allowing them to feed, grow and reproduce on their host plants. This has led to the co-evolution of a very complex interaction between the two (Mello and Silva-Filho 2002). This process of defence and counter-defence between plants and insects has been rightly termed as the co-evolutionary arms race (Whittaker and Feeny 1971). A simple model of insect infestation stress responses and how proteomics can be useful to study insect–plant interaction is outlined in Fig. 1.

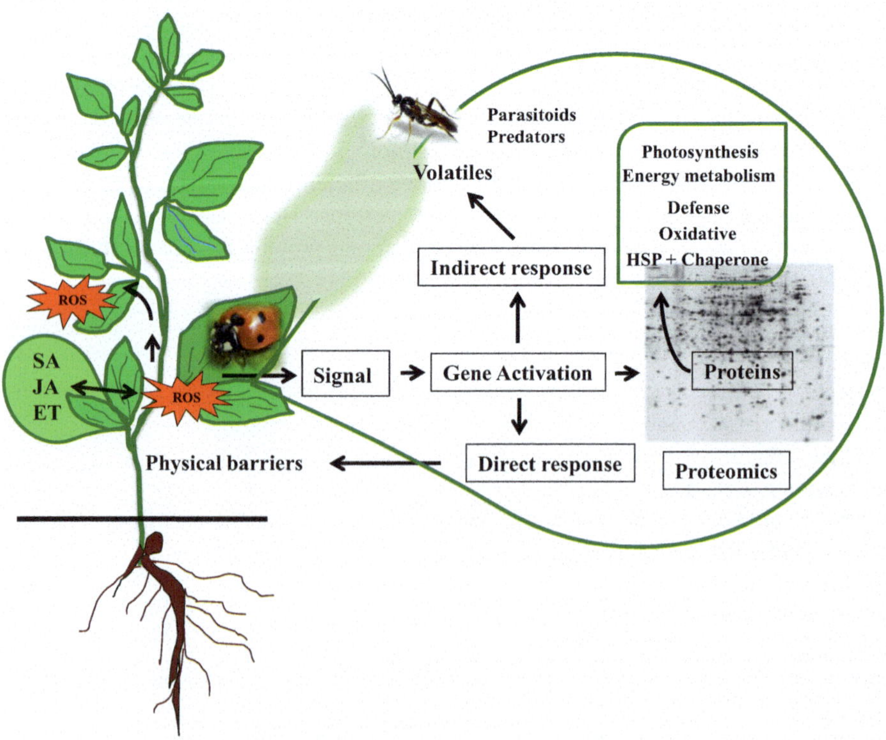

Fig. 1 Plant signalling in response to insect infection

Plants' chemicals for defence function in two different ways: firstly, to repel herbivores through direct toxicity or by making plant tissues difficult to digest and secondly, as inducible substances synthesized in response to tissue damage by herbivores (Mello and Silva-Filho 2002). Barah and Bones (2015) summed up the defence mechanisms of higher plants in response to insect attack as signal input, which include recognition of insect oral secretions, structural components of the insect mouthparts and injured plant cells, followed by signal transduction (calcium ion fluxes, phosphorylation cascades and hormonal crosstalk); signal processing which include changes in transcriptome, proteome and metabolome; and finally the production of defence compounds which can be proteins and secondary metabolites. Photosynthesis is the major source of a range of reactive oxygen species (ROS) such as singlet oxygen, superoxide and hydrogen peroxide in the chloroplasts (Kerchev et al. 2012). The ROS and phytohormone signalling also participate in the plant response to insects. Insect herbivory also impacts photosynthesis and creates shifts in the transcriptional profile of genes associated with photosynthetic reactions (Bilgin et al. 2010). A branch of science 'chemical ecology' separately studies the involvement of chemicals in interaction of organisms with each other and with their surroundings (Frederic et al. 2006).

The insects belong to one of the two categories according to strategies of defence which are a result of the evolutionary arms race: generalist insect herbivores, such as *Myzus persicae* aphid, need more complex adaptive mechanisms since they need to respond to a large array of different plant defensive chemicals (Frederic et al. 2006). The others are specialists like the chewing insect, *Plutella xylostella* (L.) (diamond-back moth); specialize in Brassicaceae family, including *Arabidopsis thaliana*; and induce differential expression of host genes/proteins (Truong et al. 2015). Another categorization is based on feeding guild and recognizes chewers versus phloem-feeders. It has been suggested often that depending on the feeding mode of an insect, different plant defence responses are induced, resulting in the activation of different plant defence mechanisms (Bidart-Bouzat and Kliebenstein 2011; Walling 2000). Many reports suggest the involvement of salicylic acid (SA) in defence against phloem-sucking insects (Walling 2000; Broekgaarden et al. 2011), whereas chewing larvae (mainly lepidopterans) cause tissue damage and jasmonic acid (JA) and ethylene (ET) induction (De Vos et al. 2005; Kessler and Baldwin 2002). Many studies directly compared chewers and suckers, and the common trend that emerged was that phloem-feeding insects induced activation of fewer genes associated with the jasmonic acid pathway, whereas the chewers induced fewer genes associated with the salicylic acid pathway. This is consistent with the prediction that phloem-feeders, like aphids, leaf-hoppers and white flies, cause only minor tissue damage and induce defence signalling pathways resembling those activated against pathogens (SA regulated) (Walling 2000; Moran and Thompson 2001; Kusnierczyk et al. 2007). In this chapter, we summarize the various proteomic approaches employable for studying this complex relationship between insects and plants. Briefly, we also summarize the different tools of proteomics which could be possibly used for this purpose.

2 Significance of Proteomics in Studying Plant Defence Against Insects

It is now known that the mechanism of plant–insect interaction is complex. Insect infestations trigger a broad range of molecular defence mechanisms, including profound alterations in gene/protein expression (Lippert et al. 2007). Current research in plant–insect interaction using tools of genomics focus mainly on late events induced by biotic stress. To better understand the process, it was found that the genomic approaches need to be complemented by qualitative and quantitative analyses of the plant at several levels including the transcriptome, proteome and metabolome (Quirino et al. 2010). Powerful techniques like microarray analysis are used to study the transcriptome, and information about which genes are involved in stress response and adaptation are reported (Renaut et al. 2006). These results, however, are not sufficient as mRNA abundance and protein level are not clearly correlated (Gygi et al. 1999), that low copy number mRNAs (potentially very important for regulation) are not measured as readily as abundant mRNAs and that gene expression studies do not provide information about either the subcellular localization of gene products or the post-translational modifications (PTMs) occurring to a protein that may be essential for its function, transport and activation (Rose et al. 2004). Phosphorylation of components is an important event in cascades involved in plant defence. Protein cleavage and degradation have also been shown to play a key role in the early events of the hypersensitive response. Proteomics technology plays an important role in studying plant defence mechanisms by mining the expression changes of proteins in response to pathogen attack (Fang et al. 2015). High-throughput quantitative proteomics studies have gained importance in plant research during the last few years to characterize proteomes and changes in them in response to biotic stresses (Barah and Bones 2015).

3 Tools of Proteomics to Study Plant–Insect Interaction

Proteomic studies aim to explain the quantity, purpose and spatial and temporal location of proteins within an organism. Proteomics approaches are either gel-based or gel-free. Many of these approaches have earlier been discussed in scientific articles and reviews (Lambert et al. 2005; Mehta et al. 2008; Lodha et al. 2013; Rustagi et al. 2018). The following are the modern proteomic tools which have been potentially been explored to unravel the complexities of plant–insect interactions, and Table 1 enlists the modern proteomic tools which have been used in the last 10 years for studying plant–inset interactions.

Table 1 Major plant–insect interaction studies using proteomics as tool in the last 10 years

S. no.	Plant	Insect	Proteomic tool used	Outcome of the study	References
1.	*Arabidopsis thaliana*	*Plutella xylostella*	2-DE analysis with mass spectrometry (MALDI-TOF-MS or LC-ESI-MS/MS)	13 differentially expressed proteins identified which were functionally associated with photosynthesis, energy metabolism, carbohydrate metabolism, lipid metabolism and amino acid metabolism	Truong et al. (2015)
2.	*Solanum lycopersicum*	*Macrosiphum euphorbiae aphid*	2-DE coupled with MALDI-TOF-MS and nanoLC-ESI-LIT-MS/MS	Identified 57 differentially expressed proteins which were functionally annotated to biological processes such as primary metabolism, cell maintenance, photosynthesis, signal transduction, stress, defence response, etc.	Coppola et al. (2013)
3.	Rice (*Oryza sativa* L.)	(*Laodelphax striatellus* Fallén (small brown planthopper; SBPH), Homoptera, Delphacidae	2D-DIGE 2-DE, MS analysis	Around 132 differentially expressed proteins were identified which were involved mainly in carbohydrate metabolic process, amino acid metabolism, stress response, photosynthesis, energy metabolism, cell wall-related proteins, transcriptional regulation and protein metabolic process	Dong et al. (2017)

(continued)

Table 1 (continued)

S. no.	Plant	Insect	Proteomic tool used	Outcome of the study	References
4.	Potato (*Solanum tuberosum* L.)	*Macrosiphum euphorbiae* Thomas. (potato aphid) *Leptinotarsa decemlineata* Say (colorado potato beetle)	2-DE, MS	Approximately 31 differentially expressed proteins were identified in response to several stress treatments as compared to healthy controls	Duceppe et al. (2012)
5.	*Arabidopsis thaliana*	*Plutella xylostella*	2D-PAGE coupled with MS/MS	Significant differences in the proteins expressed in herbivore-resistant and herbivore-susceptible lines with resistant lines showing increased production of reactive oxygen species such as H_2O_2	Collins et al. (2010)
6.	*Arabidopsis thaliana*	*Myzus persicae* Sulzer	2-DE gels followed by MALDI-TOF-MS and LC-ESI-MS/MS	Approximately 31 proteins were expressed differentially. These included proteins involved in energy metabolism, carbohydrate metabolism, amino acid metabolism, translation, photosynthesis, defence response, etc.	Truong et al. (2015)
7.	*Oryza sativa*	*Nilaparvata lugens*	Proteomic approach based on quantitative mass spectrometry	Several proteins involved in multiple pathways indicated considerable changes in response to BPH	Wei et al. (2009)

(continued)

Table 1 (continued)

S. no.	Plant	Insect	Proteomic tool used	Outcome of the study	References
				feeding. These included oxidative stress response proteins, jasmonic acid synthesis proteins, kinases,, clathrin protein, beta-glucanase protein, etc.	
8.	Sitka spruce (*Picea sitchensis*)	*Pissodes strobi* (white pine weevils)	2-D PAGE and MS/MS	A number of insect-induced proteins were identified following insect feeding such as small heat shock proteins, proteins involved in secondary metabolism, stress response proteins, oxidoreductases and a yet unreported spruce protein.	Lippert et al. (2007)

3.1 Two-Dimensional Polyacrylamide Gel Electrophoresis (2D-PAGE)

2D-PAGE separates individual proteins from a complex mixture of proteins in a polyarylamide gel matrix. The individual proteins are separated on the basis of their characteristic isoelectric point (pI) and molecular size. The separated protein spots are visualized by staining the gel (Coo- massie Brilliant Blue (CBB), silver stain, SYPRO staining, etc.)

Digital images of 2D-PAGE are analysed using specific software for quantification of proteins and characterization of molecular weight and isoelectric point and also for studying their post-translational modifications (PTMs) (Gorg et al. 2004, Wittmann-Liebold et al. 2006). The 2D-PAGE is significant in thrust research areas such as de novo sequencing and protein identification of those organisms whose genome sequences are not known, alternative detection methods for modification specific proteomics, and identification of protein isoforms and modified proteins (Rogowska-Wrzesinska et al. 2013). 2D-PAGE is a standard method for protein separation, especially useful for studying protein modifications and to find prognostic or diagnostic biomarkers in various disease states.

3.2 Fluorescent Two-Dimensional 'Difference Gel Electrophoresis' (2D-DIGE)

Two-dimensional difference gel electrophoresis (2D-DIGE) is an advanced modification of 2-D electrophoresis (2-DE). It enables the user to compare two or more protein samples simultaneously on the same gel. The 2D-DIGE has advantages over 2-DE as this technique eliminates gel-to-gel variation and achieves high resolution, sensitivity and reproducibility (Gao 2014). Since samples can be multiplexed on the same gel, this advanced technique limits the experimental variation and can quantify even a 10% difference in protein expression.

3.3 ICATs (Isotope-Coded Affinity Tags)

Isotope-coded affinity tag (ICAT) is a gel-free proteomics technique. The ICAT can provide a quantitative as well as qualitative information of protein levels under different experimental conditions. However, this technique has the limitation of selectively detecting proteins which are high in cysteine while the acidic proteins are difficult to detect (Gygi et al. 1999; Zhou et al., 2002).

3.4 iTRAQ (Isobaric Tags for Relative and Absolute Quantitation)

The iTRAQ technique, a variation of ICAT, can quantify protein obtained from different sources in a single experiment (Ross et al. 2004, Agarwal et al. 2006, Zieske 2006, Lund et al. 2007). The advantages include simplified analysis, increased analytical precision, the ability to multiplex several samples and easier quantification (Agarwal et al. 2006; Lund et al. 2007; Zieske 2006). Up to four different samples can be compared in one MS-based experiment. The iTRAQ potentially allows for the expanded coverage of the proteome by tagging tryptic peptides as all of which would possess primary amine groups, thereby eliminating the dependence on cysteine as in ICAT. (Aggarwal et al. 2006; Ross et al. 2004; Zieske 2006).

3.5 MudPIT (Multidimensional Protein Identification Technology)

MudPIT is an on-line 2D ion-exchange/reversed phase HPLC method. The MudPIT technique generates an exhaustive list of proteins which are present in a protein sample under experimentation. This technique is fast, sensitive and highly reproducible. Also, MudPIT enables to analyse the proteins of all functional and physical classes. Therefore, it is used for identification of protein complexes, large-scale cataloguing of proteins in cells and organisms, profiling of membrane and organelle proteins, determination of PTMs, protein ubiquitination in diverse plant species and quantitative analysis of protein expression (Yates et al. 2005, Cantin et al. 2006, Speers and Wu 2007, Maor et al. 2007). Though MudPIT is a relatively lengthy process as a number of fractions produced take too much time to be analysed by MS (Anguraj Vadivel 2015).

3.6 Mass Spectrometry

Mass spectrometry (MS) distinguishes molecules by their mass-to-charge ratios (m/z) and is the crucial player in proteomics world (Zhu et al. 2009, 2010). In advanced type of MS analysis, two 'soft' ionization methods, namely, electrospray ionization (ESI) (Fenn et al. 1989) and matrix-assisted laser desorption/ionization (MALDI) (Karas and Hillenkamp 1988; Tanaka et al. 1988), are used. In these two methods, non-volatile analytes such as complex proteins and peptides are ionized and then effectively analysed (Feng et al. 2015). The MS finds its application in quantification, identification of proteins, protein folding and interaction studies, protein expression cataloguing, identification of protein modification sites and detection of PTMs (Han et al. 2008; Lodha et al. 2013).

3.7 Protein Microarray

Protein microarray studies thousands of proteins simultaneously and provides information about analytes, ligands, receptors and antibody affinity-based interactions and binding partners and permits high-throughput analysis (Romanov et al. 2014). Microarrays are substantially utilized for protein–protein interaction analysis, biochemical pathway mapping, biomarker identification, host–microbe interaction studies, detection of infectious diseases, drug screening, vaccine development, enzyme–substrate profiling and immuno-profiling (Zhu et al. 2012; Romanov et al. 2014; Moore et al. 2016). Thus, protein microarray encompasses both the classical and functional proteome analysis (Lodha et al. 2013).

Both gel-based and gel-free approaches are significant in unravelling plant–insect complexities. Gel-based techniques, although widely used, have certain limitations. Insensitivity to low abundant proteins, inability to characterize the entire proteome in one gel and poor reproducibility are a few prominent limitations. The post-translational modifications may change the pI (e.g. by phosphorylations) or relative mass (e.g. by glycosylation or truncation), thereby affecting the protein mobility on a 2-DE gel. Gel-based methods are time-consuming and expensive too. The gel-free proteomics techniques could overcome most of these limitations, but they have their own shortcomings.

The gel-free proteomics technology is promising for samples in which proteins are less abundant. Liquid chromatography (LC) can efficiently separate proteins and peptides in such complex samples. Multidimensional chromatographic separation further improves the separation and identification of peptides. The advanced MS systems have boosted the identification of proteins and more sensitive and accurate protein quantitation. However, gel-free MS-based proteomics requires expensive instrumentation facility and highly skilled personnel. Both the proteomics approaches with their merits and demerits are complementary and may be used in parallel to get a more complete comprehension of protein expression and interactions.

4 Recent Proteomic Investigations to Reveal Plant–Insect Relation

Proteomics has emerged as a powerful tool of biotechnology to decipher the complex mechanism of plant–insect interaction. These tools have been efficiently harnessed to decipher the plant defence responses and molecular strategy of the insects as well during the interaction. Herein we summarize the recent studies revealing these intricacies from the last 10 years.

4.1 Plant Defence Responses

Truong et al. (2018) reported after observing existing data that chewing insects caused the reduction of photosynthesis in plants following their infestation. Their group also used a 2-DE proteome approach to detect protein expression changes in the leaves of *Arabidopsis* plants exposed to *P. xylostella* larval infestation. Thirteen proteins were successfully identified by MALDI-TOF/MS and LCESI-MS/MS, and the proteins were associated with amino acid, carbohydrate, energy, lipid metabolism and photosynthesis (Truong et al. 2018).

In order to understand the response of rice plants to small brown planthopper (*Laodelphax striatellus* Fallén, Homoptera, Delphacidae-SBPH), two rice genotypes, SBPH-resistant and SBPH-sensitive, were used. Protein expression profile analysis was carried out in the leaf sheath of SBPH-resistant and SBPH-sensitive rice lines in response to SBPH attack using 2D-DIGE, 2-DE and MS analysis. Around 132 differentially expressed proteins were identified which were found to be involved mainly in stress response, photosynthesis, protein metabolic process, carbohydrate metabolic process, energy metabolism, cell wall-related proteins, amino acid metabolism and transcriptional regulation. During SBPH attack, levels of superoxide dismutase (SOD) and glutathione (GSH) were found to be considerably higher in SBPH-resistant line as compared to SBPH-susceptible line. Also it was found that rice plants defend against the SBPH infestation by triggering the salicylic acid (SA)-dependent systemic acquired resistance pathway (Dong et al. 2017).

The defence response of *Arabidopsis thaliana* to the insect pest green peach aphid (*Myzus persicae* Sulzer) was investigated using 2-DE gels. Approximately

31 differentially expressed protein spots were observed, out of which seven and nine were identified by MALDI-TOF-MS and LC-ESI-MS/MS analysis, respectively. It was revealed that the identified proteins corresponded to carbohydrate, amino acid and energy metabolism, photosynthesis, defence response and translation, etc. Thus, it was concluded that significant changes are induced in the proteome of *Arabidopsis thaliana* following *M. persicae* infestation (Truong et al. 2015).

A detailed study to detect the response in tomato upon aphid *Macrosiphum euphorbiae* attack was carried out by Coppola et al. 2013. Using 2-DE analysis followed by MALDI-TOF-MS and nanoLC-ESI-LIT-MS/MS, a large number of differentially expressed proteins were identified found to be related to biological processes such as stress-related response, photosynthesis, signal transduction, etc. The response triggered in tomato upon aphid infestation was found to be having increased oxidative stress accompanied by the production of proteins involved in the detoxification of oxygen radicals. A model was proposed to explain the aphid elicited defence response in tomato which was based on the cross-communication of different hormone-related signalling pathways such as those related to the salicylic acid (SA), jasmonic acid (JA), ethylene and brassinosteroids.

To understand the differential response to wounding and herbivory, comparative proteomic analysis was carried out in potato plants (*Solanum tuberosum* L.*)* by two specialized insect herbivores, defoliating pest the Colorado potato beetle, *Leptinotarsa decemlineata* Say, or the potato aphid, *Macrosiphum euphorbiae* Thomas. More than 30 proteins were differentially expressed in response to different stress treatments as compared to healthy controls. The proteins identified by mass spectrometry included several defence-related proteins, like wound-inducible protease inhibitors and pathogenesis-related proteins. Therefore, differential effects of mechanical wounding, beetle leaf chewing and aphid phloem sap feeding on the defence and photosynthesis-related proteins were observed in potato leaves (Duceppe et al. 2012).

A proteomic study was conducted to understand the difference in feeding behaviour by larvae of the insect, *Plutella xylostella*, on herbivore-susceptible and herbivore-resistant *Arabidopsis thaliana* lines. 2D-PAGE analysis coupled with MS/MS revealed significant differences in the proteins expressed in herbivore-resistant and herbivore-susceptible lines with resistant lines showing increased production of reactive oxygen species (ROS). Also many proteins known to be involved in limiting ROS are damaged such as carbonic anhydrases, malate dehydrogenases, glutathione S-transferases, etc. These results led to the conclusion that increased production of ROS is one of the mechanisms of *Plutella* resistance in *Arabidopsis* (Collins et al. 2010).

Comparative analysis of protein expression profiles was carried out in the rice leaf sheaths in response to infestation by the brown planthopper (*Nilaparvata lugens* Stål, BPH). Using quantitative mass spectrometry-based proteomic approach, several proteins involved in multiple pathways showed significant changes in response to BPH feeding. These included jasmonic acid synthesis proteins, oxidative stress response proteins, beta-glucanase protein, kinases, clathrin protein, glycine cleavage system protein, photosynthesis proteins and aquaporins. Also it was observed that proteomic responses related to wounding, oxidative and pathogen stress overlapped considerably between BPH-resistant and BPH-susceptible rice lines (Wei et al. 2009).

The changes to the proteome of Sitka spruce bark (*Picea sitchensis*) tissue were examined subsequent to feeding by white pine weevils (*Pissodes strobi*) or mechanical wounding by Lippert et al. 2007. The 2-D PAGE and high-throughput MS/MS were used to examine induced changes in protein abundance and protein modification. Significant changes were observed as early as 2 h following the onset of insect feeding. Among the insect-induced proteins, a series of related small heat shock proteins, other stress response proteins, proteins involved in secondary metabolism, oxidoreductases and a novel spruce protein were significant. Comparison of protein expression and cDNA microarray profiles of induced spruce stem tissues reveals the complementary nature of transcriptome and proteome analyses and the need to apply a multifaceted approach to the large-scale analysis of plant defence systems.

4.2 Proteomics of Insect Secretions (Secretome/Regurgitome, etc.)

A major determinant of the influence of evolutionary arms races is the strategy of the insect. Generalist insect herbivores, such as *Myzus persicae* aphid, need more complex adaptive mechanisms since they need to respond to a large array of different plant defensive chemicals (Frederic et al. 2006). The signal responsible for the activation of plant defences is not only mechanical but also chemical through the action of particular molecules, commonly called elicitors (Harmel et al. 2008). The salivary effectors of herbivores have been poorly studied, although there have been similar recent approaches using transcriptomics and proteomics (Giacometti and Zavala 2016). According to Harmel et al. (2008), some elicitors in insect saliva comes in contact with plant host tissue, inducing plant defence reactions. They studied the salivary proteome of the green peach aphid, *Myzus persicae*, and discovered glucose oxidase, glucose dehydrogenase, NADH dehydrogenase, α-glucosidase and α-amylase in *M. persicae* saliva. Sap-sucking insects often inject saliva into host plants, which contains a suite of effector proteins and even microbial communities that can alter the plant defence. Lacking salivary glands, leaf-feeding beetles represent an interesting group of phytophagous insects. Feeding beetles regurgitate onto leaf surfaces, and it is thought that these oral secretions influence insect–plant interactions and even play a role in virus–vector specificity. Leaf-feeding beetles, such as *E. varivestis*, deposit regurgitant onto wounded leaves during feeding (Gedling et al. 2018). Analogous to the saliva of sap-sucking insects, it has been speculated by Gedling et al. (2018) that these oral secretions perform vital roles in the feeding process by initiating digestion and suppressing anti-herbivory host defences. They reported the first comprehensive high-throughput 'regurgitome' of a beetle species.

Frederic et al. (2006) studied the chemical ecology of *M. persicae* associated with different plant species, from Brassicaceae and Solanaceae families using a non-restrictive proteomic approach. The complex protein mixtures of the insect were separated by 2-D electrophoresis, and the related spots of proteins significantly varying were selected and identified by mass spectrometry (ESI-MS/MS) coupled with data bank investigations. Fourteen aphid proteins were found to vary according to host plant switch; ten of them were downregulated (proteins involved in glycolysis, TCA cycle and protein and lipid synthesis), while four others were overexpressed (mainly related to the cytoskeleton) (Frederic et al. 2006).

5 Conclusion and Further Prospects

Insect pests are biotic stressors and reduce yield quantity as well as quality. Plants counteract insect attacks both directly and indirectly. Many of these defences are regulated by signalling pathways where phytohormones have major roles. Plants have developed efficient mechanisms to protect them against herbivory, while insects have found diverse ways of avoiding negative effects of their host plants' defence mechanisms. Molecular genetics, genomic and proteomic technologies are now providing exciting new avenues of research in plant–insect interactions. These applications are beginning to provide in-depth information about a vast array of plantmolecular responses to insect herbivores. By using proteomics, differentially expressed proteins during insect infestation were successfully identified. These proteins participate in multiple physiological and defence processes. Functional classification analysis indicated that such proteins were associated with amino acid, carbohydrate, energy and lipid metabolism and photosynthesis. In addition, their relative abundance was upregulated or downregulated according to insect pest feeding on plant leaves.

The sequencing of the *Arabidopsis* genomes has provided the first real insights into the structure, function and location of plant–insect resistance genes. In addition, microarrays containing several thousand expressed sequences allow rapid screening of putative plant resistance-related cDNAs. Arrays for *Arabidopsis*, *Glycine* spp., barley, tomato and wheat are in use to provide genome-wide representations of plant genes involved in defence responses to insect attack. As a more complete knowledge of plant genomes and proteome develops, microarrays will provide valuable information about the identity of resistance genes and the gene products mediating their function. The better understanding of this process will allow us to achieve more effective methods for the biological control of insect pests with natural products by the development of new plant varieties with enhanced chemical defences. The knowledge about the sequences of resistance genes from different resistance sources can be used for the development of next-generation insect-resistant crop cultivars. Therefore, the cultivars with resistance genes of diverse sequence and function can be released and deployed to sustain resistance and help delay the development of virulent, resistance-breaking insects. However, proper ethical and safety guidelines have to be strictly followed during transgenic developments.

References

Agarwal K, Choe LH, Lee KH (2006) Shotgun proteomics using the iTRAQ isobaric tags. Brief Funct Genom Proteom 5(2):112–120

Anguraj-Vadivel AK (2015) Gel-based proteomics in plants: time to move on from the tradition. Front Plant Sci 6:369

Barah P, Bones AM (2015) Multidimensional approaches for studying plant defence against insects: from ecology to omics and synthetic biology. J Exp Bot 66(2):479–493

Berenbaum MR, Zangerl AR (2008) Facing the Future of Plant-Insect Interaction Research: Le Retour a' la "Raison d'Être"1. Plant Physiol 146:804–811

Bidart-Bouzat MG, Kliebenstein D (2011) An ecological genomic approach challenging the paradigm of differential plant responses to specialist versus generalist insect herbivores. Oecologia 167:677–689

Bilgin DD, Zavala JA, Zhu J, Clough SJ, Ort DR, DeLucia EH (2010) Biotic stress globally downregulates photosynthesis genes. Plant Cell Environ 33(10):1597–1613

Broekgaarden C, Voorrips RE, Dicke M (2011) Transcriptional responses of *Brassica nigra* to feeding by specialist insects of different feeding guilds. Insect Sci 18:259–272

Cantin GT, Venable JD, Cociorva D et al (2006) Quantitative phosphoproteomic analysis of the tumor necrosis factor pathway. J Proteome Res 5(1):127–134

Collins RM, Afzal M, Ward DA, Prescott MC, Sait SM et al (2010) Differential proteomic analysis of *Arabidopsis thaliana* genotypes exhibiting resistance or susceptibility to the insect herbivore, *Plutella xylostella*. PLoS One 5(4):e10103. https://doi.org/10.1371/journal.pone.0010103

Coppola V, Coppola M, Rocco M et al (2013) Transcriptomic and proteomic analysis of a compatible tomato-aphid interaction reveals a predominant salicylic acid-dependent plant response. BMC Genomics 14:515

De Vos M, Van Oosten VR et al (2005) Signal signature and transcriptome changes of *Arabidopsis* during pathogen and insect attack. Mol Plant Microbe Interact 18:923–937

Dong Y, Fang X, Yang Y, Xue G-P, Chen X, Zhang W, Wang X, Yu C, Zhou J, Mei Q, Fang W, Yan C, Chen J (2017) Comparative proteomic analysis of susceptible and resistant rice plants during early infestation by small brown planthopper. Front Plant Sci 8:1744. https://doi.org/10.3389/fpls.2017.01744

Duceppe MO, Cloutier C, Michaud D (2012) Wounding, insect chewing and phloem sap feeding differentially alter the leaf proteome of potato, *Solanum tuberosum* L. Proteome Sci 10:73

Fang X, Chen J, Dai L, Ma H, Zhang H, Yang J, Wang F, Yan C (2015) Proteomic dissection of plant responses to various pathogens. Proteomics 15(9):1525–1543

Feng X, Liu BF, Li J et al (2015) Advances in coupling microfluidic chips to mass spectrometry. Mass Spectrom Rev 34(5):535–557

Fenn JB, Mann M, Meng CK, Wong SF, Whitehouse CM (1989) Electrospray ionization for mass spectrometry of large biomolecules. Science 246(4926):64–71

Frederic F, Pascal G, Nicolas H, Gabriel M, Edwin DP, Eric H (2006) Proteomics in *Myzus persicae*: effect of aphid host plant switch. Insect Biochem Mol Biol 36:219–227

Gedling CR, Smith CM, LeMoine CMR, Cassone BJ (2018) The Mexican bean beetle (*Epilachna varivestis*) regurgitome and insights into beetle-borne virus specificity. PLoS One 13(1): e0192003

Giacometti R, Zavala JA (2016) Soybean response to pest attack. Soybeans: cultivation, nutritional properties and effects on health. Nova Science Publishers, New York, pp 35–62

Gao W (2014) Analysis of protein changes using two-dimensional difference gel electrophoresis. Molecular toxicology protocols, pp 17–30

Gorg A, Weiss W, Dunn MJ (2004) Current two-dimensional electrophoresis technology for proteomics. Proteomics 4(12):3665–3685

Gygi SP, Rochon Y, Franza BR, Aebersold R (1999) Correlation between protein and mRNA abundance in yeast. Mol Cell Biol 19:1720–1730

Han X, Aslanian A, Yates JR III (2008 Oct 1) Mass spectrometry for proteomics. Curr Opin Chem Biol 12(5):483–490

Harmel N, Letocart E, Cherqui A, Giordanengo P, Mazzucchelli G, Guillonneau F et al (2008) Identification of aphid salivary proteins: a proteomic investigation of *Myzus persicae*. Insect Mol Biol 17:165–174

Karas M, Hillenkamp F (1988) Laser desorption ionization of proteins with molecular masses exceeding 10,000 daltons. Anal Chem 60(20):2299–2301

Kerchev PI, Fenton B, Foyer CH, Hancock RD (2012) Plant responses to insect herbivory: interactions between photosynthesis, reactive oxygen species and hormonal signalling pathways. Plant Cell Environ 35(2):441–453

Kessler A, Baldwin IT (2002) Plant responses to insect herbivory: the emerging molecular analysis. Annu Rev Plant Biol 53:299–328

Kusnierczyk A, Winge P, Midelfart H, Armbruster WS, Rossiter JT, Bones AM (2007) Transcriptional responses of *Arabidopsis thaliana* ecotypes with different glucosinolate profiles after attack by polyphagous *Myzus persicae* and oligophagous *Brevicoryne brassicae*. J Exp Bot 58:2537–2552

Lambert JP, Ethier M, Smith JC, Figeys D (2005) Proteomics from gel based to gel free. Anal Chem 77(12):3771–3787

Lippert D, Chowrira S, Ralph SG, Zhuang J, Aeschliman D, Ritland C, Ritland K, Bohlmann J (2007) Conifer defense against insects: proteome analysis of Sitka spruce (*Picea sitchensis*) bark induced by mechanical wounding or feeding by white pine weevils (*Pissodes strobi*). Proteomics 7:248–270

Lodha TD, Hembram P, Tep N, Basak J (2013) Proteomics: a successful approach to understand the molecular mechanism of plant-pathogen interaction. Am J Plant Sci 4:1212–1226

Lund TC, Anderson LB, McCullar V (2007) iTRAQ is a useful method to screen for membrane-bound proteins differentially expressed in human natural killer cell types. J Proteome Res 6:644–653

Maor R, Jones A, Nühse TS et al (2007) Multidimensional protein identification technology (MudPIT) analysis of ubiquitinated proteins in plants. Mol Cell Proteomics 6(4):601–610

Mehta A, Brasileiro ACM, Souza DSL et al (2008) Plant–pathogen interactions: what is proteomics telling us?. FEBS J 275:3731–3746

Mello MO, Silva-Filho MC (2002) Plant-insect interactions: an evolutionary arms race between two distinct defense mechanisms. Braz J Plant Physiol 14(2):71–81

Moore CD, Ajala OZ, Zhu H (2016) Applications in high-content functional protein microarrays. Curr Opin Chem Biol 30:21–27

Moran PJ, Thompson GA (2001) Molecular responses to aphid feeding in *Arabidopsis* in relation to plant defense pathways. Plant Physiol 125:1074–1085

Quirino BF, Candido ES, Campos PF, Franco OL, Krüger RH (2010) Proteomic approaches to study plant–pathogen interactions. Phytochemistry 71:351–362

Renaut J, Hausman J, Wisniewski ME (2006) Proteomics and low temperature studies: bridging the gap between gene expression and metabolism. Physiol Plant 126:97–109

Rogowska-Wrzesinska A, Le Bihan MC, Thaysen-Andersen M, Roepstorff P (2013) 2D gels still have a niche in proteomics. J Proteome 2(88):4–13

Romanov V, Davidoff SN, Miles AR et al (2014) A critical comparison of protein microarray fabrication technologies. Analyst 139(6):1303–26

Rose JKC, Bashir S, Giovannoni JJ, Jahn MM, Saravanan RS (2004) Tackling the plant proteome: practical approaches, hurdles and experimental tools. Plant J 39:715–733

Ross PL, Huang YN, Marchese JN et al (2004) Multiplexed protein quantitation in *Saccharomyces cerevisiae* using amine-reactive isobaric tagging reagents. Mol Cell Proteomics 3:1154–1169

Rustagi A, Singh G, Agrawal S, Gupta PK (2018) Proteomic studies revealing enigma of Plant–Pathogen interaction. In: Singh A, Singh I (eds) Molecular aspects of plant-pathogen interaction. Springer, Singapore

Speers AE, Wu CC (2007) Proteomics of integral membrane proteins – theory and application. Chem Rev 107(8):3687–3714

Tanaka K, Waki H, Ido Y, Akita S, Yoshida Y, Yoshida T, Matsuo T (1988) Protein and polymer analyses up to *m/z* 100 000 by laser ionization time-of-flight mass spectrometry. Rapid Commun Mass Spectrom 2(8):151–153

Truong DH, Bauwens J, Delaplace P, Mazzucchelli G, Lognay G, Francis F (2015) Proteomic analysis of *Arabidopsis thaliana* (L.) Heynh responses to a generalist sucking pest (*Myzus persicae* Sulzer). Plant Biol (Stuttg) 17(6):1210–1217. https://doi.org/10.1111/plb.12363

Truong DH, Nguyen HC, Bauwens J, Mazzucchelli G, Lognay G, Francis F (2018) Plant defense in response to chewing insects: proteome analysis of *Arabidopsis thaliana* damaged by *Plutella xylostella*. J Plant Interact 13(1):30–36

Walling LL (2000) The myriad plant responses to herbivores. J Plant Growth Regul 19:195–216

Wei Z, Hu W, Lin Q, Cheng X, Tong M, Zhu L, Chen R, He G (2009) Understanding rice plant resistance to the Brown Planthopper (*Nilaparvata lugens*): a proteomic approach. Proteomics 9 (10):2798–2808. https://doi.org/10.1002/pmic.200800840

Whittaker RH, Feeny PP (1971) Allelochemics: chemical Interactions between species. Science 171 (3973):757–770

Wittmann-Liebold B, Graack HR, Pohl T (2006) Two-dimensional gel electrophoresis as tool for proteomics studies in combination with protein identification by mass spectrometry. Proteomics 6(17):4688–4703

Yates IJR, Gilchrist A, Howell KE et al (2005) Proteomics of organelles and large cellular structures. Nat Rev Mol Cell Biol 6(9):702–714

Zhou H, Ranish JA, Watts JD et al (2002) Quantitative proteome analysis by solid-phase isotope tagging and mass spectrometry. Nat Biotechnol 20(5):512–515

Zhu W, Smith JW, Huang CM (2009) Mass spectrometry- based label-free quantitative proteomics. J Biomed Biotechnol 2010. Article ID: 840581

Zhu M, Simons B, Zhu N et al (2010) Analysis of abscisic acid responsive proteins in Brassica napus guard cells by multiplexed isobaric tagging. J Proteome 73(4):790–805

Zhu N, Zhu M, Dai S et al (2012) An improved isotope-coded affinity tag technology for thiol redox proteomics. J Integr OMICS 2(1):17–23

Zieske LR (2006) A perspective on the use of iTRAQ reagent technology for protein complex and profiling studies. J Exp Bot 57(7):1501–1508

Role of miRNA in Plant Defense Against Insects

Jinu Jacob, P. Madhu, and R. Vinodh

Abstract

MicroRNAs (miRNAs) are ~20–24-nucleotide-long endogenous small RNA molecules that regulate gene expression. They are known to repress gene expression at post-transcriptional levels. In the recent years, advances in high-throughput sequencing techniques have helped in uncovering the vast population of miRNAs and their roles in plants and animals. Insects comprise the largest group of metazoans. As much as insects contribute toward the existence of plant species, they pose threat to agricultural production. Recently, multiple investigations have indicated roles of miRNAs in various processes in insects as well as insect-related responses in plants. In this chapter, we summarize the information available on miRNA discovery in insect species that are of relevance to plants, roles of miRNAs in plant-insect interactions, approaches for miRNA discovery, and application of miRNAs in developing effective crop protection strategies.

1 Introduction

The last three decades have witnessed the birth of a novel gene regulatory mechanism in eukaryotes. Small non-protein coding RNAs have emerged as significant components in the regulation of several fundamental processes in living organisms such as developmental transition and patterning, responses to biotic as well as abiotic stresses, maintenance of genome stability, response to various environmental cues,

J. Jacob (✉) · R. Vinodh
Biotechnology Division, ICAR-Indian Institute of Millets Research (IIMR), Rajendranagar, Telangana, India

P. Madhu
ICRISAT, Genomics and Trait Discovery Theme-Genetic Gains Programme, Hyderabad, India

I. K. Singh, A. Singh (eds.), *Plant-Pest Interactions: From Molecular Mechanisms to Chemical Ecology*, https://doi.org/10.1007/978-981-15-2467-7_4

and even development of diseases. The different classes of non-protein coding RNAs or ncRNAs, mainly natural antisense small interfering RNAs (nat-siRNAs), microRNAs (miRNAs), transacting small interfering RNAs (ta-siRNAs), and small interfering RNAs (siRNAs), carry out a myriad of functions in eukaryotic cells. MicroRNAs are one of the highly studied classes of ncRNAs and are ~22-nucleotide-long endogenous RNAs that influence gene regulatory networks in plants and animals. miRNAs achieve their functions by targeting mRNAs for cleavage or translational repression (Bartel 2004). In plants, miRNAs generally regulate gene expression by degrading the target mRNA, while in animals, they achieve this by blocking mRNA translation by binding with their complementary sequences in the 3′-UTR of target mRNA. miRNAs are pleiotropic in nature and an individual miRNA can have multiple targets (Selbach et al. 2008).

Existence of regulatory RNAs that are neither translated nor bear structural roles came into picture when it was identified that the gene, *lin-4*, which is involved in the control of larval developmental timing in *Caenorhabditis elegans* instead of coding for a protein, produces a pair of short RNAs (Lee et al. 1993; Wightman et al. 1993). There were multiple complementary sites for these small RNAs in the 3′-UTR of *lin-14* gene, and they regulated the generation of LIN 14 protein without affecting the mRNA content much, indicating a post-transcriptional control. Discovery of ~22-nucleotide-long regulatory RNAs encoded by *let-7* from *C. elegans* and identification of its homologues in humans, flies, and other organisms hinted toward existence of more small RNAs and brought about better clarity in understanding the roles of small RNAs (Slack et al. 2000; Reinhart et al. 2000; Pasquinelli et al. 2000). Because lin-4 and let-7 were developmentally regulated, they were referred to as "short temporal RNAs" or "stRNAs" initially. Soon several orthologues of these genes and similar genes were identified in various organisms, and scientists started calling this class of small RNAs as "microRNAs" or "miRNAs" (Lagos-Quintana et al. 2001, Lau et al. 2001). Now it is established that miRNAs regulate mRNA transcripts by binding to 3′-UTR complementary sequence in the target mRNA and either by inhibiting its translation or by degrading it.

Plants are no exception to the post-transcriptional regulation of genes by small RNAs of which siRNAs and miRNAs take the lead. miRNAs are transcribed from *MIR* genes that are found in the genome and are conserved in plant species (Reinhart et al. 2002; Nozawa et al. 2012). They are generated from a non-coding RNA molecule called pri-miRNA which is transcribed from the *MIR* genes in the genomic DNA. Pri-miRNAs are broken down into shorter precursor transcripts called pre-miRNAs that form self-complementary stem-loop structures, in a two- or multi-step process in the cytoplasm that involves an RNaseIII enzyme, DCL1. Either strand of this duplex may serve as functional miRNAs (Krol et al. 2010).

In plants, miRNAs predominantly function by target cleavage (Bartel 2004). They also employ methylation of DNA or histone to silence genes at transcriptional level (Wu et al. 2010). The abundance and diversity of miRNAs in plants are clearly indicative of the involvement of one or more of them in any biological process at some point (Djami-Tchatchou et al. 2017; D'Ario et al. 2017; Liu et al. 2018).

2 Approaches for Detection and Expression Analysis of miRNAs

miRNA detection is of paramount significance in the purview of determining its function. Bioinformatics methods are used to predict putative miRNAs from transcriptome or whole genome sequence database. Potential miRNA precursors that are capable of assuming a hairpin-like secondary structure are identified computationally (Zhang et al. 2005; Sunkar and Jagadeeswaran 2008). However, several of the pre-miRNAs elude computational detection due to the stringency of the filters used in the programs, and also, not all computationally predicted putative miRNAs are functional. Therefore, a combined approach that unites computational prediction and experimental validation is essential to discover more miRNAs in a species with comprehensive genomic information.

Northern blotting was one of the earliest techniques used for detecting miRNAs (Lee et al. 1993). This technique includes size separation of miRNA pool in a denaturing gel followed by its transfer into a membrane. The desirable labeled single miRNA probe is used to visualize the blotted hybridized miRNA. Though this technique suffers from a requirement for large quantity of RNA and low sensitivity and throughput, it is the single technique that makes quantitative visualization of miRNA possible (Hunt et al. 2015). Microarray technique allowed more massive detection of miRNAs from an RNA sample. It's a high-throughput hybridization-based technique which is semi-quantitative. They are best at comparing relative abundance of specific microRNAs between two states (e.g., treatment vs. control or disease vs. healthy), but they cannot be used for determining absolute quantities (Pritchard et al. 2012). Also this technique suffers from specificity issues between closely related sequences. Hence, generally, detection of miRNAs using microarrays needs to be coupled with quantitative reverse transcriptase-PCR (qRT-PCR) to precisely quantify gene expression.

At present, next-generation sequencing (NGS) or RNAseq is the leading technique for miRNA discovery and quantification and is the only method capable of discovering novel miRNAs (Androvic et al. 2017). It starts with construction of a small RNA cDNA library from the desired RNA sample and then sequencing the multiple reads parallelly using NGS (next-generation sequencing) methods. The greatest advantage of this method is that it facilitates discovery of known and novel miRNAs which is not possible with hybridization-based techniques. Precise identification of miRNA variants differing even by a single nucleotide is possible by RNAseq. But care should be exercised in precisely annotating millions of small RNA reads coming out of NGS into various classes of non-coding RNAs using proper bioinformatics methodologies (Yang and Qu 2012). Though high cost and requirement for huge computational infrastructure are its limitations, this technique is harnessing momentum as the most extensively used tool for miRNA discovery.

Quantitative RT-PCR is considered as the gold standard among miRNA quantification techniques (Hunt et al. 2015). It is most often used for validating the results from high-throughput miRNA discovery methods such as microarrays and next-generation sequencing. Design of miRNA primers for PCR amplification remains a challenge owing to the short length of miRNAs. It is complicated by the existence of multiple family members which differ only by a single nucleotide. Still several efficient modifications and methods have been developed to overcome these challenges (Chen et al. 2005; Jonstrup et al. 2006; Yin et al. 2012; Redshaw et al. 2013; Miao et al. 2016; Androvic et al. 2017). Thus far, several thousand plant miRNAs have been identified and uploaded in the major database of miRNA registry (miRBase). Several plant miRNAs are yet to be identified, and still large numbers are to be functionally annotated. Table 1 summarizes the most relevant literature on the miRNA population of various insects studied so far that are of relevance to plants and also mentions the miRNA species identified during various plant-insect interactions.

3 MicroRNAs in Plant-Insect Interactions

With the advent of next-generation sequencing, the small RNAome of several insects have been inventoried. Insects form the largest group of animals with complex ecological roles and are highly valuable in biological and agriculture research. MiRBase 22 release currently has miRNA inventories from 35 insect species. During the last two decades, computational and experimental approaches were successful in identifying numerous conserved and novel miRNAs from diverse plant species and insects. The progress in high-throughput sequencing technologies made the advances in unraveling the roles played by small RNAs, especially miRNAs in plant-insect interactions.

(a) **Identification of insect miRNAs**

Insects comprise the largest group of metazoans. Along with the long list of insects that are beneficial to plants and animals, there are several of them causing significant damage to crops around the world. Several attempts have been made in identifying the miRNAs controlling insect development and metabolism. Through bioinformatics techniques, direct cloning, and high-throughput sequencing, miRNA population of a few insect species that are of relevance to plants has been uncovered. The list includes honey bees (*Apis mellifera*) (Weaver et al. 2007), red flour beetle (Singh and Nagaraju 2008), locusts (Wei et al. 2009), *Nilaparvata lugens* (Chen et al. 2012; Zha et al. 2016), and others. Identifying insect miRNAs and tracking their functional roles pose greater hurdles because it's one of the largest groups of animals that are highly versatile in environmental adaptation and are endowed with unique developmental processes.

Mulberry silkworm, *Bombyx mori*, is an insect that is key to sericulture. Using a combinatorial approach consisting of computational prediction of putative miRNAs from genome sequence of *B. mori* and deep sequencing, Yu et al. (2008) initially identified 118 already existing and 151 novel miRNAs from this insect. The study brought out the fine-tuning of developmental stages by miRNAs in *B. mori*. They discovered that molting stages were characterized by significant changes in miRNA expression both in terms of types and quantities. Later on, large-scale screening for

Table 1 miRNAs identified from insects of relevance to plants

Host	Insect	Number of miRNAs identified/role of miRNAs	References
Cucumis melo	*Aphis gossypii*	Role of miRNA in aphid herbivory in melon plants identified	Sattar et al. (2012, 2016)
Chrysanthemum	*Macrosiphoniella sanborni*	303 conserved and 234 novel miRNAs	Xia et al. (2015)
Rice	*Nilaparvata lugens*	464 miRNAs identified from rice; 183 potentially novel	Wu et al. (2017b)
Rice	*Nilaparvata lugens*	41 novel miRNAs identified	Zha et al. (2016)
Solanum lycopersicum and *S. habrochaites*	*Bemisia tabaci*	44 miRNA families identified: 13 novel families	Wang et al. (2018)
Camellia sinensis	*Ectropis oblique* and mechanical wounding	Discovered 130 conserved miRNAs and 512 novel miRNAs	Jeyaraj et al. (2017)
Wheat	*Diuraphis noxia*	27 miRNAs identified	Nicolis et al. (2017)
Wheat	*Schizaphis graminum* and *Sipha flava*	72 and 56 miRNA candidates identified	Wang et al. (2017)
Wheat	*Aphis gossypii*	292 miRNAs identified	Ma et al. (2017)
Wheat	*Sogatella furcifera*	382 miRNAs (106 conserved and 276 novel)	Chang et al. (2016)
Wheat	*Bombyx mori*	354 miRNAs	Zhang et al. (2009)
Wheat	*Bombyx mori*	118 conserved and 151 novel miRNAs	Yu et al. (2008)
Wheat	*Apis mellifera*	267 novel miRNAs	Chen et al. (2010)
Wheat	*Apis mellifera*	17 novel brain-specific miRNAs	Greenberg et al. (2012)
Wheat	*Mayetiola destructor*	89 conserved and 184 novel miRNAs; 611 putative miRNAs	Khajuria et al. (2013)
Wheat	*Plutella xylostella*	384 conserved and 174 novel miRNAs	Liang et al. (2013)
Wheat	*Helicoverpa armigera*	Chitinase regulating miRNAs identified	Agrawal et al. (2013)
Wheat	*Mayetiola destructor*	921 miRNAs in total	Chen et al. (2017)
Wheat	*Manduca sexta*	164 conserved miRNAs and 22 novel miRNAs	Zhang et al. (2012a, 2014, 2015)
Wheat	*Tribolium castaneum*	118 putative miRNAs	Luo et al. (2008)

(continued)

Table 1 (continued)

Host	Insect	Number of miRNAs identified/role of miRNAs	References
Wheat	*Tribolium castaneum*	45 putative miRNAs	Singh and Nagaraju (2008)
Wheat	*Tribolium castaneum*	203 miRNAs	Marco et al. (2010)
Wheat	*Tribolium castaneum*	123 novel miRNAs	Ninova et al. (2016)
Wheat	*Tribolium castaneum*	108 conserved and 772 novel miRNAs	Wu et al. (2017a)
A. thaliana	*Heterodera schachtii*	Role of miR396 in syncytium development identified	Hewezi et al. (2012)
A. thaliana	*Heterodera schachtii*	Role of miR827 in susceptibility reaction of the host identified	Hewezi et al. (2016)
Arabidopsis	*Meloidogyne javanica*	288 miRNAs	Cabrera et al. (2016)

miRNA genes from *B. mori* was attempted by Zhang et al. (2009) through high-throughput sequencing to specifically identify their roles in insect development and metamorphosis. Existence of 354 miRNA genes was confirmed using miRNA microarrays from 3750 miRNA candidate genes that were predicted computationally. Expression analysis implied specific miRNA-mediated regulation of insect development as evidenced by the egg- and pupal-specific expression of 248 miRNAs.

Honey bee (*Apis mellifera*) is a highly social insect which is regarded as a model organism for studying social behavior. Initial studies on honey bee miRNAs concentrated on the expression validation of computationally predicted miRNAs which suggested possible role of miRNAs in age-dependent behavioral changes (Weaver et al. 2007; Behura and Whitefield 2010). Large-scale miRNA sequencing of small RNA libraries from different developmental stages of honey bees provided a basis for the role of these RNAs in regulating developmental networks and differentiation of caste (Chen et al. 2010). This study identified 267 novel miRNAs from *A. mellifera*. Functional validation of brain-specific miRNAs brought out a better picture of the division of labor in bees (Hori et al. 2011 and Greenberg et al. 2012). The study by Greenberg et al. (2012) provided the first evidence correlating neural and behavioral plasticity associated with division of labor in honey bees and variation in miRNA expression in the brain. The study identified a downregulation of several miRNAs in nurse bees in comparison to foragers, and this happened exclusively when nurses and foragers were together in the colonies, highlighting a social context-based miRNA expression pattern (Greenberg et al. 2012).

To resolve the mystery underlying the generation of phenotypes that are morphologically and reproductively diverse from the same organism genome, Ashby et al.

(2016) investigated the microRNAs that are specific to honey bee males and females as well as various castes. It was noticed that miRNA and mRNA profiles change critically in different classes of honey bees during a larva-to-pupa transition phase that is characterized by differentiation of various body parts from imaginal discs of larva. MicroRNA and transcriptional profiles differed strikingly between queens and workers that are reproductively distinct. When enriched genes belonged to metabolic pathways in queens and drones, workers had enrichment of neuronal and caste developmental genes. They also provided evidences that miRNAs target non-methylated neuronal development genes. Feeding of plant miRNA-enriched diet to honey bees resulted in the occurrence of these miRNAs in bee bread and royal jelly, with a higher proportion being in bee bread, and such bees developed more worker-like features in comparison to the control bees (Zhu et al. 2017).

One among the first agricultural pests for which miRNA profile was systematically documented was diamond back moth, *Plutella xylostella*, a worldwide pest attacking the crops of Cruciferae (Liang et al. 2013). Small RNA population from several stages of this insect, starting from the egg, larvae, pupae, and adults, were discovered by creating a pooled library of small RNAs. A total of 384 miRNAs were identified in sequencing, out of which 174 were *Plutella xylostella* specific. This study was a comprehensive report which added onto the small repertoire of information which was already existing through the works of Etebari et al. (2013). *Plutella xylostella* miRNAs were found to be highly related to those of *Bombyx mori* (used as reference genome) and two other lepidopterans, *Manduca sexta* and *Heliconius melpomene*. A microarray analysis identified developmental stage-specific expressions of 234 miRNAs suggesting their role in embryogenesis and metamorphosis.

Deep sequencing identified 72 and 56 miRNA candidates in *Schizaphis graminum* (greenbug) and *Sipha flava* (yellow sugarcane aphid) of which 14 and 8, respectively, were novel (Wang et al. 2017). 45 of them were expressed in both the species. Sbi-miR1-3p, sbimiR2- 3p, sbi-miR396-5p, sbi-miR3-5p, hvu-miR2- 5p, and sbi-miR3905p were suggested to have roles in detoxification of xenobiotics, while sbi-miR3-5p, sbi-miR4-5p, sbi-miR5076-5p, sbimiR6230-3p, sbi-miR516-3p3, sbi-miR166-3p, sbi-miR390-5p, hvumiR2-5p, and hvu-miR3-3p were targeting genes involved in digestive physiology by metabolizing starch and sucrose. miRNA repertoire of another insect having relevance to plant kingdom is a major rice plant hopper, *Sogatella furcifera* (whitebacked plant hopper). Apart from causing feeding damages on the crop, it acts as the carrier of plant viruses. 382 miRNAs and 4117 of their targets were identified from viral inoculated (viruliferous) and mock-inoculated (non-viruliferous) cultured cells of *S. furcifera* and were thoroughly characterized with respect to their conservation across species, functions, and organization within the genome (Chang et al. 2016). The miRNA pool had 106 conserved and 276 novel miRNA candidates. A key observation was that in comparison to novel miRNAs, the conserved miRNAs were more target-specific even when the putative target genes had similar functions.

Deep sequencing of a gall fly, Hessian fly (*Mayetiola destructor*) which is a wheat pest, was initially attempted in 2013 (Khajuria et al.) to dig out the miRNA population in the larval transcriptome. They identified 89 conserved and 184 novel miRNAs from the fly. Based on bioinformatics analysis of the genome sequence of Hessian fly, around 611 putative miRNA-encoding genes could be identified. They also noted a striking feature of miRNA population in Hessian fly, which were the extended members in several of the miRNA families with as many as 91 genes coding for 20 different miRNAs. The same observation was reinstated by Chen et al. 2017 when they went for a systematic analysis of developmental stage-specific miRNAs from the Hessian fly by high-throughput sequencing. Yet another special feature was the abundance of 15–16-nucleotide-long reads which constituted ~37% of the total miRNA reads. Chen et al. could identify 921 miRNAs from wheat Hessian fly.

miRNA repertoire in the tobacco hornworm (*Manduca sexta*) was comprehensively studied revealing 164 conserved and 22 novel miRNA species (Zhang et al. 2012a, 2014, 2015). As this insect is a good model for studying the biochemical processes, Zhang et al. (2015) attempted to derive clues on the roles of miRNAs in coordinating insect metabolism, immune responses, hormonal regulation and nutrient metabolism, influence of type of tissue on miRNA expression, and role of miRNA in metabolism and immune responses. They identified prospective mRNA target genes among enzymes, hormonal regulation, and nutrient metabolism pathways.

An important storage pest of agricultural products, the red flour beetle (*Tribolium castaneum*), has been identified as a model for studying the genetics and development of insects. Through scanning of whole genome sequence, 118 putative miRNAs (Luo et al. 2008) and 45 miRNAs (Singh and Nagaraju 2008) have been computationally identified in the pest. Deep sequencing of various stages of the insect transcriptome led to further identification of developmental stage-specific miRNAs (Marco et al. 2010; Ninova et al. 2016). To identify the miRNA-induced defects in metamorphosis, Wu et al. (2017a) employed RNAi to knock down the genes *Dcr-1* and *Ago-1*, and sequenced small RNAs form different life stages of *T. castaneum* resulting in the identification of 1154 unique miRNAs. There were 274 miRNAs belonging to 68 families which were conserved and 772 miRNAs that were hitherto unknown. This group for the first time identified that the putative juvenile hormone (JH) receptor gene *Met* and its downstream gene were the target of 11 and 14 miRNAs, respectively, suggesting the regulatory role for the miRNAs in JH signaling pathway.

Validation of in silico-predicted miRNAs through experiments holds importance in the study of small RNAs as the rate of false-positive results of computationally predicted miRNAs is quite high (Selbach et al. 2008). Agrawal et al. (2013) profiled the relative abundance of several microRNAs at different stages of expression of chitinase gene in *Helicoverpa armigera*. Chitinase is an enzyme catalyzing the degradation of a major component of insect cuticle which is chitin and hence having a significant role in metamorphosis in insects. A library of small RNAs was generated from various stages of larvae for studying the miRNAs that are specific to chitinase gene expression, and three miRNAs, miR-24, miR-2, and miR-131, which showed complementarity to chitinase 3′-UTR were identified. A dual luciferase assay system revealed that miR-24 regulates chitinase negatively, and when the larvae were fed on synthetic miR-24, there was significant reduction in chitinase transcript abundance and morphological aberrations resulting in molting arrest.

Yet another experimental validation of miRNA function was investigation of role of miRNAs in protease gene expression control in *Helicoverpa armigera* when exposed to plant protease inhibitors (Lomate et al. 2014). *H. armigera* is a polyphagous pest causing serious economic damage to many agricultural crops. A comparative profiling of miRNAs from *H. armigera* larvae fed on artificial diet containing protease inhibitors and control diet was carried out by constructing 12 small RNA libraries and by analyzing the expression profile of individual miRNAs. Deep sequencing using Illumina platform identified 186 unique miRNAs among which 90 were novel. miRNAs from *H. armigera* were mainly 22–25 nucleotide long, as is the case in other species. Effect of protease inhibitor diet feeding was clearly visible in terms of differences in expression of several miRNAs in the protease inhibitor-fed larvae in comparison to that fed on control diet. Target identification pointed out that the targets of several miRNAs included endo- and exo-protease genes coding for trypsin, cathepsin, chymotrypsin, amino and carboxypeptidase, etc. There were multiple target genes for many miRNAs, and many genes had multiple target sites for miRNAs.

(b) **Dynamics of host miRNAs in plant-insect interaction**

Evidence is mounting on the role of miRNAs in interaction of insects with plants. Though exact functional importance of many of the plant-specific miRNAs in herbivore gene expression networks is not clearly known and also there is still ambiguity over how feeding by an insect reprograms the miRNA expression in plants, evidences that make the picture clearer are accumulating. Large-scale molecular rearrangements including small RNA fluctuations followed by herbivory were for the first time demonstrated in *Nicotiana attenuata* exposed to *Manduca sexta* oral secretions (Pandey et al. 2008). Since then several studies were initiated that explored the role of small RNAome during plant-insect interactions.

During herbivory, insects will be ingesting large quantities of all types of RNAs including small RNAs. Since these ingested small RNAs need to go through the harsh pH environments of the insect digestive system, they were rarely thought to be evoking any change in gene expression of the host. But recent reports point to the existence of plant-derived miRNAs in considerable quantities in host bodies (Zhang et al. 2012b; Vaucheret and Chupeau 2012; Wang et al. 2017). In the publically available small RNA databases of pea, aphid, and silkworm, miR168 was the mostly occurring plant miRNA family (Zhang et al. 2012b). But when *Helicoverpa zea* (corn earworm) and *Spodoptera frugiperda* (fall armyworm) larval stages were fed on specific miRNA (including miR168)-enriched diets, deep sequencing of small RNAs could not detect miR168 abundance. Instead, the maximum detected plant miRNA in most of the insect libraries was miR1507. Interestingly, miR1507 was the predominant miRNA found even in corn-fed insects even when this miRNA doesn't exist in monocots, like corn (Zhang et al. 2012b). The authors, after thorough investigation, attributed the presence of plant miRNAs in insects/animals feeding on these plants to the experimental setup and cross-contamination. Later on, Wang et al. (2017) identified host plant-derived miRNAs in greenbugs and yellow sugarcane aphid. They could validate a subset of the miRNAs including two miRNAs that are not reported previously (sbi-miR1-3p and sbi-miR2-3p) and two already known miRNAs (sbi-miR2927a-5p and osa-miR390-5p) through RT-PCR. These miRNAs identified through deep sequencing from greenbug and yellow sugarcane aphid were expressed by sorghum (which is a host to the insects) as well as in greenbug confirming the hypothesis that miRNAs derived from plants are often ingested during feeding by insects. Cagirici et al. (2017) while studying the putative interaction between wheat and wheat sawfly, *Cephus cinctus*, showed that there are potential larval targets of wheat miRNAs. Even though elaborate studies are needed to establish the entry of small RNAs through oral feeding of insects, this mechanism can be tapped for the effective control strategies of agricultural pests.

Expression of small RNAs, which is induced by aphid feeding, has been thoroughly studied in many crops, and this was one among the first reports in the category of miRNA dynamics during interaction of plants and insects. Sattar et al. (2012) investigated the differences in expression profiles of miRNAs during insect herbivory in a resistant and susceptible interaction, taking melon (*Cucumis melo*)-aphid (*Aphis gossypii*) system (Sattar et al. 2012). In melon, a single gene governs aphid resistance, *vat* (virus aphid transmission), which is genetically dominant in nature, and *vat*$^{+}$ and *vat*$^{-}$ melon plants have different transcriptional reprogramming during aphid infestation. *Vat* imparts resistance against melon aphid as well as various mosaic viruses spread by this aphid. Most of the conserved miRNAs got expressed at higher levels during early (2 h, 4 h, 6 h) and late (8 h,10 h, 12 h after infestation) stages of aphid herbivory. miR164, miR393, miR169, miR166, miR160, miR398, miR165, miR167, miR2111, miR390, miR396, miR397, miR408, miR2911, and miR894 were increased during early stages of aphid feeding. While miR398 was the predominant miRNA during initial phases of infestation in the resistant line, its expression went down during susceptible interaction. The regulatory roles of these identified miRNAs were further understood by sequencing a degradome library from *vat*$^{+}$ melon plants (Sattar et al. 2016). For the differentially expressed miRNAs, a total of 70 miRNA/target pairs were identified, and it consisted of 28 novel combinations too. Interestingly, 11 of them were linked to phytohormone regulation. Cme-miR396, cme-miR156, cmemiR172, cme-miR394, and cme-miR159 were associated with jasmonic acid and ethylene signaling and perception and abscisic acid and gibberellic acid synthesis, and six of them were involved in auxin signaling and perception (cme-miR167, cme-miR160, cme-miR393, cme-miR319, cme-miR164, and cme-miR390). It is the transcription factors that were primary targets of majority of miRNAs. Aphid herbivory guided expression of genes that were under the regulation of such transcription factors was also studied to have a complete picture of phytohormone guided regulation of insect herbivory. Several members of such secondary genes belonging to Gretchen Hagen 3 (GH3) family got differentially expressed upon aphid feeding. Aphid feeding was found to cause a repression in expression of GH3.6, which is a target of miR160 and is under the regulation by ARF17 at two time points of infection in resistant and susceptible plants. GH3.5 was initially repressed at 6 h in *vat*$^{+}$ plants and got activated at a later point of aphid feeding, while there were no changes in its expression in the susceptible line. Irrespective of the initial expression pattern, there was a significant induction of GH3.1, 12 h after aphid feeding in both *vat*$^+$ and *vat*$^-$ leaf tissues. Levels of expression of IAA16, SAUR50 IAA13-like, and other secondary gene targets got reduced in *vat*$^+$ leaves, while that of IAA13-like has seen a rising trend in susceptible plants during 6-h and 12-h time points of aphid infection (Sattar et al. 2016).

Comparison of small RNA libraries constructed from aphid (*Macrosiphoniella sanborni*)-fed chrysanthemum leaves and control leaves revealed several differentially abundant miRNAs, especially miR159a, miR160a, and miR393a, which might be directly involved in plants' reaction to infestation by aphids (Xia et al. 2015). 24 conserved miRNAs and 37 novel miRNAs were observed in plants following aphid attack.

As there is differential transcript abundance during biotic stress in a resistant and a susceptible plant, there is differential regulation of miRNAs depending on the resistance or susceptibility of the infested plant as is shown in case of rice-BPH interaction (Wu et al. 2017b) and *Solanum*-white fly interaction (Wang et al. 2018). Brown plant hopper, BPH (*Nilaparvata lugens*), is the most devastating insect pest of modern rice cultivation system. It is a monophagous insect pest surviving on rice by feeding on phloem sap (Cha et al. 2008). Wu et al. (2017b) investigated the miRNA response of BPH feeding in rice plants that are resistant and susceptible to the pest. The resistant line P15 had *BPH15* (a BPH resistance gene) introgressed into it. The miRNA profiles of both lines were studied at three time points (0 h, 6 h, and 48 h) following attack by BPH leading to the identification of 464 miRNAs, among which 183 were not known before. BPH feeding induced more number of miRNAs to be differentially expressed in resistant rice P15 than in a susceptible line. 26 miRNAs commonly got differentially expressed upon BPH infestation irrespective of the disease response of the rice plant. The following miRNAs had a reduced expression in the resistant line after BPH feeding, and they targeted transcripts of defense-related pathways and hormone regulation, namely, miR444d (which targets a calmodulin binding protein), miR531a/b/c (targeting MAPK cascade genes), miR167a-5p (whose target is ARF16), miR399, miR156, miR1846e, and miR-3979-3p. It was also noticed that there was significant change in the expression pattern between early (6 h) and late (48 h) stages of feeding as it's the later stages where insect starts feeding on the phloem sap.

A set of miRNAs and their expression profiles were studied during four different stages of white fly feeding in two species of *Solanum* of which one is susceptible (*S. lycopersicum*) and the other one is resistant (*S. habrochaites*, a wild tomato species) to white fly (Wang et al. 2018), and miRNAs contributing to white fly resistance could be identified. The patterns of expression of miRNAs differed notably between cultivated and wild tomato species. For example, miR398, a highly conserved miRNA of higher plants which plays a key role in many plant-pathogen interactions, could be detected only in the wild species, *S. habrochaites*. miR398 showed decreased expression during the first three stages of white fly feeding hinting a role for it in resistance to the pest. miR164 showed variable expression in both the species and at different infestation stages. Computationally, several miRNAs such as miR390, miR6022, miR482, miR395, miR6027, miR530, miR6024, and miR8033, were predicted to be directed toward LRR genes, which are part of resistance proteins.

Mechanical wounding and oral secretion are the major stimulants for generating defense responses in plants in the event of pest attack (Wang et al. 2016). In order to specifically identify the miRNA population under these two conditions and to have an insight onto how insect oral feeding changes the dynamics of small RNA population over mechanical wounding, miRNA population from tea-*Ectropis oblique* system was investigated. The effect of mechanical and *E. oblique* (a geometrid insect pest of tea)-induced stress on tea plant (*Camellia sinensis*) was delineated by constructing individual small RNA libraries from tea leaves that are wounded mechanically, attacked by *E. oblique* and stress-free control tea leaves (Jeyaraj et al. 2017). Libraries revealed 130 already known miRNAs that are conserved among other plant species and 512 new miRNA species. A large number of novel (96-upregulated and 43-downregulated) and conserved (14-upregulated and 22-downregulated) miRNAs were identified, respectively, from the comparison of differential profiles of geometrid-induced/mechanical wounding libraries. All these are probable candidates of defense pathways against the geometrid pest, and several of the mRNA targets of these miRNAs belonged to defense signaling network of plants. csn-miR156a, targeting leucine-rich repeat receptor-like protein kinase (LRR-RLKs), was downregulated and csn-miR396d-3p, which targeted calcium-related proteins, was upregulated in the insect-fed leaves. Almost eight novel miRNAs were found to target reactive oxygen species (ROS) regulating enzymes such as ascorbate peroxidase, peroxidase, and glutathione sulfa-transferase which forms primary line of defense in plants during biotic stress. Most of the *E. oblique*-inducible miRNAs targeted transcription factors, namely, csn-miR160/ARF, csn-miR156/SPL, csn-miR164/NAC, csn-miR169/NFY, csnmiR171/bHLH, csn-miR172/ERF, csn-miR319/TCP, csn-miR396/GRF, and csn-miR828/MYB.

miRNAs are also implicated to have a role in metabolizing the allelochemicals secreted by plants during herbivory (Ma et al. 2017). Allelochemicals are plant-derived toxic chemicals belonging to multiple classes secreted by plants to defend herbivory. Insects, as part of herbivory, need to tackle these toxins effectively. *Aphis gossypii*, which has almost 300 host species, was studied with regard to the efficacy of its miRNA repertoire in detoxifying allelochemicals secreted by its diverse hosts (Ma et al. 2017). A major pest of cotton and cucurbit crops, *A. gossypii* was fed on artificial diet containing various plant allelochemicals such as 2-tridecanone, quercetin, gossypol, and tannic acid and also control diet (five libraries). A total of 292 miRNA species could be identified through deep sequencing the libraries. By comparing the expression of these miRNAs among the different libraries, allelochemical metabolism-related miRNAs were identified and their targets predicted. Several miRNAs from the allelochemical-treated libraries showed differential expression in comparison to control libraries. miRNAs such as Ago-miR-3191-3p, Ago-miR-8798a, Ago-miR-331-3p, Ago-miR-2179-5p, Ago-miR-1773-5p, Ago-miR-9083-2, Ago-miR-92b-5p, and Ago-miR-719 got upregulated, and

Ago-let-7-5p, Ago-miR- 100-5p, Ago-miR-44b-3p, Ago-miR-7054-3p, Ago-miR-4021-3p, Ago-miR-656a-3p, Ago-miR-4661a-3p, and Ago-miR-2238j-3p were consistently downregulated in all four of the allelochemical-treated libraries. Ago-miR-7475a-5p was found to be expressed only in the allelochemical libraries, whereas it was totally absent in the control. Several of these miRNAs' targets were identified as acetylcholinesterases, sodium channel proteins, cytochrome P450, and glutathione S-transferases which are crucial genes in handling xenobiotic stress.

4 Role of miRNAs in Plant-Nematode Interactions

miRNAs are found to be involved in plant-parasitic nematode interactions. Plants have developed a series of defense responses against nematodes which are efficiently tackled by certain sophisticated mechanisms evolved by the nematodes (Wubben et al. 2008). Both these mechanisms involve miRNAs as one of the tools. DICER-like (*dcl*) and RNA-dependent RNA polymerase (*rdr*) mutants of *Arabidopsis* were found to show decreased susceptibility to sugar beet cyst nematode, *Heterodera schachtii*. This study could identify roles for small RNAs in regulating gene expression during nematode infestation of a plant as DCL and RdRp proteins are involved in miRNA biogenesis (Hewezi et al. 2008). This research group generated two small RNA libraries from *H. schachtii*-infested *Arabidopsis* roots collected at 4 and 7 days post-inoculation (dpi) and conducted high-throughput sequencing. Out of the 861 distinct small RNA sequences that were recovered in this study, 575 sequences matched well with *Arabidopsis* genome. Real-time PCR technique was employed to know the changes in expression of these miRNAs in response to nematode infection at 4-dpi and 7-dpi time points of infection. At the 4-dpi time point, miR160, miR161, miR164, miR156, miR157, miR158a, miR167a, miR168, miR171b, miR172c, miR396a and b, miR398a, and miR775 were statistically significantly downregulated, while miR169d and miR172a were unchanged in the infested roots in comparison to the control roots. Among all miRNAs, miR156 showed the strongest upregulation and miR16, the strongest downregulation. They also observed that miRNA members of the same families with a single or a few nucleotide variations also exhibited different patterns of expression.

The same research group later established the role of miR396, a miRNA they had identified in their first study (Hewezi et al. 2008) as the key regulator of root cell reprogramming as a result of cyst nematode infection (Hewezi et al. 2012). Two members of miR396 family, miR396a and miR396b, are known to act as negative regulators of seven *growth regulating factor* (GRF) genes in *Arabidopsis* by positively controlling cell proliferation and size in leaves. This study established that the expression of miR396a and miR396b genes as well as their major target genes GRF1 and GRF3 was oppositely regulated in the syncytium upon *H. schachtii* infection, and it is this what decides cell fate specification and differentiation in the developing syncytium. The miR396-GRF regulatory system was found to alter almost half of the differentially expressed genes in *Arabidopsis* syncytium, thereby making miR396 a key regulator of the reprogramming of the root cells (Hewezi et al. 2012). Using promoter/GUS fusion assay, this group identified a role for miR827 in susceptibility of *Arabidopsis* to *H. schachtii.* miR827 targeted *nitrogen limitation adaptation* (NLA) gene encoding an ubiquitin E3 ligase enzyme (Hewezi et al. 2016). Overexpression of miR827 resulted in hyper-susceptibility to the nematode by suppressing the immune response of the host.

miRNAs were studied in root knot nematodes also. Libraries constructed from root segments of *Meloidogyne javanica*-infected (3-dpi) and control *Arabidopsis* plants revealed significant differences in their small RNA population (Cabrera et al. 2016). Altogether the libraries yielded 288 *Arabidopsis*-specific miRNAs out of which 23 each were exclusively found in the galls and in control root libraries.

5 miRNA-Based Strategies for Crop Protection

As detailed insights into the role of miRNAs started being available, new doors in gene regulation approaches have been opened. To overcome the ill effect of chemical-based pest control strategies, miRNA-based gene silencing, which is highly specific in approach, opens up new avenues.

One of the promising approaches to engineer plants that are resistant to insect pests is through the introduction of insect-specific artificial miRNAs (amiRNAs). As it is evident that miRNAs serve as important players in various developmental processes in insects, their suppression or overexpression can be used as an efficient tool in management of insect pests. Synthetic mimics and inhibitors of specific miRNAs have been used in pest control by specifically targeting certain insect genes. AmiRNA technology is more specific and hence safer than host delivered RNAi technology. The latter silences the target genes through siRNAs generated from long hairpin-dsRNA construct. RNAi technology might have many targets and might affect un-targeted insects.

Just like genes are used in pest control strategies, miRNAs could also be potentially used in integrated pest management programs as is evident by a 40% increase in larval mortality and a 70% reduction in fecundity when *Helicoverpa armigera* were orally fed by synthetic miRNA mimics (Jayachandran et al. 2013). Using har-miR-2002b mimic, *H. armigera* larval development could be successfully impaired (Jayachandran et al. 2013). *H. armigera* trypsin-like protease gene was identified as a target of har-miR-2002b, and it was observed that an enhanced supply of this miRNA affected growth of larvae and fecundity in adults. This described a novel approach in which plants are genetically modified to express miRNA inhibitors or mimics, and this serves as an effective pest control strategy. As an alternative to Bt technology, Agrawal et al. (2015) developed transgenic *Solanum lycopersicon* plants overexpressing *Helicoverpa armigera*-specific microRNAs, and the plants acquired insecticidal activity against the pest. The authors generated artificial miRNA-24 (amiR-24) targeting *H. armigera* chitinase to cause down regulation of the gene which was earlier proved (Agrawal et al. 2013). They could find that larvae fed on the transgenic plants failed to molt and eventually died. In a similar line, Guo et al. (2014) also employed amiRNA technology to develop *Myzus persicae*-resistant transgenic tobacco plants. Using amiRNAs targeting acetylcholinesterase gene (*Ace*), coding for a key enzyme in the insect central nervous system, exhibited good aphid resistance. In yet another promising example, host delivered silencing of a predominant isoform of *Ace* gene of *H. armigera* was achieved in *Arabidopsis* transgenic plants (Saini et al. 2018). Effectiveness of amiRNAs was explored in soybean cyst nematode (SCN) resistance by targeting three nematode fitness genes and successfully achieving significant reductions in SCN population densities (Tian et al. 2016).

Still, the use of miRNAs in gene silencing applications should be attempted with caution and after thorough functional validation as it is known that miRNAs are pleiotropic in nature and can have multiple targets.

6 Conclusion

Plant-insect interaction is a highly complex and regulated process that is controlled at multiple levels. The discovery of small RNAs has helped in understanding the mechanisms underlying their regulation better. Plant miRNAs have been identified on a large scale, and many of them have been implicated in biotic stress responses. Yet insects fall far behind other animals and plants in miRNA-related research. Identification of more miRNAs and elucidation of their function are essential for clearly deciphering how the interaction between plants and insects happens at the molecular level. Until then, significant challenges remain to be addressed before miRNA-based RNAi technology can be successfully employed for pest control in the field.

References

Agrawal N, Sachdev B, Rodrigues J, Sree KS, Bhatnagar RK (2013) Development associated profiling of chitinase and microRNA of *Helicoverpa armigera* identified chitinase repressive microRNA. Sci Rep 3:2292

Agrawal A, Rajamani V, Reddy VS, Mukherjee SK, Bhatnagar RK (2015) Transgenic plants over-expressing insect-specific microRNA acquire insecticidal activity against *Helicoverpa armigera*: an alternative to Bt-toxin technology. Transgenic Res. https://doi.org/10.1007/s11248-015-9880-x

Androvic P, Valihrach L, Elling J, Sjoback R, Kubista M (2017) Two-tailed RT-qPCR: a novel method for highly accurate miRNA quantification. Nucleic Acids Res 45(15):e144

Ashby R, Foret S, Searle I, Maleszka R (2016) MicroRNAs in honey bee caste determination. Sci Rep 6:18794. https://doi.org/10.1038/srep18794

Bartel DP (2004) MicroRNAs:genomics, biogenesis, mechanism, and function. Cell 116:281–297

Behura SK, Whitfield CW (2010) Correlated expression patterns of microRNA genes with age-dependent behavioural changes in honeybee. Insect Mol Biol 19:431–439

Cabrera J, Barcala M, Garcıa A, Rio-Machın A, Medina C, Jaubert-Possamai S, Favery B, Maizel A, Ruiz-Ferrer V, Fenoll C et al (2016) Differentially expressed small RNAs in Arabidopsis galls formed by *Meloidogyne javanica*: a functional role for miR390 and its TAS3-derived tasiRNAs. New Phytol 209:1625–1640

Cagirici HB, Biyiklioglu S, Budak H (2017) Assembly and annotation of transcriptome provided evidence of miRNA mobility between wheat and wheat stem sawfly. Front Plant Sci 8:1653. https://doi.org/10.3389/fpls.2017.01653

Cha YS, Ji H, Yun DW, Ahn BO, Lee MC, Suh SC, Lee CS, Ahn EK, Jeon YH, Jin ID, Sohn JK, Koh HJ, Eun MY (2008) Fine mapping of the rice Bph1 gene, which confers resistance to the brown planthopper (*Nilaparvata lugens* stal), and development of STS markers for marker-assisted selection. Mol Cell 26:146–151

Chang Z-X, Tang N, Lin W, Zhang L-Q, Akinyemi IA, Wu Q-F (2016) Identification and characterization of microRNAs in the white-backed planthopper, *Sogatella furcifera*. Insect Sci 23:452–468. https://doi.org/10.1111/1744-7917.12343

Chen C et al (2005) Real-time quantification of microRNAs by stem-loop RT-PCR. Nucleic Acids Res 33(20):e179. https://doi.org/10.1093/nar/gni178

Chen X, Yu X, Cai Y, Zheng H, Yu D, Liu G, Zhou Q, Hu S, Hu F (2010) Next-generation small RNA sequencing for microRNAs profiling in the honey bee *Apis mellifera*. Insect Mol Biol 19:799–805

Chen Q, Lu L, Hua H, Zhou F, Lin Y (2012) Characterization and comparative analysis of small RNAs in three small RNA libraries of the brown planthopper (*Nilaparvata lugens*). PLoS One 7: e32860

Chen D, Whitworth JR, Chen MS (2017) MicroRNA variants, expression, and putative target genes in the gall midge *Mayetiola destructor*. J Mol Biol Tech 1(1):103

D'Ario M, Griffith-Jones S, Ki M (2017) Small RNAs: big impact on plant development. Trends Plant Sci 22(12):1056–1068

Djami-Tchatchou AT, Sanan-Mishra N, Ntushelo K, Dubery IA (2017) Functional roles of microRNAs in agronomically important plants—potential as targets for crop improvement and protection. Front Plant Sci 8:378. https://doi.org/10.3389/fpls.2017.00378

Etebari K, Hussain M, Asgari S (2013) Identification of microRNAs from *Plutella xylostella* larvae associated with parasitization by *Diadegma semiclausum*. Insect Biochem Mol Biol 43 (4):309–318

Greenberg JK, Xia J, Zhou X, Thatcher SR, Gu X, Ament SA, Newman TC, Green PJ, Zhang W, Robinson GE, Ben-Shahar Y (2012) Behavioral plasticity in honey bees is associated with differences in brain microRNA transcriptome. Genes Brain Behav. https://doi.org/10.1111/j.1601-183X.2012.00782.x

Guo H, Song X, Wang G, Yang K, Wang Y et al (2014) Plant-generated artificial small RNAs mediated aphid resistance. PLoS One 9(5):e97410. https://doi.org/10.1371/journal.Pone.0097410

Hewezi T, Howe P, Maier TR, Baum TJ (2008) Arabidopsis small RNAs and their targets during cyst nematode parasitism. Mol Plant-Microbe Interact 21:1622–1634

Hewezi T, Maier TR, Nettleton D, Baum TJ (2012) The Arabidopsis microRNA396-GRF1/GRF3 regulatory module acts as a developmental regulator in the reprogramming of root cells during cyst nematode infection. Plant Physiol 159:321–335

Hewezi T, Piya S, Qi M, Balasubramaniam M, Rice JH, Baum TJ (2016) Arabidopsis miR827 mediates post-transcriptional gene silencing of its ubiquitin E3 ligase target gene in the syncytium of the cyst nematode *Heterodera schachtii* to enhance susceptibility. Plant J 88:179–192

Hori S, Kaneko K, Saito T, Takeuchi H, Kubo T (2011) Expression of two microRNAs, *ame*-mir-276 and -1000, in the adult honeybee (*Apis mellifera*) brain. Apidologie 42:89–102

Hunt EA, Broyles D, Head T, Deo SK (2015) MicroRNA detection: current technology and research strategies. Annu Rev Anal Chem 8:217–237

Jayachandran B, Hussain M, Asgari S (2013) An insect trypsin-like serine protease as a target of microRNA: utilization of microRNA mimics and inhibitors by oral feeding. Insect Biochem Mol Biol 43:398–406

Jeyaraj A, Liu S, Zhang X, Zhang R, Shangguan M, Wei C (2017) Genome-wide identification of microRNAs responsive to *Ectropis oblique* feeding in tea plants (*Camellia sinensis* L.). Sci Rep 7:13634

Jonstrup SP, Koch J, Kjems J (2006) A microRNA detection system based on padlock probes and rolling circle amplification. RNA 12(9):1747–1752. https://doi.org/10.1261/rna.110706

Khajuria C, Williams CE, El Bouhssini M, Whitworth RJ, Richards S, Stuart JJ, Chen MS (2013) Deep sequencing and genome-wide analysis reveals the expansion of microRNA genes in the gall midge *Mayetiola destructor*. BMC Genomics 14:187. https://doi.org/10.1186/1471-2164-14-187

Krol J, Loedige I, Filipowicz W (2010) The widespread regulation of microRNA biogenesis, function and decay. Nat Rev Genet 11:597–610

Lagos-Quintana M, Rauhut R, Lendeckel W, Tuschl T (2001) Identification of novel genes coding for small expressed RNAs. Science 294:853–858

Lau NC, Lim LP, Weinstein EG, Bartel DP (2001) An abundant class of tiny RNAs with probable regulatory roles in *Caenorhabditis elegans*. Science 294:858–862

Lee RC, Feinbaum RL, Ambros V (1993) The *C. elegans* heterochronic gene *lin-4* encodes small RNAs with antisense complementarity to *lin-14*. Cell 75:843–854

Liang P, Feng B, Zhou XG, Gao XW (2013) Identification and developmental profiling of microRNAs in diamondback moth, *Plutella xylostella* (L.). PLoS One 8:e78787

Liu H, Yu H, Tang G et al (2018) Small but powerful:function of microRNAs in plant development. Plant Cell Rep 37:515

Lomate PR, Mahajan NS, Kale SM, Gupta VS, Giri AP (2014) Identification and expression profiling of *Helicoverpa armigera* microRNAs and their possible role in the regulation of digestive protease genes. Insect Biochem Mol Biol 54:129–137

Luo Q, Zhou Q, Yu X, Lin H, Hu S, Yu J (2008) Genome-wide mapping of conserved microRNAs and their host transcripts in *Tribolium castaneum*. J Genet Genomics 35:349–355

Ma K-S, Li F, Liu Y, Liang P-Z, Chen X-W, Gao X-W (2017) Identification of microRNAs and their response to the stress of plant allelochemicals in *Aphis gossypii* (Hemiptera: Aphididae). BMC Mol Biol 18:5. https://doi.org/10.1186/s12867-017-0080-5

Marco A, Hui JH, Ronshaugen M, Griffiths-Jones S (2010) Functional shifts in insect microRNA evolution. Genome Biol Evol 2:686–696

Miao J, Wang J, Guo J, Gao H, Han K, Jiang C, Miao P (2016) A plasmonic colorimetric strategy for visual miRNA detection based on hybridization chain reaction. Sci Rep 6:32219

Nicolis VF, Greyling SM, Venter E (2017) Isolation of early responsive microRNA from *Diuraphis noxia* (Hemiptera-Aphididae) resistant wheat. J Econ Entomol 110(3):1298–1306

Ninova M, Ronshaugen M, Griffiths-Jones S (2016) MicroRNA evolution, expression, and function during short germband development in *Tribolium castaneum*. Genome Res 26:85–96

Nozawa M, Miura S, Nei M (2012) Origins and evolution of miRNA genes in plant species. Genome Biol Evol 4:230–239

Pandey SP, Shahi P, Gase K, Baldwin IT (2008) Herbivory-induced changes in the small-RNA transcriptome and phytohormone signaling in *Nicotiana attenuata*. PNAS 105(12):4559–4564

Pasquinelli AE, Reinhart BJ, Slack F, Martindale MQ, Kuroda M, Maller B, Srinivasan A, Fishman M, Hayward D, Ball E et al (2000) Conservation across animal phylogeny of the sequence and temporal regulation of the 21 nucleotide *let-7* heterochronic regulatory RNA. Nature 408:86–89

Pritchard CC, Cheng HH, Tewari M (2012) MicroRNA profiling: approaches and considerations. Nat Rev Genet 13:358–369

Redshaw N, Wilkes T, Whale A, Cowen S, Huggett J, Foy CA (2013) A comparison of miRNA isolation and RT-qPCR technologies and their effects on quantification accuracy and repeatability. BioTechniques 54(3):155–164. https://doi.org/10.2144/000114002

Reinhart BJ, Slack FJ, Basson M, Bettinger JC, Pasquinelli AE, Rougvie AE, Horvitz HR, Ruvkun G (2000) The 21 nucleotide *let-7* RNA regulates developmental timing in *Caenorhabditis elegans*. Nature 403:901–906

Reinhart BJ, Weinstein EG, Rhoades MW, Bartel B, Bartel DP (2002) MicroRNAs in plants. Genes Dev 16:1616–1626

Saini RP, Raman V, Dhandapani G, Malhotra EV, Sreevathsa R, Kumar PA et al (2018) Silencing of *HaAce1* gene by host-delivered artificial microRNA disrupts growth and development of *Helicoverpa armigera*. PLoS One 13(3):e0194150. https://doi.org/10.1371/journal.pone.0194150

Sattar S, Song Y, Anstead JA, Sunkar R, Thompson GA (2012) *Cucumis melo* microRNA expression profile during aphid herbivory in a resistant and susceptible interaction. Mol Plant-Microbe Interact 25:839–848. https://doi.org/10.1094/MPMI-09-11-0252

Sattar S, Addo-Quaye C, Thompson GA (2016) miRNA-mediated auxin signaling repression during Vat-mediated aphid resistance in *Cucumis melo*. Plant Cell Environ 39:1216–1227. https://doi.org/10.1111/pce.12645

Selbach M, Schwanhausser B, Thierfelder N, Fang Z, Khanin R, Rajewsky N (2008) Widespread changes in protein synthesis induced by microRNAs. Nature 455:58–63

Singh J, Nagaraju J (2008) *In silico* prediction and characterization of microRNAs from red flour beetle (*Tribolium castaneum*). Insect Mol Biol 17:427–436

Slack FJ, Basson M, Liu Z, Ambros V, Horvitz HR, Ruvkun G (2000) The *lin-41* RBCC gene acts in the *C. elegans* heterochronic pathway between the *let-7* regulatory RNA and the LIN-29 transcription factor. Mol Cell 5:659–669

Sunkar R, Jagadeeswaran G (2008) *In silico* identification of conserved microRNAs in large number of diverse plant species. BMC Plant Biol 8:37

Tian B, Li J, Oakley TR, Todd TC, Trick HN (2016) Host-derived artificial microRNA as an alternative method to improve soybean resistance to soybean cyst nematode. Genes 7:122. https://doi.org/10.3390/genes7120122

Vaucheret H, Chupeau Y (2012) Ingested plant miRNAs regulate gene expression in animals. Cell Res 22:3–5

Wang YN, Tang L, Hou Y, Wang P, Yang H, Wei CL (2016) Differential transcriptome analysis of leaves of tea plant (*Camellia sinensis*) provides comprehensive insights into the defense responses to *Ectropis oblique* attack using RNA-Seq. Funct Integr Genomics 16:383–398

Wang H, Zhang C, Dou Y, Yu B, Liu Y, Heng-Moss T, Lu G, Wachholtz M, Bradshaw JD, Twigg P, Scully E, Palmer N, Sarath G (2017) Insect and plant-derived miRNAs in greenbug (*Schizaphis graminum*) and yellow sugarcane aphid (*Sipha flava*) revealed by deep sequencing. Gene 599:68–77

Wang K, Su X, Cui X, Du Y, Zhang S, Gao J (2018) Identification and characterization of microRNA during *Bemisia tabaci* infestations in *Solanum lycopersicum* and *Solanum habrochaites*. Hortic Plant J. https://doi.org/10.1016/j.hpj.2018.03.002

Weaver DB, Anzola JM, Evans JD, Reid JG, Reese JT, Childs KL, Zdobnov EM, Samanta MP, Miller J, Elsik CG (2007) Computational and transcriptional evidence for microRNAs in the honey bee genome. Genome Biol 8(6):R97

Wei Y, Chen S, Yang P, Ma Z, Kang L (2009) Characterization and comparative profiling of the small RNA transcriptomes in two phases of locust. Genome Biol 10(1):R6

Wightman B, Ha I, Ruvkun G (1993) Post-transcriptional regulation of the heterochronic gene *lin-14* by *lin-4* mediates temporal pattern formation in *C. elegans*. Cell 75:855–862

Wu L, Zhou H, Zhang Q, Zhang J, Ni F, Liu C, Qi Y (2010) DNA methylation mediated by a microRNA pathway. Mol Cell 38:465–475

Wu W, Xiong W, Li C, Zhai M, Li Y, Ma F, Li B (2017a) MicroRNA-dependent regulation of metamorphosis and identification of microRNAs in the red flour beetle, *Tribolium castaneum*. Genomics 109(5–6):362–373

Wu Y, Lv W, Hu L, Rao W, Zeng Y, Zhu L, He Y, He G (2017b) Identification and analysis of brown planthopper-responsive microRNAs in resistant and susceptible rice plants. Sci Rep 7:8712. https://doi.org/10.1038/s41598-017-09143-y

Wubben MJ, Jin J, Baum TJ (2008) Cyst nematode parasitism of *Arabidopsis thaliana* is inhibited by salicylic acid (SA) and elicits uncoupled SA independent pathogenesis-related gene expression in roots. Mol Plant-Microbe Interact 21:424–432

Xia X, Shao Y, Jiang J, Du X, Sheng L, Chen F et al (2015) MicroRNA expression profile during aphid feeding in Chrysanthemum (*Chrysanthemum morifolium*). PLoS One 10(12):e0143720. https://doi.org/10.1371/journal.pone.0143720

Yang JH, Qu LH (2012) DeepBase: annotation and discovery of microRNAs and other noncoding RNAs from deep-sequencing data. Methods Mol Biol 822:233–248

Yin BC, Liu YQ, Ye BC (2012) One-step, multiplexed fluorescence detection of microRNAs based on duplex-specific nuclease signal amplification. J Am Chem Soc 134(11):5064–5067. https://doi.org/10.1021/ja300721s

Yu X, Zhou Q, Li S-C, Luo Q, Cai Y et al (2008) The silkworm (*Bombyx mori*) microRNAs and their expressions in multiple developmental stages. PLoS One 3(8):e2997. https://doi.org/10.1371/journal.pone.0002997

Zha W et al (2016) Characterization and comparative profiling of the small RNA transcriptomes in the Hemipteran insect *Nilaparvata lugens*. Gene 595:83–91

Zhang BH, Pan XP, Wang QL, Cobb GP, Anderson TA (2005) Identification and characterization of new plant microRNAs using EST analysis. Cell Res 15:336–360

Zhang Y, Zhou X, Ge X, Jiang J, Li M, Jia S, Yang X, Kan Y, Miao X, Zhao G, Li F, Huang Y (2009) Insect-specific microRNA involved in the development of the silkworm *Bombyx mori*. PLoS One 4(3):e4677. https://doi.org/10.1371/journal.pone.0004677

Zhang X, Zheng Y, Jagadeeswaran G, Ren R, Sunkar R, Jiang H (2012a) Identification and developmental profiling of conserved and novel microRNAs in *Manduca sexta*. Insect Biochem Mol Biol 42:381–395

Zhang Y, Wiggins BE, Lawrence C, Petrick J, Ivashuta S, Heck G (2012b) Analysis of plant-derived miRNAs in animal small RNA datasets. BMC Genomics 13:381

Zhang X, Zheng Y, Jagadeeswaran G, Ren R, Sunkar R, Jiang H (2014) Identification of conserved and novel microRNAs in *Manduca sexta* and their possible roles in the expression regulation of immunity-related genes. Insect Biochem Mol Biol 47:12–22

Zhang X, Zheng Y, Cao X, Ren R, Yu X-Q, Jiang H (2015) Identification and profiling of *Manduca sexta* microRNAs and their possible roles in regulating specific transcripts in fat body, hemocytes, and midgut. Insect Biochem Mol Biol 62:11–22

Zhu K, Liu M, Fu Z, Zhou Z, Kong Y, Liang H, Lin Z, Luo J, Zheng H, Wan P, Zhang J, Zen K, Chen J, Hu F, Zhang C-Y, Ren J, Chen X (2017) Plant microRNAs in larval food regulate honeybee caste development. PLoS Genet 13(8):e1006946. https://doi.org/10.1371/journal.pgen.1006946

Role of Mapks During Plant-Insect Interaction

Kaur Manjeet and Sunita Yadav

Abstract

Mitogen-activated protein kinases (MAPKs) are components of one of the early signaling events during plant-insect interactions. MAPK cascade proceeds in three steps where MAPKKK phosphorylates MAPKK, which further phosphorylates MAPK. MAPK activation further triggers downstream cascade of events that include alteration in the levels of plant hormones, reshaping the transcriptome and proteome, leading to plant defense against insect. In this chapter, we examined different MAPKs with special attention to their roles in triggering defense responses in various plants in response to insect attack. We discuss the role of known MAPKs, which have been identified and characterized from various plant species till date, specifically during plant-insect interaction. However, there was limited information available regarding the molecular mechanisms and genes encoding receptors during insect attack. We focused more on the three-tiered MAPK, their interaction leading to altered hormone levels finally resulting in defense responses in plants.

Keywords

Plant-insect interactions · Mitogen-activated protein kinases (MAPKs) · Mitogen-activated protein kinase kinases (MAPKKs) · Mitogen-activated protein kinase kinase kinases (MAPKKKs) · Herbivore-associated molecular patterns (HAMPs) · Jasmonic acid (JA) biosynthesis · Systemin

K. Manjeet
Department of Plant Sciences, School of Life Sciences, University of Hyderabad, Hyderabad, Telangana, India

S. Yadav (✉)
Department of Botany, Sri Venkateswara College, University of Delhi, New Delhi, India

I. K. Singh, A. Singh (eds.), *Plant-Pest Interactions: From Molecular Mechanisms to Chemical Ecology*, https://doi.org/10.1007/978-981-15-2467-7_5

1 Introduction

To resist herbivory, plants protect themselves in two ways: either by preformed structures such as modifications of epidermal tissue system like trichomes, thorns, and thick cuticles or by using secondary metabolites. Some secondary metabolites are expressed constitutively and are found at almost steady concentrations irrespective of insect attack. While some secondary metabolites are either present in low concentrations or are absent before insect attack, however, the concentration rapidly increases after insect infestation; such defenses are referred to as inducible (Hettenhausen et al. 2015). The compounds released during plant defense are expensive as the production consumes energy and nutrients. The inducible production is thus considered more cost-effective (Steppuhn and Baldwin 2007) compared to the constitutive production. Inducible defenses are regulated by signal transduction events. To withstand the attacking insects, plants should be able to recognize the attacking enemy. The immediate recognition, decoding of signals and adequate response, is the key to successful defense by plants. The cascade of plant-insect interaction can be categorized into two categories, the early events that are initiated within first seconds to minutes and the late events depend mainly on genomics and proteomics (Maffei et al. 2007). The early events during plant-insect interactions include perception of molecular patterns and defense effectors, ionic imbalance which changes the transmembrane potential, Ca^{2+} – signaling, MAPK activation and protein phosphorylation, the activation of NADPH oxidase and production of reactive oxygen and nitrogen species (ROS and RNS), and phytohormones and their cross talk which finally lead to gene expression (Zebelo and Maffei 2015).

As the herbivore attacks the plant and MAPK signaling is activated, the levels of plant hormones alter, reshaping the transcriptome and proteome, hence providing defense against insect attack. Ethylene (ET), jasmonic acid (JA), and salicylic acid (SA) play an important role among the downstream signaling pathways. SA, JA, and ET interact and render plant the best defense strategy (Glazebrook 2005). Among all, JA occupies a key position in activating direct and indirect defense responses (War et al. 2012). The signaling pathways of SA, ABA (Abscisic acid), and ET work in a cross talk network with JA signaling pathway (Bodenhausen and Reymond 2007). Production of defensive proteins reduces the capacity of the insect herbivore to digest the plant and is thus an important event in downstream signaling during plants-insect interaction (Fürstenberg-Hägget al. 2013).

Aphids and whiteflies feed on phloem sap and change the profile of secondary metabolites (Walling 2000). It is interesting to note that like plant-pathogen interaction, an R gene known as Mi-1 (resistance to *Meloidogyne incognita*) provides resistance to aphids, whiteflies, and nematodes (Kaloshian 2004). Another R gene, Bph14, provides resistance to brown plant hoppers (*Nilaparvata lugens*) in *Oryza sativa*. According to Du et al. 2009, Mi-1 gene and Bph 14 both code for a coiled coil, nucleotide-binding site, and leucine-rich repeat protein. Both aphids and brown plant hoppers (phloem sap suckers) are capable of inducing SA signaling. The involvement of these R-genes in herbivore resistance is still unknown (Wu and Baldwin 2010).

In this chapter we focus on MAPKs, which are activated immediately after the herbivore attack. Biochemical and genetic evidences proved that MAPKs play a vital role in plant-herbivore interactions. To understand plant defense responses, we should know that there is always an overlapping of different plant stress responses at different stages of signal transduction (Stratmann 2008). In spite of having different sources of stress, MAPKs act as a common sink and regulate different cellular responses (Fig. 1).

MAPKs consist of 11 modules (I–XI), and the dual phosphorylation of threonine and tyrosine residues in a TXY (threonine-X-tyrosine, in submodule VIII) motif allows its activation (Hanks et al. 1988). The phosphorylation of conserved serine and/or threonine groups in the activation loop allows the activation of MAPK kinases (MAPKKs or MAP 2 K or MEKs or MKKs) through MAPKK kinases (MAPKKKs or MAP 3 K or MEKKs). Further, there is phosphorylation of downstream substrates like transcription factors/enzymes with the help of these activated MAPKs. The stress-related responses are then triggered downstream to substrate phosphorylation (Hazzalin and Mahadevan 2002). Different herbivore-associated molecular patterns (HAMPs) have already been known to activate MAPK signaling. For example, OS of *Schistocerca gregaria* activate MPK3/6 in *Arabidopsis thaliana* (Schafer et al. 2011), and fatty acid conjugates identified in various lepidopteran insects also trigger MAPK signaling (Wu et al. 2007). It is considered that much information regarding elicitors triggering specific defense responses, activation of plant signaling cascades, and further defense responses during plant-insect interaction is still unraveled.

2 Mapk Signaling in Host Cells

The MAPK signaling events occur in all eukaryotes including yeasts. In plants, MAPK signal transduction is involved in various responses against biotic and abiotic stresses. Plant MAPKs and mammalian extracellular signal-regulated kinase (ERK) were found to be more closely related to each other. Enormous number of MAPKs is found in plant genomes; however the number of MAPKKs is reduced to half. According to Champion et al. (2004), the number of MAPKKKs that are functional in plant genome is still under dispute. Twenty MAPKs, 10 MAPKKs, and 60 MAPKKKs were reported in *Arabidopsis* genome (Ichimura MAPK Group 2002).

2.1 MAPKs

MAPKs phosphorylate cytosolic as well as nuclear proteins which act as substrates for MAPKs; these substrates **include** transcription factors and enzymes. In comparison with yeast and humans (having 6 and 10 MAPKs, respectively), plants have expanded families of MAPKs (20, 15, and 21 in Arabidopsis, rice, and poplar genomes, respectively), MAPKKs, and MAPKKKs as revealed through genome

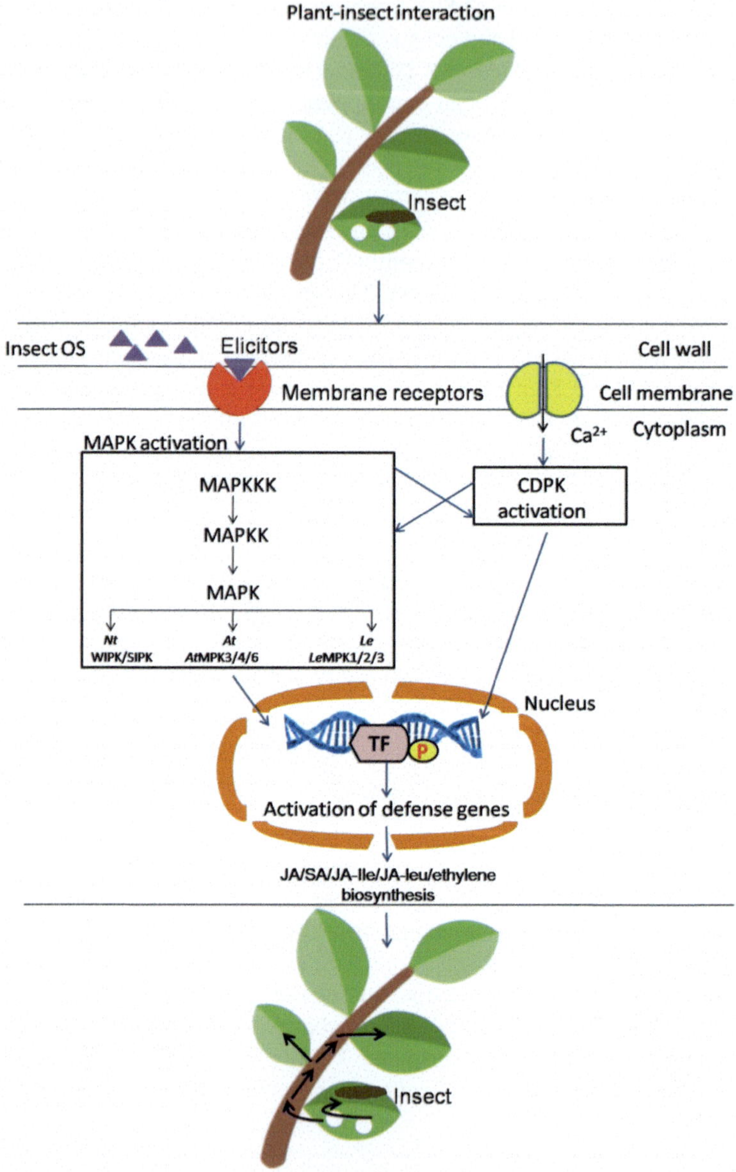

Fig. 1 Defense responses in systemic and local leaves

sequencing projects (Hamel et al. 2006). The immense amount of MAPKs in plants suggests that due to the presence of these diverse MAP kinases, plants adapt themselves to various environmental stresses. Some of the plant MAPKs are similar to the mammalian ERK MAPKs in having a Thr-Glu-Tyr (TEY) activation motif, while others have a Thr-Asp-Tyr (TDY) activation motif, which is unique to plant MAPKs.

Four categories of plant MAPKs (A, B, C, and D) have been formed on the basis of phylogenetic analysis of amino acid sequences and TXY phosphorylation motifs. A major clade is formed by MAPKs of subfamilies A, B, and C whose members have a well-conserved TEY motif. The other clade includes MAPKs of subfamily D that possess a TDY motif. In addition, the evolutionarily conserved C-terminal CD domain which functions as a docking site for MAPKKs, phosphatases, and protein substrates is not found in the subfamily D members. However, they are characterized through a long C-terminal sequence (Hettenhausen et al. 2015).

Some of the well-characterized MAPKs include MPK3, MPK4, and MPK6. These are activated by a diversity of other stimuli like abiotic stresses, pathogens, and oxidative stress. The MPK3 and 6 positively regulate the defense responses, whereas MPK4 negatively regulates biotic stress signaling. Plant MAPKs that are extensively studied include MAPKs from tobacco, i.e., the wound-induced MAPK (WIPK) and SA-induced MAPK (SIPK). Also, their orthologs from *Arabidopsis thaliana* (*At*MPK3 and *At*MPK6), alfalfa {SAMK (stress-activated MAPK) and SIMK (salt stress-inducible MAPK)}, and parsley (*Pc*MPK6) have been the major investigated class of plant MAPKs. *Le*MPK1, *Le*MPK2, and *Le*MPK3 are the MAPKs in tomato that have been extensively studied.

Till date, maize and rice are the only monocots where orthologs of WIPK have been characterized. There are reports showing expression of maize *Zm*MPK5 (a close relative of *Nt*SIPK) in senescent leaves and in response to low temperature, but there are no reports showing their response against wounding or JA application. The *Os*MSRMK2 from *Oryza sativa* is closer to *Nt*WIPK rather than *Nt*SIPK. Another salicylic acid-induced (wound-inducible) MAPK termed as *Os*SIPK is an ortholog of *Nt*SIPK (Rakwal and Agrawal 2003).

Further, we will discuss the role of MAPKs from *Nicotiana tabacum* (SIPK and WIPK), *Lycopersicon esculentum* (*Le*MPK1, *Le*MPK2, *Le*MPK3), and *Arabidopsis thaliana* MAPKs (MPKs) in response to different stress conditions, with special reference to plant-insect interactions.

2.1.1 Role of SIPK and WIPK in Plant-Insect Interactions

The involvement of MAPKs in plant-herbivore interaction was first reported by Seo et al. (1995) in *N. tabacum*. In response to wounding in *N. tabacum*, an immediate activation of WIPK and SIPK is observed consequently leading to the accumulation of jasmonic acid. WIPK silencing indicated that WIPK is involved in the production of wound-induced JA (Seo et al. 2007). A reduction in the activity of WIPK and SIPK is observed upon silencing of a tobacco receptor-like protein kinase, and a simultaneous reduction in JA accumulation is found upon wounding (Takabatake et al. 2006). Additionally, reduced levels of JA and defense-related genes were observed after antisense expression of WIPK. Various signaling pathways including WIPK and SIPK influence the biosynthesis of JA in wounded tobacco plants. Thus, SIPK/WIPK regulates the synthesis of an unknown wound-induced inhibitor, which either alone or together with JA suppresses the synthesis of SA.

An augmentation in the activity of SIPK is observed as a retort to almost all biotic and abiotic responses (Nakagami et al. 2005). In response to wounding and other stress signals, transcriptional and posttranslational regulation of WIPK and orthologs is observed (Stratmann 2008). However, SIPK and orthologs without significant changes in transcript levels are activated at posttranslational levels (Higgins et al. 2007; Ichimura et al. 2000; Zhang and Klessig 1998). Wounding and insect attack on plants lead to an increased systemic MAPK activity (Seo et al. 1999; Stratmann and Ryan 1997). A report by Wu et al. (2000) revealed fatty acyl conjugates (FACs) in *Manduca sexta* OS as the accountable elicitors. Further, they reported that after application of *Manduca sexta* OS to puncture wounds in leaves of *N. attenuata*, there was a dramatic amplification in SIPK activity and WIPK transcripts (Wu et al. 2007). Wounding and OS-elicitor-mediated biosynthesis of JA, salicylic acid (SA), JA-Ile/JA-Leu conjugate, and ethylene are regulated through upstream SIPK and WIPK signaling machinery. As mentioned earlier, protein kinases work as a network, and there is always a complicated transcriptional cross talk occurring among them. Transcriptional analysis further revealed that SIPK and WIPK not only regulate each other's transcript accumulation in response to wounding and OS-elicitation; they also lead to an increased production of many defense-related genes like MAPKs and CDPKs.

A rapid activation of SIPK orthologs in tomato was observed upon treatment of leaves or suspension cells of *Lycopersicon esculentum* with systemin (Ryan 2000). An increase in the levels of JA was also reported in *Oryza sativa* plants upon the overexpression of an ortholog of WIPK, i.e., MK1 (Lee et al. 2004).

JA biosynthesis involves at least eight different enzymes (Wasternack 2007). There are reports stating that SIPK and WIPK carry out different functions; however, the mode of action is still not known. Linolenic acid (18:3) is released from chloroplast membranes through SIPK-regulated lipase. However, WIPK alter the action of allene oxide synthase (Kallenbach et al. 2010). There are reports indicating that only SIPK is involved in ethylene production in *M. sexta-N. attenuata* interaction (Wu et al. 2007).

2.1.2 Role of *Le*MPK1, *Le*MPK2, and *Le*MPK3 in Tomato Plants

Systemin, a wound signaling peptide, mediates wound response in tomato. Antisense inhibition and pro-systemin overexpression experiments indicated that systemin is the most important component leading to resistance of tomato plants against insects (Chen et al. 2005; McGurl et al. 1994). Systemin amplify as well as prolong the MAPK activity. Systemin-regulated MAPKs activation depends on receptor. During wounding, systemin stimulate *Le*MPK1, *Le*MPK2, and *Le*MPK3 (the *Lycopersicon esculentum* MAPKs). The orthologs of the tobacco SIPK, Ntf4, and WIPK are, respectively, depicted by LeMAPK1, 2, and 3 (Higgins et al. 2007; Holley et al. 2003). *Le*MPK1 and 2 share 95% sequence identity, and both are stimulated by stress-related stimuli like wounding, oligosaccharide elicitors, and fusicoccin (Higgins et al. 2007; Holley et al. 2003). The cytosolic enzymes involved in ethylene synthesis, i.e., 1-aminocyclopropane-1-carboxylate synthase 2 and 6 (ACS2, ACS6) are the known substrates of *Le*MPK1/2 ortholog*At*MPK6 (Liu and Zhang 2004). Various responses generating from stress signals converge at *Le*MPK1 and 2 (Holley et al. 2003; Higgins et al. 2007; Pearce et al. 2001; Stratmann et al. 2000). Virus-induced gene silencing (VIGS) of *Le*MPK1/2, *Le*MPK3, and *Le*MKK2 was used to assess the role of MAPK(K)s in aphid resistance. The aphid population was resistant while feeding on Mi-1-expressing tomato leaves suggesting the involvement of a MAPK cascade in Mi-1-mediated resistance to aphids (Li et al. 2006).

Prosystemin boosts plant resistance against insects. When prosystemin gene was overexpressed in transgenic tomato and plants were fed with *M. sexta,* an activation of tomato homologues of WIPK (MPK3) and SIPK (MPK1 and MPK2) was observed within 10 min. Moreover, a better growth of *M. sexta* larvae was observed on MPK1/2-co-silenced plants as these plants have reduced expression of JA biosynthetic genes (Kandoth et al. 2007).

2.1.3 Role of MAPKs in *Arabidopsis thaliana*

As mentioned earlier, based on phylogenetic analysis, MAPK have been categorized into four groups (A/B/C and D). Environment- and hormone-mediated responses are known to be modulated by group A of MAPKs in *A. thaliana*. And many environmental stresses are known to activate MPK6 and its orthologs in species (Kovtun et al. 2000; Nühse et al. 2000; Ichimura et al. 2000; Desikan et al. 2001; Yuasa et al. 2001). MPK3 gene expression is induced and activated by environmental as well as oxidative stress (Kovtun et al. 2000). Tobacco WIPK shares high sequence similarity with MPK3. Schafer et al. (2011) proved that applying grass hopper OS to wounded Arabidopsis leaves highly promotes MAPK activity (MPK3 and MPK6) in addition to OPDA, JA, JA-Ile, and ET and cytosolic calcium ($[Ca^{2+}]_{cyt}$) levels (Schafer et al. 2011).

Group B MAPKs are involved in both environmental stress responses and cell division. These MAPKs have not been studied in detail. Biochemical analysis using a MPK4-specific antibody indicated that MPK4 gets activated during both biotic and abiotic stress (Ichimura et al. 2000; Desikanet al. 2001). A distinct cluster known as B2 is formed within group B by alfalfa MMK3, tobacco *Ntf*6, and MPK13, which are activated in a cell cycle-dependent manner (Calderini et al. 1998; Bögre et al. 1999).

Group C MPKs represent the least studied group of MPKs. Schaffer et al. (2001) used microarray analysis for detection of circadian rhythm regulated expression of MPK7.

Group D MPKs are distinguished by the presence of TDY motif in their T-loop. This group includes eight members of the *Arabidopsis* MPKs. At N-terminus, MPK 8, 9, and 15 possess short stretches of ~60–80 amino acids; MPK8 and 9 possess a serine-rich and a glutamic acid-rich region, respectively. Group D genes are induced by blast fungus (e.g., rice BWMK1) and wounding (e.g., alfalfa TDY1) (He et al. 1999; Schoenbecket al. 1999).

As discussed earlier, MPKs in Group A and B possess CD domain which is conserved throughout the evolution (Tanoue et al. 2000). CD domain is a docking site for MAPKKs, phosphatases, and protein substrates. This domain possesses [LH]

[LHY]Dxx[DE]xx[DE]EPxC (where x represents any amino acid) sequence which includes two adjacent D and E acidic residues. These residues are essential for interaction with a cluster of K and R basic amino acids present in MAPKKs (Tanoue et al. 2000). On similar lines, the bulky L, H, and Y hydrophobic residues in the CD domain of MAPK have a binding affinity in the MAPKK docking site for their respective LxLxL residues. However, there seems to be some modification in the CD domain of group C MPKs. Additionally, none of the domain versions is found in the group D sequences.

2.2 MAPKKs

The direct activators of MAPKs present upstream in the MAPK signaling cascade are MAP kinase kinases (MAPKKs). MAPKKs have numerous MAPK targets as indicated through lesser number of MAPKKs than MAPKs in corresponding genomes. The exact function of MAPKKs in plant-insect interactions is not yet completely unraveled, although attempts are continuously being made in this direction (Kiegerl et al. 2000; Gao et al. 2008; Liu et al. 2004). MAPKKs in plants have been classified into four groups based on the similarities in the sequences of these proteins.

According to Yang et al. (2001), an enhanced activation of SIPK and WIPK is observed when a constitutively active MAPKK known as MEK2 (in *N. tabacum*) is strongly expressed. It is likely that the common MAPKK for SIPK and WIPK is NtMEK2. Similar reports have also been found in Arabidopsis protoplasts where MPK6 and MPK3 were found to be activated by the homologues of MEK2, i.e., MKK4 and MKK5 (Asai et al. 2002). As a defense mechanism against tobacco hornworm (*Manduca sexta*), the functions of MAPKKs like MEK1, MEK2, MKK1, SIPKK, and NPK2 were examined in *N. attenuata* by Heinrich et al. (2011a, b). However, only one MAPKK, i.e., MEK2, was found to be necessary for triggering SIPK and WIPK after wounding, although it might be possible that some unknown MAPKKs might be involved in the complete activation of SIPK and WIPK. The activity levels of trypsin proteinase inhibitors were found to be affected by all of the abovementioned MAPKKs, suggesting a pathway devoid of jasmonic acid signaling and SIPK and WIPK (MAPKs) signaling in regulating plant resistance to herbivores.

2.3 MAPKKKs

The MAPKKK family has more members and a greater variety of primary structures and domain composition as compared to MAPKs and MAPKKs (Ichimura MAPK Group 2002).

There are two large subfamilies of the highly divergent plant MAPKKKs:

1. MEKK-like kinases (12 members in Arabidopsis)

2. Raf-like kinases (around 50 members in Arabidopsis)

Several biochemical and genetic data support the involvement of MEKK-like members (MEKK1, MAPKKKα, YDA) functioning as MAPKKKs upstream of MAPKKs in plant MAPK cascades (Asai et al. 2002; del Pozoet al. 2004; Meng et al. 2012; Qiu et al. 2008; Ren et al. 2008; Wang et al. 2007). However, there is no evidence to support any Raf-like kinase member (the two functionally characterized CTR1 and EDR1) as probable MAPKKKs. Recently, Qiao et al. (2012) and Ju et al. (2012) have demonstrated that CTR1 directly phosphorylates EIN2, confirming that CTR1 is not a MAPKKK. The *Arabidopsis* MAPKKKs such as MP3K6/MP3K7, YODA, and ANP2/ANP3 play a crucial role in plant developmental processes (pollen viability, plant cell division, stomatal development and in controlling extra-embryonic fate) (Chaiwongsar et al. 2006; Krysan et al. 2002; Lukowitz et al. 2004; Bergmann et al. 2004), and MEKK1 and ANP1 act in the environmental stress response (Ichimura et al. 2006; Nakagami et al. 2006; Suarez-Rodriguez et al. 2007), while CTR1 has a pivotal role in ethylene signaling (Kendrick and Chang 2008).

3 Mapk Signaling in Systemic Tissues

Accumulation of defensive metabolites prior to attack by insects leads to a beneficial response as observed in case of systemic-induced responses. First instance for systemic defense was observed in tomato, whereby proteinase inhibitors (small peptides present in the insect midgut which inhibit digestive proteinases) were induced in the wounded/local leaves as well as in distal untreated/systemic leaves (Hettenhausen et al. 2015).

Systemic activation of MAPKs is a significant step in the signaling events, as it leads to appropriate defenses in intact unharmed tissues. Upon local and systemic infection of tobacco leaves, an increase in the amount of WIPK mRNA was observed within a minute (Seo et al. 1995). Further, there was an immediate increase in the expression of WIPK in leaves positioned near the already cut tobacco stems (Seo et al. 1999). Systemic activation of MAPK has been observed in response to wounding in soybean (Lee et al. 2001) and tomato (Stratmann and Ryan 1997). Analysis of both local and systemic activities of SIPK after simulated *M. sexta* feeding in *N. attenuata* by Hettenhausen et al. (2015) pointed to the fact that SIPK was activated only in some specific systemic leaves. Even after 30 min of local elicitation, any increase in SIPK activity was not observed in *N. attenuata* systemic leaves. Therefore, SIPK activity could not be triggered by wounding alone; however, simulated herbivory is required for systemic elevated activity of SIPK. Thus, systemic SIPK signaling is activated after specific recognition events between *M. sexta* and *N. attenuata*. According to Wu et al. (2007), some specific unharmed regions of herbivore-attacked leaves also show SIPK activity. Further studies suggested that to activate MAPK signaling, there is transport of mobile signal from the attacked leaf to specific areas of the same or systemic leaves. Further evidences highlight the involvement of electric signals formed through GLUTAMATE RECEPTOR-LIKE genes in systemic signaling. An active form of jasmonate, i.e., jasmonyl-isoleucine is sufficiently produced in leaves through these electric signals (Mousavi et al. 2013).

4 Future Perspectives

Protein phosphorylation is an important cellular regulatory mechanism. Many enzymes and receptors are activated/deactivated by phosphorylation and dephosphorylation events through the specific kinases and phosphatases, respectively. And around one-third of all proteins are phosphorylated at any particular time in a eukaryotic cell. Knowing the importance of MAPKs in animals and yeasts, much focus has been shifted to know the roles played by them in plants as well. Plant-pathogen interaction has been studied in greater details because of an underdeveloped challenge that remains in genetically modifying insects.

Multiple MAPK signaling pathways have been studied in detail to explore MAPKKs, MAPKs, and direct MAPK substrates (Kong et al. 2012). But still a lot of questions remain unanswered on signaling mechanisms operative through MAPKs during plant-insect interactions. For example, how specificity of MAPK is achieved in terms of substrate recognition? And how plants mediate the specific responses after recognition of different insect herbivores? The precise molecular mechanism is again unanswered. Which genes are coding for receptors for elicitors (insect-derived)? How MAPKs govern the buildup of plant hormones and restructure the transcriptome is still unknown. Enriching our knowledge in this field will help us to generate tools for plant resistance against insects.

Induced resistance could be taken into consideration when developing crop cultivars that produce the inducible response once mildly infested. This could be exploited as one of the pivotal components of integrated pest management for sustainable crop production. Understanding the mechanisms of induced resistance to build up the natural defense system against injury caused by herbivores, the elicitors of induced responses can be applied on crop plants. Induced resistance can therefore be considered as a pivotal component for pest management.

References

Asai T, Tena G, Plotnikova J, Willmann MR, Chiu WL, Gomez-Gomez L, Boller T, Ausubel FM, Sheen J (2002) MAP kinase signalling cascade in Arabidopsis innate immunity. Nature 415:977–983

Bergmann DC, Lukowitz W, Somerville CR (2004) Stomatal development and pattern controlled by a MAPKK kinase. Science 304:1494–1497

Bodenhausen N, Reymond P (2007) Signaling pathways controlling induced resistance to insect herbivores in Arabidopsis. Mol Plant Microbe Interact 20:1406–1420

Bögre L, Calderini O, Binarova P, Mattauch M, Till S, Kiegerl S, Jonak C, Pollaschek C, Barker P, Huskisson NS, Hirt H, Heberle-Bors E (1999) A MAP kinase is activated late in plant mitosis and becomes localized to the plane of cell division. Plant Cell 11:101–113

Calderini O, Bögre L, Vicento O, Binarova P, Heberle-Bors E, Wilson C (1998) A cell cycle regulated MAP kinase with a possible role in cytokinesis in tobacco cells. J Cell Sci 111:3091–3100

Chaiwongsar S, Otegui MS, Jester PJ, Monson SS, Krysan PJ (2006) The protein kinase genes MAP3K epsilon 1 and MAP3K epsilon 2 are required for pollen viability in Arabidopsis thaliana. Plant J 48:193–205

Champion A, Picaud A, Henry Y (2004) Reassessing the MAP3K and MAP4K relationships. Trends Plant Sci 9:123–129

Chen H, Wilkerson CG, Kuchar JA, Phinney BS, Howe GA (2005) Jasmonate-inducible plant enzymes degrade essential amino acids in the herbivore midgut. Proc Natl Acad Sci USA 102:19237–19242

delPozo O, Pedley KF, Martin GB (2004) MAPKKKα is a positive regulator of cell death associated with both plant immunity and disease. EMBO J 23:3072–3082

Desikan R, Hancock JT, Ichimura K, Shinozaki K, Neill SJ (2001) Harpin induces activation of the Arabidopsis mitogen-activated protein kinases AtMPK4 and AtMPK6. Plant Physiol 126:1579–1587

Du B, Zhang W, Liu B, Hu J, Wei Z, Shi Z, He R, Zhu L, Chen R, Han B, He G (2009) Identification and characterization of Bph14, a gene conferring resistance to brown planthopper in rice. Proc Natl Acad Sci USA 106:22163–22168

Fürstenberg-Hägg J, Zagrobelny M, Bak S (2013) Plant defense against insect herbivores. Int J Mol Sci 14:10242–10297

Gao M, Liu J, Bi D, Zhang Z, Cheng F, Chen S, Zhang Y (2008) MEKK1, MKK1/MKK2 and MPK4 function together in a mitogen-activated protein kinase cascade to regulate innate immunity in plants. Cell Res 18:1190–1198

Glazebrook J (2005) Contrasting mechanisms of defense against biotrophic and necrotrophic pathogens. Annu Rev Phytopathol 43:205–227

Hamel LP, Nicole MC, Sritubtim S, Morency MJ, Ellis M, Ehlting J, Beaudoin N, Barbazuk B, Klessig D, Lee J, Martin G, Mundy J, Ohashi Y, Scheel D, Sheen J, Xing T, Zhang S, Seguin A, Ellis BE (2006) Ancient signals: comparative genomics of plant MAPK and MAPKK gene families. Trends Plant Sci 11:192–198

Hanks SK, Quinn AM, Hunter T (1988) The protein kinase family: conserved features and deduced phylogeny of the catalytic domains. Science 241:42–52

Hazzalin CA, Mahadevan LC (2002) MAPK-regulated transcription: a continuously variable gene switch? Nat Rev Mol Cell Biol 3:30–40

He C, Fong SHT, Yang D, Wang GL (1999) BWMK1, a novel MAP kinase induced by fungal infection and mechanical wounding in rice. Mol Plant-Microbe Interact 12:1064–1073

Heinrich M, Baldwin IT, Wu J (2011a) Three MAPK kinases, MEK1, SIPKK, and NPK2, are not involved in activation of SIPK after wounding and herbivore feeding but important for accumulation of trypsin proteinase inhibitors. Plant Mol Biol Report 30:731–740

Heinrich M, Baldwin IT, Wu J (2011b) Two mitogen-activated protein kinase kinases, MKK1 and MEK2, are involved in wounding- and specialist lepidopteran herbivore Manduca sexta-induced responses in Nicotianaattenuata. J Exp Bot 62:4355–4365

Hettenhausen C, Schuman MC, Wu J (2015) MAPK signaling: a key element in plant defense response to insects. Insect Sci 22:157–164

Higgins R, Lockwood T, Holley S, Yalamanchili R, Stratmann JW (2007) Changes in extracellular pH are neither required nor sufficient for activation of mitogen-activated protein kinases (MAPKs) in response to systemin and fusicoccin in tomato. Planta 225:1535–1546

Holley SR, Yalamanchili RD, Moura DS, Ryan CA, Stratmann JW (2003) Convergence of signaling pathways induced by systemin, oligosaccharide elicitors, and ultraviolet-B radiation at the level of mitogen-activated protein kinases in Lycopersicon peruvianum suspension-cultured cells. Plant Physiol 132:1728–1738

Ichimura MAPK Group (2002) Mitogen-activated protein kinase cascades in plants: a new nomenclature. Trends Plant Sci 7:301–308

Ichimura K, Mizoguchi T, Yoshida R, Yuasa T, Shinozaki K (2000) Various abiotic stresses rapidly activate Arabidopsis MAP kinases ATMPK4 and ATMPK6. Plant J 24:655–666

Ichimura K, Casais C, Peck SC, Shinozaki K, Shirasu K (2006) MEKK1 is required for MPK4 activation and regulates tissue-specific and temperature-dependent cell death in Arabidopsis. J Biol Chem 281:36969–36976

Ju C, Yoon GM, Shemansky JM, Lin DY, Ying ZI, Chang J, Garrett WM, Kessenbrock M, Groth G, Tucker ML, Cooper B, Kieber JJ, Chang C (2012) CTR1 phosphorylates the central regulator EIN2 to control ethylene hormone signaling from the ER membrane to the nucleus in Arabidopsis. Proc Natl Acad Sci USA 109:19486–19491

Kallenbach M, Alagna F, Baldwin IT, Bonaventure G (2010) Nicotiana attenuate SIPK, WIPK, NPR1, and fatty acid-amino acid conjugates participate in the induction of jasmonic acid biosynthesis by affecting early enzymatic steps in the pathway. Plant Physiol 152:96–106

Kaloshian I (2004) Gene-for-gene disease resistance: bridging insect pest and pathogen defense. J Chem Ecol 30:2419–2438

Kandoth PK, Ranf S, Pancholi SS, Jayanty S, Walla MD, Miller W, Howe GA, Lincoln DE, Stratmann JW (2007) Tomato MAPKs LeMPK1, LeMPK2, and LeMPK3 function in the systemin-mediated defense response against herbivorous insects. Proc Natl Acad Sci USA 104:12205–12210

Kendrick MD, Chang C (2008) Ethylene signaling: new levels of complexity and regulation. Curr Opin Plant Biol 11:479–485

Kiegerl S, Cardinale F, Siligan C, Gross A, Baudouin E, Liwosz A, Eklof S, Till S, Bogre L, Hirt H, Meskiene I (2000) SIMKK, a mitogen-activated protein kinase (MAPK) kinase, is a specific activator of the salt stress-induced MAPK, SIMK. Plant Cell 12:2247–2258

Kong Q, Qu N, Gao M, Zhang Z, Ding X, Yang F, Li Y, Dong OX, Chen S, Li X, Zhang Y (2012) The MEKK1-MKK1/MKK2-MPK4 kinase cascade negatively regulates immunity mediated by a mitogen-activated protein kinase kinase kinase in Arabidopsis. Plant Cell 24:2225–2236

Kovtun Y, Chiu WL, Tena G, Sheen J (2000) Functional analysis of oxidative stress-activated mitogen-activated protein kinase cascade in plants. Proc Natl Acad Sci USA 97:2940–2945

Krysan PJ, Jester PJ, Gottwald JR, Sussman MR (2002) An Arabidopsis mitogen-activated protein kinase kinasekinase gene family encodes essential positive regulators of cytokinesis. Plant Cell 14:1109–1120

Lee S, Hirt H, Lee Y (2001) Phosphatidic acid activates a wound-activated MAPK in Glycine max. Plant J 26(5):479–486

Lee DE, Lee IJ, Han O, Baik MG, Han SS, Back K (2004) Pathogen resistance of transgenic rice plants expressing mitogen-activated protein kinase 1, MK1, from Capsicum annuum. Mol Cell 17:81–85

Li Q, Xie QG, Smith-Becker J, Navarre DA, Kaloshian I (2006) Mi-1-mediated aphid resistance involves salicylic acid and mitogen-activated protein kinase signaling cascades. Mol Plant-Microbe Interact 19:655–664

Liu Y, Zhang S (2004) Phosphorylation of 1-aminocyclopropane-1-carboxylic acid synthase by MPK6, a stress-responsive mitogen-activated protein kinase, induces ethylene biosynthesis in Arabidopsis. Plant Cell 16:3386–3399

Liu Y, Schiff M, Dinesh-Kumar SP (2004) Involvement of MEK1 MAPKK, NTF6 MAPK, WRKY/MYB transcription factors, COI1 and CTR1 in N-mediated resistance to tobacco mosaic virus. Plant J 38:800–809

Lukowitz W, Roeder A, Parmenter D, Somerville C (2004) A MAPKK kinase gene regulates extra-embryonic cell fate in Arabidopsis. Cell 116:109–119

Maffei ME, Mithofer A, Boland W (2007) Before gene expression: early events in plant–insect interaction. Trends Plant Sci 12:310–316

McGurl B, Orozco-Cardenas ML, Pearce G, Ryan CA (1994) Overexpression of the prosystemin gene in transgenic tomato plants generates a systemic signal that constitutively induces proteinase inhibitor synthesis. Proc Natl Acad Sci USA 91:9799–9980

Meng X, Wang H, He Y, Liu Y, Walker JC, Torii KU, Zhang S (2012) A MAPK cascade downstream of ERECTA receptor-like protein kinase in regulating Arabidopsis inflorescence architecture by promoting localized cell proliferation. Plant Cell 24:4948–4960

Mousavi SA, Chauvin A, Pascaud F, Kellenberger S, Farmer EE (2013) GLUTAMATE RECEPTOR-LIKE genes mediate leaf-to-leaf wound signalling. Nature 500(7463):422–426

Nakagami H, Pitzschke A, Hirt H (2005) Emerging MAP kinase pathways in plant stress signalling. Trends Plant Sci 10:339–346

Nakagami H, Soukupova H, Schikora A, Zarsky V, Hirt H (2006) A mitogen-activated protein kinase kinase kinase mediates reactive oxygen species homeostasis in arabidopsis. J Biol Chem 281:38697–38704

Nühse TS, Peck SC, Hirt H, Boller T (2000) Microbial elicitors induce activation and dual phosphorylation of the Arabidopsis thaliana MAPK 6. J Biol Chem 275(11):7521–7526

Pearce G, Moura D, Stratmann J, Ryan CA (2001) Production of multiple plant hormones from a single polyprotein precursor. Nature 411:817–820

Qiao H, Shen Z, Huang SS, Schmitz RJ, Urich MA, Briggs SP, Ecker JR (2012) Processing and subcellular trafficking of ER-tethered EIN2 control response to ethylene gas. Science 338:390–393

Qiu JL, Zhou L, Yun BW, Nielsen HB, Fiil BK, Petersen K, Mackinlay J, Loake GJ, Mundy J, Morris PC (2008) Arabidopsis mitogen-activated protein kinase kinases MKK1 and MKK2 have overlapping functions in defense signaling mediated by MEKK1, MPK4, and MKS1. Plant Physiol 148:212–222

Rakwal R, Agrawal GK (2003) Wound signaling-coordination of the octadecanoid and MAPK pathways. Plant Physiol Biochem 41:855–861

Ren D, Liu Y, Yang KY, Han L, Mao G, Glazebrook J, Zhang S (2008) A fungal-responsive MAPK cascade regulates phytoalexin biosynthesis in Arabidopsis. PNAS USA 105:5638–5643

Ryan CA (2000) The systemin signaling pathway: differential activation of plant defensive genes. Biochim Biophys Acta 1477:112–121

Schafer M, Fischer C, Meldau S, Seebald E, Oelmuller R, Baldwin IT (2011) Lipase activity in insect oral secretions mediates defense responses in Arabidopsis. Plant Physiol 156:1520–1534

Schaffer R, Landgraf J, Accerbi M, Vernadette S, Larson M, Wisman E (2001) Microarray analysis of diurnal and circadian-regulated genes in Arabidopsis. Plant Cell 13:113–123

Schoenbeck MA, Samac DA, Fedorova M, Gregerson RG, Gantt JS, Vance CP (1999) The alfalfa (Medicago sativa) TDY1gene encodes a mitogen-activated protein kinase homolog. Mol Plant–Microbe Interact 12:882–893

Seo S, Okamoto M, Seto H, Ishizuka K, Sano H, Ohashi Y (1995) Tobacco MAP kinase: a possible mediator in wound signal transduction pathways. Science 270:1988–1992

Seo S, Sano H, Ohashi Y (1999) Jasmonate-based wound signal transduction requires activation of WIPK, a tobacco mitogen-activated protein kinase. Plant Cell 11:289–298

Seo S, Katou S, Seto H, Gomi K, Ohashi Y (2007) The mitogen-activated protein kinases WIPK and SIPK regulate the levels of jasmonic and salicylic acids in wounded tobacco plants. Plant J 49:899–909

Steppuhn A, Baldwin IT (2007) Resistance management in a native plant: nicotine prevents herbivores from compensating for plant protease inhibitors. Ecol Lett 10:499–511

Stratmann J (2008) MAP kinases in plant responses to herbivory. In: Schaller A (ed) Induced plant resistance to herbivory. Springer, Berlin. https://doi.org/10.1007/978-1-4020-8182-8_16

Stratmann JW, Ryan CA (1997) Myelin basic protein kinase activity in tomato leaves is induced systemically by wounding and increases in response to systemin and oligosaccharide elicitors. Proc Natl Acad Sci U S A 94:11085–11089

Stratmann JW, Stelmach BA, Weiler EW, Ryan CA (2000) UVB/UVA radiation activates a 48 kDa myelin basic protein kinase and potentiates wound signaling in tomato leaves. Photochem Photobiol 71:116–123

Suarez-Rodriguez MC, Adams-Phillips L, Liu Y, Wang H, Su SH, Jester PJ, Zhang S, Bent AF, Krysan PJ (2007) MEKK1 is required for flg22-induced MPK4 activation in Arabidopsis plants. Plant Physiol 143:661–669

Takabatake R, Shigemi S, Ito N, Gotoh Y, Mitsuhara I, Ohashi Y (2006) Involvement of wound-induced receptor-like protein kinase in the wound signal transduction in tobacco plant. Plant J 47:249–257

Tanoue T, Adachi M, Moriguchi T, Nishida E (2000) A conserved docking motif in MAP kinases common to substrates, activators and regulators. Nat Cell Biol 2:110–116

Walling LL (2000) The myriad plant responses to herbivores. J. Plant Growth Regul 19:195–216

Wang H, Ngwenyama N, Liu Y, Walker JC, Zhang S (2007) Stomatal development and patterning are regulated by environmentally responsive mitogen-activated protein kinases in Arabidopsis. Plant Cell 19:63–73

War AR, Paulraj MG, Ahmad T, Buhroo AA, Hussain B, Ignacimuthu S, Sharma HC (2012) Mechanisms of plant defense against insect herbivores. Plant Signal Behav 7:1306–1320

Wasternack C (2007) Jasmonates: an update on biosynthesis, signal transduction and action in plant stress response, growth and development. Ann Bot 100:681–697

Wu J, Baldwin IT (2010) New insights into plant responses to the attack from insect herbivores. Annu Rev Genet 44:1–24

Wu WJ, Hettenhausen C, Meldau S, Baldwin IT (2000) Herbivory rapidly activates MAPK signaling in attacked and unattacked leaf regions but not between leaves of Nicotianaattenuata. Plant Cell 19:1096–1122

Wu J, Hettenhausen C, Meldau S, Baldwin IT (2007) Herbivory rapidly activates MAPK signaling in attacked and unattacked leaf regions but not between leaves of *Nicotiana attenuata*. Plant Cell 19:1096–1122

Yang KY, Liu Y, Zhang S (2001) Activation of a mitogen-activated protein kinase pathway is involved in disease resistance in tobacco. Proc Natl Acad Sci U S A 98:741–746

Yuasa T, Ichimura K, Mizoguchi T, Shinozaki K (2001) Oxidative stress activates AtMPK6, an Arabidopsis homologue of MAP kinase. Plant Cell Physiol 42:1012–1016

Zebelo SA, Maffei ME (2015) Role of early signalling events in plant–insect interactions. J Exp Bot 66:435–448

Zhang S, Klessig DF (1998) The tobacco wounding-activated mitogen-activated protein kinase is encoded by SIPK. Proc Natl Acad Sci USA 95:7225–7230

Thioredoxins as Molecular Players in Plants, Pests, and Pathogens

Pragati Kumari, Arvind Gupta, and Saurabh Yadav

Abstract

Throughout their life cycle, plants are exposed to various kinds of biotic and abiotic stress conditions. Plant responds to such stressors through the orchestrated machinery at the molecular level and exerts its defense. Plants have sophisticated mechanisms to sense environmental cues, which guard them to mount a protective response. The antioxidant machinery in the plants comprises enzymatic and non-enzymatic proteins. The enzymatic proteins include glutaredoxins, thioredoxins, and many others. Thioredoxin (Trx) are multifunctional small redox proteins found in every living organism, and various Trxs have been identified in plants. The two reactive cysteine residues are found in the conserved motif of thioredoxins. They play post-translational regulatory role in number of cellular processes such as oxidative stresses and plant pathogen interactions. Trxs are reduced by NADP-thioredoxin reductase (NTR) in the presence of NADPH. In model plant, *Arabidopsis thaliana*, At Trxs are pathogen-inducible and contribute towards plant defense via expression of the defense responsive pathogenesis-related (PR) genes. The most important family of thioredoxin proteins is Trxh, having their role in many versatile processes including defense responses in plants. We present upcoming, prospective roles of thioredoxin proteins of plants, insects as well as pathogens and their role in biological interactions. This chapter would assist plant scientists in further exploring the

Authors Pragati Kumari and Arvind Gupta have equally contributed to this chapter.

P. Kumari
Scientist Hostel-S-02, Srinagar Garhwal, Uttarakhand, India

A. Gupta · S. Yadav (✉)
Department of Biotechnology, Hemvati Nandan Bahuguna Garhwal (Central) University, Srinagar Garhwal, Uttarakhand, India

I. K. Singh, A. Singh (eds.), *Plant-Pest Interactions: From Molecular Mechanisms to Chemical Ecology*, https://doi.org/10.1007/978-981-15-2467-7_6

crucial role of thioredoxins in response to attack by the insects causing losses to economically important plants.

Keywords

Herbivory · Redox regulations · Thioredoxins · Oxidative stress · TRXs

1 Introduction

Survival of plants and crop yield depend on the plant's ability for adaptation to many different types of environmental stresses, which generally induce oxidative stress. The environmental stress modulates the accumulation of reactive oxygen species (ROS) in plant cells and may induce oxidative damage, thereby limiting plant growth and development (Cassells and Curry 2001). Acting as signal transducers, ROS gets involved in many growth and developmental processes, biotic and abiotic response to environmental stimuli, and programmed cell death (Berglund and Ohlsson 1995; Shao et al. 2019). Environmental stress causes increased accumulation of ROS in plants. Many factors like stress, phytohormones stimulate ROS production, which may induce other routes or directly acts as defense compounds (Bailey-Serres and Mittler 2006). The ROS homoeostasis between its production and scavenging may get disturbed by plethora of biotic and abiotic factors (Apel and Hirt 2004; Wang and Hai 2016). When ROS level increases and exceeds the appropriate amount, it results in oxidative stress (Scandalios 2005; Sharma et al. 2012). It is well-known that high ROS concentrations are highly harmful to the organisms and irreversible damage occurs while causing loss of physiological activity and leading to cell death. Therefore, oxidative damage causes defense mechanisms to get mounted for regulating toxic levels of ROS (Mullineaux and Baker 2010).

ROS belongs to the group of many reactive species, free radicals, etc. which are derived from oxygen. Singlet oxygen, hydrogen peroxide, superoxide radical, and hydroxyl radical are the most common ROS. These are very toxic to the cellular biological macromolecules and lead to oxidative damage to the cell (Apel and Hirt 2004; Mittler et al. 2004). In chloroplasts, mitochondria, and peroxisomes, ROS generation is more due to high metabolic activity of these compartments (Mittler et al. 2004). Mehler reaction in chloroplasts, mitocondrial electron transport, and peroxisomal photorespiration are the main sources of ROS generation. ROS causes damage to macromolecules like nucleic acids, lipids, protein, and carbohydrates and affects many cellular functions (Scandalios 2005; Sharma et al. 2012). Thus, to protect the cellular machinery, the balance between production and ROS scavenging must be tightly regulated. To maintain growth, metabolism, and overall development, it is necessary to scavenge ROS and control the damage caused by ROS molecules.

Enzymatic and non-enzymatic antioxidants are responsible for maintaining the balance between the detoxification and production of ROS (Mittler et al. 2004; Nareshkumar et al. 2020). Antioxidants of non-enzymatic nature includes glutathione (GSH), redox buffers ascorbate (AsA), carotenoids, tocopherols, and phenolic compounds (Scandalios 2005; Mittler et al. 2004; Gratão et al. 2005). Enzymatic components comprise superoxide dismutase, catalase, peroxiredoxins, glutathione reductase, glutathione peroxidase, ascorbate peroxidase, guaiacol peroxidase, monodehydroascorbate reductase, and dehydroascorbate reductase (Caverzan et al. 2016). Trx comprises a major component of enzymatic components of ROS machinery.

2 Thioredoxins: A General Account

Trxs are small proteins having -Trp-Cys-Gly-Pro-Cys- amino acids within active site and redox active disulfide bridge. Thioredoxins are evolutionarily conserved in bacteria, fungi, plants, and animals (Wang et al. 2013: Japelaghi et al. 2012) with approximately 12 kDa molecular mass. Two types of TRX sequence are found in bacteria and animals, but in plants multiple forms are known, viz., chloroplastic (two), cytoplasmic (one), and mitochondrial (one). Even though the thioredoxins are of similar size and have related redox properties, they deal with different and specific functions (Fig. 1). The structural complementarity helps in the close interaction specifically between the different thioredoxins and putative targets (Pasternak et al. 2020). To regulate target protein's function and signaling in plants, these thioredoxins help in disulfide bond reduction by their redox-active dithiols in active site (Holmgren 1985). Thioredoxins in plants belong to a large gene family, and

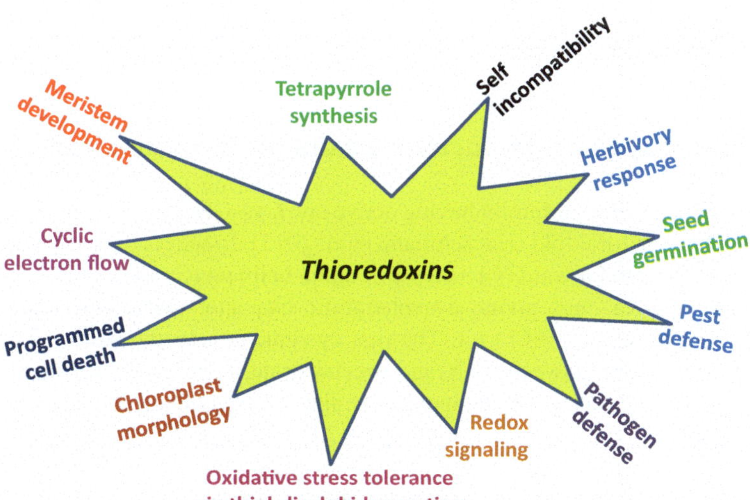

Fig. 1 Diverse roles of plant thioredoxin (TRX) proteins

20 different thioredoxin isoforms are found in *Arabidopsis thaliana*, viz., Trxs f1–2, m1–4, x, y1–2 and z (plastid) and Trx o1–2 (mitochondria and nucleus) and eight Trx h (cytosol, nucleus, endoplasmatic reticulum and mitochondria) (Delorme-Hinoux et al. 2016; Meyer et al. 2012). There are many different thioredoxin reduction systems in plants (Meyer et al. 2012; Li et al. 2011). NADPH via A (NTRA) and NTRB outside chloroplasts while in chloroplast Fdx reduces thioredoxins via Fdx-Trx reductase (FTR). An unusual type of NADPH-Trx reductase called as NTRC is present in chloroplast.

3 Mechanisms of Action of Thioredoxins

Thioredoxins are characterized by its structural features and also by its catalytic activity. These are stable proteins 110 amino acids in length, and two cysteine residues are in conserved catalytic site WC[G/P]PC and thus classified as dithiol thioredoxins. In monothiol Trxs, an unusual catalytic site has C-terminal cysteine residue replaced by serine (CXXS). In dithiol mechanism, both cysteine residues do catalytic functions of the enzyme, and target protein is reduced. The X–X dipeptide sequence present in the thioredoxin superfamily help in interactions with the target proteins and determine its redox potential (Krimm et al. 1998; Lin and Chen 2004; Jeong et al. 2004). Several thioredoxins-h proteins have unusual catalytic site WCPPC, and out of the five isoforms, three isoforms of subgroup I with WCPPC site is found in *A. thaliana* (Meyer et al. 2002). Partial complementation of sulfate assimilation phenotype was observed by modification of AtTrxh3 active site (WCPPC to WCGPC) using yeast heterologous complementation assays (Bréhelin et al. 2000). In *Populus trichocarpa*, the mutation (WCGPC to WCPPC) of the PtTrxh3 active site modifies the protein conformation, and it was analyzed that the prolyl residue modifies active site and does the interactions specifically with target proteins (Gelhaye et al. 2003a, b).

4 How Cysteine Reduction Occurs by Thioredoxins?

Trxs are of the TRX superfamily having active site Cys-Gly-Pro-Cys, and this forms characteristic structural fold (Hanschmann et al. 2013). In the redox active motif, the first cysteine is found toward N-terminus of alpha helix which forms dipole moment which reduces pKa, thus giving a nucleophilic character (Mata-Perez and Spoel 2018; Kortemme et al. 1996) to it. The first cysteine attacks the oxidized thiols in substrate targets and between TRX and the substrate, forming the covalent mixed disulfide bond. The second cysteine of active site resolves the mixed disulfides, thus reducing the substrate (Mata-Perez and Spoel 2018; Meyer et al. 2008; Jacquot et al. 2002). In plants, two thioredoxin reductases identified are NADPH-dependent thioredoxin reductase (NTR) and Ferredoxin-thioredoxin reductase (FTR) (Saurav et al. 2019; Nikkanen et al. 2017). NTR and FTR reduce oxidized TRX to recycle it back to its reduced enzymatic state (Mata-Perez and Spoel 2018; Meyer et al. 2008;

Holmgren and Bjornstedt 1995; Vieira Dos Santos and Rey 2006). In sulfenic acids and disulfides, there are cysteine modifications which are reduced by TRX-NTR/FTR systems (Tarrago et al. 2010; Meyer et al. 2009).

5 Plant Immunity and Redox-Based Post-Translational Protein Modifications

When a plant encounters stress, the signal needs to be transferred to those proteins having role in mounting a defense response, and ROS and reactive nitrogen species (RNS) have the potential to destroy the cellular machinery. But, these toxic molecules also serve as sensors by exploiting the redox-sensitive cysteine residues which acts as OFF and ON buttons during signaling. Thus, cysteine residues which are redox-sensitive act as ideal signaling switches in redoxins protein like thioredoxin proteins which harbor them. The cysteine moiety of Trx has the potential to perform multiple different reversible redox-based modifications like S-sulfenation (SeOH), S-nitrosylation (SeNO), and S-thiolation (SeS), which aid in defense response against biotic interactions. The S-thiolation includes disulfide bonds and glutathione attached covalently to cysteine residues in S-glutathionylation (Huang et al. 2018; Trost et al. 2017). S-Nitrosothiol-containing proteins (protein-SNO) are related to disease resistance. Elevated levels of free NO or NO donor molecules correlates well with high amounts of protein-SNO and have disturbed immune signaling, in mutant plants (Mata-Perez and Spoel 2018; Wang et al. 2014; Feechan et al. 2005; Rustérucci et al. 2007). Salicylic acid (SA) is a signaling phytohormone which is increased under pathogen attack. Later the elevated protein-SNO levels suppress this salicylic acid levels, and disease susceptibility is seen. The S-nitrosoglutathione (GSNO) also triggers S-glutathionylation on receptive thiols and potent NO donor for NO bioactivity and later gets reduced by GSNO Reductase 1 (GSNOR1) (Feechan et al. 2005). Many immune signaling is regulated by S-nitrosothiols. S-Glutathionylation plays important role in protein function and prevents irreversible thiol oxidation which causes protein dysfunction (Dalle-Donne et al. 2009). S-Nitrosylation of cysteines are important for the suppression of plant immune system.

The activity of enzymes is regulated by S-nitrosothiols which gets accumulated upon infection during pathogen effector triggered immunity (Romero-Puertas et al. 2008; Yun et al. 2011). The SA-binding activity and carbonic anhydrase activity of salicylic acid binding protein (SABP3), which is a positive regulator of effector-triggered immunity and S-nitrosylation activity, suppress the function (Wang et al. 2009). S-Nitrosothiols regulate enzymatic activity and finally are involved in plant immunity. NPR1 (nonexpressor of pathogenesis-related genes 1) is an important molecule which is SA-responsive and undergoes conformational changes upon oxidative modifications under conformational signaling, during plant defense (Fig. 2). When a pathogen attacks, SA level increases, and NPR1 disulfide bonds are quickly reduced, and NPR1 monomers are translocated to the nucleus, thereby activating SA-responsive genes of defense response (Ahn and Thiele 2003; Kinkema

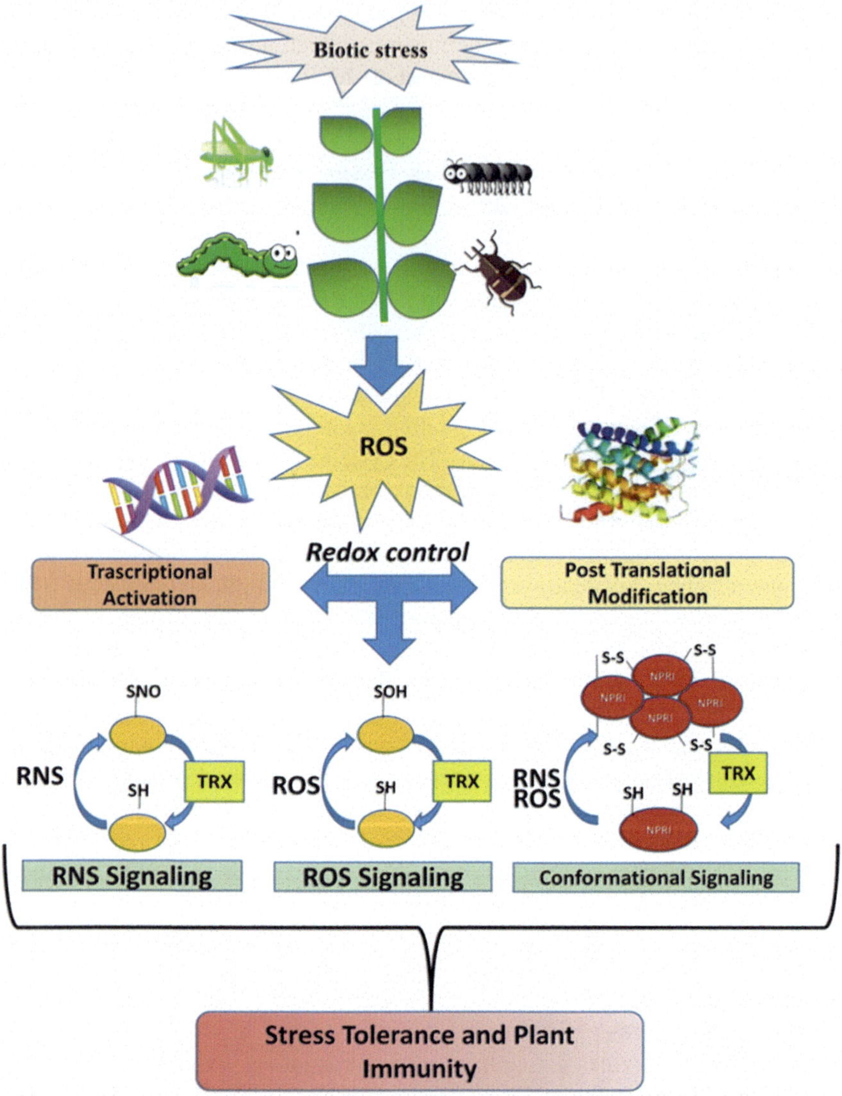

Fig. 2 Thioredoxin signaling in plants and role in plant immunity. The biotic factors act as stress and insects cause oxidative stress in plants. There are ROS, RNS, and conformational signaling, and the redox regulation is via transcriptional regulation and post-translational modifications. The thioredoxins reduce the oxidized proteins and help in stress tolerance

et al. 2000). NPR1 protein interacts with TGA transcription factors, and TGA1 & TGA4 contains conserved cysteine residues, which plays a crucial role under oxidative stress generated by plant pathogen interactions (Fig. 2). These well-documented findings allow us to conclude correlation between redox-based cysteine modifications and help in expression of genes to get involved in plant immunity.

Thioredoxin (Trx) system utilizes selenol and thiol groups to create a reducing cellular environment for fighting the oxidative as well as nitrosative stress (Ren et al. 2017). The glutathione (GSH) in reduced form and cellular thioredoxin system helps in denitrosylation of S-nitrosylated proteins (Ren et al. 2017). Thus, the role of thioredoxin-induced molecular changes upon attack by pathogens, insect, etc. finds special relevance in the future research.

6 Involvement of Different Thioredoxin Activities to Regulate Plant Immunity

Upcoming roles of different TRX enzymes shed light on the diverse roles for regulating plant immune signaling. The first report about the TRX in plant immunity was reported in the study by a protein of tomato known as Cf-9 protein that regulates resistance by effector-triggered immunity in both tomato and *Nicotiana benthamiana* (Mata-Perez and Spoel 2018; Rivas et al. 2004). The thioredoxin interacting with Cf-9 resistance protein is called Cf-9-interacting TRX (CITRX) but now christened as TRX-z (Arsova et al. 2010). In absence of a pathogen, CITRX/TRX z blocks interaction of Cf-9 and other interacting partner, so Cf-9 is prevented from inducing immunity. Under the attack by pathogen, the negative regulation of Cf-9 was eliminated by downregulating the CITRX/TRX because of which there is more accumulation of ROS leading to phosphorylation of kinases and induction of defense-related genes and activation of effector-triggered immunity (Rivas et al. 2004). Trxs play very prominent role in local and systemic signaling in immune response under the effect of the phytohormone salicylic acid. Pathogenic elicitors, wounding, abscission, and senescence induce TRXh5 gene expression. Out of the eight members of TRX-h group, only TRXh5 was upregulated under such stresses (Tada et al. 2008; Laloi et al. 2004; Reichheld et al. 2002), and it promotes SA-induced gene expression when the conformational state of NPR1, which is a SA-responsive co-activator, changes (Kinkema et al. 2000). NPR1 in oligomeric form having disulfide cross-linking between conserved cysteine residues remains in the cytoplasm. Under salicylic acid signaling response, the pathogen-inducible TRXh5 and TRXh3 which are constitutively expressed enzymes help in reducing disulfide bonds of NPR1. Now, monomeric NPR1 activates SA-inducible defense-related genes (Tada et al. 2008). The thioredoxin TRXh5 reduces protein-SNO molecule of S-nitrosylated NPR1 (Mata-Perez and Spoel 2018). By the foregoing discussion, we can see different TRX which are redox-based key immune signaling proteins in regulating post-translational modifications. Apart from canonical thioredoxin enzymes, the superfamily of TRX has members with a typical domain structures. In different plants, TRX are used in various signaling pathways of immunity but may get overpowered by certain plant pathogens. *Ralstonia solanacearum*, a pathogen, produces the RipAY, a putative γ-glutamyl cyclotransferase (effector protein) involved in glutathione breakdown. *R. solanacearum* exploits the host thioredoxin system to degrade RipAY activity, which results in inhibition of the pattern-triggered immunity (Mata-Perez and Spoel

2018; Mukaihara et al. 2016). There is substrate specificity observed as cytosolic TRX-h specifically activates RipAY but not by TRXs of chloroplasts under in vitro conditions. It was also observed that RipAY activity was independent of TRX-h active site, thus proving the thiol-independent signaling of TRX-h enzymes (Mukaihara et al. 2016). *Cochliobolus victoriae*, a necrotrophic fungus, is causal agent of Oat Victoria blight, and its victorin toxin induces apoptosis in susceptible plants, thus having role in pathogenicity (Sweat and Wolpert 2007; Mata-Perez and Spoel 2018; Lorang et al. 2012). At the active site 39th position of cysteine residues, the Victorin toxin binds to thioredoxin-h5, thereby inhibiting its activity. This again activates a NB-LRR resistance protein, and there is oxidative burst and programmed cell death (PCD) by the LOV1 (Locus Orchestrating Victorin effects1)-dependent resistance response (Lorang et al. 2012; Sweat and Wolpert 2007). *C. victoriae* exploits cytotoxic nature of LOV1 which gives victorin sensitivity (Lorang et al. 2012; Sweat and Wolpert 2007). The *C. victoriae* overpowers TRX h5-dependent resistance signaling pathway which causes PCD and confers host susceptibility by exploiting this response (Lorang et al. 2007; Sweat and Wolpert 2007).

7 How Cellular Redox Changes During Pathogen Challenge?

The molecular mechanism which explains the manner in which cellular redox alters upon pathogen challenge and leads to transcriptional reprogramming which culminates in disease resistance was analyzed in earlier studies. It was proposed that the redox-related signals are sensed and transmitted through protein-SNO and thioredoxins present in cytosol. The NPR1 oligomers to monomer exchange are catalyzed by cytosolic TRXs. Under the attack by pathogens, salicylic acid induces TRX-h5 to catalyze the instant release of NPR1 monomer. NPR1 oxidation is induced by salicylic acid, and reduction is done sequentially after the application of systemic induced response (SAR), which results in transient oxidative fluctuations (Mou et al. 2003). This observation was seen after treating the plants with a mixture of cycloheximide, the proteasome inhibitor MG115, and salicylic acid (Tada et al. 2008). When there is no protein synthesis, the monomeric NPR1 accumulates after SA application and is highest 12 h after treatment and monomers again oligomerize later after 16 h. Thioredoxins help in the conversion of oligomeric NPR1 to monomeric forms (Balmer et al. 2003; Yamazaki et al. 2004), which is very critical requirement for plant defense.

Myrosinase is an enzyme related to ABA signaling and is differentially expressed during stress conditions (Lambrix et al. 2001). In response to the biotic stress, the enzyme myrosinase-2 modulates its expression, and in the thioredoxin mutants, i.e., trxf1, trxf2, and trxm3, the enzyme myrosinase was downregulated (Lambrix et al. 2001). The epithiospecifier protein was upregulated in thioredoxin m3 mutant (trxm3) (Lambrix et al. 2001). **E**ndothelial cell **S**pecific **M**olecule**1** (ESM1) is a GDSL esterase/lipase and gets upregulated in mutants of trxm1, trxm2, and trxm4. This was also seen to be related to plants response to insect-infestation (Zhang et al. 2006). Regarding the role of thioredoxins in abiotic stress, overexpression of

Arabidopsis AtTrx-h3 caused heat resistance (Lee et al. 2009; Park et al. 2009), and Thioredoxin h3 from rice (OsTRXh) were found to be modulated by salt and cold stress (Xie et al. 2009).

8 Thioredoxin (from Plants) as a Signaling Component of *Chickpea-Helicoverpa* Interaction

The h-type of TRX proteins, i.e., (Trx-h), are the largest TRX superfamily (Atkinson and Babbitt 2009). Recently, an inducible Trx h was characterized during the *Helicoverpa-Cicer arietinum* interaction and designated as CaHaTrx-h (Singh et al. 2018) which contains "CGFS" kind of thioredoxin having 113 amino acids from chickpea plant (Singh et al. 2018). Productivity of chickpea and soybean is limited by biotic stresses including herbivory by *Helicoverpa armigera* (Singh et al. 2018; Wang et al. 2017). During chickpea-*Helicoverpa* interactions, there are many defense-related proteins that are modulated. The transcript level of a thioredoxin homolog Trx h5 modulated during herbivory and after treatment with signaling molecules like salicylic acid, jasmonic acid, ABA, etc. Since there was moderate CaHa-Trx-h expression seen under normal growth conditions, it proved its role in the normal growth and development process (Singh et al. 2018). After the *Helicoverpa* attack or feeding on the chickpea plants, the expression of CaHa-Trx-h instantly within 3 h reached to 12-fold (Singh et al. 2018). When mechanical wounding was induced, the transcript increase was not up to the extent of insect feeding. In response to phytohormones/signaling compound treatments also, change in gene expression level was observed. Hence, it can be concluded that there is critical role of thioredoxins during biotic stress especially during insect plant interactions.

9 Thioredoxins (from Insect) from Common Cutworm (*Spodoptera litura*)

Genes encoding thioredoxins have been known from insects such as fruit fly (*Drosophila melanogaster*), silkworm (*Bombyx mori*), Western honeybee (*Apis mellifera*), Indian-meal moth (*Plodia interpunctella*), and Chinese honeybee (*Apis cerana*) (Pellicena-Palle et al. 1997; Kim et al. 2007; Corona and Robinson 2006; Hoflehner et al. 2012; Yao et al. 2013). During interaction with plants, the insects and plants have to get involved under oxidative stress atmosphere. OMICS studies were instrumental in analysis of the thioredoxins from different insects and have proven their role in insect-plant interaction. After their upregulation upon biotic and abiotic stress exposure, they act as key players for protection against ROS damage (Kang et al. 2015; Yao et al. 2013). *Spodoptera litura*, a tobacco caterpillar, is economically important insect pest, which attacks the agricultural crops. *Spodoptera litura* is found throughout the tropical and temperate Asia, Australasia, and Pacific Islands (Gaur and Mogalapu 2018; Feakin 1973; Kranz et al. 1977). *Spodoptera litura* is a tobacco caterpillar and is one of the most economically important insect pests which attack the agricultural crops. *Spodoptera litura* is found throughout the tropical and temperate Asia, Australasia, and Pacific Islands (Feakin 1973; Kranz et al. 1977). Since *S. litura* is a polyphagous insect (Holloway 1989), it has huge potential to reach and invade new geographical areas and gets adapted to new climatic conditions. Looking into the economical importance of *S. litura*, investigations on this voracious feeder continued and Kang et al. (2015) cloned and functionally characterized thioredoxins from *Spodoptera litura.* There are two different types of thioredoxin genes which have full-length open reading frame (ORF), viz., SlTrx1 and SlTrx2 that were differentially localized. There was protein level similarity of 94% with sequence of important insect *Helicoverpa armigera.* There was constitutive expression of SlTrx1 and SlTrx2 transcripts in many tissues although with different expression levels (Kang et al. 2015). The sixth-instar larval stage showed significant changes in the expression of SlTrx 1 and SlTrx 2 in the tissues, epidermis and midgut (Kang et al. 2015). When *S. litura* larva were exposed to hydrogen peroxide, cumene hydroperoxide, or metaflumizone on first day of sixth-instar, the transcript levels of SlTrx1 and SlTrx2 were measured using quantitative PCR (qRT-PCR). SlTrx1 and SlTrx2 levels were significantly elevated in the midgut, epidermis, and fat body after treatment, as compared with the control. Kang et al. (2015) also reported the differential expression of SlTrx1 and SlTrx2 on exposure to pro-oxidants like hydrogen peroxide, cumene hydroperoxide, indoxacarb, and metaflumizone (Kang et al. 2015). During plant-insect interactions, oxidative stress is generated, and the ROS molecules act as the signaling factors, which culminate in mounting of defense response. This potentiates the fact that such upregulation may play important role in response toward oxidative stress. Trxs have role in the defense response against abiotic stresses and have been earlier reported in a variety of insects, viz., fruit fly (*D. melanogaster*) (Svensson et al. 2003). Kang et al. (2015) reported *S. litura*'s genes SlTrx1 and SlTrx2 to be of thioredoxin family and also reported the first comparative study of two thioredoxins. Two thioredoxins systems are found in mammalian cells, viz., cytosolic Trx1 and the mitochondrial Trx2 systems. Phylogenetically SlTrx1 showed similarity to cytosolic Trx1s, while SlTrx2 showed similarity to mitochondrial Trx2 (Kang et al. 2015; Lu and Holmgren 2014). The functional roles of SlTrx1 and SlTrx2 during the oxidative stress and molecular basis imply their role as antioxidants in cutworms. The biological control of *S. litura* and control methods will be explored in the future based on the research done. Thus, the plant-insect interaction when in harmful mode needs the newer transgenic technologies and CRISPR-CAS9 for efficient control of crop damage.

10 Molecular Characterization of Two Thioredoxin Genes (from Insect) of *Grapholita molesta*

Grapholita molesta (Lepidoptera: Tortricidae), oriental fruit moth, acts as primary pest of stone fruits which results in major economic damage. Only in few insects, there are reports related to the role of thioredoxins, e.g., *Drosophila melanogaster,*

Bombyx mori, and *Apis cerana cerana*. In *D. melanogaster*, there are three thioredoxin genes (Trx1, Trx2, and TrxT) having their role in oxidative defense system (Svensson et al. 2003). During high temperature and microbial infection, thioredoxins from *B. mori* involve itself in resisting the oxidative stress (Kim et al. 2007). Similarly in *Apis mellifera*, the thioredoxin proteins act in antioxidant defense by scavenging mitochondrial ROS (Corona and Robinson 2006). The HaTrx2, thioredoxin protein from *Helicoverpa armigera*, was responsible for providing antioxidant defense (Shen et al. 2018; Kang et al. 2015; Zhang et al. 2015). These are the data about thioredoxins from insects, which prove their role in maintaining redox homeostasis, thus giving oxidative stress resistance in insects. In *Grapholita molesta*, there was no any report of role of thioredoxins during plant insect interaction. *Grapholita molesta* is a pest, which severely affects the plants worldwide grown in most of the temperate fruit cultivating areas (Kirk et al. 2013). Since, the organophosphates insecticide resistance was observed in *G. molesta* (Kanga et al. 2003), the question was to search for the underlying molecular mechanisms related to the genes, which may be involved in the insect's defense mechanism. Shen et al. (2018) hypothesized that the oxidative stress-inducible thioredoxins, i.e., Trx2 and Trx-like1, may provide resistance to the insecticides in *G. molesta*. They had a clue about the possible role of Trxs in *G. molesta* on the basis of their expression patterns observed under different types of stress treatments. For functional genomics study, overexpression and downregulation of candidate genes are employed. The RNAi lines provided downregulation of GmTrx2, and GmTrx-like1, which upregulated the oxidative stress-related enzymes, suggesting role of the thioredoxins in oxidative stress defense. The results of such studies help the researchers to undertake research related to the protein structures, determination of active sites, etc. of thioredoxins in *G. molesta*, which will be helpful in devising strategies for targeted pest control. The foregoing discussion about possible role of TRXs will be useful for delineating the antioxidant defense processes of thioredoxin system in insects, which is less researched.

11 Thioredoxin 2 and Thioredoxin Reductase 1 (from Insect) of *Helicoverpa armigera*

Throughout Asia, cotton bollworm (*Helicoverpa armigera*), a lepidopteran pest, devastates and incurs economic losses in the economically important plants, viz., cotton, chickpea, corn, vegetable, etc. (Wu et al. 2008). In *Spodoptera litura* thioredoxin peroxidase gives resistance to biocontrol fungus *Nomuraea rileyi* (Zhang et al. 2015; Chen et al. 2014). Zhang et al. (2015) thought that thioredoxin and thioredoxin reductase may help in providing resistance to the infection of pathogenic microbes in insects. In the knockdown larvae of HaTrx2- or HaTrxR1-lines and normal larvae, the ROS generation and lipid peroxidation level correlated TRX function with the resistance to nucleopolyhedrovirus (NPV). The work of Zhang et al. (2015) related to thioredoxins and thioredoxin reductases in insects will help in making good insecticides. The HaTrx2 from *Helicoverpa* was found to be similar to Trx sequences from other selected insect species. HaTrx2 and HaTrxR1 transcripts were significantly upregulated after NPV infection (Zhang et al. 2015). The protective functions of HaTrx2 and HaTrxR1 were observed, against the oxidative stress and NPV infection. In *H. armigera,* it was shown that both HaTrx2 and HaTrxR1 have redox active sites and are of thioredoxins and thioredoxin reductase families, respectively (Zhang et al. 2015). Possibly, HaTrx2 and HaTrxR1 remove elevated ROS levels, caused by NPV attack and *H. armigera* infestation. It is believed that this may be the possible mechanism how Trx and TrxR provide antioxidant defense. So, there are enough evidences that both HaTrx2 and HaTrxR1 protect from oxidative stress and enhance our knowledge about insect's TRX system (Zhang et al. 2015).

12 Role of Thioredoxin-Like Protein (from Pathogen) of *Bemisia tabaci*

Among viruses, *Begomovirus* (family *Geminiviridae*) is the largest genus (Gutierrez 1999; Zerbini et al. 2017), and around the globe severe crop losses are due to geminiviruses (Hanley-Bowdoin et al. 2013). *Begomovirus, Turncurtovirus, Mastrevirus, Topocuvirus, Becurtovirus, Curtovirus, Eragrovirus, Capulavirus,* and *Grablovirus* are few genera which fall under geminiviruses, on the basis of type of insect vector and their genome organization (Hanley-Bowdoin et al. 2013; Varsani et al. 2014, 2017). In India, *Cotton leaf curl virus* (CLCuV) and *Tomato leaf curl New Delhi virus (*ToLCNDV) are major begomoviruses, and during feeding they get transmitted through sole vector, *Bemisia* (Hogenhout et al. 2008). The pathway of transmission of begomoviruses is through stylet, esophagus, filter chamber, midgut, hemolymph, and salivary glands, and then it enters into another healthy plant (Czosnek et al. 2002; Ghanim et al. 2001). Very less information exists about the begomoviral proteins and their interacting partners, in *B. tabaci.* There exist many reports that the begomoviruses interact with different molecules/proteins in the alimentary canal of insect during transmission (Götz et al. 2012; Ohnesorge and Bejarano 2009). The coat protein (CP), which is the exclusive protein and an important molecular player, interacts insect vector's protein and is responsible for the viral specificity in the insect vector (Briddon et al. 1990; Höfer et al. 1997; Noris et al. 1998). Now it was imperative by the researchers involved in this characterization to understand the complete pathway. So, protein-protein interaction was employed using yeast-two hybrid (Y2H) technique to isolate and identify putative interacting proteins (Saurav et al. 2019). Thus, coat proteins of both CLCuV and ToLCNDV were used for Y2H analysis. It is already known that upon *Tomato yellow leaf curl China virus* (TYLCCNV) infection in *B. tabaci*, there was change in the gene expression profile of the insect. Out of the various interacting partners, one clone for which 5′ RACE was performed that showed homology with thioredoxin-like protein (TLP). TLP is a member of thioredoxin-like superfamily (Saurav et al. 2019). Thioredoxin-like protein (TLP) interacted with coat protein of both CLCuV and ToLCNDV. It is well-known that the thioredoxin system is evolutionary conserved across taxa and found in various organisms. Thioredoxins has role in host-pathogen interaction also (Chibani et al. 2009a, b), and it can be assumed that the interaction of TLP (a thioredoxin-like superfamily protein) with coat protein of both CLCuV and ToLCNDV has significant role in the virus transmission. There is not much data as to how the virus transmits by *B. tabaci*, and mechanism is also not deciphered. The thioredoxin-like protein, which interacts with coat protein of both CLCuV and ToLCNDV, has potential role in plant-insect interactions. But the exact role in the transmission of virus and its route in *B. tabaci* must be known for protecting crop damage.

13 Thioredoxins in Plant Immunity Against TMV and CMV

The disease in agricultural crops are mostly the result of plant-pathogen and plant-insect interactions, and most of the plant viruses like CMV and TMV which cause plant diseases rely on various biotic vectors for the transmission and survival in host (Whitfield et al. 2015). Mostly insects are the plant virus-transmitting vectors and are whiteflies, aphids, leafhoppers, planthoppers, etc. (Bragard et al. 2013). The oxidative stress studies and the redoxin proteins from plant-insect interactions shed light on the molecular mechanism. The thioredoxin protein, i.e., TRX-h functions in the redox regulation (Atkinson and Babbitt 2009) and defense response (Jiang et al. 2010). The plants are protected from oxidative stress damage by involvement of thiol-disulfide interchanges, and the Trx-h proteins help in transfer of reducing equivalents (Dal Piaz et al. 2010) to relieve hydrogen peroxide toxicity and other reactive oxygen species (ROS) involved (Blokhina and Fagerstedt 2009). Beyond the inherent characteristic feature as antioxidant activities, now the Trx-h function as signaling molecules are also being explored (Bashandy et al. 2010). The thioredoxins, Trx-h, play important role in redox-related defense response due to their redox-related cellular activity (Kandasamy et al. 2009). After this, pathogenesis-related (PR) genes get expressed, and systemic acquired response develops (Laloi et al. 2004; Tada et al. 2008). In *Arabidopsis*, the regulation of NPR1 and SAR by AtTRX-h3 and AtTRXh5 shows the functional relationship between Trx-h and defense responses in plants. In the response to a bacterial pathogen in *Arabidopsis*, the expression of PR1 genes was characterized by the regulation of NPR1 (Laloi et al. 2004; Tada et al. 2008), but the roles of NPR1 and PR genes are not known to be active during CMV and TMV infection. The defense functions of Trx-h family members are very less known. In *Nicotiana*, three Trx-h homologues -NtTRXh1, NtTRXh2, and TRXh-like genes have been identified. The function of NtTRXh3 (Trx-h like gene) was deciphered in plants during virus infection by overexpression and silencing of this gene in tobacco.

14 Conclusion

Today the present-day world is busy in finding sustainable methods to combat against many plant diseases, and this has led researchers to delve deep into defense mechanisms of plants. Enzymatic redoxins, like thioredoxins, are an important class of oxidative stress-related proteins and are ubiquitously present in many organisms. Mostly the proteins are in reduced form, but when level of oxidized proteins is increased, they get reduced by thioredoxins. The transcript level of TRXs helped in understanding the molecular event inside the cell during plant-insect interactions. The observations from various studies of *H. armigera* enriches role of the TRX system and oxidative stress, in insects. Therefore, further research is needed for the development of microbial pesticides and new chemicals for targeting *TRX* systems. The protein interactions between plant viruses and their corresponding insect vectors is a key molecular interface, which determines how infected host plant transmits to new hosts. This specificity of plant-insect interactions opens up avenues for possible control. The advent of "omics" technologies proves as facilitating our current knowledge and significantly contributes in this domain. The emerging techniques like imaging, molecular biotechnology, high throughput, gene editing will better tailor the mechanisms in favor of the plants, and the harmful insects will be easily overpowered.

References

Ahn SG, Thiele DJ (2003) Redox regulation of mammalian heat shock factor 1 is essential for Hsp gene activation and protection from stress. Genes Dev 17:516–528

Apel K, Hirt H (2004) Reactive oxygen species: metabolism, oxidative stress, and signal transduction. Annu Rev Plant Biol 55:373–399

Arsova B et al (2010) Plastidial thioredoxin z interacts with two fructokinase-like proteins in a thiol-dependent manner: evidence for an essential role in chloroplast development in Arabidopsis and Nicotiana benthamiana. Plant Cell 22:1498–1515

Atkinson HJ, Babbitt PC (2009) Glutathione transferases are structural and functional outliers in the thioredoxin fold. Biochemistry 48:11108–11116

Bailey-Serres J, Mittler R (2006) The roles of reactive oxygen species in plants cells. Plant Physiol 141(2):311

Balmer Y, Koller A, del Val G, Manieri W, Schürmann P, Buchanan BB (2003) Proteomics gives insight into the regulatory function of chloroplast thioredoxins. Proc Natl Acad Sci 100 (1):370–375

Bashandy T, Guilleminot J, Vernoux T, Caparros-Ruiz D, Ljung K, Meyer Y, Reichheld JP (2010) Interplay between the NADP-linked thioredoxin and glutathione systems in Arabidopsis auxin signaling. Plant Cell 22:376–391

Berglund T, Ohlsson AB (1995) Defensive and secondary metabolism in plant tissue cultures, with special reference to nicotinamide, glutathione and oxidative stress. Plant Cell Tissue Organ Cult 43:137–145

Blokhina O, Fagerstedt KV (2009) Reactive oxygen species and nitric oxide in plant mitochondria: origin and redundant regulatory systems. Physiol Plant 138:447–462

Bragard C, Caciagli P, Lemaire O, Lopez-Moya JJ, MacFarlane S, Peters D, Susi P, Torrance L (2013) Status and prospects of plant virus control through interference with vector transmission. Annu Rev Phytopathol 51:177–201. https://doi.org/10.1146/annurev-phyto-082712-102346

Bréhelin C, Mouaheb N, Verdoucq L, Lancelin JM, Meyer Y (2000) Characterization of determinants for the specificity of Arabidopsis thioredoxins h in yeast complementation. J Biol Chem 275:31641–31647

Briddon RW, Pinner MS, Stanley J, Markham PG (1990) Geminivirus coat protein gene replacement alters insect specificity. J Virol 177(1):85–94. https://doi.org/10.1016/0042-6822(90)90462-Z

Cassells AC, Curry RF (2001) Oxidative stress and physiological, epigenetic and genetic variability in plant tissue culture: implications for micropropagators and genetic engineers. Plant Cell Tissue Organ Cult 64:145–157

Caverzan A, Casassola A, Patussi Brammer SP (2016) Reactive oxygen species and antioxidant enzymes involved in plant tolerance to stress. In: Shanker A (ed) Abiotic and biotic stress in plants—recent advances and future perspectives. InTech, Rijeka, pp 463–480. https://doi.org/10.5772/61368. ISBN 978-953-51-2250-0

Chen H et al (2014) Thioredoxin peroxidase gene is involved in resistance to biocontrol fungus *omuraea rileyi* in *Spodoptera litura*: gene cloning, expression, localization and function. Dev Comp Immunol 44(1):76–85

Chibani K, Wingsle G, Jacquot J-P, Gelhaye E, Rouhier N (2009a) Comparative genomic study of the thioredoxin family in photosynthetic organisms with emphasis on *Populus trichocarpa*. Mol Plant 2(2):308–322. https://doi.org/10.1093/mp/ssn076

Chibani K, Couturier J, Selles B, Jacquot JP, Rouhier N (2009b) The chloroplastic thiol reducing systems: dual functions in the regulation of carbohydrate metabolism and regeneration of antioxidant enzymes, emphasis on the poplar redoxin equipment. Photosynth Res 104:75–99

Corona M, Robinson GE (2006) Genes of the antioxidant system of the honeybee: annotation and phylogeny. Insect Mol Biol 15:687–701. https://doi.org/10.1111/j.1365-2583.2006.00695.x

Czosnek H, Ghanim M, Ghanim M (2002) The circulative pathway of begomoviruses in the whitefly vector *Bemisia tabaci* insights from studies with tomato yellow leaf curl virus. Ann Appl Biol 140(3):215–231. https://doi.org/10.1111/j.1744-7348.2002.tb00175.x

Dal Piaz F, Braca A, Belisario MA, De Tommasi N (2010) Thioredoxin system modulation by plant and fungal secondary metabolites. Curr Med Chem 17:479–494

Dalle-Donne I, Rossi R, Colombo G, Giustarini D, Milzani A (2009) Protein S-glutathionylation: a regulatory device from bacteria to humans. Trends Biochem Sci 34:85–96

Delorme-Hinoux V et al (2016) Nuclear thiol redox systems in plants. Plant Sci 243:84–95

Feakin SD (1973) Pest control in groundnuts, PANS manual no. 2. ODA, London

Feechan A et al (2005) A central role for S-nitrosothiols in plant disease resistance. Proc Natl Acad Sci U S A 102:8054–8059

Gaur N, Mogalapu S (2018) Pests of soybean. In: Pests and their management. Springer, Singapore, pp 137–162

Gelhaye E, Rouhier N, Vlamis-Gardikas A, Girardet JM, Sautière PE, Sayzet M et al (2003a) Identification and characterization of a third thioredoxin h in poplar. Plant Physiol Biochem 41:629–635

Gelhaye E, Rouhier N, Jacquot JP (2003b) Evidence for a subgroup of thioredoxin h that requires GSH/Grx for its reduction. FEBS Lett 555:443–448

Ghanim M, Rosell RC, Campbell LR, Czosnek H, Brown JK, Ullman DE (2001) Digestive, salivary, and reproductive organs of *Bemisia tabaci* (Gennadius) (Hemiptera: Aleyrodidae) B type. J Morphol 248(1):22–40. https://doi.org/10.1002/jmor.1018

Götz M, Popovski S, Kollenberg M, Gorovitz R, Brown JK, Cicero JM, Czosnek H, Winter S, Ghanim M (2012) Implication of *Bemisia tabaci* heat shock protein 70 in begomovirus-whitefly interactions. J Virol 86:13241–13252. https://doi.org/10.1128/JVI.00880-12

Gratão PL, Polle A, Lea PJ, Azevedo RA (2005) Making the life of heavy metal-stressed plants a little easier. Funct Plant Biol 32(6):481–494

Gutierrez C (1999) Geminivirus DNA replication. Cell Mol Life Sci 56(3–4):313–329. https://doi.org/10.1007/s000180050433

Hanley-Bowdoin L, Bejarano ER, Robertson D, Mansoor S (2013) Geminiviruses: masters at redirecting and reprogramming plant processes. Nat Rev Microbiol 11(11):777–788. https://doi.org/10.1038/nrmicro3117

Hanschmann EM, Godoy JR, Berndt C, Hudemann C, Lillig CH (2013) Thioredoxins, glutaredoxins, and peroxiredoxins – molecular mechanisms and health significance: from cofactors to antioxidants to redox signaling. Antioxid Redox Signal 19:1539–1605

Höfer P, Bedford ID, Markham PG, Jeske H, Frischmuth T (1997) Coat protein gene replacement results in whitefly transmission of an insect nontransmissible geminivirus isolate. Virology 236 (2):288–295. https://doi.org/10.1006/viro.1997.8751

Hoflehner E, Binder M, Hemmer W, Mahler V, Panzani RC, Jarisch R, Wiedermann U, Duchêne M (2012) Thioredoxin from the Indianmeal moth Plodia interpunctella: cloning and test of the allergenic potential in mice. PLoS One 7:e42026

Hogenhout SA, Ammar E-D, Whitfield AE, Redinbaugh MG (2008) Insect vector interactions with persistently transmitted viruses. Annu Rev Phytopathol 46:327–359. https://doi.org/10.1146/annurev.phyto.022508.092135

Holloway JD (1989) The moths of Borneo: family Noctuidae, trifine subfamilies: Noctuinae, Heliothinae, Hadeninae, Acronictinae, Amphipyrinae, Agaristinae. Malay Nat J 42 (2–3):57–228

Holmgren A (1985) Thioredoxin. Annu Rev Biochem 54:237–271

Holmgren A, Bjornstedt M (1995) Thioredoxin and thioredoxin reductase. Methods Enzymol 252:199–208

Huang J, Willems P, Van Breusegem F, Messens J (2018) Pathways crossing mammalian and plant sulfenomic landscapes. Free Radic Biol Med 122:193–201

Jacquot JP et al (2002) Thioredoxins and related proteins in photosynthetic organisms: molecular basis for thiol dependent regulation. Biochem Pharmacol 64:1065–1069

Japelaghi RH, Haddad R, Garoosi G (2012) Isolation, identification and sequence analysis of a thioredoxin h gene, a member of subgroup III of h-type Trxs from grape (*Vitis vinifera* L. cv. Askari). Mol Biol Rep 39:3683–3693. https://doi.org/10.1007/s11033-011-1143-1

Jeong W, Yoon HW, Lee SR, Rhee SG (2004) Identification and characterization of TRP14, a thioredoxin-related protein of 14 kDa. New insights into the specificity of thioredoxin function. J Biol Chem 279:3142–3150

Jiang H, Song W, Li A, Yang X, Sun D (2010) Identification of genes differentially expressed in cauliflower associated with resistance to Xanthomonas campestris pv. campestris. Mol Biol Rep 38(1):621–629

Kandasamy S, Loganathan K, Muthuraj R, Duraisamy S, Seetharaman S, Thiruvengadam R, Ponnusamy B, Ramasamy S (2009) Understanding the molecular basis of plant growth promotional effect of *Pseudomonas fluorescens* on rice through protein profiling. Proteome Sci 7:47

Kang T, Wan H, Zhang Y, Shakeel M, Lu Y, You H et al (2015) Comparative study of two thioredoxins from common cutworm (Spodoptera litura): cloning, expression, and functional characterization. Comp Biochem Physiol 182:47–54. https://doi.org/10.1016/j.cbpb.2014.12.004

Kanga LHB, Pree DJ, van Lier JL, Walker GM (2003) Management of insecticide resistance in oriental fruit moth (Grapholita molesta; Lepidoptera: Tortricidae) populations from Ontario. Pest Manag Sci 59:921–927. https://doi.org/10.1002/ps.702

Kim YJ, Lee KS, Kim BY, Choo YM, Sohn HD, Jin BR (2007) Thioredoxin from the silkworm, *Bombyx mori*: cDNA sequence, expression, and functional characterization. Comp Biochem Physiol B 147:574–581

Kinkema M, Fan W, Dong X (2000) Nuclear localization of NPR1 is required for activation of PR gene expression. Plant Cell 12:2339–2350

Kirk H, Dorn S, Mazzi D (2013) Worldwide population genetic structure of the oriental fruit moth (Grapholita molesta), a globally invasive pest. BMC Ecol 13:12. https://doi.org/10.1186/1472-6785-13-12

Kortemme T, Darby NJ, Creighton TE (1996) Electrostatic interactions in the active site of the N-terminal thioredoxin-like domain of protein disulfie isomerase. Biochemistry 35:14503–14511

Kranz J, Schumutterer H, Koch W (eds) (1977) Diseases pests and weeds in tropical crops. Verlag Paul Parley, Berlin/Hamburg

Krimm I, Lemaire S, Ruelland E, Miginiac-Maslow M, Jacquot JP, Hirasawa M et al (1998) The single mutation Trp35 Ala in the 35–40 redox site of *Chlamydomonas reinhardtii* thioredoxin h affects its biochemical activity and the pH dependence of C36-C39 1H–13C NMR. Eur J Biochem 255:185–195

Laloi C, Mestres-Ortega D, Marco Y, Meyer Y, Reichheld JP (2004) The Arabidopsis cytosolic thioredoxin h5 gene induction by oxidative stress and its W-box-mediated response to pathogen elicitor. Plant Physiol 134:1006–1016

Lambrix V, Reichelt M, Mitchell-Olds T, Kliebenstein DJ, Gershenzon J (2001) The Arabidopsis epithiospecifier protein promotes the hydrolysis of glucosinolates to nitriles and influences Trichoplusiani herbivory. Plant Cell 13:2793–2807. https://doi.org/10.1105/tpc.13.12.2793

Lee JR, Lee SS, Jang HH, Lee YM, Park JH, Park SC, Moon JC, Park SK, Kim SY, Lee SY (2009) Heat-shock dependent oligomeric status alters the function of a plant-specific thioredoxin-like protein, AtTDX. Proc Natl Acad Sci 106(14):5978–5983

Li H, Lu H, Guo D et al (2011) Molecular characterization of a thioredoxin h gene (HbTRX1) from *Hevea brasiliensis* showing differential expression in latex between self-rooting juvenile clones and donor clones. Mol Biol Rep 38:1989–1994. https://doi.org/10.1007/s11033-010-0321-x

Lin TY, Chen TS (2004) A positive charge at position 33 of thioredoxin primarily affects its interaction with other proteins but not redox potential. Biochemistry 43:945–952

Lorang JM, Sweat TA, Wolpert TJ (2007) Plant disease susceptibility conferred by a "resistance" gene. Proc Natl Acad Sci U S A 104:14861–14866

Lorang J et al (2012) Tricking the guard: exploiting plant defense for disease susceptibility. Science 338:659–662

Lu J, Holmgren A (2014) The thioredoxin antioxidant system. Free Radic Biol Med 66:75–87

Mata-Perez C, Spoel SH (2018) Thioredoxin-mediated redox signalling in plant immunity. Plant Sci 279:27–33

Meyer Y, Vignols F, Reichheld JP (2002) Classification of plant thioredoxins by sequence similarity and intron position. Methods Enzymol 347:394–402

Meyer Y et al (2008) Glutaredoxins and thioredoxins in plants. Biochim Biophys Acta 1783:589–600

Meyer Y, Buchanan BB, Vignols F, Reichheld JP (2009) Thioredoxins and glutaredoxins: unifying elements in redox biology. Annu Rev Genet 43:335–367

Meyer Y et al (2012) Thioredoxin and glutaredoxin systems in plants: molecular mechanisms, crosstalks, and functional significance. Antioxid Redox Signal 17:1124–1160

Mittler R, Vanderauwera S, Gollery M, Van Breusegem F (2004) Reactive oxygen gene network of plants. Trends Plant Sci 9(10):490–498

Mou Z, Fan W, Dong X (2003) Inducers of plant systemic acquired resistance regulate NPR1 function through redox changes. Cell 113:935

Mukaihara T, Hatanaka T, Nakano M, Oda K (2016) Ralstonia solanacearum type III effctor RipAY is a glutathione-degrading enzyme that is activated by plant cytosolic thioredoxins and suppresses plant immunity. MBio 7:e00359–e00316

Mullineaux PM, Baker NR (2010) Oxidative stress: antagonistic signaling for acclimation or cell death. Plant Physiol 154(2):521–525. https://doi.org/10.1104/pp.110.16140

Nareshkumar A, Subbarao S, Vennapusa AR et al (2020) Enzymatic and non-enzymatic detoxification of reactive carbonyl compounds improves the oxidative stress tolerance in cucumber, tobacco and rice seedlings. J Plant Growth Regul. https://doi.org/10.1007/s00344-020-10072-w

Nikkanen L, Toivola J, Diaz MG, Rintamaki E (2017) Chloroplast thioredoxin systems: prospects for improving photosynthesis. Philos Trans R Soc Lond Ser B Biol Sci 372:20160474

Noris E, Vaira AM, Caciagli P, Masenga V, Gronenborn B, Accotto GP (1998) Amino acids in the capsid protein of tomato yellow leaf curl virus that are crucial for systemic infection, particle formation, and insect transmission. J Virol 72:10050–10057

Ohnesorge S, Bejarano ER (2009) Begomovirus coat protein interacts with a small heat-shock protein of its transmission vector (*Bemisia tabaci*). Insect Mol Biol 18(6):693–703. https://doi.org/10.1111/j.1365-2583.2009.00906.x

Park SK, Jung YJ, Lee JR, Lee YM, Jang HH, Lee SS, Park JH, Kim SY, Moon JC, Lee SY (2009) Heat-shock and redox-dependent functional switching of an h-type Arabidopsis thioredoxin from a disulfide reductase to a molecular chaperone. Plant Physiol 150(2):552–561

Pasternak Y, Ohana M, Biron-Shental T et al (2020) Thioredoxin, thioredoxin interacting protein and transducer and activator of transcription 3 in gestational diabetes. Mol Biol Rep 47:1199–1206. https://doi.org/10.1007/s11033-019-05221-8

Pellicena-Palle A, Stitzinger SM, Salz HK (1997) The function of the Drosophila thioredoxin homolog encoded by the deadhead gene is redox-dependent and blocks the initiation of development but not DNA synthesis. Mech Dev 62:61–65

Reichheld JP, Mestres-Ortega D, Laloi C, Meyer Y (2002) The multigenic family of thioredoxin h in *Arabidopsis thaliana*: specifi expression and stress response. Plant Physiol Biochem 40:685–690

Ren X, Zou L, Zhang X, Branco V, Wang J, Carvalho C, Holmgren A, Lu J (2017) Redox signaling mediated by thioredoxin and glutathione systems in the central nervous system. Antioxid Redox Signal 27:989–1010. https://doi.org/10.1089/ars.2016.6925

Rivas S et al (2004) CITRX thioredoxin interacts with the tomato Cf-9 resistance protein and negatively regulates defence. EMBO J 23:2156–2165

Romero-Puertas MC et al (2008) Proteomic analysis of S-nitrosylated proteins in *Arabidopsis thaliana* undergoing hypersensitive response. Proteomics 8:1459–1469

Rustérucci C, Espunya MC, Díaz M, Chabannes M, Martínez MC (2007) S-nitrosoglutathione reductase affords protection against pathogens in Arabidopsis, both locally and systemically. Plant Physiol 143:1282–1292

Saurav GK, Rana VS, Popli S et al (2019) A thioredoxin-like protein of *Bemisia tabaci* interacts with coat protein of begomoviruses. Virus Genes 55:356. https://doi.org/10.1007/s11262-019-01657-z

Scandalios JG (2005) Oxidative stress: molecular perception and transduction of signal triggering antioxidant gene defenses. Braz J Med Biol Res 38(7):995–1014

Shao Y, Guo M, He X et al (2019) Constitutive H2O2 is involved in sorghum defense against aphids. Rev Braz Bot 42:271–281

Sharma P, Jha AB, Dubey RS, Pessarakli M (2012) Reactive oxygen species, oxidative damage, and antioxidative defense mechanism in plants under stressful conditions. J Bot 2012:1–26

Shen Z-J, Liu Y-J, Gao X-H, Liu X-M, Zhang S-D, Li Z, Zhang Q-W, Liu X-X (2018) Molecular identification of two thioredoxin genes from *Grapholita molesta* and their function in resistance to emamectin benzoate. Front Physiol 9:1421. https://doi.org/10.3389/fphys.2018.01421

Singh A, Tyagi C, Nath O, Singh IK (2018) Helicoverpa-inducible Thioredoxin h from *Cicer arietinum*: structural modeling and potential targets. Int J Biol Macromol 109:231–243. https://doi.org/10.1016/j.ijbiomac.2017.12.079

Svensson MJ, Chen JD, Pirrotta V, Larsson J (2003) The thioredoxinT and deadhead gene pair encode testis- and ovary-specific thioredoxins in *Drosophila melanogaster*. Chromosoma 112:133–143

Sweat TA, Wolpert TJ (2007) Thioredoxin h5 is required for victorin sensitivity mediated by a CC-NBS-LRR gene in Arabidopsis. Plant Cell 19:673–687

Tada Y et al (2008) Plant immunity requires conformational changes of NPR1 via Snitrosylation and thioredoxins. Science 321:952–956

Tarrago L et al (2010) Plant thioredoxin CDSP32 regenerates 1-cys methionine sulfoxide reductase B activity through the direct reduction of sulfenic acid. J Biol Chem 285:14964–14972

Trost P, Fermani S, Calvaresi M, Zaffgnini M (2017) Biochemical basis of sulphenomics: how protein sulphenic acids may be stabilized by the protein microenvironment. Plant Cell Environ 40:483–490

Varsani A, Navas-Castillo J, Moriones E, Hernández-Zepeda C, Idris A, Brown JK, Zerbini FM, Martin DP (2014) Establishment of three new genera in the family Geminiviridae: Becurtovirus, Eragrovirus and Turncurtovirus. Arch Virol 159(8):2193–2203. https://doi.org/10.1007/s00705-014-2050-2

Varsani A, Roumagnac P, Fuchs M, Navas-Castillo J, Moriones E, Idris A, Briddon RW, Rivera-Bustamante R, Zerbini FM, Martin DP (2017) Capulavirus and Grablovirus: two new genera in the family Geminiviridae. Arch Virol 162(6):1819–1831. https://doi.org/10.1007/s00705-017-3268-6

Vieira Dos Santos C, Rey P (2006) Plant thioredoxins are key actors in the oxidative stress response. Trends Plant Sci 11:329–334

Wang X, Hai C (2016) Novel insights into redox system and the mechanism of redox regulation. Mol Biol Rep 43:607–628

Wang YQ et al (2009) S-nitrosylation of AtSABP3 antagonizes the expression of plant immunity. J Biol Chem 284:2131–2137

Wang Q, Hou Y, Qu J et al (2013) Molecular cloning, expression, purification and characterization of thioredoxin from Antarctic Sea-ice bacteria Pseudo alteromonas sp. AN178. Mol Biol Rep 40:6587–6591. https://doi.org/10.1007/s11033-013-2771-4

Wang C et al (2014) Free radicals mediate systemic acquired resistance. Cell Rep 7:348–355

Wang X, Lu J, Chen H et al (2017) Comparative analyses of transcriptome and proteome in response to cotton bollworm between a resistant wild soybean and a susceptible soybean cultivar. Plant Cell Tissue Organ Cult 129:511–520

Whitfield AE, Falk BW, Rotenberg D (2015) Insect vector-mediated transmission of plant viruses. Virology 479:278–289. https://doi.org/10.1016/j.virol.2015.03.026

Wu KM, Lu YH, Feng HQ, Jiang YY, Zhao JZ (2008) Suppression of cotton bollworm in multiple crops in China in areas with Bt toxin-containing cotton. Science 321(5896):1676–1678

Xie G, Kato H, Sasaki K, Imai R (2009) A cold-induced thioredoxin h of rice, OsTrx23, negatively regulates kinase activities of OsMPK3 and OsMPK6 in vitro. FEBS Lett 583(17):2734–2738

Yamazaki D, Motohashi K, Kasama T, Hara Y, Hisabori T (2004) Target proteins of the cytosolic thioredoxins in *Arabidopsis thaliana*. Plant Cell Physiol 45(1):18–27

Yao P, Lu W, Meng F, Wang X, Xu B, Guo X (2013) Molecular cloning, expression and oxidative stress response of a mitochondrial thioredoxin peroxidase gene (AccTpx-3) from *Apis cerana* cerana. J Insect Physiol 59:273–282

Yun BW et al (2011) S-nitrosylation of NADPH oxidase regulates cell death in plant immunity. Nature 478:264–268

Zerbini FM, Briddon RW, Idris A, Martin DP, Moriones E, Navas-Castillo J, Rivera-Bustamante R, Roumagnac P, Varsani A, ICTV Report Consortium (2017) ICTV virus taxonomy profile: Geminiviridae. J Gen Virol 98:131–133. https://doi.org/10.1099/jgv.0.000738

Zhang Z-Y, Ober JA, Kliebenstein DJ (2006) The gene controlling the quantitative trait locus EPITHIOSPECIFIER MODIFIER1 alters glucosinolate hydrolysis and insect resistance in Arabidopsis. Plant Cell 18:1524–1536. https://doi.org/10.1105/tpc.105.039602

Zhang S et al (2015) Sequence analysis, expression profiles and function of thioredoxin 2 and thioredoxin reductase 1 in resistance to nucleopolyhedrovirus in *Helicoverpa armigera*. Sci Rep 5:15531. https://doi.org/10.1038/srep15531

Plant Volatiles and Their Role in Insect Olfaction

Muhammad Binyameen, Qasim Ali, Amit Roy, and Fredrik Schlyter

Abstract

Plant volatiles are the invisible players in the plant-insect co-evolutionary arms race. They are involved in various plant-mediated tri-trophic interactions within the ecosystem. Volatiles, emitted from different parts of the plant, serves as a cue for the host-seeking herbivores. Interestingly, insects perceive and process such complex environmental cues with their advanced olfactory system. During herbivory, plants also produce defensive volatiles recognized as herbivore-induced plant volatiles (HIPVs) that often serve a dual purpose by attracting the natural enemies and giving an alarming signal to the plants nearby. Insects tend to avoid the host plants emitting such defensive volatiles and non-host plants, releasing non-host volatiles (NHVs). Understanding the dynamics of host selection through plant volatile recognition by pest insects is highly important to develop eco-friendly pest management practices employing strategies such as push-pull

M. Binyameen (✉)
Department of Entomology, Faculty of Agricultural Sciences and Technology, Bahauddin Zakariya University, Multan, Pakistan

Faculty of Forestry and Wood Sciences, EXTEMIT-K, Czech University of Life Sciences, Prague, Czech Republic
e-mail: mbinyameen@bzu.edu.pk

Q. Ali · A. Roy
Faculty of Forestry and Wood Sciences, EXTEMIT-K, Czech University of Life Sciences, Prague, Czech Republic
e-mail: aliq@fld.czu.cz; roy@fld.czu.cz

F. Schlyter
Faculty of Forestry and Wood Sciences, EXTEMIT-K, Czech University of Life Sciences, Prague, Czech Republic

Division of Chemical Ecology, Department of Plant Protection Biology, Swedish University of Agricultural Sciences, Alnarp, Sweden
e-mail: schlyter@fld.czu.cz

I. K. Singh, A. Singh (eds.), *Plant-Pest Interactions: From Molecular Mechanisms to Chemical Ecology*, https://doi.org/10.1007/978-981-15-2467-7_7

strategy. In the present chapter, we have reviewed different aspects of plant volatile production, insect olfactory system organization, the contribution of plant volatiles in host-seeking behavior of insects, and finally their potential use in formulating insect pest management strategies.

Keywords

Insect olfaction · Host selection · Plant volatiles · Push-pull strategy · Odourant receptor (OR) · Non-host volatiles (NHVs)

1 Introduction

Plants and insects are coexisting since the last 350 million years, and during that period, different forms of interactions have evolved between them. Plant volatiles are the primary agents for such interactions. For instance, these interactions may assist a plant by insect-mediated pollination or damage by insect herbivory (Schiestl 2010). Volatile organic compounds (VOCs) are emanated in small amounts from various parts of a healthy plant, i.e., leaves, flowers, stem, branches, roots, etc. However, under any stress (biotic or abiotic) or upon damage, the rate of emission increases to many folds (Das et al. 2013). More than 1700 blends of plant volatiles have been identified so far from different plant species belonging to more than 90 plant families. However, these volatile blends are comprised of only 1% of the plant secondary metabolites (Holopainen and Blande 2012; Dudareva et al. 2006). These blends of volatiles have messages encrypted for the one or different species, and the translation of such messages are receiver species-specific (Tumlinson 2014). The chemical composition and intensity of volatiles also depend on the plant species and/or stress to which they have been subjected to. Naturally, VOCs are emitted from vegetative and reproductive parts of a plant. Usually, vegetative parts produce less VOCs; however, in case of mechanical damage or pest attack, the amount of volatiles increases substantially. Plant parts, such as the bark of conifers and leaves of basil and peppermint, produce a high quantity of VOCs (Niinemets et al. 2013).

Insects have a striking capability to distinguish between the VOCs released from the host and non-host plants, and they try to avoid the habitat with VOCs from non-host plants (Schlyter 2012). Herbivore-induced plant volatiles (HIPVs) are released in response to herbivory from the plants to help the attacked plants either by the release of oviposition and feeding deterrents or by calling the natural enemies for assistance to get rid of the herbivore. The latter is categorized as an indirect defense by plants, which is often more important than the direct chemical defense (Tumlinson 2014). Plants release a variety of HIPVs, which could be different, both qualitatively and quantitatively, from VOCs released during normal conditions (Aartsma et al. 2017). In several plant species, HIPVs, namely, terpenes, sesquiterpenes, homoterpenes, and some other compounds, are produced due to the regulation of specific biosynthetic pathways as a counteraction to insect herbivory (Tumlinson 2014). By volatile emissions, an herbivore-attacked plant not only builds up a defense against invaders but also helps neighboring plants through associational resistance or priming defenses. Thus, HIPVs not only participate in the tri-trophic interaction by attracting natural enemies of the herbivores but also deliver an "alarm" signal to neighboring plants (Zakir et al. 2013; Himanen et al. 2010).

During last 2–3 decades, significant advancement has been made in analytical tools for studying the dynamics of plant-insect interactions, which gives us an in-depth knowledge of the molecular mechanism underlying volatile synthesis, regulation, function, and their role in various interactions with insects (Stam et al. 2014; Mithöfer et al. 2009; Tabata 2018; Renwick 1989; Bernays and Chapman 2007; Dicke 2000). In this book chapter, we have reviewed the recent advancements made in the field of insect chemical ecology by touching the basics for plant volatile synthesis and their detection by insects. Insect olfactory system, signal transduction, and translation into behaviors and host plant selection mechanisms are also explained. We also discussed the plant defense against herbivores using VOCs. This latest information will help the readers, especially young scientists, to understand the basic concepts of insect and plant interactions, latest techniques being used to analyze plant chemistry and insect responses ranging from landscape to the level of a gene, and devise plant protection strategies using natural plant-derived volatile compounds.

1.1 Plant Volatiles: Synthesis and Release of Different Types of Plant Volatiles

Plant volatiles are synthesized through a series of biochemical pathways resulting in the production of diverse types [Fig. 1]. Plant volatiles mostly belong to the different functional groups of chemicals such as terpenes, phenylpropanoids, volatile fatty acid derivatives and amino acid volatile derivatives, etc.

Terpenes are synthesized in plants by consecutive condensation of isopentenyl diphosphate (IPP) and its isomer, dimethylallyl diphosphate (DMAPP), which is five-carbon precursor (Baldwin 2010). IPP and DMAPP are synthesized in two different pathways: in the cytosol, through Mevalonic acid (MVA) pathway (Kuzuyama 2002), while in plastid through methyl-erythritol-phosphate (MEP) pathway from pyruvate and glyceraldehyde-3-phosphate (Rohmer 1999; Eisenreich et al. 1998). Enzymatic property of prenyltransferases transfers the IPP and DMAPP into farnesyl diphosphates (FPP) in cytosol and plastids. Geranyl pyrophosphate synthase (GPPS) action on IPP and DMAPP produces geranyl pyrophosphate (GPP), a C_{10} molecule and precursor of monoterpenes (Ogura and Koyama 1998). In the presence of farnesyl pyrophosphate synthase (FPPS) in the cytosol, two IPP and one DMAPP molecules produce farnesyl pyrophosphate (FPP), which is a C_{15} molecule and precursor for sesquiterpenes (McGarvey and Croteau 1995).

Similarly, in the presence of geranylgeranyl pyrophosphate synthase (GGPPS), three IPP molecules and one DMAPP produce geranylgeranyl pyrophosphate (GGPP), a C20 molecule and precursor of diterpenes (Ogura and Koyama 1998), while irregular, acyclic C11- and C16-homoterpenes are derived from the enzymatic

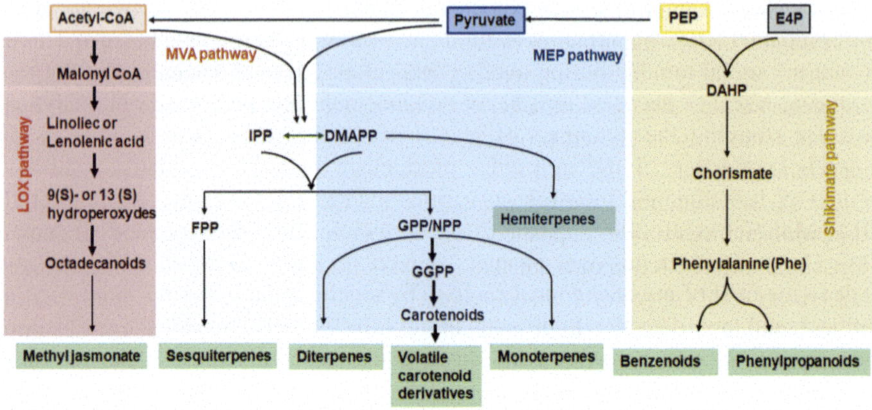

Fig. 1 Scheme of biosynthetic pathways producing plant volatile organic compounds (VOCs). Precursors for VOCs originate from primary metabolism of the plant (enclosed at the top). The shikimate, the methylerythritol phosphate (MEP), the mevalonic acid (MVA), and lipoxygenase (LOX) pathways are the four primary metabolic pathways leading to the emission of monoterpenes, diterpenes, hemiterpenes, sesquiterpenes, benzenoids, phenylpropanoids, volatile carotenoid derivatives, methyl jasmonate, and green leaf volatiles. VOCs are highlighted with green-text color. Abbreviations: DMAPP, dimethylallyl pyrophosphate; DAHP, 3-deoxy-d-arabinoheptulosonate-7 phosphate; FPP, farnesyl pyrophosphate; GGPP, geranylgeranyl pyrophosphate; NPP, neryl pyrophosphate; PEP, erythrose 4-phosphate (E4P); Phe, phenylalanine; IPP, isopentenyl pyrophosphate; GPP, geranyl pyrophosphate; phospho*enol*pyruvate. (Adopted with modification from Dudareva et al. 2013)

action on FPP and GGPP, respectively, and further oxidized by cytochrome P450 monooxygenase (Tholl et al. 2011). From the enzymatic action of terpene synthases on FPP, GPP, and GGPP, a large number of terpene volatiles are released.

Phenylpropanoids are produced from L-phenylalanine through Shikimate pathway (Vergoz et al. 2007). The catalyzing activity of L-phenylalanine ammonia-lyase in phenylpropanoid pathway converts the L-phenylalanine into *trans*-cinnamic acid (Moreno et al. 2014). By hydroxylation and methylation, *trans*-cinnamic acid converts into hydroxycinnamic acid esters and then into alcohols, aldehydes, and hydroxycinnamic acids (Humphreys and Chapple 2002). Benzenoids are also synthesized during phenylpropanoid pathway from *trans*-cinnamic acid. Specially, benzenoids are produced by cleavage of the *trans*-cinnamic acid at C_2 unit by either the CoA-dependent–β-oxidative pathway or CoA-independent–non-β-oxidative pathway. Sometimes, a combination of both of these pathways are involved. In CoA-dependent–β-oxidative pathway, oxidation of fatty acids produces the benzenoids, while in CoA-independent–non-β-oxidative pathway; *trans*-cinnamic acid's hydration and degradation produce the benzaldehyde that is converted to benzoic acid by an enzyme called benzaldehyde dehydrogenase (Boatright et al. 2004). Similar to pathways of terpene productions, little is known about the enzyme actions during the synthesis of phenylpropanoids; however, considerable advancement has been made on this topic.

Fatty Acid Volatile Derivatives are another large group of plant volatile having *trans*-2-hexenal, *cis*-3-hexenol, and methyl jasmonate plant volatiles. For their production, unsaturated fatty acid (C_{18}) is deoxygenated by lipoxygenases (LOX) at C_9, or C_{13} and enzyme is named as its location activity, LOX-9 or LOX-13. Fatty acids produce 9-hydroperoxy and 13-hydroperoxy derivatives by these enzyme actions, and these two derivatives are catalyzed further. Allene oxide synthase (AOS) catalyze 13-hydroxyperoxy derivatives into 12,13-epoxy octadecatrienoic acid, and further series of reaction leads the synthesis of jasmonic acid. Jasmonic acid is converted into a volatile ester, methyl-jasmonate, by the enzymatic action of jasmonic acid carboxyl methyltransferase (Song et al. 2005; Feussner and Wasternack 2002). For the hydroxy and peroxy derivatives, breakdown action of hydroperoxide lyase produces fewer carbon molecules which are catalyzed again by alcohol dehydrogenases for the production of aldehydes (1-hexanal, *trans*-2-hexenal, nonanal) and alcohols (*cis*-3-hexenol). These aldehydes and alcohols are produced in green organs, as a result of wounding (Dudareva et al. 2013).

Amino Acid Volatile Derivatives besides phenylpropanoids production, serve as the precursor of alcohols, aldehydes, acids, and esters. Flowers and fruit aromas are the primary sources of these volatiles, where nitrogen and sulfur are the main components (Baldwin 2010). Deamination and transamination of amino acids by aminotransferases lead to the formation of α-keto acid (Gonda et al. 2010). This acid goes to decarboxylation and then into reductions, oxidations, and esterification for the production of aldehydes, acids, alcohols, and esters, respectively (Reineccius 2016). Amino acids can also act as a precursor of acyl-CoAs, which is essential for alcohol esterification in the presence of alcohol acyltransferases catalyst (Gonzalez et al. 2009).

1.2 Plant Fitness Cost for Volatile Production

Naturally, VOCs arise from different parts of a plant during functions of growth and development. Plants need insects for pollination purpose, but they try to avoid the interaction with herbivore. Some plants have evolved the mechanism for floral scent emission to do so (Haverkamp et al. 2018; Pattrick et al. 2017; Lev-Yadun 2016). Floral scents are primary metabolites; however, under stress, the plant also produces secondary metabolites. For producing metabolites, plants have to pay some cost (Bennett and Wallsgrove 1994; Neilson et al. 2013). Cost of volatile synthesis from an induced plant can be estimated by comparing the plant growth and fecundity to another healthy plant (Strauss et al. 2002; Cipollini and Heil 2010). Mostly, volatile production cost has been measured in the non-woody herbaceous and agricultural plant. The production of terpenes volatiles has been found to be the most costly one (Paré and Tumlinson 1999). However, the cost of secondary volatile compounds relies on the intensity of the stress. For instance, tobacco plants, affected by low herbivory, produce less seed as compared to the healthy plants undergoing no herbivory. Interestingly, plants suffering intermediate herbivory with no effect on the foliage part are documented to have more fitness than a plant having insects feeding on leaves (Baldwin 1998). There is considerable attention toward the fitness cost assessment for the shorter-lived annual and biennial field plants, whereas there is a lack of information on the fitness of woody plants due to their longer lifespan.

1.3 Dispersion of Plant Volatiles in the Air

VOCs of plants are released into the air and detected by other plants and insects. Insects use their olfactory system to detect these volatile compounds. Air plays a crucial role in transporting these volatile compounds from the site of release to the olfactory organs of an insect. Naturally, odours or floral scents arising from a plant dispersed into the air and traveled toward the direction of the air flow. On the site of emission, the concentration of the volatiles within an odour plume is higher, but its surface area is small. Structure of odour plume changes as it goes away from the plume emission point. Size of an odour plume and distance between the odour molecules within an odour plume increase with an increase in traveling distance from the source while its concentration decreases. This movement is like the movement of smoke coming from the upper mouth of a chimney, which is visible when coming out and gradually with time, its visibility reduces, and its area increases in the air. Movement of the odour plume in the air stretches the concentration of odours. Insects in the surrounding environment detect the volatiles and respond accordingly (Murlis et al. 1992; Cardé and Willis 2008). Mixing of odour with other sources depends on the distance between two sources and distance of odour molecule from its source (Myrick et al. 2009).

The site of odour synthesis and its quantity is vital for its movement as the larger site of production, and higher concentration increases the traveling distance of odour molecules (Cardé and Willis 2008; Farré-Armengol et al. 2016). When air carries volatiles to the antennae, herbivores first obtain the information about their origin (host or non-host plant). Then, they try to analyze the direction and speed of the volatile, which help them to measure the distance of the odour source from its current position. If herbivore is interested in the volatile, they start moving toward the plant emitting the volatiles. It means that an herbivore has to fly or walk toward the opposite route of the air flow to reach the odour source. The concentration of the volatile increases with the movement serving as an indicator to the herbivore for correct orientation toward the origin (Murlis et al. 1992; Beyaert and Hilker 2014). Depending on environmental conditions, odour plume can alter with time as well as distance. The odour of low volatility will travel less distance than the odour of higher volatility (Beyaert and Hilker 2014).

2 Insect Olfactory System and Functions

Insects have a sophisticated olfactory system that is responsible for the detection and differentiation of diverse odour stimuli. Insect olfactory receptors, which are the basic units in an olfactory system and expressed on the dendritic membrane of olfactory sensory neurons (OSNs), empower the insects to sense and recognize volatile compounds for the selection of host plant and mate. Moreover, the olfactory system also plays an essential role in avoidance from non-host plants, predators, and parasitoids (Carraher et al. 2015). Most of the work done so far to study the morphology and physiology of the insect olfactory system is subjected upon *Drosophila* (Hallem et al. 2006); however, many studies have also been performed on moths, bark beetles, and mosquitoes (Hansson and Stensmyr 2011; de Fouchier et al. 2017; Park et al. 2018; Andersson et al. 2015; Rinker et al. 2013; Zhang and Schlyter 2004).

2.1 Olfactory Organs: Types and Structure

Size and shape of an insect change during different developmental stages (metamorphosis); therefore, the olfactory organs also differ at each stage. This diversity is due to habitat and survival competency. For example, in many insect species, larvae and adults of the same species have different habitats and foods, such as in mosquitoes and dragonflies, immatures live and feed in the water, while adults fly in the air and visit pollens for food. So, both stages have a different structure of olfactory organs according to their habits and needs (Szyszka and Galizia 2015).

In insects, main organs used for olfaction are antennae, maxillary palpi, and labial palpi; they all are present on the head (Missbach et al. 2014; Wicher 2015). Insect antennae are more involved in olfaction than maxillary and labial palpi. Antenna detects different kinds of pheromones, host and non-host odourants. Maxillary palpi in mosquito are responsible for the detection of CO_2, and mosquito uses the information detected by maxillary palpi during host-seeking behavior for the blood (Majeed et al. 2014). While maxillary palpi in *Drosophila* are found responsible for the detection of different odours from fruits, microbes, and feces (Dweck et al. 2016), labial palpi are also involved for CO_2 and other odour detection in many insect species (Ning et al. 2016; Hansson and Stensmyr 2011).

Larvae have simpler olfactory organs than adults, which mostly encounter with the host plant on which larvae feed due to feeding habits and less mobility. Maxillary palpi are almost similar to adult and have no apparent difference. Head of the immature has antennal extensions, which house sensilla. In *Drosophila* larvae, olfactory organs are consisting of the central olfactory dome and peripheral sensilla. Insect antennae are predominantly equipped with olfactory sensilla for olfactory signals, but gustatory sensilla are also present on the antennae (Joseph and Carlson 2015; Depetris-Chauvin et al. 2015). With reaching the adult stage, a number of sensilla also increased (Kostromytska et al. 2015; Hallberg et al. 2012). Usually, in adult insects, antennae are the core olfactory organs for capturing and processing of odour stimuli. An antenna can freely move in all directions and contains three parts, namely, scape, pedicel, and flagellum. Morphology of the antenna, i.e., shape and size, varies in different insect species, i.e., short in housefly and dragonfly, clubbed in butterflies, pectinate in mostly moths, elbow type in ants and weevils, while beetles have a broad diversity in the antenna (Table 1). Hair-like porous structures, called sensilla, are present on the antennae, maxillary-palpi, and labial palpi, filled with sensillum lymph (Song et al. 2017; Krishnan and Wairkar 2018; Ando et al. 2019).

2.2 Peripheral Olfactory System

Olfactory sensilla that are present on olfactory organs show considerable variations in shape and sizes depending on the species of insects as well as within a species, i.e., sexual dimorphism. With regard to external morphology, insect olfactory sensilla have been classified into different types. For example, in *Spodoptera littoralis*, which is a polyphagous model moth species, six different types of olfactory sensilla, such as trichoid (hair-like), basiconic (hair-like but short), coeloconic (peg type in a pit), grooved peg (like a peg on antennal surface), and auricilic (rabbit hair-like) (Fig. 2), have been identified (Binyameen et al. 2012). There are pores on the surface of these olfactory sensilla serving as gates for the entrance of the air carrying volatile compounds are hydrophobic in nature. These hydrophobic odour molecules pass from the hydrophilic medium of sensillum lymph with the help of olfactory binding proteins (OBPs) to reach at the site of a receptor, which is expressed at the dendritic membrane of olfactory sensory neurons (OSNs) (Pelosi et al. 2018; Leal 2013; Suh et al. 2014). OBPs can be either like pheromone binding proteins (PBPs) that bind and transport pheromone specifically or general (non-specific), which can bind and transport any odourant molecules from a host or non-host plant to the receptor site. Hence, the interaction of odour molecules with OBPs is considered as the first step for the activation of the insect olfactory system. Usually, there is a specific type of OBP and OSNs for each type (or structurally similar type) of odours (Fig. 3). So, irrelevant odour molecules are not detected by the system. Thus, OBPs play a dual role: discrimination of odours (an OBP binds only specific odour molecules which can fit the binding site) and transportation of odourants to the site of ORs. There has been ample evidence that OBPs release odourants at the site of an OR, and odourants directly activate the OR, but a study on *Drosophila* also demonstrated that a specific OBP could activate the OR while binding the odourant (Leal 2013).

2.2.1 Redrafted from Cseke, Kaufman et al. (2007)

ORs are also dependent on a collaborator called (Orco), a co-receptor (Suh et al. 2014). However, Orco cannot recognize natural odourants. It builds a heteromeric complex with the OR specific for odour and helps in signal transduction (Larsson et al. 2004). Once the odour molecules are attached to OR and activate it, it is essential to free the site for incoming odour molecules, and this is done by different odour degrading enzymes (ODE) (Leal 2013; Brito et al. 2016). For example, in

Table 1 Different types of insect antenna

Antenna	Insect	Key structural features	Reference
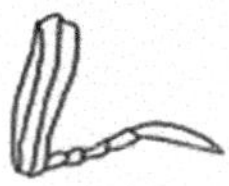 Lamellate	Beetles, i.e., (*Aegus jansoni*)	Leaf-like surfaces at the antennal club	(Beutel et al. 2013)
 Serrate	Beetles, i.e., (*Mordella leuco*)	Saw-like in appearance	
 Pectinate	Beetles, i.e., (*Pedilus flabellatus*, male)	Comb-like in appearance	
 Flabellate	Beetles, i.e., (*Sandalus niger*, male)	Long folded fan-like appearance at the terminal portion	
 Clavate	Butterflies, i.e., (*Danaus genutia*)	Gradual thickening from base to the tip of the antenna	
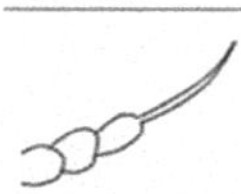 Stylate	Robber flies, i.e., (*Mallophora ruficauda*)	A long, slender point at final segment of the antenna	
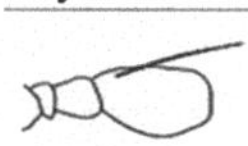 Aristate	House fly, i.e., (*Musca domestica*)	Pouch-like antennae having a lateral bristle	
 Plumose	Mosquito, i.e., (*Aedes aegypti*)	Feather or dense hair on antennae	
 Moniliform	Termite i.e., (*Coptotermes formosanus*)	String of bead appearance	
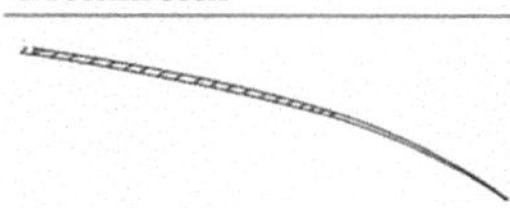 Setaceous	Cockroach i.e., (*Gryllotalpa gryllotalpa*)	Hair like and gradually taper from base to the tip	
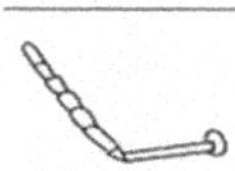 Geniculate	Ants i.e., (*Oecophylla longiceps*)	Bend like knee or elbow	

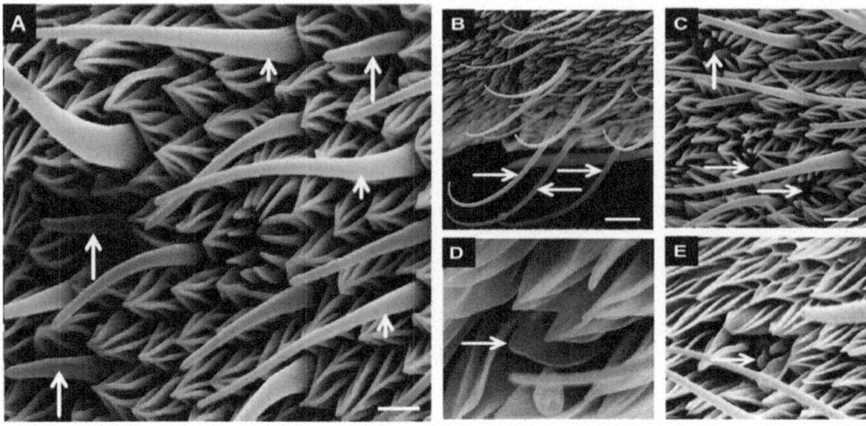

Fig. 2 Different antennal olfactory sensilla morphology in female *Spodoptera littoralis*. (**a**) Basiconic sensilla (long arrows), short trichoid (short arrows). (**b**) Long-trichoid sensilla. (**c**) Coeloconic sensilla (arrows). (**d**) Auricilic sensilla (arrow). (**e**) Grooved peg sensilla (arrow). (Adapted from Binyameen et al. 2012)

Antheraea polyphemus Cramer, ODE is an esterase (Vogt and Riddiford 1981), whereas in the case of *M. sexta* L. this is an aldehyde oxidase. Usually, ODEs can make more than one million receptor sites free in less than 0.5 s (Rybczynski et al. 1989). Sensory neuron membrane proteins (SNMPs) are linked to the CD36 gene family, which is responsible for the production of membrane-embedded glycoproteins. These glycoproteins help in the transportation of fatty acids, lipoproteins, and oxidized phospholipids (Stengl 2010; Silverstein and Febbraio 2009). SNMPs have two subfamilies: SNMP1 and SNMP2. SNMPs might lower the energy barrier of pheromone from pheromone receptors for association and dissociation (Li et al. 2014). OSNs are bipolar; dendrites sink in the sensilla, while axons project into the antennal lobes (ALs), the primary olfactory hubs of the insect brain. Attachment of odour molecule to the dendrites creates potential in the dendritic transmembrane, known as receptor potential. Development of this potential makes the membrane depolarizes and opens the voltage-gated ion channels for the movement of ions.

Bidirectional ions move through the membrane which creates an action potential (AP), which travels in the membrane called a nerve impulse carrying information about the odour molecule. Strength and frequency of the AP depend on the strength of odour reception on the antenna (Gu et al. 2009). OSNs in larvae are like adults OSNs in the structure; however, the main difference is the number of OSNs; more in adults than in larvae. The difference in numbers of OSNs justifies the significance of OSNs in the adults as they have to capture odourant from a long distance for locating food and a potential mating partner as well as oviposition site for laying eggs (Kreher et al. 2005; Fishilevich et al. 2005) (Fig. 3).

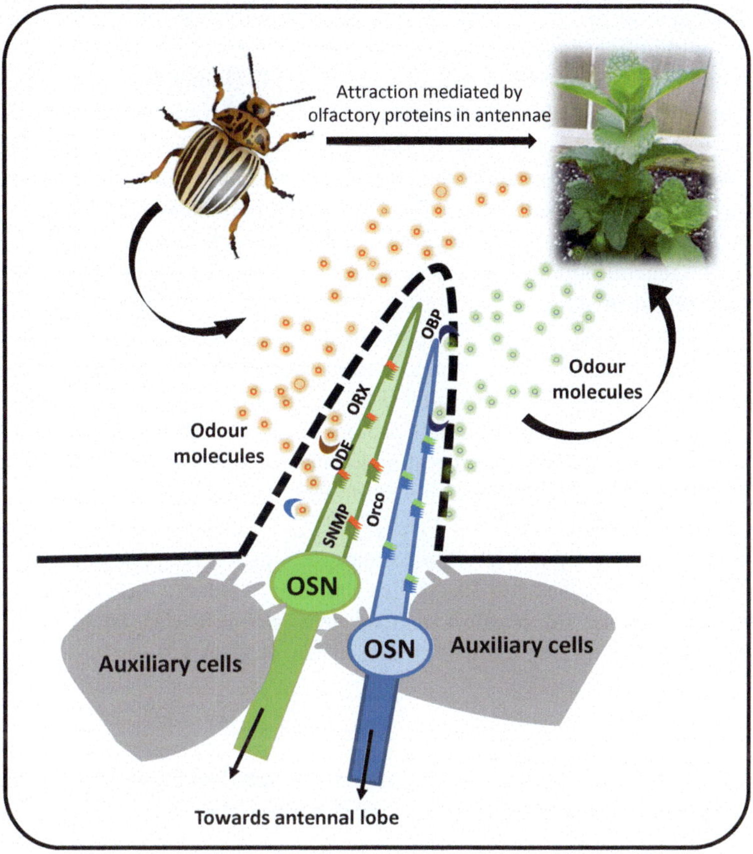

Fig. 3 Olfactory proteins present in insect antennae involved during host plant selection process in insects. OBPs bind to the hydrophobic host odour molecules and aid their transportation to odourant sensory neuron (OSN) membranes. Movement of odour molecules toward chemoreceptors is facilitated by sensory neuron membrane proteins (SNMPs). Response to a particular odour is mediated by odourant receptor complex. Once the desired signal is conveyed, odourant degrading enzymes (ODEs) degrades the odour molecules. OSNs are surrounded by auxiliary cells, which produces OBPs and lymph. (Adopted with modification from Gonzalez 2017)]

2.3 Central Nervous System

All OSNs have a projection of axon that forms a synaptic connection with the antennal lobes (ALs). ALs have both sensory and motor neurons. They are responsible for antennal movement and response against stimuli. ALs are spherical condensed neuropil structures, known as glomeruli, and their numbers vary in each species like 60 in *Bombyx mori* L., and *M. sexta*, 50 in *Drosophila melanogaster* Meigen, 70 in *Mamestra brassicae* L., 60 in *P. brassicae* have, and 1000 in *Locusta migratoria* L. (Visser 1986; Renou 2014). However, intraspecifically, number, location, and size of all glomeruli are similar (Hu et al.

2011; Watanabe et al. 2010). The functional specialization of glomeruli is quite interesting within Lepidoptera, where strong sexual dimorphism is observed in ALs. A group of identified glomeruli is making the macro glomerulus complex (MGC), which entirely tuned to the components of the sex pheromone blend while other small glomeruli are for general odour blends (Galizia and Rössler 2010; Ian et al. 2017). OSNs carry the information about general odourants and project into ordinary glomeruli (Sharma and Matsunami 2014). Local interneurons (LNs) connect the glomeruli, and the majority of these LNs are inhibitory (Chou et al. 2010; Reisenman et al. 2008). General odour OSNs show selectivity and their reception range permit the categorization of odourants according to their functional groups. A functional types of the general odour OSNs is perceived by specific glomeruli in the AL to provide information about the odour. Within glomeruli, OSNs make a synaptic connection with the projection neurons (PNs), which transmit the AP to the higher regions of the brain. The higher brain is present in insect protocerebrum and consists of mushroom bodies (MBs), the lateral and superior protocerebrum, and the lateral horn (LH). PNs synapse with thousands of intrinsic neurons, the Kenyon cells, in the calyces in the MBs (Galizia 2014; Wicher 2015). Decision-making, learning, and memory signals are going to MBs while innate and instinctive behavioral signals are integrated with LHs (Clark and Ray 2016; Sato and Touhara 2008). Subsequently, the brain analyzes the information, and the response is displayed in the form of behavior by the insect (Takemura et al. 2017; Aso et al. 2014).

2.4 Functional Characterization of Insect Olfactory Receptors Using Different Analytical and Molecular Approaches

In insects, it is known that ORs are accountable for the detection of odourants. From the initial discovery of ORs in *Drosophila melanogaster* (Gao and Chess 1999; Clyne et al. 1999; Vosshall et al. 1999), ORs of insects are subjected to in-depth research, both for basic information about insect sensory systems and for their enormous prospect as novel targets for the development of products that could influence behaviors of harmful crop pests and disease vectors. After the OMICs advancement in current years, more profound studies on ORs have been performed resulting characterization of many ORs each year. Many methodologies for functional characterization of insect ORs have developed, both in vivo and in vitro, which has made the deorphanization of ORs possible (Montagné et al. 2015; de Fouchier et al. 2017; Mathew et al. 2013; Wang et al. 2010).

Single sensillum recording (SSR) is a powerful tool for the in vivo study of ORs. This technique enables us to understand the translation of olfactory signals into the electrical pulse and how insect brain receives them. Precisely, an electrode is entered in the insect sensillum that measures the action potential (AP) difference between the reference electrode and sensillum lymph (Pellegrino et al. 2010; Olsson and Hansson 2013). By screening active volatiles from an odour source having high specificity and sensitivity, OSNs for the different olfactory signal can be distinguished and

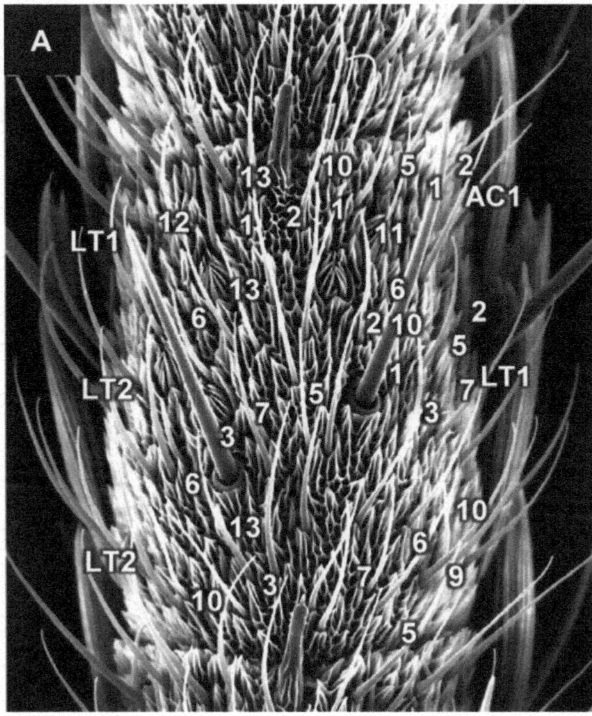

Fig. 4 Single sensillum recording based positional mapping for functional classes of OSNs on the antennae of female moth, *Spodoptera littoralis*. Adapted from Binyameen et al. (2012)

mapped on the insect antenna to develop a functional-morphological antennal map (Fig. 4).

Single sensillum recording has been used extensively to develop functional morphological maps mainly in Dipteran (*Drosophila*, mosquitoes), Lepidopteran (moths), and Coleopteran (bark beetles) insects (Binyameen et al. 2012; Vosshall et al. 1999; Ghaninia et al. 2014; Andersson et al. 2009; Hallem et al. 2004).

Among the "in vivo" expression systems for functional characterization of ORs, the most well-known is the *Drosophila* "*Empty Neuron System* (*ENS*)." ENS is a sophisticated expression system, where a target OR can be selectively expressed in *D. melanogaster* antennae in place of the endogenous DmelOR22a (Gonzalez et al. 2016; de Fouchier et al. 2017). This replacement is made through a driver line (GAL4) containing the Or22a promoter in the Δ halo background and a *Drosophila* fly line with UAS–'OR gene' transgene [Fig. 5]. In this manner, the desired OR gene from any source is introduced next to UAS-promoter to be expressed in *Drosophila* "empty/mutant" ab3A (basiconic sensilla) neuron. However, in Or67d^{GAL4} knock-in system, mutant alleles are generated by replacement of the open reading frame of Or67d with GAL4, and subsequently an independent UAS–"OR gene" transgene insertion is performed into the Or67d^{GAL4} line allowing for the expression of the desired "OR gene" in the antennal trichoid sensilla (at1) (Kurtovic et al. 2007).

For electrical responses of the modified OSNs, SSR technique can be used for the recording of exogenous candidate OR spectrum. This technique offers the authentic

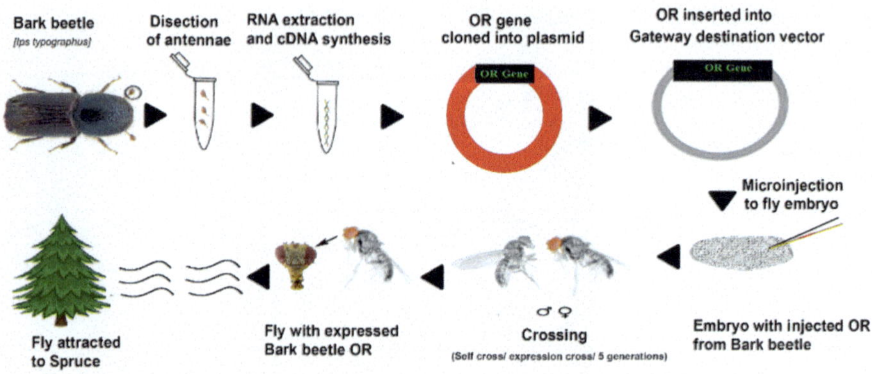

Fig. 5 Graphic representation of the heterologous expression of Bark beetle OR in *Drosophila* empty neuron system (ENS). (For details procedure, refer to Gonzalez et al. 2016)

cellular machinery to process ORs, permitting their appropriate transferring. For the recording of response, stimulus (odourant) is applied in vapor form like natural condition. Expressed OR is positioned in basiconic ab3 sensilla of the *D. melanogaster* antennae (Montagné et al. 2015). In vitro expression strategies are preferable due to their speed and suitability for the characterization of ORs functionally. Among in vitro strategies, heterologous expression of ORs in *Xenopus* oocytes and human embryonic kidney (HEK) cells are the most abundantly used for ORs studies. Other in vitro systems being used for ORs functional characterization are S2, *Sf9*, high-five insect cells, and cell-free expression systems (Montagné et al. 2015). *Xenopus* oocytes are suitable cell type for the study of insect ORs due to their large size and strength. Sensibly, it was used to deorphanize the first insect OR, a *Drosophila* OR (Wetzel et al. 2001). The oocytes turn out to be ligand-responsive upon expression of an OR via injection of their RNA into the cell and can be measured for responsiveness one at a time, in a liquid medium, using the two-electrode voltage-clamp technique (Nakagawa et al. 2005; Sakurai et al. 2004; Fleischer et al. 2018). Cultured cell lines, such as mammalian HEK293 (Fleischer et al. 2018) and insect *Sf9* (Hattori et al. 2016) cells, are also used for the heterologous expressions. Maintaining the cultured cell lines in the laboratory and no need for rearing alive insects are the main benefits of these heterologous expression systems. Moreover, they have stable transform lines, and by an inducing agent, OR expression can be studied (Corcoran et al. 2014). Lately, cell-free expression systems have been developed to produce recombinant insect ORs (Tegler et al. 2015; Gonzalez et al. 2016). These expression systems have the potential for formulating new functional approaches aiding the in-depth study of the insect ORs.

3 Role of Insect Olfaction in Host Plant Selection

Host plant and herbivore interaction are regulated by selection and acceptance of host plant for herbivory and oviposition needs. Phytophagous insects, usually living in a complex habitat consisting of both host and non-host plants, feed on a wide range of host plants from different families. Generally, they tend to be closer to the host plant dominated habitat with less density and odours from non-host plants. To meet this basic criterion, insects use stimuli associated with the host and non-host plants during the process of host plant selection (Renwick 1989; Raffa et al. 2016; Binyameen et al. 2013). Insects are well-equipped with sensory receptors, which enable them to perceive these stimuli to search the desired habitat. A series of selection process begins after habitat selection that ends with the finding of a suitable host plant in the habitat and then rewarding host site [Fig. 6]. Series of these selections (detailed below) are finalized by chemical and visual cues, in which olfaction plays a key role (Bernays and Chapman 2007).

3.1 Habitat Selection

Food and reproduction are the motivating factors while searching for a suitable habitat by a plant-feeding insect. Suitable habitat searching is infrequent by the presence of unwanted plants, host tree defense, intra- and interspecific resource competitions, presence of natural enemies, and environmental factors. Selecting the wrong habitat results in less survivorship and reproduction afterward. Volatiles

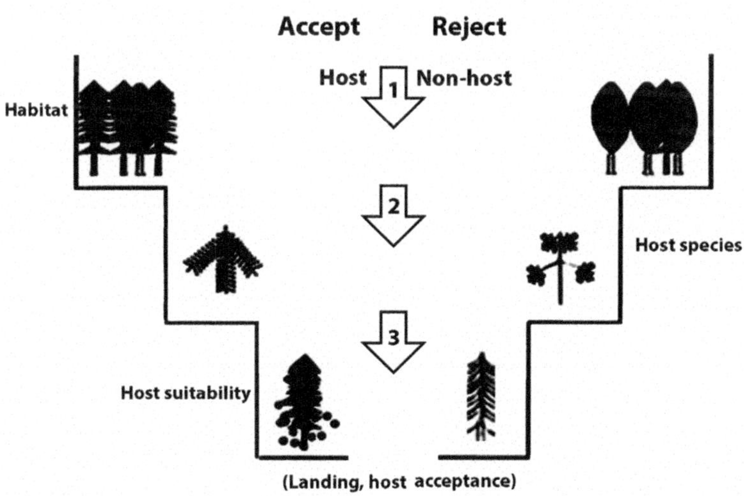

Fig. 6 Illustration of the host tree selection process in a conifer feeding bark beetle. (**1**) Habitat selection. (**2**) Selection of a preferred host species. (**3**) Choice of a suitable host plant before landing on a host plant and picking a site for feeding or oviposition. (Adopted from Zhang and Schlyter 2004)

from the host plants are the main drivers for suitable habitat selection. Host plants, from the nearest habitat, spread their odours from their different body parts, e.g., flower, fruit or leaves, into the air. Olfactory receptors present on the insect's antenna help in detecting these odours over a long distance.

Insect locomotion (walk or flight) get it closer to its nearest available habitat through capturing odours of the plants present in the air (Raffa et al. 2016). Insect moves toward the odours of the host plant, and as it gets closer, odour strength becomes stronger. Chemical olfactory cues for habitat search vary for different insect species. For example, *Plutella xylostella* L. and other crucivores detect the habitat from glucosinolates odours. Hydrolysis of glucosinolates results in a decreased searching ability (Renwick 2002). Adults of *Bembidiono btusidens* Fall also find their habitat by the odour reception arising from the algal microorganisms from the soil near the roots of the host plant (Evans 1983). Olfaction may help an insect to reject a habitat if non-host plant cues are present with host plant odours (Binyameen et al. 2013). For example, cabbage root flies tend to avoid habitats surrounded by other non-host plants due to the presence of the non-host plant volatiles that repels the insect and disturbs the regular feeding and other activities (Kostal and Finch 1994). Moreover, chemical cues from conspecific species are also influencing the habitat selection. In nature, herbivore perceives the pheromones and plant volatiles as an ensemble rather that is from the host or non-host plant. Further decision about habitat acceptance and rejection is taken by the insect brain (Trona et al. 2013). Insect decision is also based on prior feeding (i.e., at the larval stage); however, in some species like *S. littoralis*, previous experience about the habitat is exploited for the better habitat selection. This plasticity in behavior enhances the chance to accept and establish a new habitat for resources or to avoid unfavorable conditions (Lhomme et al. 2017). So, habitat selection is the foundation of the complex insect-host plant interaction.

3.2 Host Plant Selection

Phytophagous insects range from specialists to the generalists. For them, locating a suitable host plant is a challenging duty which is completed in two phases, choosing and selecting. Choosing is done on a far distance with olfactory and visual cues, while the selection phase includes the contact with the host to confirm its acceptability by gustatory cues (Carrasco et al. 2015). Both, olfactory and gustatory senses, integrated work to deal with the plant odours and nutritional value in the CNS while physical characters of the plant are also considered for host suitability. When a right combination of all cues are perceived, the plant is accepted as a suitable host, or in case of the wrong perception, avoidance behavior is elicited (Bruce et al. 2005). Host suitability is in term of host quality and food quantity, among the available host plant species in that habitat. Selection of host plant satisfies different ecological needs of an insect including adult nectar-feeding, oviposition, eggs and larvae protection, and larval fitness. The strength of volatile from host plant indicates the vigor of the plant and its suitability as plants releasing a low concentration of volatile are less preferred by the insect (Bruce et al. 2005). Role of olfaction in host plant selection is proposed by two different hypotheses: species-specific odour recognition that deals with finding host plant from few chemical compounds arising from a particular species or closely related plant species and ratio-specific odour recognition and finding host plant from ubiquitous volatiles and their specific ratios. Supporting evidence is mostly in favor of ratio-specific odour recognition (Morawo and Fadamiro 2016; Pashalidou et al. 2015). *Dryocosmus kuriphilus* Yasumatsu (chestnut gall wasp) was attracted to the intact or old twigs; however, the 1 h old mechanically damaged twigs were found unattractive for the wasp due to change in the ratio of the perceiving compounds (Germinara et al. 2011). Moreover, *Papilio polyxenes* (Stoll) made more landings and laid more eggs on a substrate having the odour of host plant and contact stimulus than the substrate having only contact stimulus. This confirms that an insect mostly relies on its olfactory system as compared to the visual during host-seeking (Feeny et al. 1989). When an insect gets in contact with the host plant, chemical and physical conditions of the plant also come under consideration. For instance, gravid female butterflies move their foreleg chemosensilla on the host plant surface (known as drumming) to examine the suitability of the plant because sometimes immobile neonates are unable to feed due to the phytochemistry of the plant (Ozaki et al. 2011; Ômura 2018).

3.3 Site Selection on a Host Plant

A plant has different phenological states ranging from vegetative to reproductive parts, including flowers and fruits. Most adult insects while looking for nectars visit flower buds or the plant parts secreting extra-floral nectaries. Phytophagous female insects, looking for oviposition site, either land on plant leaves, flowers, fruits, or stems depending on their oviposition preferences and larval feeding habits using their olfactory senses and visual cues. After landing, gustation or mechanoreception gets involved in making a final decision to accept or reject that plant (Webster and Cardé 2017). Volatile odourants released by a plant can act as an attractant for one insect species and deterrent for the other. For example, in brassicaceae crops, glucosinolates, having a pungent smell, serve as an attractant for the different aphid species; *Brevicoryne brassicae* L., *Myzus persicae* Sulzer, and *Lipaphis erysimi* (Kalt). Moreover, the concentration of glucosinolates is higher in leaves than in other parts, so these aphid species prefer to stay and reproduce on the leaves (Khan et al. 2011; Bruce 2014). Likewise, with the senescence of plant, feeding preferences of aphid shift from old outer leaves to the inner younger leaves that have more concentration of glucosinolates. Conversely, bird-cherry-oat aphid, *Rhopalosiphum padi,* or other aphid species that feed on non-brassica host plants do not feed on brassicaceae crops (Bruce and Pickett 2011). Similarly, female of *S. littoralis* (Boisduval) prefers to lay eggs on leaves that are healthy and not releasing any oviposition deterrents (Rojas et al. 2003), because site selection is crucial to reduce the risks for the upcoming generation (Knolhoff and Heckel 2014). Aggregation behavior of herbivores observed in the response of conspecific attraction that is carrying social information for attracted conspecific indicating high-quality resource sites; however, it may also cause competition at high densities (Raitanen et al. 2013). Attraction preferences of an herbivore may change with time ranging from a few minutes to hours, i.e., female of *S. littoralis* switch its attraction from floral parts for the nectar to the leaves for oviposition after few hours of mating (Saveer et al. 2012) . However, male preference for flower does not change after mating (Kromann et al. 2015). Hence, such preferences are often gender and situation-specific.

While searching for a plant site, an herbivore insect always considers the nutritional value of the plant and its oviposition preferences and protection of the eggs and neonates from the natural enemies. For instance, insect feed or lay the eggs mostly on the lower side of leaves; this reduces the chances for encountering the natural enemies. Leafhoppers and other sap-sucking insects try to feed on a rolled leaf to hide their visual presence (Arocha et al. 2005; Stafford et al. 2012). Although the majority of chewing insects primarily feed on leaves, but often at later plant stages, they switch to reproductive parts because reproductive parts such as flower, pod, boll, and fruit are more nutritious and also provide an optimum hiding place from predators and parasitoids. So, insects select the best site according to their perception, plant response, and environmental conditions.

4 Role of Plant Volatiles in Pest Management

Plant volatiles are responsible for mediating interaction with herbivores, predators, pollinators, and even with neighboring plants (Aartsma et al. 2017). Considerable improvements have been made in understanding the functional mechanisms underlying such complex ecological networks. The fundamental reason behind releasing volatiles by plants is to attract pollinators. Plants also release volatiles for their defense to inhibit attraction or repel herbivorous insects (Mithöfer and Boland 2012; Fürstenberg-Hägg et al. 2013). Plants often use volatiles to attract natural enemies to escape the herbivores attack (Kersch-Becker et al. 2017) and to alarm other plants for the plausible upcoming stress (Schuman and Baldwin 2018).

4.1 Oviposition-Induced Plant Volatiles as Oviposition Deterrents

The egg is the most fragile stage in insect life cycle due to its immobility and exposure to natural enemies. Hence, it is essential for the egg to go through embryonic development deprived of maltreatment. Mostly, herbivore female lays its eggs on the plant part after choosing a specific host plant carefully. Oviposition by female insects is the warning of upcoming herbivory, and hence it induces the defense response in plants. Plants synthesize and release oviposition-induced plant volatiles (OIPVs), which induce negative impact to the deposited eggs, repel ovipositing females, or serve as a signal to the natural enemies (Reymond 2013).

Earlier, it was thought that only females produce oviposition deterrents to avoid the oviposition by conspecific females on the same site or plant. However, the first report about OIPVs was reported by Blaakmeer (1994), where found OIPVs from the cabbage leaves carrying recent oviposition of *P. brassicae*.

After recognition of egg-specific elicitors, plant activates its direct and indirect defense against deposited eggs. Direct defense comprises necrosis of the tissues around deposited eggs, neoplasm development, development of tissues to crush the eggs, and production of toxic compounds. The indirect defense comprises a bouquet of volatiles, which attract the natural enemies (Hilker and Fatouros 2016; Tamiru et al. 2015). The general concept is that plant recognizes deposited eggs as microbial pathogen invasion, and in response, innate immune is activated (Reymond 2013). Moreover, the emission of OIPVs was stopped after the removal of eggs from the leaves. More than 20 insect species have been already reported causing OIPVs release by the host plant. However, the synthesis or release of OIPVs has been found in insect species-specific. When eggs were deposited by *P. brassicae,* a specialist herbivore, and *M. brassicae*, a generalist herbivore, on black mustard, only the eggs of *P. brassicae* were accountable for the production of OIPVs (Pashalidou et al. 2013). Similarly, *Sogatella furcifera* (Horváth) eggs on rice plant resulted in an increased production of benzyl benzoate to damage the eggs (Yang et al. 2013).

For calling the natural enemies, plants produce different volatile cues for guiding the natural enemies to the deposited eggs (Reymond 2013; Hilfiker et al. 2014). Oviposition induced terpenoids are also crucial for attracting natural enemies (Hilker and Meiners 2011). Eggs of *S. furcifera* on rice induce changes in the concentration of fatty acids on leaf wax (high concentration of tetratriacontanoic acid and low concentration of tetracosanoic acid) that aids the parasitoids in finding the eggs (Blenn et al. 2012). Essential oils or water extracts from different non-host plants can act as an oviposition deterrent for the many insect pests of agriculture and human health importance (Wallingford et al. 2016; Lamy et al. 2017). So, egg deposition cause changes in the chemistry of the host plant that may induce the species-specific defense mechanisms, directly or indirectly.

4.2 Herbivore-Induced Volatile Organic Compounds and Their Role in Herbivore-Plant Interaction

Herbivore-induced plant volatiles are emitted in response to herbivory and play a crucial role in the interaction with herbivores, natural enemies, and unattacked neighboring plants. Synthesis of HIPVs depends on the stress (insect or disease attack) and species of plants.

HIPVs and Herbivorous Insects' Interaction Many plant species counteract to herbivory with an increased emission of VOCs. However, the response of herbivorous insects against HIPVs is also species-specific. For example, a plant producing HIPVs may be less attractive for some species of aphid (Bernasconi et al. 1998) and spider mites but more attractive for other herbivore insects such as beetles (Kalberer et al. 2001; Halitschke et al. 2008; Bolter et al. 1997), moths (Rojas 1999; Anderson and Alborn 1999), and some other species of aphids (Van Tol et al. 2009). A herbivore insect can obtain many encrypted messages from HIPVs; the indication of defense mechanism activation or overcome of plant defense or presence of other species, i.e., HPVs from aphid infested cabbage plants, is avoided by *M. brassicae* (Rojas 1999). Although herbivore attack is adequate for the production of HIPVs, insect-induced elicitors can influence HIPVs qualitatively as well as quantitatively (Chen et al. 2006; Holopainen and Gershenzon 2010).

HIPVs and Carnivorous Insects' Interaction HIPVs are cues for the natural enemies to find the herbivore. Profile of the HIPVs differs with the herbivore species and quantity depends on the density of the herbivores. Natural enemies can locate their prey or host, i.e., herbivore by using the smell released from the body of a herbivore, but HIPVs released from the plant attacked by the herbivore are much stronger signals than the odour of the herbivore itself. Focusing detectability on HIPVs, the efficiency of the natural enemies is better optimizing (McCormick et al. 2012). HIPVs production from a damaged plant is thought to be an invitation for the carnivorous insects to feed on the herbivore attacking that plant. Many spidery mites and carnivorous insects detect the volatile cues from an herbivore infested plant and perform the role of natural enemies (Arimura et al. 2005). Such insects belong to five orders, namely, Hymenoptera, Coleoptera, Heteroptera, Diptera, and Neuroptera (Dicke and Baldwin 2010). HIPVs attract specialist parasitoids, e.g., HIPVs emitted from the maize leaves infested with *S. littoralis* attracted the *parasitic wasps, Cotesia marginiventris* (Cresson), and *Microplitisru fiventris Kokujev* (D'Alessandro et al. 2006). *The plant*-mediated attraction of natural enemies (parasitoids and predators) helps the infested plant by enhancing parasitism and predation pressure on the feeding herbivores (Kelly et al. 2014; Davidson et al. 2015; Tamiru and Khan 2017). Attraction and discrimination of the natural enemies based on the nutritional value of the herbivore is less documented. Usually, they are attracted toward the HIPVs from a healthy plant than a plant having infestation signals for other herbivores and pathogens (Pierre et al. 2011; Snoeren et al. 2010).

HIPVs and Neighbor Plant Interaction HIPVs transfer message within and between neighboring plants. For example, synthesis of HIPVs from a damaged plant induces the direct (constitutive) or indirect (induced) defense in plants against the herbivore. In addition to these defense mechanisms, HIPVs can induce another indirect defense strategy called associational resistance in conspecific neighboring plants by inducing secretion of extrafloral nectaries (Kost and Heil 2006). For HIPVs production, plants identify the herbivore attack by damage-associated molecular patterns (DAMPs) and herbivore-associated molecular patterns (HAMPs). DAMPs are plant-based products that indicate tissue damage, while HAMPs are herbivore products that come in plant contact during feeding (Turlings and Erb 2018). During the direct defense, a plant prevents herbivore feeding by producing chemicals such as alkaloids, terpenoids, phenylpropanoids, and fatty acids. Defensive proteins such as protease inhibitors also have their role in the direct defense. These proteins target and inhibit the digestive proteins in herbivores (Zhu-Salzman and Zeng 2015). The first evidence of associational resistance among plants was reported in the 1980s (Baldwin and Schultz 1983). Later laboratory and field studies have revealed that HIPVs have plenty of phytohormones (Arimura et al. 2002) and defense-related metabolites such as proteinase inhibitors, terpenoids and phenolic compounds (Kessler 2015; Bruce et al. 2008; Erb et al. 2015; Engelberth et al. 2004).

5 Conclusion and Future Directions

Plant volatiles are synthesized through various biosynthetic pathways under normal and/or stress conditions and act as a medium of communication between plants and arthropods, especially insects. Insects have a well-organized and sophisticated olfactory system for the detection of plant volatiles. Substantial advancement has been reached in understanding the insect olfactory sensory system and its mode of action using advanced physiological and molecular techniques, which have helped to understand the insect-plant interactions. Usually, insects move toward the volatiles of interest. However, often, some learning from previous experience can help, i.e., non-host plant volatile, natural enemies smell, toxic chemical presence, etc. Such behavioral learning and NHVs can be applied to develop sustainable pest management practices. However, the practical feasibility of such applications in agriculture as well as forestry warrants further investigations. While mix cropping and spraying of plant essential oils can be useful in agricultural fields, but in the forest, a frequent spray of volatiles is not very realistic. Hence, mixed forests are plausibly the best way to protect them from pest insects.

Acknowledgments MB, AR, FS is supported by the EU project "EXTEMIT – K," No. CZ.02.1.01/0.0/0.0/15_003/0000433 financed by OP RDE. AR and QA also supported by Internal Grant Agency (IGA) of Czech University of Life Sciences Prague for financial support through Project IGA D_01_18 and IGA C_06_18 respectively. BZU is acknowledged for facilitating MB.

References

Aartsma Y, Bianchi FJ, Werf W, Poelman EH, Dicke M (2017) Herbivore-induced plant volatiles and tritrophic interactions across spatial scales. New Phytol 216(4):1054–1063

Anderson P, Alborn H (1999) Effects on oviposition behaviour and larval development of *Spodoptera littoralis* by herbivore induced changes in cotton plants. Entomol Exp Appl 92 (1):45–51

Andersson MN, Larsson MC, Schlyter F (2009) Specificity and redundancy in the olfactory system of the bark beetle *Ips typographus*: single-cell responses to ecologically relevant odours. J Insect Physiol 55(6):556–567

Andersson MN, Löfstedt C, Newcomb RD (2015) Insect olfaction and the evolution of receptor tuning. Front Ecol Evol 3:53

Ando T, Sekine S, Inagaki S, Misaki K, Badel L, Moriya H, Sami MM, Itakura Y, Chihara T, Kazama H (2019) Nanopore formation in the cuticle of an insect olfactory sensillum. Curr Biol 29(9):1512–1520

Arimura G, Ozawa R, Nishioka T, Boland W, Koch T, Kühnemann F, Takabayashi J (2002) Herbivore-induced volatiles induce the emission of ethylene in neighboring lima bean plants. Plant J 29(1):87–98

Arimura G-i, Kost C, Boland W (2005) Herbivore-induced, indirect plant defences. BBA-Mol Cell Biol L 1734(2):91–111

Arocha Y, López M, Fernández M, Piñol B, Horta D, Peralta E, Almeida R, Carvajal O, Picornell S, Wilson M (2005) Transmission of a sugarcane yellow leaf phytoplasma by the delphacid planthopper *Saccharosydne saccharivora*, a new vector of sugarcane yellow leaf syndrome. Plant Pathol 54(5):634–642

Aso Y, Hattori D, Yu Y, Johnston RM, Iyer NA, Ngo T-T, Dionne H, Abbott L, Axel R, Tanimoto H (2014) The neuronal architecture of the mushroom body provides a logic for associative learning. elife 3:e04577

Baldwin IT (1998) Jasmonate-induced responses are costly but benefit plants under attack in native populations. Proc Natl Acad Sci 95(14):8113–8118

Baldwin IT (2010) Plant volatiles. Curr Biol 20(9):R392–R397

Baldwin IT, Schultz JC (1983) Rapid changes in tree leaf chemistry induced by damage: evidence for communication between plants. Science 221(4607):277–279

Bennett RN, Wallsgrove RM (1994) Secondary metabolites in plant defence mechanisms. New Phytol 127(4):617–633

Bernasconi ML, Turlings TC, Ambrosetti L, Bassetti P, Dorn S (1998) Herbivore-induced emissions of maize volatiles repel the corn leaf aphid, *Rhopalosiphum maidis*. Entomol Exp Appl 87(2):133–142

Bernays EA, Chapman RF (2007) Host-plant selection by phytophagous insects, vol 2. Springer, New York

Beutel RG, Friedrich F, Yang X-K, Ge S-Q (2013) The orders of Hexapoda. In: Insect morphology and phylogeny: a textbook for students of entomology. De Gruyter, Berlin/New York, pp 174–479

Beyaert I, Hilker M (2014) Plant odour plumes as mediators of plant–insect interactions. Biol Rev 89(1):68–81

Binyameen M, Anderson P, Ignell R, Seada MA, Hansson BS, Schlyter F (2012) Spatial organization of antennal olfactory sensory neurons in the female *Spodoptera littoralis* moth: differences in sensitivity and temporal characteristics. Chem Senses 37(7):613–629

Binyameen M, Hussain A, Yousefi F, Birgersson G, Schlyter F (2013) Modulation of reproductive behaviors by non-host volatiles in the polyphagous Egyptian cotton leafworm, *Spodoptera littoralis*. J Chem Ecol 39(10):1273–1283

Blaakmeer A, Hagenbeek D, Van Beek T, De Groot A, Schoonhoven L, Van Loon J (1994) Plant response to eggs vs. host marking pheromone as factors inhibiting oviposition by *Pieris brassicae*. J Chem Ecol 20(7):1657–1665

Blenn B, Bandoly M, Küffner A, Otte T, Geiselhardt S, Fatouros NE, Hilker M (2012) Insect egg deposition induces indirect defense and epicuticular wax changes in *Arabidopsis thaliana*. J Chem Ecol 38(7):882–892

Boatright J, Negre F, Chen X, Kish CM, Wood B, Peel G, Orlova I, Gang D, Rhodes D, Dudareva N (2004) Understanding in vivo benzenoid metabolism in petunia petal tissue. Plant Physiol 135 (4):1993–2011

Bolter CJ, Dicke M, Van Loon JJ, Visser J, Posthumus MA (1997) Attraction of Colorado potato beetle to herbivore-damaged plants during herbivory and after its termination. J Chem Ecol 23 (4):1003–1023

Brito NF, Moreira MF, Melo AC (2016) A look inside odourant-binding proteins in insect chemoreception. J Insect Physiol 95:51–65

Bruce T (2014) Glucosinolates in oilseed rape: secondary metabolites that influence interactions with herbivores and their natural enemies. Ann Appl Biol 164(3):348–353
Bruce TJ, Pickett JA (2011) Perception of plant volatile blends by herbivorous insects–finding the right mix. Phytochemistry 72(13):1605–1611
Bruce TJ, Wadhams LJ, Woodcock CM (2005) Insect host location: a volatile situation. Trends Plant Sci 10(6):269–274
Bruce TJ, Matthes MC, Chamberlain K, Woodcock CM, Mohib A, Webster B, Smart LE, Birkett MA, Pickett JA, Napier JA (2008) Cis-Jasmone induces *Arabidopsis* genes that affect the chemical ecology of multitrophic interactions with aphids and their parasitoids. Proc Natl Acad Sci 105(12):4553–4558
Cardé RT, Willis MA (2008) Navigational strategies used by insects to find distant, wind-borne sources of odour. J Chem Ecol 34(7):854–866
Carraher C, Dalziel J, Jordan MD, Christie DL, Newcomb RD, Kralicek AV (2015) Towards an understanding of the structural basis for insect olfaction by odourant receptors. Insect Biochem Mol Biol 66:31–41
Carrasco D, Larsson MC, Anderson P (2015) Insect host plant selection in complex environments. Curr Opin Insect Sci 8:1–7
Chen H, Jones AD, Howe GA (2006) Constitutive activation of the jasmonate signaling pathway enhances the production of secondary metabolites in tomato. FEBS Lett 580(11):2540–2546
Chou Y-H, Spletter ML, Yaksi E, Leong JC, Wilson RI, Luo L (2010) Diversity and wiring variability of olfactory local interneurons in the *Drosophila* antennal lobe. Nat Neurosci 13 (4):439
Cipollini D, Heil M (2010) Costs and benefits of induced resistance to herbivores and pathogens in plants. Plant Sci Rev 5:1–25
Clark JT, Ray A (2016) Olfactory mechanisms for discovery of odourants to reduce insect-host contact. J Chem Ecol 42(9):919–930
Clyne PJ, Warr CG, Freeman MR, Lessing D, Kim J, Carlson JR (1999) A novel family of divergent seven-transmembrane proteins: candidate odourant receptors in *Drosophila*. Neuron 22(2):327–338
Corcoran JA, Jordan MD, Carraher C, Newcomb RD (2014) A novel method to study insect olfactory receptor function using HEK293 cells. Insect Biochem Mol Biol 54:22–32
Cseke LJ, Kaufman PB, Kirakosyan A (2007) The biology of essential oils in the pollination of flowers. Nat Prod Commun 2(12):1317–1336
D'Alessandro M, Held M, Triponez Y, Turlings TC (2006) The role of indole and other shikimic acid derived maize volatiles in the attraction of two parasitic wasps. J Chem Ecol 32 (12):2733–2748
Das A, Lee S-H, Hyun TK, Kim S-W, Kim J-Y (2013) Plant volatiles as method of communication. Plant Biotechnol Rep 7(1):9–26
Davidson M, Nielsen MC, Butler R, Castañé C, Alomar O, Riudavets J, Teulon D (2015) Can semiochemicals attract both western flower thrips and their anthocorid predators? Entomol Exp Appl 155(1):54–63
de Fouchier A, Walker WB III, Montagné N, Steiner C, Binyameen M, Schlyter F, Chertemps T, Maria A, Francois M-C, Monsempes C (2017) Functional evolution of Lepidoptera olfactory receptors revealed by deorphanization of a moth repertoire. Nat Commun 8:15709
Depetris-Chauvin A, Galagovsky D, Grosjean Y (2015) Chemicals and chemoreceptors: ecologically relevant signals driving behavior in *Drosophila*. Front Ecol Evol 3:41
Dicke M (2000) Chemical ecology of host-plant selection by herbivorous arthropods: a multitrophic perspective. Biochem Syst Ecol 28(7):601–617
Dicke M, Baldwin IT (2010) The evolutionary context for herbivore-induced plant volatiles: beyond the 'cry for help'. Trends Plant Sci 15(3):167–175
Dudareva N, Negre F, Nagegowda DA, Orlova I (2006) Plant volatiles: recent advances and future perspectives. Crit Rev Plant Sci 25(5):417–440

Dudareva N, Klempien A, Muhlemann JK, Kaplan I (2013) Biosynthesis, function and metabolic engineering of plant volatile organic compounds. New Phytol 198(1):16–32
Dweck HK, Ebrahim SA, Khallaf MA, Koenig C, Farhan A, Stieber R, Weißflog J, Svatoš A, Grosse-Wilde E, Knaden M (2016) Olfactory channels associated with the *Drosophila* maxillary palp mediate short-and long-range attraction. elife 5:e14925
Eisenreich W, Schwarz M, Cartayrade A, Arigoni D, Zenk MH, Bacher A (1998) The deoxyxylulose phosphate pathway of terpenoid biosynthesis in plants and microorganisms. Chem Biol 5(9):R221–R233
Engelberth J, Alborn HT, Schmelz EA, Tumlinson JH (2004) Airborne signals prime plants against insect herbivore attack. Proc Natl Acad Sci 101(6):1781–1785
Erb M, Veyrat N, Robert CA, Xu H, Frey M, Ton J, Turlings TC (2015) Indole is an essential herbivore-induced volatile priming signal in maize. Nat Commun 6:6273
Evans W (1983) Habitat selection in the Carabidae. Coleopt Bull 37:164–167
Farré-Armengol G, Peñuelas J, Li T, Yli-Pirilä P, Filella I, Llusia J, Blande JD (2016) Ozone degrades floral scent and reduces pollinator attraction to flowers. New Phytol 209(1):152–160
Feeny P, Städler E, Åhman I, Carter M (1989) Effects of plant odour on oviposition by the black swallowtail butterfly, *Papilio polyxenes* (Lepidoptera: Papilionidae). J Insect Behav 2 (6):803–827
Feussner I, Wasternack C (2002) The lipoxygenase pathway. Annu Rev Plant Biol 53(1):275–297
Fishilevich E, Domingos AI, Asahina K, Naef F, Vosshall LB, Louis M (2005) Chemotaxis behavior mediated by single larval olfactory neurons in *Drosophila*. Curr Biol 15 (23):2086–2096
Fleischer J, Pregitzer P, Breer H, Krieger J (2018) Access to the odour world: olfactory receptors and their role for signal transduction in insects. Cell Mol Life Sci 75:1–24
Fürstenberg-Hägg J, Zagrobelny M, Bak S (2013) Plant defense against insect herbivores. Int J Mol Sci 14(5):10242–10297
Galizia CG (2014) Olfactory coding in the insect brain: data and conjectures. Eur J Neurosci 39 (11):1784–1795
Galizia CG, Rössler W (2010) Parallel olfactory systems in insects: anatomy and function. Annu Rev Entomol 55:399–420
Gao Q, Chess A (1999) Identification of candidate *Drosophila* olfactory receptors from genomic DNA sequence. Genomics 60(1):31–39
Germinara GS, De Cristofaro A, Rotundo G (2011) Chemical cues for host location by the chestnut gall wasp, *Dryocosmus kuriphilus*. J Chem Ecol 37(1):49–56
Ghaninia M, Olsson SB, Hansson BS (2014) Physiological organization and topographic mapping of the antennal olfactory sensory neurons in female hawkmoths, *Manduca sexta*. Chem Senses 39(8):655–671
Gonda I, Bar E, Portnoy V, Lev S, Burger J, Schaffer AA, Ya T, Gepstein S, Giovannoni JJ, Katzir N (2010) Branched-chain and aromatic amino acid catabolism into aroma volatiles in *Cucumis melo* L. fruit. J Exp Bot 61(4):1111–1123
Gonzalez F (2017) Smells of sociality. Diss. (sammanfattning/ summary) Alnarp: Sveriges lantbruksuniv., Acta Universitatis agriculturae Sueciae, 1652–6880; 2017:39; eISBN 978-91-576-8852-1
Gonzalez M, Gaete-Eastman C, Valdenegro M, Figueroa CR, Fuentes L, Herrera R, Moya-León MA (2009) Aroma development during ripening of *Fragaria chiloensis* fruit and participation of an alcohol acyltransferase (FcAAT1) gene. J Agric Food Chem 57(19):9123–9132
Gonzalez F, Witzgall P, Walker WB III (2016) Protocol for heterologous expression of insect odourant receptors in Drosophila. Front Ecol Evol 4:24
Gu Y, Lucas P, Rospars J-P (2009) Computational model of the insect pheromone transduction cascade. PLoS Comput Biol 5(3):e1000321
Halitschke R, Stenberg JA, Kessler D, Kessler A, Baldwin IT (2008) Shared signals–'alarm calls' from plants increase apparency to herbivores and their enemies in nature. Ecol Lett 11(1):24–34

Hallberg E, Hansson BS, Löfstedt C (2012) Sensilla and proprioceptors. In: Volume 2: morphology, physiology, and development, vol 4. De Gruyter, Berlin/Boston, pp 267–288
Hallem EA, Ho MG, Carlson JR (2004) The molecular basis of odour coding in the *Drosophila* antenna. Cell 117(7):965–979
Hallem EA, Dahanukar A, Carlson JR (2006) Insect odour and taste receptors. Annu Rev Entomol 51:113–135
Hansson BS, Stensmyr MC (2011) Evolution of insect olfaction. Neuron 72(5):698–711
Hattori T, Nakanishi K, Mori T, Tomita M, Tsumoto K (2016) The method used to culture host cells (Sf9 cells) can affect the qualities of baculovirus budding particles expressing recombinant proteins. Biosci Biotechnol Biochem 80(3):445–451
Haverkamp A, Hansson BS, Baldwin IT, Knaden M, Yon F (2018) Floral trait variations among wild tobacco populations influence the foraging behavior of hawkmoth pollinators. Front Ecol Evol 6:19
Hilfiker O, Groux R, Bruessow F, Kiefer K, Zeier J, Reymond P (2014) Insect eggs induce a systemic acquired resistance in Arabidopsis. Plant J 80(6):1085–1094
Hilker M, Fatouros NE (2016) Resisting the onset of herbivore attack: plants perceive and respond to insect eggs. Curr Opin Plant Biol 32:9–16
Hilker M, Meiners T (2011) Plants and insect eggs: how do they affect each other? Phytochemistry 72(13):1612–1623
Himanen SJ, Blande JD, Klemola T, Pulkkinen J, Heijari J, Holopainen JK (2010) Birch (*Betula spp.*) leaves adsorb and re-release volatiles specific to neighbouring plants – a mechanism for associational herbivore resistance? New Phytol 186(3):722–732
Holopainen JK, Blande JD (2012) Molecular plant volatile communication. In: Sensing in nature. Springer, New York, pp 17–31
Holopainen JK, Gershenzon J (2010) Multiple stress factors and the emission of plant VOCs. Trends Plant Sci 15(3):176–184
Hu J-H, Wang Z-Y, Sun F (2011) Anatomical organization of antennal-lobe glomeruli in males and females of the scarab beetle *Holotrichia diomphalia* (Coleoptera: Melolonthidae). Arthropod Struct Dev 40(5):420–428
Humphreys JM, Chapple C (2002) Rewriting the lignin roadmap. Curr Opin Plant Biol 5 (3):224–229
Ian E, Kirkerud NH, Galizia CG, Berg BG (2017) Coincidence of pheromone and plant odour leads to sensory plasticity in the heliothine olfactory system. PLoS One 12(5):e0175513
Joseph RM, Carlson JR (2015) *Drosophila* chemoreceptors: a molecular interface between the chemical world and the brain. Trends Genet 31(12):683–695
Kalberer NM, Turlings TC, Rahier M (2001) Attraction of a leaf beetle (*Oreina cacaliae*) to damaged host plants. J Chem Ecol 27(4):647–661
Kelly JL, Hagler JR, Kaplan I (2014) Semiochemical lures reduce emigration and enhance pest control services in open-field predator augmentation. Biol Control 71:70–77
Kersch-Becker MF, Kessler A, Thaler JS (2017) Plant defences limit herbivore population growth by changing predator–prey interactions. Proc R Soc Lond B Biol Sci 284(1862):20171120
Kessler A (2015) The information landscape of plant constitutive and induced secondary metabolite production. Curr Opin Insect Sci 8:47–53
Khan M, Ulrichs C, Mewis I (2011) Water stress alters aphid-induced glucosinolate response in *Brassica oleracea var. italica* differently. Chemoecology 21(4):235–242
Knolhoff LM, Heckel DG (2014) Behavioral assays for studies of host plant choice and adaptation in herbivorous insects. Annu Rev Entomol 59:263–278
Kost C, Heil M (2006) Herbivore-induced plant volatiles induce an indirect defence in neighbouring plants. J Ecol 94(3):619–628
Kostal V, Finch S (1994) Influence of background on host plant selection and subsequent oviposition by the cabbage root fly (*Delia radicum*). Entomol Exp Appl 70(2):153–163
Kostromytska O, Scharf ME, Buss EA (2015) Types and functions of mole cricket (Orthoptera: Gryllotalpidae) antennal and palpal sensilla. Fla Entomol 98(2):593–605

Kreher SA, Kwon JY, Carlson JR (2005) The molecular basis of odour coding in the *Drosophila* larva. Neuron 46(3):445–456

Krishnan B, Wairkar YP (2018) Electroantennograms (EAGs) and Electroretinograms (ERGs) in the genetic dissection of synaptic function in *Drosophila melanogaster*. In: Extracellular recording approaches. Springer, New York, pp 157–184

Kromann SH, Saveer AM, Binyameen M, Bengtsson M, Birgersson G, Hansson BS, Schlyter F, Witzgall P, Ignell R, Becher PG (2015) Concurrent modulation of neuronal and behavioural olfactory responses to sex and host plant cues in a male moth. Proc R Soc Lond B Biol Sci 282 (1799):20141884

Kurtovic A, Widmer A, Dickson BJ (2007) A single class of olfactory neurons mediates behavioural responses to a Drosophila sex pheromone. Nature 446:542–546

Kuzuyama T (2002) Mevalonate and nonmevalonate pathways for the biosynthesis of isoprene units. Biosci Biotechnol Biochem 66(8):1619–1627

Lamy FC, Poinsot D, Cortesero A-M, Dugravot S (2017) Artificially applied plant volatile organic compounds modify the behavior of a pest with no adverse effect on its natural enemies in the field. J Pest Sci 90(2):611–621

Larsson MC, Domingos AI, Jones WD, Chiappe ME, Amrein H, Vosshall LB (2004) Or83b encodes a broadly expressed odourant receptor essential for *Drosophila* olfaction. Neuron 43 (5):703–714

Leal WS (2013) Odourant reception in insects: roles of receptors, binding proteins, and degrading enzymes. Annu Rev Entomol 58:373–391

Lev-Yadun S (2016) Plants are not sitting ducks waiting for herbivores to eat them. Plant Signal Behav 11(5):e1179419

Lhomme P, Carrasco D, Larsson M, Hansson B, Anderson P (2017) A context-dependent induction of natal habitat preference in a generalist herbivorous insect. Behav Ecol 29(2):360–367

Li Z, Ni JD, Huang J, Montell C (2014) Requirement for *Drosophila* SNMP1 for rapid activation and termination of pheromone-induced activity. PLoS Genet 10(9):e1004600

Majeed S, Hill SR, Ignell R (2014) Impact of elevated CO2 background levels on the host-seeking behaviour of *Aedes aegypti*. J Exp Biol 217(4):598–604

Mathew D, Martelli C, Kelley-Swift E, Brusalis C, Gershow M, Samuel AD, Emonet T, Carlson JR (2013) Functional diversity among sensory receptors in a *Drosophila* olfactory circuit. Proc Natl Acad Sci 110(23):E2134–E2143

McCormick AC, Unsicker SB, Gershenzon J (2012) The specificity of herbivore-induced plant volatiles in attracting herbivore enemies. Trends Plant Sci 17(5):303–310

McGarvey DJ, Croteau R (1995) Terpenoid metabolism. Plant Cell 7(7):1015

Missbach C, Dweck HK, Vogel H, Vilcinskas A, Stensmyr MC, Hansson BS, Grosse-Wilde E (2014) Evolution of insect olfactory receptors. elife 3:e02115

Mithöfer A, Boland W (2012) Plant defense against herbivores: chemical aspects. Annu Rev Plant Biol 63:431–450

Mithöfer A, Boland W, Maffei ME (2009) Chemical ecology of plant-insect interactions. In: Molecular aspects of plant disease resistance. Wiley-Blackwell, Chirchester, pp 261–291

Montagné N, de Fouchier A, Newcomb RD, Jacquin-Joly E (2015) Advances in the identification and characterization of olfactory receptors in insects. In: Progress in molecular biology and translational science, vol 130. Elsevier, Amsterdam, pp 55–80

Morawo T, Fadamiro H (2016) Identification of key plant-associated volatiles emitted by Heliothis virescens larvae that attract the parasitoid, *Microplitis croceipes*: implications for parasitoid perception of odour blends. J Chem Ecol 42(11):1112–1121

Moreno A, Palacios I, Blanc S, Fereres A (2014) Intracellular salivation is the mechanism involved in the inoculation of cauliflower mosaic virus by its major vectors *Brevicoryne brassicae* and *Myzus persicae*. Ann Entomol Soc Am 98(6):763–769

Murlis J, Elkinton JS, Carde RT (1992) Odour plumes and how insects use them. Annu Rev Entomol 37(1):505–532

Myrick AJ, Park KC, Hetling JR, Baker TC (2009) Detection and discrimination of mixed odour strands in overlapping plumes using an insect-antenna-based chemosensor system. J Chem Ecol 35(1):118–130

Nakagawa T, Sakurai T, Nishioka T, Touhara K (2005) Insect sex-pheromone signals mediated by specific combinations of olfactory receptors. Science 307(5715):1638–1642

Neilson EH, Goodger JQ, Woodrow IE, Møller BL (2013) Plant chemical defense: at what cost? Trends Plant Sci 18(5):250–258

Niinemets Ü, Kännaste A, Copolovici L (2013) Quantitative patterns between plant volatile emissions induced by biotic stresses and the degree of damage. Front Plant Sci 4:262

Ning C, Yang K, Xu M, Huang L-Q, Wang C-Z (2016) Functional validation of the carbon dioxide receptor in labial palps of *Helicoverpa armigera* moths. Insect Biochem Mol Biol 73:12–19

Ogura K, Koyama T (1998) Enzymatic aspects of isoprenoid chain elongation. Chem Rev 98 (4):1263–1276

Olsson SB, Hansson BS (2013) Electroantennogram and single sensillum recording in insect antennae. In: Pheromone signaling. Springer, Totowa, pp 157–177

Ômura H (2018) Plant secondary metabolites in host selection of butterfly. In: Chemical ecology of insects. CRC Press, Boca Raton, pp 3–27

Ozaki K, Ryuda M, Yamada A, Utoguchi A, Ishimoto H, Calas D, Marion-Poll F, Tanimura T, Yoshikawa H (2011) A gustatory receptor involved in host plant recognition for oviposition of a swallowtail butterfly. Nat Commun 2:542

Paré PW, Tumlinson JH (1999) Plant volatiles as a defense against insect herbivores. Plant Physiol 121(2):325–332

Park KC, Lee JA, Suckling DM (2018) Antennal olfactory sensory neurones responsive to host and nonhost plant volatiles in gorse pod moth *Cydia succedana*. Physiol Entomol 43(2):86–99

Pashalidou FG, Lucas-Barbosa D, van Loon JJ, Dicke M, Fatouros NE (2013) Phenotypic plasticity of plant response to herbivore eggs: effects on resistance to caterpillars and plant development. Ecology 94(3):702–713

Pashalidou FG, Gols R, Berkhout BW, Weldegergis BT, van Loon JJ, Dicke M, Fatouros NE (2015) To be in time: egg deposition enhances plant-mediated detection of young caterpillars by parasitoids. Oecologia 177(2):477–486

Pattrick JG, Shepherd T, Hoppitt W, Plowman NS, Willmer P (2017) A dual function for 4-methoxybenzaldehyde in Petasites fragrans? Pollinator-attractant and ant-repellent. Arthropod Plant Interact 11(5):623–627

Pellegrino M, Nakagawa T, Vosshall LB (2010) Single sensillum recordings in the insects *Drosophila melanogaster* and *Anopheles gambiae*. J Vis Exp 17(36):e1725

Pelosi P, Zhu J, Knoll W (2018) Odourant-binding proteins as sensing elements for odour monitoring. Sensors 18(10):3248

Pierre PS, Dugravot S, Ferry A, Soler R, van Dam NM, Cortesero AM (2011) Aboveground herbivory affects indirect defences of brassicaceous plants against the root feeder *Delia radicum* Linnaeus: laboratory and field evidence. Ecol Entomol 36(3):326–334

Raffa K, Andersson MN, Schlyter F (2016) Host selection by bark beetles: playing the odds in a high-stakes game. In: Advances in insect physiology, vol 50. Elsevier, Amsterdam, pp 1–74

Raitanen J, Forsman JT, Kivelä SM, Mäenpää MI, Välimäki P (2013) Attraction to conspecific eggs may guide oviposition site selection in a solitary insect. Behav Ecol 25(1):110–116

Reineccius G (2016) Flavor chemistry and technology. CRC Press, Boca Raton

Reisenman CE, Heinbockel T, Hildebrand JG (2008) Inhibitory interactions among olfactory glomeruli do not necessarily reflect spatial proximity. J Neurophysiol 100(2):554–564

Renou M (2014) Pheromones and general odour perception in insects. In: Neurobiology of chemical communication. CRC Press, Boca Raton, pp 23–56

Renwick J (1989) Chemical ecology of oviposition in phytophagous insects. Experientia 45 (3):223–228

Renwick JAA (2002) The chemical world of crucivores: lures, treats and traps. Entomol Exp Appl 104(1):35–42

Reymond P (2013) Perception, signaling and molecular basis of oviposition-mediated plant responses. Planta 238(2):247–258

Rinker DC, Zhou X, Pitts RJ, Rokas A, Zwiebel LJ (2013) Antennal transcriptome profiles of anopheline mosquitoes reveal human host olfactory specialization in *Anopheles gambiae*. BMC Genomics 14(1):749

Rohmer M (1999) The discovery of a mevalonate-independent pathway for isoprenoid biosynthesis in bacteria, algae and higher plants. Nat Prod Rep 16(5):565–574

Rojas JC (1999) Influence of host plant damage on the host-finding behavior of *Mamestra brassicae* (Lepidoptera: Noctuidae). Environ Entomol 28(4):588–593

Rojas JC, Virgen A, Cruz-López L (2003) Chemical and tactile cues influencing oviposition of a generalist moth, *Spodoptera frugiperda* (Lepidoptera: Noctuidae). Environ Entomol 32 (6):1386–1392

Rybczynski R, Reagan J, Lerner MR (1989) A pheromone-degrading aldehyde oxidase in the antennae of the moth *Manduca sexta*. J Neurosci 9(4):1341–1353

Sakurai T, Nakagawa T, Mitsuno H, Mori H, Endo Y, Tanoue S, Yasukochi Y, Touhara K, Nishioka T (2004) Identification and functional characterization of a sex pheromone receptor in the silkmoth *Bombyx mori*. Proc Natl Acad Sci 101(47):16653–16658

Sato K, Touhara K (2008) Insect olfaction: receptors, signal transduction, and behavior. In: Chemosensory systems in mammals, fishes, and insects. Springer, Berlin, pp 203–220

Saveer AM, Kromann SH, Birgersson G, Bengtsson M, Lindblom T, Balkenius A, Hansson BS, Witzgall P, Becher PG, Ignell R (2012) Floral to green: mating switches moth olfactory coding and preference. Proc R Soc Lond B Biol Sci 279(1737):2314–2322. rspb20112710

Schiestl FP (2010) The evolution of floral scent and insect chemical communication. Ecol Lett 13 (5):643–656

Schlyter F (2012) Semiochemical diversity in practice: Antiattractant semiochemicals reduce bark beetle attacks on standing trees – a first meta-analysis. Psyche: J Entomol 2012:1–10

Schuman MC, Baldwin IT (2018) Field studies reveal functions of chemical mediators in plant interactions. Chem Soc Rev 47:5338–5353

Sharma R, Matsunami H (2014) Mechanisms of olfaction. In: Bioelectronic nose. Springer, Dordrecht, pp 23–45

Silverstein RL, Febbraio M (2009) CD36, a scavenger receptor involved in immunity, metabolism, angiogenesis, and behavior. Sci Signal 2(72):re3–re3

Snoeren TA, Mumm R, Poelman EH, Yang Y, Pichersky E, Dicke M (2010) The herbivore-induced plant volatile methyl salicylate negatively affects attraction of the parasitoid *Diadegma semiclausum*. J Chem Ecol 36(5):479–489

Song MS, Kim DG, Lee SH (2005) Isolation and characterization of a jasmonic acid carboxyl methyltransferase gene from hot pepper (*Capsicum annuum* L.). J Plant Biol 48(3):292–297

Song L-M, Wang X-M, Huang J-P, Zhu F, Jiang X, Zhang S-G, Ban L-P (2017) Ultrastructure and morphology of antennal sensilla of the adult diving beetle *Cybister japonicus* Sharp. PLoS One 12(3):e0174643

Stafford CA, Walker GP, Ullman DE (2012) Hitching a ride: vector feeding and virus transmission. Commun Integr Biol 5(1):43–49

Stam JM, Kroes A, Li Y, Gols R, van Loon JJ, Poelman EH, Dicke M (2014) Plant interactions with multiple insect herbivores: from community to genes. Annu Rev Plant Biol 65:25–64

Stengl M (2010) Pheromone transduction in moths. Front Cell Neurosci 4:133

Strauss SY, Rudgers JA, Lau JA, Irwin RE (2002) Direct and ecological costs of resistance to herbivory. Trends Ecol Evol 17(6):278–285

Suh E, Bohbot JD, Zwiebel LJ (2014) Peripheral olfactory signaling in insects. Curr Opin Insect Sci 6:86–92

Szyszka P, Galizia CG (2015) Olfaction in insects. In: Handbook of olfaction and gustation. Wiley, Hoboken, pp 531–546

Tabata J (2018) Chemical ecology of insects: applications and associations with plants and microbes. CRC Press, Boca Raton

Takemura S-y, Aso Y, Hige T, Wong A, Lu Z, Xu CS, Rivlin PK, Hess H, Zhao T, Parag T (2017) A connectome of a learning and memory center in the adult *Drosophila* brain. elife 6:e26975

Tamiru A, Khan Z (2017) Volatile semiochemical mediated plant defense in cereals: a novel strategy for crop protection. Agronomy 7(3):58

Tamiru A, Khan ZR, Bruce TJ (2015) New directions for improving crop resistance to insects by breeding for egg induced defence. Curr Opin Insect Sci 9:51–55

Tegler LT, Corin K, Hillger J, Wassie B, Yu Y, Zhang S (2015) Cell-free expression, purification, and ligand-binding analysis of *Drosophila melanogaster* olfactory receptors DmOR67a, DmOR85b and DmORCO. Sci Rep 5:7867

Tholl D, Sohrabi R, Huh J-H, Lee S (2011) The biochemistry of homoterpenes–common constituents of floral and herbivore-induced plant volatile bouquets. Phytochemistry 72 (13):1635–1646

Trona F, Anfora G, Balkenius A, Bengtsson M, Tasin M, Knight A, Janz N, Witzgall P, Ignell R (2013) Neural coding merges sex and habitat chemosensory signals in an insect herbivore. Proc R Soc Lond B Biol Sci 280(1760):20130267

Tumlinson JH (2014) The importance of volatile organic compounds in ecosystem functioning. J Chem Ecol 40(3):212–213

Turlings TC, Erb M (2018) Tritrophic interactions mediated by herbivore-induced plant volatiles: mechanisms, ecological relevance, and application potential. Annu Rev Entomol 63:433–452

Van Tol R, Helsen H, Griepink F, De Kogel W (2009) Female-induced increase of host-plant volatiles enhance specific attraction of aphid male *Dysaphis plantaginea* (Homoptera: Aphididae) to the sex pheromone. Bull Entomol Res 99(6):593–602

Vergoz V, Roussel E, Sandoz J-C, Giurfa M (2007) Aversive learning in honeybees revealed by the olfactory conditioning of the sting extension reflex. PLoS One 2(3):e288

Visser J (1986) Host odour perception in phytophagous insects. Annu Rev Entomol 31(1):121–144

Vogt RG, Riddiford LM (1981) Pheromone binding and inactivation by moth antennae. Nature 293 (5828):161–163

Vosshall LB, Amrein H, Morozov PS, Rzhetsky A, Axel R (1999) A spatial map of olfactory receptor expression in the *Drosophila* antenna. Cell 96(5):725–736

Wallingford AK, Connelly HL, Dore Brind'Amour G, Boucher MT, Mafra-Neto A, Loeb GM (2016) Field evaluation of an oviposition deterrent for management of spotted-wing drosophila, *Drosophila suzukii*, and potential nontarget effects. J Econ Entomol 109(4):1779–1784

Wang G, Carey AF, Carlson JR, Zwiebel LJ (2010) Molecular basis of odour coding in the malaria vector mosquito *Anopheles gambiae*. Proc Natl Acad Sci 107(9):4418–4423

Watanabe H, Nishino H, Nishikawa M, Mizunami M, Yokohari F (2010) Complete mapping of glomeruli based on sensory nerve branching pattern in the primary olfactory center of the cockroach *Periplaneta americana*. J Comp Neurol 518(19):3907–3930

Webster B, Cardé RT (2017) Use of habitat odour by host-seeking insects. Biol Res 92 (2):1241–1249

Wetzel CH, Behrendt H-J, Gisselmann G, Störtkuhl KF, Hovemann B, Hatt H (2001) Functional expression and characterization of a *Drosophila* odourant receptor in a heterologous cell system. Proc Natl Acad Sci 98(16):9377–9380

Wicher D (2015) Olfactory signaling in insects. In: Progress in molecular biology and translational science, vol 130. Elsevier, Amsterdam, pp 37–54

Yang J-O, Nakayama N, Toda K, Tebayashi S, Kim C-S (2013) Elicitor (s) in *Sogatella furcifera* (Horvath) causing the Japanese rice plant (Oryza sativa L.) to induce the ovicidal substance, benzyl benzoate. Biosci Biotechnol Biochem 77(6):1258–1261

Zakir A, Sadek MM, Bengtsson M, Hansson BS, Witzgall P, Anderson P (2013) Herbivore-induced plant volatiles provide associational resistance against an ovipositing herbivore. J Ecol 101 (2):410–417

Zhang QH, Schlyter F (2004) Olfactory recognition and behavioural avoidance of angiosperm nonhost volatiles by conifer-inhabiting bark beetles. Agric For Entomol 6(1):1–20

Zhu-Salzman K, Zeng R (2015) Insect response to plant defensive protease inhibitors. Annu Rev Entomol 60:233–252

Direct and Indirect Defence Against Insects

Anupama Razdan Tiku

Abstract

To obtain nutrients from their hosts, insects feeding on plants use a variety of approaches. Instead of being submissive victims during such interactions, plants respond to these herbivores by producing a variety of defensins in the form of toxins, post-defence peptides, etc. so that the physiological or metabolic processes in the insect could be targeted (direct defence mechanisms) or emit volatile organic compounds to attract natural enemies of these herbivores (indirect defence). Direct and indirect defence mechanisms in plants may act together in concurrence or independent of each other. Both types of defence mechanisms could be either constitutive (always present in the plant) or induced (produced in response to damage or stress caused by herbivores). Evolution of induced chemical defences is linked with formation of chemical substances which are not involved in either photosynthetic or metabolic activities or growth development and reproduction of plants, but these organic compounds also known as secondary metabolites or allelochemicals are only required for defence against herbivores. These metabolites/toxins interfere with insect's metabolism by blocking specific biochemical reactions. These natural defence mechanisms in plants (specially the induced one) could be exploited and used as an important tool in the field of pest management where one can restrict or cut down the use of large amount of insecticides or pesticides in crop fields.

1 Introduction

Plant and insect herbivores have existed together from more than 400 million years ago. During the period of long-term interactions, both of them have coevolved and have also developed various mechanisms to fight against each other. Plants having

A. R. Tiku (✉)
Department of Botany, Ramjas College, University of Delhi, Delhi, India

I. K. Singh, A. Singh (eds.), *Plant-Pest Interactions: From Molecular Mechanisms to Chemical Ecology*, https://doi.org/10.1007/978-981-15-2467-7_8

huge biomass and less mobility protect themselves from insects and predators by producing various chemicals and possessing different types of morphological features or physical traits (Fürstenberg-Hägg et al. 2013). Insect herbivores are challenged by plants either directly by affecting predators' biology through various protective structures present on the surface of plants (thorns, spines, trichomes, thicker leaves, etc.) or by producing lethal allelochemicals such as phenols, quinones, terpenoids, anthocyanins and alkaloids which could kill or slow down the development of insects (War et al. 2012) or indirectly by releasing mixture of volatile substances which specifically attract the predators of these insect herbivores. They can also invite the natural enemies of their herbivores by providing food in the form of extra floral nectar or by providing shelter to them.

Plant-herbivore interaction is a multidisciplinary field of biology which involves various disciplines to explain chemical and ecological processes taking place during this interaction or influencing it. After understanding the nature of gene expression of the plant defensive traits it could be easily applied for designing crop plants with better protection from predators, limiting the usage of harmful pesticides for insect control. Still this race continues as all of them coexist and coevolve. Having ample knowledge of these complex allelochemicals involved in plant-herbivore interaction would help in the production of new resistant varieties of crops.

Pre-formed / Constitutive defense mechanism which include the physical and chemical barriers exist in the plants even before insect attacks the plant whereas inducible defense mechanism gets activated only after insect attacks the plant. Inducible defence occurs in three steps: recognition of pests/pathogens, signal transduction and then production of defence-related chemicals (Dangl and McDowell 2006; Ferry et al. 2004). Inducible system is more important and brings upon less metabolic cost for the host plant. Being a pest-specific mechanism, this defence system has become the focus of research for the last few decades. Inducible defences are mainly of two types: direct and indirect defence mechanisms.

2 Direct Defence Against Insect Herbivores

Plants' morphological traits like thorns, trichomes, cell wall thickness, epidermal cuticular wax and lignification produce the first level of physical barrier for insect herbivores, whereas allelochemicals acting as toxins affect growth and development and reduce digestion in herbivores and act as the next barrier to defend the plants from its predators. These toxic chemicals are more effective when acting together in a synergistic manner whereas they show less effect when working separately. In *Lycopersicon esculentum* protein inhibitors (PIs), phenolics, alkaloids and oxidative enzymes are more effective on insects when working together synergistically, causing difficulty for insect in ingestion, digestion and metabolism (Steppuhn and Baldwin 2007). Another example seen is the case of *Nicotiana* species where nicotine, trypsin and PIs when working together give the best defence response to *Spodoptera exigua* (Duffey and Stout 1996).

2.1 Structural Defences in Plants

In plants, structural defence is provided by change in morphological or anatomical trait which confers fitness advantage for that plant by directly preventing herbivores from feeding on it. These include various types of spines, prickles and thorns (spinescens), hairlike trichomes, toughened and hardened leaves (sclerophylly), formation of cuticular wax, thickness in cell wall due to suberisation and lignification, granular mineral deposition in plant tissues (druses, cystoliths, phytoliths mainly composed of silica and calcium oxalate/carbonate) and "divaricated branching" in some plants where shoots with wiry stems are produced at wide axillary angles so that they interweave and are widespread (Schoonhoven et al. 2005). In sclerophylly, leaves become hard, therefore reducing the palatability and digestibility of the tissues, making it difficult for insects to ingest and digest. Layer of hairs extends from the epidermal cells (trichomes) of the aboveground plant parts (pubescence) such as leaves, stem, fruits, etc. and occurs in various forms and shapes (spiral, stellate, hooked, glandular, etc.) (Hanley et al. 2007).

2.1.1 Role of Trichomes

Trichomes play an important role in defending plants from different types of insect pests in both toxic and preventive ways (Handley et al. 2005). Higher density of trichomes has a negative effect over the feeding, ovipositional behaviour and larval nutrition of insect pests. They also interfere with movement of arthropods and insects on the plant surface, resulting in the limited access to the leaf epidermal surface of these herbivores (Agrawal et al. 2009). Trichomes are either glandular or non-glandular with diverse shapes depending upon their functions. Glandular trichomes provide better defence by secreting mixture of allelochemicals such as alkaloids, terpenoids and flavonoids which are toxic and repulsive and even trap insects, therefore providing a combination of chemical and structural defence. In a few plants, it has been reported that the density of trichomes increases in new leaves after the insect damages or attacks the plant, e.g. in *Salix cinerea* L.

After the damage caused due to attack by Coleopteran or adult leaf beetles (Pbratora vulgatissima L.) and in case of Alnus incana by beetles density of trichomes increases in new leaves. In case of black mustard, both density of trichomes and glucosinolate level increase when attacked by Pieris rapae (Dalin and Björkman 2003). Increased trichome density varies from 25 to 1000% in response to herbivory. In a few cases, herbivore attack induces change in the proportion of glandular and non-glandular trichomes (Agrawal 1999). In case of spider mites, most of their eggs are consumed by predatory mites (*Phytoseiulus persimilis*). When the spider mites attack the plants it leads to the increase in the trichome density of the host plant. This increase in trichome density provides shelter to the predatory mites (indirect defense).

2.1.2 Role of Epicuticular Waxes

The cuticle portion of many vascular plants gets covered with a layer made up of hydrophobic substances, and these waxy coatings therefore play a very important role in protecting the plants from dehydration, plant microbe invasion and insect herbivory. The thickness of these wax layers varies from plant to plant, but the thicker the wax layers are, better the plants are protected from herbivores as it has been studied in various cases where insects experience difficulties in eating leaves covered with wax crystals, for example, epicuticular wax crystals present on the surface of young leaves of *Eucalyptus globulus* make the leaves' surface slippery due to which insect herbivores (psyllids) are unable to adhere to the leaf surface, decreasing the survival rate of these insects as they die due to starvation (Chen 2018).

2.1.3 Role of Leaf Toughness

The cell wall of plants is strengthened due to deposition of complex polymers such as cellulose, lignin, callose and suberin along with sclerenchymatous tissue, making the plant's surface tough and resistant to mechanical injury caused by tearing action of insect mandibles and penetration of piercing-sucking mouthparts of insect herbivores. *Spodoptera exigua* (beet armyworm) takes thrice more time swallowing celery leaves (*Apium graveolens*) than eating *Chenopodium murale* leaves because the former have 1.5 times tough surface than the latter (Hanley et al. 2007).

3 Plant Defensive Compounds/Allelochemicals

Plants produce organic compounds (secondary metabolites) as by-products during the synthesis of primary metabolites, and these compounds have no role in plants' growth and development. They perform various physiological functions such as UV protectants, as tools for storage and transport of nitrogen and as attractants for pollinating and seed-dispersing animals, though their main function is to work as plant defence chemicals against pathogens and herbivores (Mao et al. 2007; Tiku 2018). Structures of more than 50,000 plant defence chemicals have been studied and this amount is only a small fraction of the total secondary metabolites/defensins existing in nature. These metabolites or defensive chemicals are known as phytoanticipins if they are constitutive in nature and phytoalexins if they are induced in response to insect or pathogen attack. When herbivores attack, beta-glycosidase activates phytoanticipins, resulting in the release of various biocidal aglycones. Examples of some common phytoanticipins are benzoxazinoids (BXs) present in members of Gramineae. BX glycosides are hydrolysed by beta-glycosidase, leading to the production of biocidal aglycone during tissue damage caused by insect herbivores. Glycosinolate is another phytoanticipin that is hydrolysed by myrosinases during disruption of tissues and at the time of herbivory. Induced secondary metabolites/phytoalexins include alkaloids, terpenoids, isoflavonoids, etc. (Table 1). In *Zea mays*, resistance to corn earworm (*Helicoverpa zea*) occurs only in the presence of secondary metabolites like C-glycosyl flavone maysin and chlorogenic acid which is a phenylpropanoid product (Nuessly et al. 2007). In *Sorghum bicolor*, 4, 4-dimethyl cyclooctene is formed to protect it from *A. soccata* (shoot fly) (Chamarthi et al. 2011). Secondary metabolites are usually responsible for generating direct defence mechanisms in plants, but it is required to

Table 1 Role of plant secondary metabolites against insect herbivores

Class of the chemical	Mode of action	Plant/insect system
Tannins (polyphenol)	?	Oak/multiple insects
Maysin and apimaysin (flavones)	?	Corn/corn earworm (*Helicoverpa zea*)
Isoorientin (flavonoid)	Toxicity	Corn/corn earworm (*Helicoverpa zea*)
Various flavonoids	Toxicity (Ca2C-ATPase inhibitor?)	Roostertree/bean weevil (*Callosobruchus chinensis*)
Glyceollin (isoflavonoid)	?	Soybean/coleopterans
Lignin	Anti-nutrition?	Aspen/gypsy moth (*Lymantria dispar*)
Sinalbin (thioglucosides)	Repellent	Crucifers/multiple insects
Sinigrin (thioglucosides)	Repellent	Crucifers/multiple insects
Dhurrin (cyanogenic glucoside)	Toxicity (hydrogen cyanide, HCN)	*Arabidopsis*/flea beetle (*Phyllotreta nemorum*)
DIMBOA (hydroxamic acid)	Toxicity and repellent	Corn/multiple insects
DIBOA (hydroxamic acid)	Toxicity and repellent	Rye/multiple insects
Monoterpenes and diterpenes	Toxicity	Spruce/white pine weevil (*Pissodes strobi*)
Saponin tomatine (glycosylated triterpenes)	Toxicity (disrupting cell membrane)	Tomato/multiple insects
Sesquiterpenes	Repellent	Tobacco/multiple insects

reveal their unidentified or emerging signalling pathways. Scientific techniques such as mass spectrometry, chemical profiling and high-throughput sequencing used for gene expression analysis could be used to identify new signalling molecules which are involved in providing resistance to plants from insects and other herbivores. These techniques could be used in identification of genes and enzymes present in plant cells which are involved in biosynthesis of defence-related chemicals in plants. These bioactive specialised compounds target biological system (nervous, digestive and endocrine) of insect herbivores and also provide specific order, taste and colour to plants (Engelberth 2006). They act as repellents for general insects and attractants for specific insects.

3.1 Role of Alkaloids

They are bioactive natural products well-known for their metabolic effects in mammals, for example, nicotine, morphine, caffeine, cocaine, etc. (Howe and Jander 2008), and work as deterrents for insect herbivores. They are synthesised from amino acids in the roots, and around 15,000 different alkaloids have been reported in 20% of all the vascular plants. Alkaloids derived from quinolizidine (sparteine and cytisine) are more effective against insects, for example, "demissine" (Fig. 1) *present* in nightshade potato (*Solanum demissum*) protects it from Colorado beetle

Fig. 1 Demissine (alkaloid)

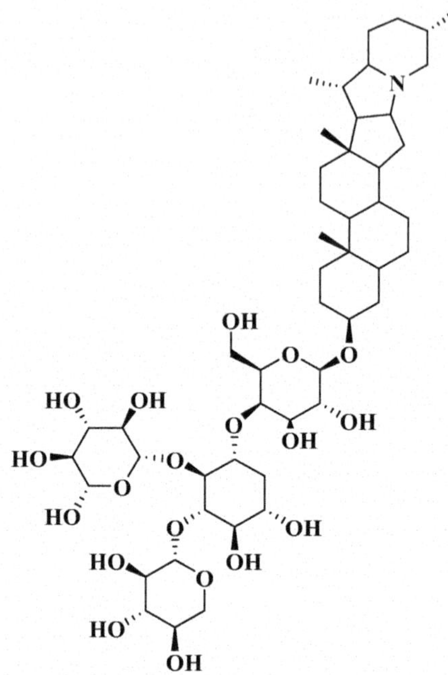

Fig. 2 DIBOA (benzoxazinoide)

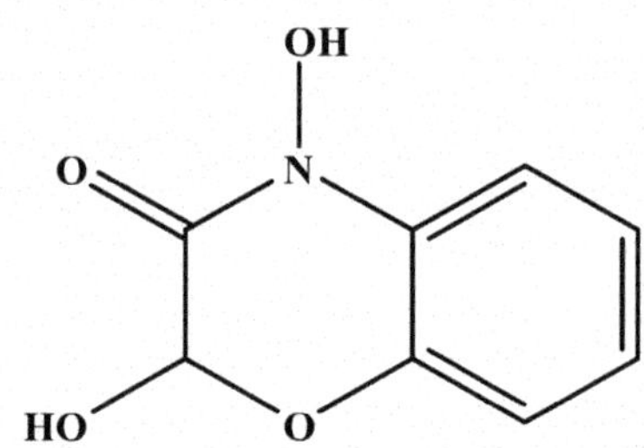

(*Leptinotarsa decemlineata*) and potato leafhopper (*Empoasca fabae)* (Ziegler and Facchini 2008). PAs (pyrrolizidine alkaloids) are synthesised from arginine and ornithine amino acids occurring naturally in many plants as non-toxic N-oxides (Fattorusso and Taglialatela-Scafati 2007).

Once they reach the digestive tract of insect herbivores under alkaline conditions, they are quickly reduced and form uncharged, toxic and hydrophobic alkaloids, which can easily pass through membranes. For example, PAs are extremely toxic to milkweed bug (*Oncopeltus fasciatus)* and to aphid (*Rhopalosiphum padi*) (Petterson et al. 1991).

3.2 Role of Benzoxazinoides

They are another group of defence-related bioactive compounds like DIBOA-Glc (2,4-dihydroxy-1,4-benzoxazin-3-one-glucoside) (Fig. 2) and DIMBOA-Glc

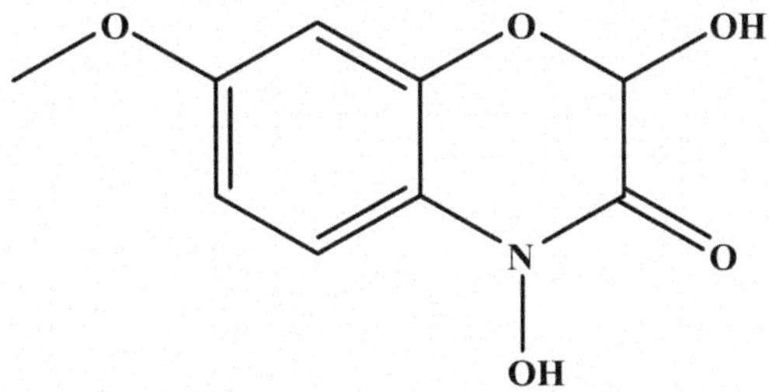

Fig. 3 DIMBOA (benzoxazinoide)

(2,4-dihydroxy-7-methoxy-1,4-benzoxazin-3-one-glucoside (Fig. 3) which are present in species of family Poaceae and synthesised from indole-3-glycerol phosphate. BX1 cleaves off glycerol phosphate and BX1-BX9 catalyses the conversion. Formation of DIBOA is catalysed by BX2-BX5 and glycosyl, and glycosyl group is added to stabilise the reaction by BX8/BX9. DIBOA-Glc is converted to DIMBOA-Glc in the presence of BX6-BX7 (Dutartre et al. 2012). In maize, formation of free indoles is catalysed by indole-3-glycerol phosphatase lyase which is a homologue of BX1 and then activated by volicitin (Møller and Seigler 1991). DIMBOA works against several insect pests like maize plant louse (*Rhopalosiphum maidis*) and first-brood European corn borer (*Ostrinia nubilalis*). DIMBOA-Glc also gets converted into HDMBOA-Glc (2-beta-d-glycopyranosyloxy-4,7-dimethoxy-1,4-benzoxazin-3-one) due to the action of jasmonic acid-induced 4-O-methyltransferase, and it acts as a strong deterrent for *S. littoralis* and *S. frugiperda* (Glauser et al. 2011).

3.3 Role of Cyanogenic Glycosides

They are present in more than 2600 species belonging to more than 150 families and 550 genera including maximum classes of vascular plants. Cyanogenic glycosides are glucosides derived from amino acids which are aromatic or branched-chain, for example, tyrosine from which "dhurrin" (Fig. 4) is derived in *Sorghum bicolor* (Dustan and Henry 1902), isoleucine and valine giving rise to *linamarin* and *lotaustralin* glucosides in lotus (*Lotus japonicus*) *and cassava (Manihot esculenta)* (McMahon et al. 1995; Forslund et al. 2004) and phenylalanine forming *amygdalin* and *prunasin in the family Rosaceae.* Before herbivory in intact plant tissues, they are stored in vacuoles of plant cells. Once the plant tissues are damaged due to attack of insects/herbivores CNglcs get exposed to β-glycosidases present in plastids or apoplasts, leading to the formation of sugar and cyanohydrin due to hydrolysis. Both of them decompose spontaneously into toxic hydrogen cyanide and aldehyde or ketone. Volatile toxic HCN is an inhibitor of cytochrome C oxidases acting in mitochondrial respiratory pathway (Brattsten et al. 1983; Davis and Nahrstedt 1985). In *M. esculenta*, tubers presence of CNglcs increases resistance towards cassava burrower bug (*Cyrtomenus bergi). Amygdalin* and *prunasin* are present in bitter almonds, making them resistant to larvae of flat headed wood borer (*Capnodis tenebrionis*) (Malagón and Garrido 1990).

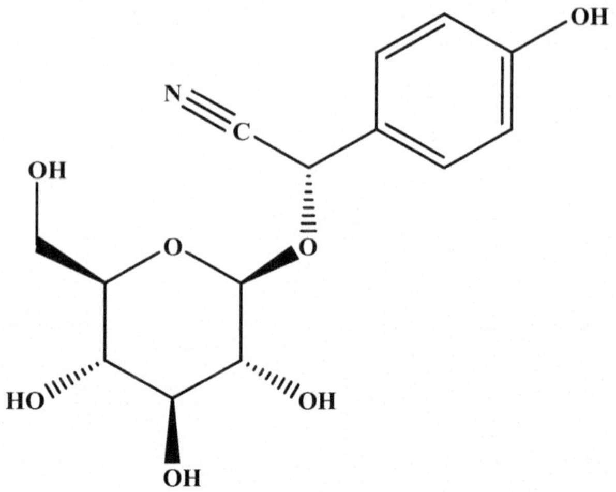

Fig. 4 Dhurrin (cyanogenic glucose)

3.4 Role of Glucosinolates (GSL)

They are sulphur- and nitrogen-containing compounds specific to Cruciferae and Capparales. GSL and glucosides derived from amino acids are divided into four groups depending upon amino acid side chain. Fifty per cent are aliphatic GSL derived from indole and methionine, 10% indole GSL formed from tryptophan and 10% aromatic GSL synthesised either from tyrosine or phenylalanine, and the rest (30%) have unknown biosynthetic origin. GSLs are also located in the vacuole of plant cells where they are protected from myrosinases (thioglucosidases). After insect attack due to tissue damage, myrosinase hydrolyses GSL, giving rise to toxic by-products like isothiocyanates (R-N=C=S), thiocyanates (R-S-C ≡ N) and nitriles. These break-down products are the part of vegetable oils (mustard oil) and give flavour to vegetable foods and act as toxins/feeding repellents for insects and other herbivores (Bennett et al. 1994), for example, in *Brassica napus* (rapeseed), presence of GSL protects them from field slug (*Deroceras reticulatum*). In old cotyledons of *Sinapis alba* (white mustard), there is less quantity of *sinalbin* (GSL) (Fig. 5), and these parts are more prone to attack by flea beetle (*Phyllotreta cruciferae*) (Bodnaryk 1991).

3.5 Role of Nonprotein Amino Acids

Plants mainly members of Leguminosae produce high concentrations of toxic nonprotein amino acids containing arginine analogue canavanine (Fig. 6) which along with its break-down product canaline are effective substrates for enzymes utilising arginine and ornithine amino acids. In most of the organisms, arginyl-tRNA synthetase is unable to distinguish between arginine and canavanine so that the latter gets incorporated into proteins, leading to deleterious effects (Rosenthal 1991).

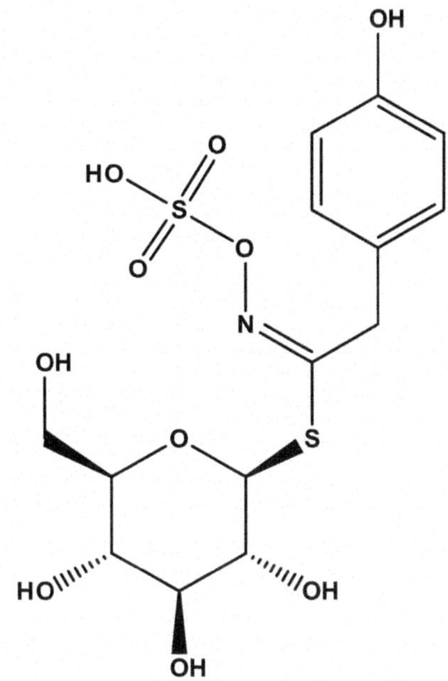

Fig. 5 Sinalbin (glucosinolate)

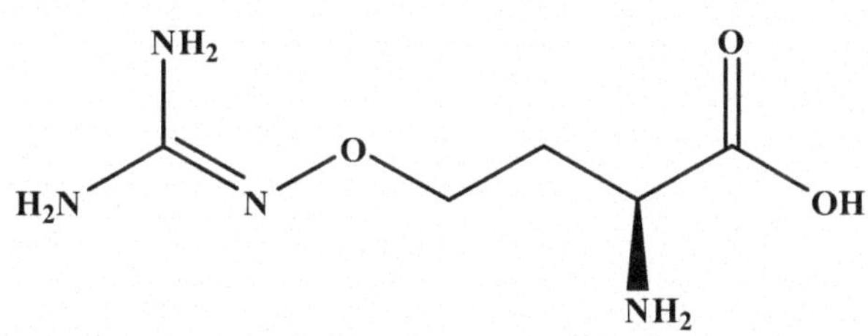

Fig. 6 Canavanine (nonprotein amino acid)

Those nonprotein amino acids are non-toxic for the plants producing it but work as deterrent for their predators.

3.6 Role of Phenolics

Phenolics are the most important secondary metabolites produced by most of the plants, and they are the widespread group of defensive compounds which protects them from insects and other herbivores. An example of a phenolic heteropolymer is lignin which is an important defensin that protects plants from insects. Lignin deposition increases leaf toughness and decreases nutritional content on the leaf. In some plants, it has been marked that lignin synthesis is induced during insect or pathogen attack, and its rapid deposition restricts the further growth of herbivores. At this time, lignin-associated genes like CAD/CAD-like genes are overexpressed in the plant tissues which are attacked by herbivores (Schmelz et al. 2007a, b). Phenols

Fig. 7 Salicylic acid (phenol)

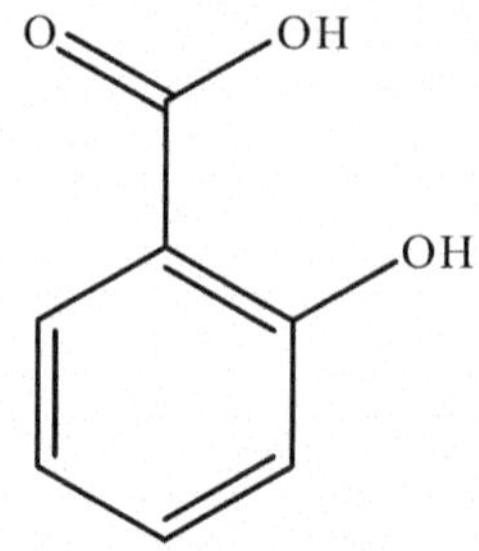

get oxidised into quinones, and the reaction is catalysed by two enzymes: PPO (polyphenol oxidase) and POD (peroxidase). Quinones being toxic in nature work as plant defence chemicals by binding covalently to the leaf proteins and hinder the digestion of such leaves by insect herbivores or exhibit direct toxicity to insects (Bhonwong et al. 2009). After getting alkylated, amino acids of leaf proteins lose their nutritional value, hence negatively affecting insect growth and development. Phenols also take part in cyclic reduction of ROS (reactive oxygen species), for example, H_2O_2, hydroxide radicals, superoxide anion and singlet oxygen, and these reduced chemicals then activate chain/series of reactions inside the plant cells, leading to activation of enzymes involved in plant defence mechanism (Doss et al. 2000). In the leaves of *Salix* plant, salicylates (simple phenolics) (Fig. 7) work as deterrent for herbivores like *Operophtera brumata* L. (Treutter 2006). Salicylates (derived from benzoic acid) produced in *Salix* species leaves restrict the growth and development of *Operophtera brumata* (oak moth) (Ruuhola et al. 2001). Catechol-based phenolics are present in leaves of *Fragaria* (strawberry) and provide protection from two-spotted spider mite (*Tetranychus urticae*) (Luczynski et al. 1990). These phenolics bind covalently to the mites' digestive enzymes inactivating them. Gossypol is a phenolic present in cotton plant and is deleterious to many insects like *Heliothis virescens* (tobacco budworm), *H. zea* (bollworm), etc. (Maxwell et al. 1965).

3.7 Role of Flavonoids

They are derivatives of simple phenols, and their chemical structure is based on 15-carbon skeleton in which two aromatic rings are connected by three-carbon bridge (C6-C3-C6). They have been classified under different subgroups, i.e. flavonols, flavones, chalcones, flavandiols, proanthocyanidins and their derivatives (anthocyanidins) as well as condensed tannins. Whenever there is any injury or infection in plant tissue, synthesis of flavonoids increases. More than 6000 flavonoid compounds have been reported in plants. Their main role is to protect plants from various biotic and abiotic stresses (UV radiation, insect pests and microbes) and also during plant-environment interactions. Flavonoids and isoflavonoids both influence the behaviour, growth and development of insect herbivores as they are toxic to insect cells and interact with different cellular enzymes through complexation. They also hunt for the free radicals like ROS and supress their formation chelating the metals (Treutter 2006). The toxic nature of some flavones and flavonoids (5-methoxyisoron chocarpin and 5-hydroxyisoderricin) has been reported in *Tephrosia vogelii* Hook, *T. purpurea* L. and *T. villosa* specially affecting the feeding herbivores like *Spodoptera* species. Transcription factors controlling flavonoid production if overexpressing can also increase the resistance in plants, and it has been reported in *Arabidopsis* where overexpression makes the plant resistant to *Spodoptera frugiperda* (Johnson and Dowd 2004). In different plants, a range of flavonoids have been reported (licoisoflavonA8B; Angustone A, B and C; luteone; and wighteone) that act as feeding deterrent to insect herbivores and other pathogens (Lane et al. 1987). Similarly in chickpea (wild variety), isoflavonoids like judaicin, 2-methoxy-judaicin, judaicin-7-0-glucoside and maackiain act as defensins against *Helicoverpa armigera*. Maackiain and judaicin also restrict feeding by *Spodoptera frugiperda* and *S. littoralis*, whereas alliarinoside and cyanopropenyl glycoside were found to be deterrent to *Pieris napioleraceae* L. (American butterfly) (Simmonds and Stevenson 2001). Isovitexin-6-D-β-glucopyranoside which is a flavone glycoside is a feeding deterrent to late instars (Renwick et al. 2001).

3.8 Role of Tannins

They are astringent and bitter polyphenolic compounds found throughout the plant kingdom, occurring mainly in woody plants and forest tress, and act as feeding deterrent to many insect herbivores. Their deterrent function is mainly due to their ability to precipitate proteins and to be oxidatively active. Tannins are chemically diverse but have the ability to bind proteins which is an important factor in defence against insect pests. They also bind with the digestive enzymes of herbivores by hydrogen bonding or covalent bonding of proteins-N4 groups and precipitate them (Sharma and Agarwal 1983; Barbehenn and Peter Constabel 2011). Tannins can also chelate the metal ions required to activate the digestive enzymes of herbivores. Their toxic chemical nature is mainly responsible for protecting plants from insect herbivores. Inside the gut of insects, tannins oxidise due to high pH and form quinones, semiquinone radicals and reactive oxygen species which cause toxicity in insects. Dismutation of semiquinones to quinones leads to accumulation of hydrogen peroxide (stable oxidant), which causes oxidative damage to epithelial cells (Barbehenn and Peter Constabel 2011). Once taken in by the insects, tannins adversely affect the gut wall permeability (Clausen et al. 1992). Recent research studies have shown that condensed tannins perform better role in plant defence and are more deleterious to herbivores. They are oligomeric or polymeric flavonoids, which are also known as proanthocyanidins.

Tannin production is induced in *Populus tremuloides* Michx. When the leaves get damaged due to herbivory, there is transcriptional activation of flavonoid pathway. Genes responsible for synthesis of tannins have been identified, and their activity is mainly due to expression of PtMYB134 (a condensed tannin regulatory gene), which gets activated after tissue damage (Mellway et al. 2009). Leaves of *Quercus robur* L. produce series of condensed tannins ((+)-catechin, (+)-gallocatechin and vanillin) and get protected by *Operophtera brumata* (winter moth) larvae (Feeny 1968). Groundnut (*Arachis* sp.) leaves produce procyanidin polymers which are found to be toxic for *Aphis craccivora* (Grayer et al. 1992).

3.9 Role of Terpenoids

They are synthesised from acetyl-CoA or glycolytic intermediates. Their classification is based upon the number of isoprene units or five-carbon elements (CH_3-CH_2-CH-$(H_3C)_2$), monoterpenes (10-C), sesquiterpenes (15-C), diterpenes (20-C), sesterterpenes (25-C), triterpenes (30-C) and terpenes with "n" number of isoprene units known as polyterpenes (Von Dahl et al. 2007). They have a major role in plants' defence mechanism as they act as antifeedant, toxins, repellents and modifiers of insect development. In plants, terpenes are mainly present in the form of essential oils (mixture of mono- and sesquiterpenes), having repellent and toxic effect on insects. "Limonene" terpenoid (Fig. 8) is present in citrus plants, and it repels leafcutter ant, *Atta cephalotes* (Engelberth 2006). In conifers like fir and pine, "monoterpenes" are produced which are toxic to many insects including different species of bark beetle (Trapp and Croteau 2001). Conifers also produce "oleoresins" which are mixture of terpenoids and phenolics stored in ducts (high pressurised intercellular spaces). During the damage caused by a herbivore, the duct breaks and the resin comes out trapping the insect (stem-boring bark beetle). On exposure to air, highly volatile mono- and sesquiterpenes present in resin get evaporated, leaving insect trapped in solid resin acids (Phillips and Croteau 1999). Terpenoids provide better defence to plants when they act synergistically, for example, binary mixture of trans anethole and thymol with citronellal have ten times stronger effect against *Spodoptera litura* (tobacco cutworm) (Hummelbrunner and Isman 2001).

3.10 Role of Plant Defensive Proteins

Nutritional requirements of insects are the same as that of animals, and any indigestion caused by utilisation of plant proteins will show drastic effects on insects' physiology; therefore, they are always in search of healthy host plants that can provide them proper food for nutrition, mating and oviposition and also fulfil the requirements of offspring. Many plant proteins ingested by insects are stable and

Fig. 8 Limonene (terpenoid)

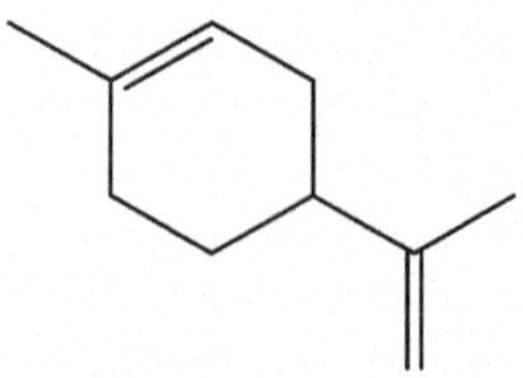

remain intact in the midgut crossing the gut wall into haemolymph. If there is any alteration in the amino acid structure of these proteins, it also alters its function, and this change occurs due to stress caused by insect attack which brings quantitative and qualitative changes in protein, which in turn play an important role in signal transduction and oxidative defence (Table 2). Complete knowledge of toxicity and mechanism of plant resistance proteins (PRPs) is possible after understanding protein structure and post-translational modification. Latest research in the field of proteomics, metabolomics and microarray have revealed information regarding a wide spectrum of PRPs involved in plant defence against herbivores. Diverse feeding habits of insects trigger multiple signalling pathways (jasmonic acid, salicylic acid, ethylene) to regulate the production of insect-inducible proteins. Few defence proteins reduce the ability of insect herbivore to digest the plant, while anti-digestive proteins slow down the rate of enzymatic conversion of ingested food, and anti-nutritive proteins limit the utilisation of food by changing its physical availability or chemical identity.

4 Classification of Defence Proteins

4.1 Protease Inhibitors (PIs)

PIs are very important elements for plant defence from insects and are regulated by signal transduction pathways that are initiated by predation and transduced as a response to wounding. Injury in plant tissues releases local and systemic extracellular inducers of the signal pathways, for instance, inhibitors which are produced through octadecanoid pathway, where breaking down of linolenic acid which in turn induces expression of PI genes, for example, in soybean and jasmonic acid and its methyl esters, increases mRNA levels of other wound responsive-genes like chalcone synthase (Koiwa et al. 1997). They are widely distributed in plants and are mainly present in storage organs (seeds and tubers). These inhibitors bind with digestive enzymes (gut proteinases) affecting protein digestion inside guts, resulting in shortage of amino acids which affects the development of insects and can even cause death due to starvation. Inside the midgut region of the insect's digestive tract, four different types of proteinases/endopeptidases are found that are used by the herbivore to cleave internal peptide bonds of plant proteins, for example, **serine proteases** present in orthopteran, coleopteran and lepidopteran groups, having alkaline pH of their midgut lumen. SPs are further divided into trypsin-like, elastase-like and chymotrypsin-like proteases. **Cysteine** and **aspartic** proteases are present in Hemiptera, Diptera *and* Coleoptera classes with acidic gut pH. The last and smallest class is **metalloproteinases**. Plants can produce protease inhibitors (PIs) for all the four different types of proteinases which can inhibit the proteolytic activity of midgut enzymes, therefore decreasing the availability of amino acids for insect herbivores. Insect-damaged tissue shows more production of inhibitors. Trypsin inhibitors are produced in damaged tissue of *Glycine max* when attacked by *Tribolium confusum* (Lipke et al. 1954). On the basis of their action mechanisms,

Table 2 Role of plant defensive proteins against insect herbivores

Putative defence protein	Plant species	Insect species
PIs		*Schizaphis graminum*
		Manduca sexta
	Sorghum bicolor	*Helicoverpa armigera*
	Tomato	*Manduca sexta*
	Gossypium hirsutum	*Spodoptera littoralis*
	Solanum nigrum	*Spodoptera exigua*
	Nicotiana attenuata	*Spodoptera exigua*
	Transgenic *Arabidopsis*/oilseed rape	*Plutella xylostella*
	Transgenic *Arabidopsis*/ tobacco	*Mamestra brassicae*
		Spodoptera littoralis
LOXs	*Cucumis sativus*	*Spodoptera littoralis*
	Nicotiana attenuata	*Bemisia tabaci*
	Alnus glutinosa	*Agelastica alni*
	Wheat	*Sitobion avenae*
	Tomato	*Macrosiphum euphorbiae Myzus persicae*
	Nicotiana attenuata	*Myzus nicotianae*
Peroxidases	*Alnus glutinosa*	*Agelastica alni*
	Arabidopsis	*Bemisia tabaci* (whitefly)
	Buffalograss	*Blissus occiduus*
	Poplar	*Lymantria dispar*
	Medicago sativa	*Aphis medicaginis*
	Corn	*Spodoptera littoralis*
	Rice	*Spodoptera frugiperda*
PPOs		*Manduca sexta*
	Tomato	*Blissus occiduus*
	Buffalograss	*Spodoptera frugiperda*
	Tomato	*Helicoverpa armigera*
Chitinases	*Sorghum bicolor*	*Schizaphis graminum*
Hevein-like protein	*Arabidopsis*	*Bemisia tabaci*
Catalase	Bufallograsses	*Blissus occiduus*
SOD	Medicago sativa	*Aphis medicaginis*
Lectin	Tobacco	Aphids
	Chickpea	*Aphis craccivora*
		Mayetiola destructor
	Wheat	*Anagasta kuehniella*
	Tobacco	*Zabrotes subfasciatus*
		Callosobruchus maculates
		Aphids
	Rice	*Nilaparvata lugens*
	Wheat	Aphids

(continued)

Table 2 (continued)

Putative defence protein	Plant species	Insect species
	Arabidopsis	*Pieris rapae*
		Spodoptera littoralis
	Tobacco	*Spodoptera littoralis, Manduca sexta*
		Acyrthosiphon pisum
	Arum	*Lipaphis erysimi, Aphis craccivora*

PIs are further classified as suicide inhibitor, transition state inhibitor and protein protease inhibitor and chelating agents. PI databases like Swiss-Prot have information of more than 5000 inhibitors along with numerous iso-inhibitors present in different plant systems (De Leo et al. 2001). Some other PI inhibitors are squash inhibitors (present in squash family and only in those plants that are active against serine proteases), for example, *Cyclanthera pedata* (wild cucumber) and *Benincasa hispida (*wax gourd) (Atiwetin et al. 2006); cereal trypsin inhibitor (BRI) (present in Poaceae family of cereals and in plants showing amylase inhibitory activity), for example, *Secale cereale* (rye), *Festuca arundinacea* (fescue), *Hordeum vulgare* (barley), etc.; Bowman-Birk inhibitors (BBIs) (present in legumes like soybean); and mustard trypsin inhibitor (MSI) (present in members of family Brassicaceae), for example, *Sinapis alba* (white mustard) and *Brassica napus* (rye).

PPI genes which are found in the insect gut and work against a common protease have been transferred in plants via genetic transformation technique so that the level of insect tolerance could be enhanced in such transformed plants (Table 3). PI-II and carboxypeptidase inhibitor (PCI) genes isolated from potato leaves are transferred to tomato in order to enhance resistance from *Liriomyza trifolii* and *Heliothis obsoleta* (Abdeen et al. 2005). Similarly NA-PI and β-hordothionin (β-HTH) genes from barley were transferred to tobacco and the leaves of these transgenic tobacco when ingested by *Helicoverpa armigera*, i.e. tobacco budworm, showed slower larvae development and increased death rate of the insect as compared to larvae feeding on non-transformed tobacco (Charity et al. 1999). Products of WCI2 and WCI5 genes extracted from *Psophocarpus* species (winged bean) have shown strong insecticidal effects on various insects feeding on transgenics, transformed by these two genes (Telang et al. 2009).

There are still various strains of insects which show negligible effect of PPIs, while others evolve and try to adapt a survival strategy to beat the effect of PIs. Hence, to overcome such adaptive mechanisms, various successful strategies have been developed for efficient pest management, and they are production of transgenes by using the technique of "gene stacking".

RNAi technique is used for producing desired resistant transgenic. In "gene stacking" technique, transgenic plants expressing combination of transgenes are produced, for instance, CRY toxins or inhibitors from bacteria *Bacillus thuringiensis*. Transgenic cotton plants expressing CRY1Ac gene (trypsin inhibitor

Table 3 Role of proteinaceous inhibitors (with anti-insect activity) in transgenic plants

Source and name of inhibitor	Transgenic plants	Targeted herbivore
Cowpea trypsin inhibitor	Tobacco	Tomato hornworm, *Manduca sexta*
Tobacco trypsin protease inhibitor	Tobacco	Tomato hornworm, *Manduca sexta*
Potato proteinase inhibitor 1	Tobacco	Tomato hornworm, *Manduca sexta*
Cowpea trypsin inhibitor	Tobacco	Cluster caterpillar, *Spodoptera litura*
Soybean Kunitz inhibitor	Tobacco	Cluster caterpillar, *Spodoptera litura*
Sweet potato trypsin inhibitor	Tobacco	Cluster caterpillar, *Spodoptera litura*
Soybean Kunitz inhibitor	Tobacco	Egyptian cotton worm, *Spodoptera littoralis*
Potato trypsin inhibitor	Tobacco	Green looper, *Chrysodeixis eriosoma*
Tobacco trypsin protease inhibitor	Tobacco	American bollworm, *Helicoverpa armigera*
Cowpea trypsin inhibitor	Potato	Tomato moth, *Lacanobia oleracea*
Soybean Kunitz inhibitor	Potato	Egyptian cotton worm, *Spodoptera littoralis*
Cowpea trypsin inhibitor	Cotton	American bollworm, *Helicoverpa armigera*
Manduca (insect) protease inhibitor	Cotton	Sweet potato whitefly, *Bemisia tabaci*
Cowpea trypsin inhibitor	Apple	Multiple insects
Tobacco multi-domain inhibitor	Apple	Light-brown apple moth, *Epiphyas postvittana*
Soybean Kunitz proteinase inhibitor	Poplar	Polyphagous moth, *Lymantria dispar*
Soybean Kunitz proteinase inhibitor	Poplar	Poplar tip moth, *Clostera anastomosis*
Rice cysteine protease inhibitor	Poplar	Poplar leaf beetle, *Chrysomela tremulae*
Cowpea trypsin inhibitor	Strawberry	Multiple insects
Cowpea trypsin inhibitor	Strawberry	Vine weevil, *Otiorhynchus sulcatus*
Sweet potato trypsin inhibitor	Cauliflower	*Pieris conidia*
Sweet potato trypsin inhibitor	Cauliflower	Diamondback moth, *Plutella xylostella*
Cowpea trypsin inhibitor	Cauliflower	Small cabbage white butterfly, *Pieris rapae*
Tobacco trypsin protease inhibitor	Pea	American bollworm, *Helicoverpa armigera*
Potato proteinase inhibitor	Rice	Rice stem borer, *Chilo suppressalis*
Winged bean trypsin inhibitor	Rice	Rice stem borer, *Chilo suppressalis*
Cowpea trypsin inhibitor	Rice	Rice stem borer, *Chilo suppressalis*
Cowpea trypsin inhibitor	Rice	Pink stem borer, *Sesamia inferens*
Potato proteinase inhibitor 2	Rice	Pink stem borer, *Sesamia inferens*
Soybean Kunitz trypsin inhibitor	Rice	Brown planthopper, *Nilaparvata lugens*
Barley trypsin inhibitor Cme	Rice	Rice weevil, *Sitophilus oryzae*

(continued)

Table 3 (continued)

Source and name of inhibitor	Transgenic plants	Targeted herbivore
Cowpea trypsin inhibitor	Wheat	Angoumois grain moth, *Sitotroga cerealella*
Barley trypsin inhibitor Cme	Wheat	Angoumois grain moth, *Sitotroga cerealella*

gene) transformed from *Bacillus thuringiensis* and CPT1 gene from cowpea show strong inhibitory effects towards *Helicoverpa armigera* as compared to normal plants (Cui et al. 2011; Macedo et al. 2015). Similarly another transgenic line of cotton was developed expressing *Nicotiana alata* protease inhibitor gene (NAPI) and showed better resistance against *H. punctigera* (Dunse et al. 2010). In transgenic lives of *Arabidopsis thaliana* plant, resistance against *Tetranychus urticae* (spider mite) was enhanced due to expression of cystatin Icy6 and Itr1 (trypsin Inhibitor) genes transferred from barley.

RNAi has also been used to develop resistance in plants to insects. This approach was used to attenuate the expression of Mysp (a serine protease) in *A. thaliana* against aphid *Myzus persicae*. Aphids feeding on transgenes of *A. thaliana* expressing dsRNA of MYsp showed reduced growth and survival rate in larvae of these insects (Bhatia et al. 2012).

4.2 Role of Lectins

They are sugar-binding glycoproteins present in storage organs and protective structure of some plants, are abundant in nature and provide protection against a range of pests. Lectins can easily survive inside the digestive tract of herbivores; therefore, there potential could be utilised as naturally occurring insecticides against insect pests (Table 2) (Vandenborre et al. 2011). They behave either as anti-nutritive or toxic substance for the herbivores (insects) by binding to the glycosyl groups of membranes lining the digestive tract, followed by a series of harmful systemic reactions. Molecular structures of lectins remain intact even under fluctuating pH conditions of the digestive tract and therefore are able to damage the luminal epithelial membrane of pests, hindering nutrient digestion and absorption. Disruption caused by improper metabolism of proteins, lipids and carbohydrates leads to enlargement or atrophy of key tissues changing the hormonal and immunological status, stopping growth and development in insects (Vandenborre et al. 2009).

Lectin could be classified into five families on the basis of comparing carbohydrate recognition domains (CRD): cereal lectins, legume lectins and C-, P- and S-type of lectins. Out of these five, only the first two are found in plants. The first lectin discovered in plants was glucose-/mannose-specific concanavalin A, ConA, present in jack bean (*Canavalia ensiformis*) (Edelman et al. 1972), and it was later transferred in *S. tuberosum where* it expresses to retard the growth of *L. oleracea* and

M. persicae (Gatehouse and Gatehouse 1998). In *Phaseolus vulgaris*, PHA (Phaseolus vulgaris agglutinin) which act on *C. maculatus* and "acrelin", which show toxicity for *Zabrotes subfasciatus* (Been weevil) are produced (Osborn et al. 1988). Wheat germ agglutinin (WGA) isolated from corn when used in different concentrations (2 mM–25 mM) changed the growth pattern and mortality of blowfly (*Lucilia cuprina*) larvae from 50 to 100%. GNA (*Galanthus nivalis* agglutinin) also known as snow drop lectin reduces the growth and development of foxglove aphid, *Aulacorthum solani*, in vivo and also when it expresses in transgenic *S. tuberosum* (Down et al. 1996). In rice plant, GNA works against *Nilaparvata lugens* (planthopper) by binding to the luminal surface of midgut epithelial cells of the insect after recognising the carbohydrate moieties of glycoproteins present on cell surface or carbohydrate moieties of glycol-conjugates present inside the gut. GNAs could also lead to systemic toxic effect, once they cross the midgut epithelial barrier and then move into the insect's circulatory system. Plant lectins are induced by elicitors in response to various stresses, for example, jasmonic acids in leaves of *Nicotiana* induce the expression of NICTABA lectin gene affecting insect herbivores like *S. littoralis, Manduca sexta* L. *and Tetranychus urticae* (Vandenborre et al. 2009). HFR1, HFR2 and HFR3 are mannose-binding jacalin-like lectins induced by the larvae of *M. destructor* (Hessian fly) in wheat plant. Different feeding behaviours of insects result in expression of different lectins, for example, *Rhopalosiphum padi Koch* (bird cherry-oat aphid) which is a phloem-feeding insect, and it induces the production of HFR3 and HFR2 one after the other, and larvae of *S. frugiperda* induce the production of only HFR2 in monocots (Giovanini et al. 2007; Puthoff et al. 2005). Several other jasmonate-induced lectins are also produced in leaf tissues of various monocots like wheat, rye, rice, barley, maize, etc.

4.3 Role of Enzymes

Crop production is greatly influenced by various types of environmental stresses including insect herbivory. Feeding by insects triggers numerous plant biochemical processes specifically involved in tolerance mechanism. It has been shown, through diverse studies, the role of plant oxidative enzymes in protecting the plant from biotic stresses induced by herbivores. These enzymes disrupt nutrient uptake by insects due to formation of electrophiles including polyphenol oxidases (PPOs), peroxidases (PODs) and other peroxidase formed by oxidising mono- or dihydroxyphenols, leading to formation of reactive O-quinones that polymerise and form covalent adducts with nucleophilic groups of insect proteins (due to their electrophilic nature), e.g. -SH or e-NH_2 of lysine (Gill et al. 2010). Other examples of oxidative enzymes are lipoxygenases (LOX), phenylalanine, ammonia lyase, superoxide dismutase, etc. This defence mechanism in plants has received considerable attention in recent years.

4.4 Role of Polyphenol Oxidases (PPOs)

PPOs are metalloenzymes induced due to attack by insect herbivores, and they catalyse conversion of monophenols and O-diphenols (chlorogenic acid) to highly reactive quinones. Being highly reactive intermediate compounds, quinones readily polymerise to react with nucleophilic side chain of amino acids and cross-link the proteins reducing their nutritive value for insects. When the conditions are acidic, quinones convert into semiquinone radicals which latter form ROS and in basic conditions; quinines react with cellular nucleophiles. PPO-generated quinones and ROS are more toxic to herbivores (particularly arthropods) than the original phenols. They also enhance melanin formation which increases cell wall resistance to insect and herbivores. Signalling molecules and injury occurring in plant tissues due to herbivory induce the expression of PPO genes, and the enzyme oxidase accumulates in leaves, stems, flowers and roots of the plants particularly in young tissues which are more prone to insect attack. Foliage and fruit of *Lycopersicon esculentum* (tomato) contain PPO and POX, compartmentally separated from orthodihydroxyphenolic substrate. When leaf tissues are damaged due to insect feeding, both the enzyme and the substrate mix with each other, leading to rapid oxidation of phenolics to ortho-quinones affecting the herbivore *H. zea* (tomato fruit worm) or *Spodoptera exigua* feeding on tomato foliage (Felton and Tumlinson 2008). In *Brassica oleracea* var. *capitata*, PPO defend the plant from phloem-sucking aphid (*Brevicoryne brassicae* L.) (Khattab 2007). Potato PPO genes are transferred into tomato plants, and the transgenics were expressing PPO genes both ways (sense and antisense orientation), making the plant resistant to *Spodoptera litura* F. (cutworm) (Mahanil et al. 2008).

4.5 Peroxidase (POD)

PODs are monomeric hemoproteins and their distribution within the cells is as soluble membrane-bound and cell wall-bound proteins. These are the enzymes catalysing oxidoreduction reaction between hydrogen peroxide and reductant.

$$H_2O_2 + AH \rightarrow 2H_2O + A$$

$$ROOR' + e^{-}donor(2e^{-}) + 2H^{+} \rightarrow ROH + R'OH$$

Peroxidases are present in animals, plants and microorganisms. On the basis of structural and catalytic properties, they are divided into three superfamilies (Welinder 1986).

Class III plant peroxidases: They are mainly glycoproteins made up of single polypeptide chain containing ferriprotoporphyrin 1x as a prosthetic group. They catalyse oxidation of compounds using H_2O_2 as an oxygen acceptor and also participate in multiple physiological functions such as lignification, auxin catabolism and suberisation and wound healing, cross-linking of cell wall structural proteins, senescence and salt tolerance which play direct or indirect role in defence against herbivores (Hiraga et al. 2001). As they are involved in cross-linking of wall proteins, this phenomenon creates mechanical barrier and suppresses penetration of herbivores, for example, the cell wall of mature periderm of *Zea mays* L. gets resistant from *Sitophilus zeamais* (maize weevil) after the action of peroxidases (García-Lara et al. 2004).

Ascorbate peroxidases: Located in chloroplast, cytoplasm, mitochondria, peroxisomes and apoplast of the plant cell catalysing oxidoreduction reaction between H_2O_2 and ascorbic acid (substrate). They are responsible for modulation of ROS and removal of H_2O_2.

$$AA + H_2O_2 \rightarrow DHA + H_2O$$

Glutathione peroxidases: They also protect the cell from the effect of H_2O_2. Previous studies have indicated that the level of POD increases when insects attack or damage the leaf tissue and produce oxidative metabolites and promote lignification. They also produce phenoxy and other oxidative radicals in association with phenols and reduce the digestibility of plant products (leaf, fruit, stem, etc.), which leads to nutrient deficiency in insects affecting their growth and development (Zhang et al. 2008). PODs can also cause direct toxicity in guts of insect herbivores.

4.6 Role of Lipoxygenases (LOXs)

They are another group of anti-oxidative enzymes which play an important role in plant defence mechanism via octadecanoid pathway (Bruinsma et al. 2009). LOXs are family of iron-containing enzymes that can catalyse deoxygenation of polyunsaturated fatty acids present in lipids (containing cis, cis-1,4-penta-diene structure) to unsaturated fatty acid peroxide which are chemically or enzymatically degraded to unstable and highly reactive aldehydes, epoxides, Y ketols and ROS (singlet oxygen, superoxide ion, hydroxyl radicals, acyl and carbon-centred radicals) (Bruinsma et al. 2010).

$$F.Acids + O_2 \rightarrow F\ Acid\ hydro\ peroxides$$

LOX triggers the synthesis of acyclic or cyclic compounds known as oxylipins which are produced after fatty acid oxidation and play numerous important functions in the plant defence against insect herbivores. Linolenic and linoleic acids are the common substrates for LOX and are produced by plants after the cells are damaged due to insect feeding. LOX catalyses oxidation of these fatty acids to form hyperoxides, superoxide anions and peroxy radicals (Gardner 1980). These free radicals being highly reactive promote lignification which act as mechanical barrier for the insects. LOX oxidises linolenic acid in JA signalling pathway to produce oxidative enzymes and protease inhibitors (direct defence) or volatile organic compounds (VOC) that attract natural enemies of insect pests (indirect defence)

(Bruinsma et al. 2010). Hypo-oxides are further metabolised into JA and traumatin. Soybean lipoxygenases retarded the growth of larvae of *Manduca sexta* L. when its 4% concentration was mixed into larvae's artificial diet. Red bean and soybean LOX restricts the growth of *Helicoverpa zea* (Felton et al. 1994).

5 Indirect Defence Against Insect Herbivores

It includes those characters that themselves do not affect herbivores but attract their natural enemies and thus reduce plant loss. After recognising the elicitors (sulphur-containing amino acids, fatty acid-amino acid conjugates, peptides, esters, enzymes and fragments of cell wall) associated with herbivores, damaged plant tissues release a mixture of volatiles (terpenes, indoles and nitrogenous compounds) that can attract predators and natural enemies of those herbivores (Aljbory and Chein 2016). Few constitutive characters which are already present in plants (domatia, food bodies and extrafloral nectars) can also invite natural enemies of the plant herbivores or can provide food and shelter to the predators' enemies. Induced indirect defences have better impact on herbivores; therefore, they have attained more attention recently and are studied on genetic, biochemical, physiological and ecological levels (Maffei 2010). Plants produce mixture of volatile and non-volatile compounds in order to attract enemies of their predators' oviposition deterrent, also known as HIPVs (herbivore-induced plant volatiles) (Hagenbucher et al. 2013; Xu et al. 2015). They are lipophilic compounds having high vapour pressure and are released through leaves, flowers and fruits into the air or from roots in to the soil, whenever there is an attack from herbivores.

5.1 Role of Elicitors

The first step involved in plant defence is through recognition of insect attack by plants via insect- or plant-derived-specific compounds known as elicitors which are produced by various mechanisms (Howe and Jander 2008). Few elicitors are first produced in attacking insect herbivores and then injected in to plant tissues in the form of oral secretions and ovipositional fluids of insects (Erb 2009; Diezel et al. 2009). Examples of some well-characterised elicitors which could induce indirect defence mechanism in plants are as follows.

5.1.1 Fatty Acid-Amino Acid Conjugates

They are made up of saturated and unsaturated fatty acids which conjugated to amino acids (α-glutamic acid or α-glutamine). Out of various fatty acid-amino acid conjugates identified in different insect species, volicitin (Fig. 9) is the most important and well-characterised compound (N-(17-hydroxylinolenoyl)-Gln). It was first identified in *Spodoptera exigua* inducing production of volatiles in *Zea mays* plants (Alborn et al. 1997). Comparative studies done in various types of elicitors showed that volicitin possesses the widest range of phytohormones and volatile-inducing

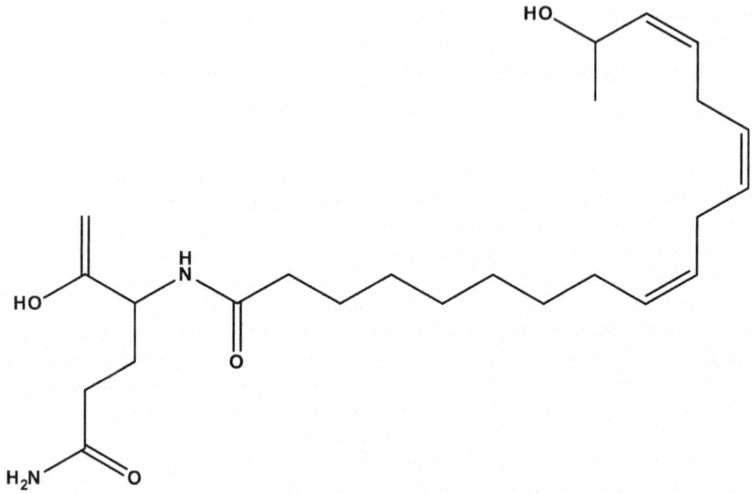

Fig. 9 Volicitin (FA amino acid conjugate)

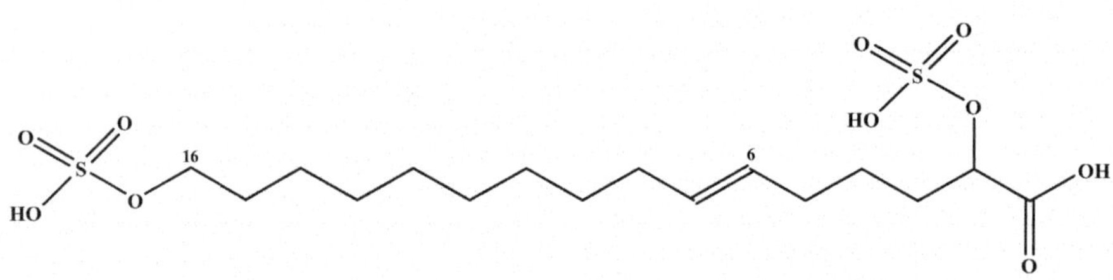

Fig. 10 Caeliferin (sulfooxy fatty acids)

activity in different plant species such as eggplant, soybean and maize (Schmelz et al. 2009). In *Zea mays*, plasma membrane protein binds with volicitin which initiates defence mechanism in plants (Truitt et al. 2004). The relationship between volicitin and other amino acid conjugates is still to be assessed for eliciting the release of specific set of volatiles attracting natural enemies of specific plants.

5.1.2 Sulfooxy Fatty Acids

They are also known as sulphur-containing fatty acids and are characterised as caeliferins (O'Doherty et al. 2011). Structure of caeliferins constitutes saturated and monosaturated sulphated α-hydroxyl fatty acids in which ω-carbon is functionalised either with a sulphated hydroxyl or carboxyl conjugated to glycine with the help of an amide bond (Schäfer et al. 2011a, b). Active role of caeliferins as an elicitor was first recognised in corn plants from the regurgitant of *Schistocerca gregaria* (grasshopper), and other species of suborder Orthoptera contain caeliferin A: 16 O (Schäfer et al. 2011a, b). Artificial synthesis of caeliferin A: 16 (Fig. 10) has been done which would induce the production of JA and ET (ethylene) in

Arabidopsis when the same concentration was delivered as same present in saliva of grasshopper (Peng et al. 2016).

5.1.3 Esters

Another group of elicitors involved in plant defence reactions are 3-hydroxypropanoate esters of long-chain α,ω-diols (Doss et al. 2000). They are also known as bruchins (Fig. 11), and unlike the previously studied elicitors (produced in insects and delivered into plant tissues through oral secretions), they are only delivered into plant tissues via eggs of insects (Oliver et al. 2000). Till date they have been only reported in *Bruchus pisorum* (pea weevil) and *Callosobruchus maculatus* (cowpea weevil). *Pieris brassicae* and *P. rapae* (butterflies) lay eggs on plants, therefore depositing the elicitors which could bring changes in plant tissues and then attracting egg predators *Trichogramma* (Fatouros et al. 2009).

Bruchins trigger uncontrolled division of cells at the sites of egg attachment, resulting in formation of tumour-like outgrowths beneath the egg, which prevents entry of larvae inside the pod (Doss et al. 2000) and exposes them to predators, parasites and desiccation (Fatouros et al. 2008). They are involved in indirect defence mechanism but without induction of volatiles to attract the natural enemies of their predators (Bruessow et al. 2010).

5.1.4 Peptides

Few of the proteins secreted by insects also work as elicitors and activate defence mechanism in plants as they are capable of performing functions such as plant manipulation, partial digestion of pure ingested food, etc. (Kessler and Baldwin

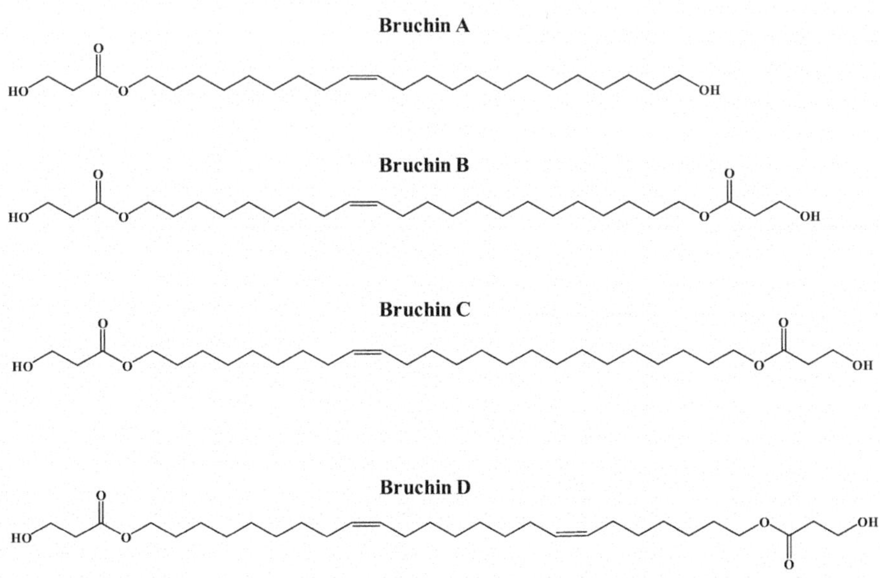

Fig. 11 Bruchins (esters)

2002). Inceptin is one of the important peptide secreted by insects, and it serves as an elicitor to activate indirect defence mechanism (Schmelz et al. 2006). It is a disulphide-bridged peptide derived from chloroplastic ATP synthase g-subunits present in plants, and after ingestion inside the insect gut, it gets converted into inceptin due to proteolytic cleavage and then delivered back into plant tissues through insect's oral secretion (Felton 2008; Schmelz et al. 2009). Inceptin triggers increased secretion of salicylic acid (SA), JA and ethylene (ET) which further releases plant volatiles like terpenoids (Schmelz et al. 2007a, b).

5.1.5 Enzymes

Oral secretions of insects contain mixture of enzymes (proteases, oxidases, glucosidases, etc.) which are not only involved in breaking down of preingested food but also act as elicitors to activate indirect defence mechanism in plants (Xu et al. 2015). For example, glucose oxidases catalyse oxidation of D-glucose to hydrogen peroxide and D-glucono-lactone, and it is produced inside labial glands of lepidopteran species (*Helicoverpa zea*, *S. exigua*, *M. sexta*) and several other insect species like *Apis mellifera* (honey bee), *Myzus persicae* (aphid) and *S. americana* (grasshopper). In tomato, indirect defence is provided due to glucose oxidases present in the salivary gland of insect predator (Louis et al. 2014). Similarly β-glucosidase triggers formation of volatile terpenes in cabbage, corn plants and lima bean (Bonaventure et al. 2011). Plants recognise both the enzyme proteins in a ligand receptor interaction as in the case of other elicitors. Glucose oxidase releases hydrogen peroxide which might serve as a signal for volatile production, and β-glucosidase releases glucose which triggers the production of elicitors. Therefore both enzymes are themselves not elicitors but producers of elicitors.

5.1.6 Fragments of Cell Walls

They are the elicitors formed due to damage of plant cells as a result of herbivore attack and are known as DAMPs (damage-associated molecular patterns) (Erb et al. 2012), for example, oligogalacturonides (oligomers of alpha-1,4-linked galacturonosyl residues) released from damage plant cell walls due to partial degradation of homogalacturonan (Ferrari et al. 2013; Benedetti et al. 2015). Oligogalacturonides activate the defence mechanism in plants via octadecanoid pathway, and JA are involved in functioning of these elicitors (John et al. 1997).

Researchers are still on the initial phase of identifying elicitors and their potential signal transduction components. But more research with improved techniques could invent new types of better elicitors. Already known elicitors (bruchins, volicitin, etc.) could be manufactured industrially and then applied artificially to plants even before invasion of herbivores to control the damage.

5.1.7 Role of Plant Hormones

They are the signal molecules regulating growth and development of plants and making them respond to various environmental stresses (Verma et al. 2016). Researchers have also investigated the role of different plant growth regulators in inducing defence mechanism when plants are attacked by pathogens and predators

(Wei et al. 2014; Vos et al. 2015). When the plant is protecting itself from the natural enemies, the level of these hormones automatically increases to a certain level in order to initiate a cascade of signal transduction. Some of the plant hormones involved in indirect defence mechanism in plants include JA, SA and ET (Zhang et al. 2013).

5.1.8 Jasmonic Acid (JA)

It is one of the important plant hormones involved in inducing indirect plant defence (Hettenhausen et al. 2013). When there is attack of herbivores, JA is synthesised in plants from linolenic acid due to action of multiple enzymes (phospholipase, lipooxygenases, allene oxide cyclases and allene oxide synthases) (Wasternack and Hause 2013). Herbivore-associated elicitors that are involved in inducing JA pathway have been reported in many species, for example, *S. frugiperda*, *S. gregaria*, *M. sexta*, etc. (Schmelz et al. 2009; Schäfer et al. 2011a, b). In plants, JA is modified into various active forms such as MeJA (methyl jasmonates), JA-Ile and OPDA (12-oxophytodienoic acid) which is a precursor for JA (Woldemariam et al. 2012). Those plants which are exposed to MeJA show increased emission of volatiles and attract predators of parasites better than the plants elicited with herbivore elicitors (Heil 2004; Bruinsma et al. 2009). EFN produced by JA is used as an alternate food by the enemies of insect herbivores.

5.1.9 Salicylic Acid

It is a monohydroxybenzoic acid derived from cinnamate which is produced from phenylalanine due to activity of phenylalanine ammonia lyases (Chen et al. 2009). SA is an important endogenous plant growth regulator regulating various metabolic and physiological processes in plants such as defence, growth and development (Pieterse and Van Loon 2004). SA-related signalling pathways are involved in emission of volatiles which are involved in direct defence mechanism of plant (Diezel et al. 2009). SA gets directly modified into its bioactive derivative known as MeSA. Higher levels of MeSA attract natural predators of insect herbivores in lima bean and tomato plants when attacked by spider mite (Ozawa et al. 2000; Dicke and Sabelis 1988), in pear plants on infestation of spider mite and psyllid (De Boer et al. 2004) and in potato plants due to attack of Colorado potato beetle herbivore (Bolter et al. 1997). Leaves of poplar tree were treated with MeSA, leading to expression of defence-related genes and over-emission of volatile compounds (Arimura et al. 2000). MeSA can therefore attract natural enemies of insect herbivores and also induce emission of other volatile compounds for indirect defence of plants. JA provides defence against leaf-chewing herbivore, whereas SA induces defence against piercing and sucking insects (Zhao et al. 2009). Synergistic interaction between JA and SA pathways also induces indirect defence in plants against herbivores (Bari and Jones 2009). When plants are attacked by chewing insects, then JA pathway gets activated and SA molecules are converted to MeSA volatiles via methylation process and are emitted as attractants for natural enemies of insects, resulting in synergistic effects between the two hormones for indirect defence mechanism in plants.

5.1.10 Ethylene

ET is produced in plants when herbivores attack and is synthesised from methionine due to sequential enzymatic activity of S-adenosylomethionine synthetases, 1-amino cyclopropane-1-carboxylic acid (ACC) synthases and ACC oxidases (Wang et al. 2002). Insects' oral secretion and elicitors induce overproduction of ET, and it has been observed in many plant species such as pine trees and tomato plants. Blockage of ET with 1-MCP diminishes herbivory-induced volatile emissions, thus affecting the defence mechanism of plants. JA and ET signalling pathways are mostly operated synergistically to activate the expression of few or defence-related genes in plants like tomato and *Arabidopsis* (Lorenzo et al. 2003).

5.2 Role of Volatiles

Major volatiles produced in different plants after the attack by different herbivores show remarkable similarity, though their compositions differ in different cases. Examples of few volatiles and their functions are as follows.

5.3 Nitrogenous Compounds

Nitrogenous volatiles commonly emitted from herbivore-damaged plants are nitriles and oximes (Irmisch et al. 2014). Though they are emitted in minute amounts, they still play a crucial role in attracting predators of the herbivores. Proportion of these compounds in a volatile mix is also specific for different predators. Biosynthetic pathways of nitrogenous compounds are different in different plant systems, for instance, glucosinolates produce nitrogenous volatiles in crucifers (Hopkins et al. 2009) and amino acid derivatives catalysed by cytochrome P450 enzyme produce NVs in plant species such as gerbera, lima bean and cucumber (Irmisch et al. 2014). Nitrogen-containing compounds such as nitriles, aldoxime, benzyl cyanides, methylbutyraldooximes, etc. have been detected in volatile form in many species of cotton wood such as *Populus nigra*, *P. trichocarpa* and *P. canescens* in response to attack from caterpillars and gypsy moth.

5.4 Volatile Indoles

These are indole and indole-alkaloid molecules containing nitrogen attached to pyrrole ring. Presence of indoles has been detected in many plant species such as maize, rice, gerbera, lima bean and cotton after the attack of insect herbivores (Frey et al. 2000; Zhuang et al. 2012). Under laboratory conditions, mixture of indoles and terpenoids is released by damaged plants that are attacked by lepidopteran larvae attracting *Cotesia marginiventris* which is their natural enemy (Turlings and Berney 2016). Alone indoles can attract natural enemies of big-eyed bug (*Geocoris*) (James 2005). In maize plant, indole formation regulates emission of other volatiles and triggers the synthesis of mono- and homoterpenes in insect-attacked plants, preparing other plant tissues and neighbouring plants against incoming herbivores (Erb et al. 2015).

5.5 Green Leaf Volatiles

FA derivatives are common volatiles in plants imparting typical odour of green leaves and therefore are also named as green leaf volatiles (Christensen et al. 2013). Their emission is at peak when the leaf tissues are damaged/disrupted and C18 unsaturated fatty acids (linoleic and linolenic acids) are released from plasma membrane of damaged cells. These fatty acids are oxidised and decarboxylated by hydroperoxide lyases, leading to production of volatiles such as C_6-alcohols and C_6-aldehydes and esters ((Z)-3-hexon-1-yl acetate) (Pare and Tumlinson 1999). In some plants, FA derivatives can also be produced from lipoxygenase via oxylipin pathway (Melan et al. 1993). Unlike other volatiles, they are released immediately after damage caused by herbivory and also provide quick information regarding quick location of a feeding herbivore (Yu et al. 2008). Earlier emission of GLVs plays an important role in inducing emission of other herbivore-induced plant volatiles and participates in intra- and inter-plant signalling (Allmann and Baldwin 2010), for example, rapid production of JA and emission of sesquiterpene in maize after exposing to synthetic green leaf volatiles (Ton et al. 2007). Secondly, it activates transcription of the genes responsible for biosynthesis of ethylene in lima bean and also triggers emission of local and systemic terpenes in tomato plants (Arimura et al. 2002; Farag and Pare 2002). Genetic transformation could help in transferring genes which would express to develop plants with high level of volatiles to attract natural enemies of their predators, but higher level of volatiles can also have negative impact on physiology of the plant. Alternatively, mixture of volatiles could be prepared artificially in industries and applied on plants before the outbreak of insect pests (Sharma et al. 2015), but for that complete knowledge regarding the structure, function and effect of these volatiles on environment is required.

5.6 Role of Inducible Constitutive Traits

5.6.1 Extrafloral Nectar

It is an aqueous solution made of sugars and amino acids secreted from extrafloral nectary organs (Koptur 1992). Most of the plant species produce it and are found on different parts of plants such as shoots, leaves and inflorescence. It can attract both predators and their natural enemies. Plants can produce them as such regardless of herbivory, but in the presence of herbivores due to damage of tissues, secretion is more. Extrafloral nectar could be more effective when used in combination with volatiles and JA (Heil 2004). In lima bean plants, treatment with mixture of volatiles showed elevated levels of EFN.

5.6.2 Domatia

They are small morphogenetic hollow structures present on plants, providing shelter for mites and other arthropods. In return, they benefit the plant by removing spores of pathogens and preying on insect herbivores (O'Connell et al. 2015; Tempfli et al. 2015). Most of the arthropods living in domatia are ants, mites and thrips.

6 Conclusion

Since the last 40 years in plant sciences after the discovery of systemic signalling in tomato and digestion-reducing PIs, many other mechanisms for plant defence against insect herbivores have been identified. Various studies done in the field of microarray have shown ample number of genes present in plants and expressing only after the attack from insect herbivores (Howe and Jander 2008). Studies on defence strategy of plants show that herbivory causes change from growth and development mode towards defence-oriented metabolism, which is also an evolutionary aspect. Despite so much research done in this field, this mechanism in plants to defend themselves still remains a mystery. Lots of efforts have been done to know the signalling events involved in defence mechanism of plants, but many of its components are yet to be discovered, for example, the order of the early events like calcium flux and phosphorylation cascades. To fulfil the purpose, screening of more and more novel components and knockout mutant studies and mapping of defence-related genes are required to have a complete knowledge of these defence-related pathways and have a comprehensive information about signalling events involved in defence against herbivores. Induced resistance has gained appropriate momentum and also has attracted the attention of many researchers working in the field of evolutionary ecology, entomology, plant physiology and defence-related techniques such as genetic transformation. Some of the defence responses might also have negative effects on humans as well as surrounding environment as they involve toxic bioactive natural products and proteins that reduce digestibility of plant material. For instance, elicitors can have negative effects if one does not have complete knowledge of the chemical changes that they induce in plants and alterations occurring in plant growth and yield. Still if they reduce the usage of synthetic insecticides by developing crop plants resistant to insect herbivores, it would be an economical gain for food and production industry and also best at environmental level. Researchers working in the field of biology and ecology have attained great help due to advancement in technologies of genomics, proteomics, chromatography, mass spectroscopy, purification and testing of various chemicals, monitoring systems and online communication. All these advanced technologies will open the pathway for systemic and integrated research on various aspects of direct and indirect mechanisms of plant defence against insect herbivores and will provide necessary conditions to utilise the valuable information for pest management.

References

Abdeen A, Virgos A, Olivella E, Villanueva J, Aviles X, Gabarra R, Prat S (2005) Multiple insect resistance in transgenic tomato plants overexpressing two families of plant proteinase inhibitors. Plant Mol Biol 57(2):189–202

Agrawal AA (1999) Induced responses to herbivory in wild Radish: effects on several herbivores and plant fitness. Ecology 80:1713–1723. https://doi.org/10.1890/0012-9658(1999)080[1713:IRTHIW]2.0.CO;2

Agrawal AA, Fishbein M, Jetter R, Saltines JP, Goldstein JB, Freitag AE, et al (2009) Phylogenetic ecology of leaf surface traits in the milkweeds (*Asclepias* spp.): chemistry, ecophysiology, and insect behaviour. New Phytol 183:848–867; PMID: 19522840; https://doi.org/10.1111/j.1469-8137.2009.02897.x

Alborn HT, Turlings TCJ, Jones TH, Stenhagen G, Loughrin JH (1997) An elicitor of plant volatiles from beet armyworm oral secretions. Science 276:945–949

Allmann S, Baldwin IT (2010) Insects betray themselves in nature to predators by rapid isomerization of green leaf volatiles. Science (New York, N.Y.) 329:1075–1078

Aljbory Z, Chen M-S (2016) Indirect plant defense against insect herbivores: a review. Insect Sci 25:2–23. https://doi.org/10.1111/1744-7917.12436

Arimura G, Ozawa R, Shimoda T, Nishioka T, Boland W, Takabayashi J (2000) Herbivory-induced volatiles elicit defence genes in lima bean leaves. Nature 406:512–515

Arimura GI, Ozawa R, Nishioka T, Boland W, Koch T, Kühnemann F, Takabayashi J (2002) Herbivore-induced volatiles induce the emission of ethylene in neighboring lima bean plants. Plant J 29:87–98

Atiwetin P, Harada S, Kamei K (2006) Serine protease inhibitor from Wax Gourd (Benincasa hispida [Thumb] seeds). Biosci Biotech Biochem 70(3):743–745

Barbehenn RV, Peter Constabel C (2011) Tannins in plant herbivore interactions. Phytochemistry 72:1551–1565; PMID: 21354580; https://doi.org/10.1016/j.phytochem.2011.01.040

Bari R, Jones J (2009) Role of plant hormones in plant defence responses. Plant Mol Biol 69:473–488. https://doi.org/10.1007/s11103-008-9435-0

Benedetti M, Pontiggia D, Raggi S, Cheng Z, Scaloni F, Ferrari S, Ausubel FM, Cervone F, De Lorenzo G (2015) Plant immunity triggered by engineered *in vivo* release of oligogalacturonides, damage-associated molecular patterns. Proc Natl Acad Sci U S A 112:5533–5538

Bennett RN, Wallsgrove RM (1994) Secondary metabolites in plant defence mechanisms. New Phytol 127(4):617–633

Bhatia V, Bhattacharya R, Uniyal PL, Singh R, Niranjan RS (2012) Host generated siRNAs attenuate expression of serine protease gene in Myzus persicae. PLoS One 7(10):e46343. https://doi.org/10.1371/journal.pone.0046343. Epub 2012 Oct 10

Bhonwong A, Stout MJ, Attajarusit J, Tantasawat P (2009) Defensive role of tomato polyphenol oxidases against cotton bollworm (Helicoverpa armigera) and beet armyworm (Spodoptera exigua). J Chem Ecol 35:28–38; PMID: 19050959; https://doi.org/10.1007/s10886-008-9571-7

Bodnaryk RP (1991) Developmental profile of sinalbin in mustard seedlings, *Sinapis alba* L., and its relationship to insect resistance. J Chem Ecol 17:1543–1556

Bolter CJ, Dicke M, van Loon JJ, Visser JH, Posthumus MA (1997) Attraction of Colorado potato beetle to herbivore – damaged plants during herbivory and after its termination. J Chem Ecol 23:1003–1023

Bonaventure G, van Doorn A, Baldwin IT (2011) Herbivore-associated elicitors: FAC signaling and metabolism. Trends Plant Sci 16:294–299

Brattsten LB, Samuelian JH, Long KY, Kincaid SA, Evans CK (1983) Cyanide as a feeding stimulant for the southern armyworm, Spodoptera eridania. Ecol Entomol 8:125–132

Bruessow F, Gouhier-Darimont C, Buchala A, Metraux J-P, Reymond P (2010) Insect eggs suppress plant defence against chewing herbivores. Plant J 62:876–885

Bruinsma M, Posthumus MA, Mumm R, Mueller MJ, van Loon JJA, Dicke M (2009) Jasmonic acid-induced volatiles of *Brassica oleracea* attract parasitoids: effects of time and dose, and comparison with induction by herbivores. J Exp Bot 60:2575–2587; PMID:19451186; https://doi.org/10.1093/jxb/erp101

Bruinsma M, van Broekhoven S, Poelman EH, Posthumus MA, Müller MJ, van Loon JJA, Dicke M (2010) Inhibition of lipoxygenase affects induction of both direct and indirect plant defences against herbivorous insects. Oecologia 162(2):393–404

Chamarthi SK, Sharma HC, Vijay PM, Narasu LM (2011) Leaf surface chemistry of sorghum seedlings influencing expression of resistance to sorghum shoot fly, *Atherigona soccata*. J Plant Biochem Biotechnol 20:211–216. https://doi.org/10.1007/s13562-011-0048-3

Charity JAA, Merilyn AB, Dennis JW, Malcom Higgins TJV (1999) Transgenic tobacco and peas expressing a proteinase inhibitor from Nicotiana alata have increased insect resistance. Mol Breed 5:357–365

Chen MS (2018) Inducible direct plant defense against insect herbivores: a review article. Insect Sci 25:2–23. https://doi.org/10.1111/j.1744-7917.2008.00190.x

Chen Z, Zheng Z, Huang J, Lai Z, Fan B (2009) Biosynthesis of salicylic acid in plants. Plant Signal Behav 4:493–496

Christensen SA, Nemchenko A, Borrego E, Murray I, Sobhy IS, Bosak L, DeBlasio S, Erb M, Robert CAM, Vaughn KA et al (2013) The maize lipoxygenase, ZmLOX10, mediates green leaf volatile, jasmonates and herbivore-induced plant volatile production for defense against insect attack. Plant J 74:59–73

Clausen TP, Reichardt PB, Bryant JP, Provenza F (1992) Condensed tannins in plant defense: a perspective on classical theories. In: Hemingway RW, Laks PE (eds) Plant polyphenols. Basic life sciences, vol 59. Springer, Boston

Cui J, Luo J, Werf WVD, Ma Y, Xia J (2011) Effect of pyramiding Bt and CpTI genes on resistance of cotton to Helicoverpa armigera (Lepidoptera: Noctuidae) under laboratory and field conditions. J Econ Entomol 104:673–684

Dalin P, Björkman C (2003) Adult beetle grazing induces willow trichome defence against subsequent larval feeding. Oecologia 134:112–118; PMID: 12647188; https://doi.org/10.1007/s00442-002-1093-3

Dangl JL, McDowell JM (2006) Two modes of pathogen recognition by plants. Proc Natl Acad Sci U S A 103:8575–8576

Davis RH, Nahrstedt A (1985) Cyanogenesis in insects. In: Kerkut GA, Gilbert LI (eds) Comprehensive insect physiology, biochemistry pharmacology. Pergamon Press, Oxford, pp 635–654

De Boer JG, Posthumus MA, Dicke M (2004) Identification of volatiles that are used in discrimination between plants infested with prey or nonprey herbivores by a predatory mite. J Chem Ecol 30:2215–2230

De Leo F, Bonadé-Bottino M, Ceci LR, Gallerani R, Jouanin L (2001) Effects of a mustard trypsin inhibitor expressed in different plants on three lepidopteran pests. Insect Biochem Mol Biol 31:593–602; PMID:11267898; https://doi.org/10.1016/S0965-1748(00)00164-8

Dicke M, Sabelis MW (1988) How plants obtain predatory mites as bodyguards. Netherlands J Zool 38:149–165

Diezel C, von Dahl CC, Gaquerel E, Baldwin IT (2009) Different lepidopteran elicitors account for cross-talk in herbivory-induced phytohormone signaling. Plant Physiol 150:1576–1586

Doss RP, Oliver JE, Proebsting WM, Potter SW, Kuy SR, Clement SL, Williamson RT, Carney JR, DeVilbiss ED (2000) Bruchins: insect-derived plant regulators that stimulate neoplasm formation. Proc Natl Acad Sci U S A 97:6218–6223

Down RE, Gatehouse AMR, Hamilton WDO, Gatehouse JA (1996) Snowdrop lectin inhibits development and decreases fecundity of the glasshouse potato aphid when administered *in vitro* and via transgenic plants both in laboratory and glasshouse trials. J Insect Physiol 42:11–12

Duffey SS, Stout MJ (1996) A nutritive and toxic components of plant defense against insects. Arch Insect Biochem Physiol 137. https://doi.org/10.1002/(SICI)1520-6327(1996)32:1<3::AIDARCH2>3.0.CO;2-1

Dunse KM, Stevens JA, Lay FT, Gaspar YM, Heath RL, Anderson MA (2010) Coexpression of potato type I and II proteinase inhibitors gives cotton plants protection against insect damage in the field. Proc Natl Acad Sci U S A 107:15011–15015; PMID: 20696895; https://doi.org/10.1073/pnas.1009241107

Dustan WR, Henry TA (1902) Cyanogenesis in plants, II. The great millet, *Sorghwtin vulgare*. Chem News 85:399–410

Dutartre L, JHilliou F, Feyereisen R (2012) Phylogenomics of the benzoxazinoid biosynthetic pathway of poaceae: gene duplications and origin of the *Bx* cluster. BMC Evol Biol 12:64. Int. J. Mol. Sci. 2013, 14 10286

Edelman GM, Cunningham BA, Reeke GN, Becker JW, Waxdal MJ, Wang JL (1972) The covalent and three-dimensional structure of concanavalin A. Proc Natl Acad Sci U S A 69:2580–2584

Engelberth J (2006) Secondary metabolites and plant defense. In: Taiz L, Zeiger E (eds) Plant physiology, vol 4. Sinauer Associates, Sunderland, pp 315–344

Erb M (2009) Modification of plant resistance and metabolism by above- and belowground herbivores. Ph.D. Dissertation, Université de Neuchâtel, Neuchâtel, Swizerland

Erb M, Glauser G, Robert CAM (2012) Induced immunity against belowground insect herbivores—activation of defenses in the absence of a jasmonate burst. J Chem Ecol 38:629–640

Erb M, Veyrat N, Robert CA, Xu H, Frey M, Ton J, Turlings TC (2015) Indole is an essential herbivore-induced volatile priming signal in maize. Nat Commun 6:6273

Farag MA, Pare PW (2002) C6-green leaf volatiles trigger local and systemic VOCemissions in tomato. Phytochemistry 61:545–554

Fatouros NE, Broekgaarden C, Bukovinszkine'Kiss G, van Loon JJ, Mumm R, Huigens ME, Dicke M, Hilker M (2008) Male-derived butterfly anti-aphrodisiac mediates induced indirect plant defense. Proc Natl Acad Sci U S A 105:10033–10038

Fatouros NE, Pashalidou FG, Aponte Cordero WV, van Loon JJA, Mumm R, Dicke M, Hilker M, Huigens ME (2009) Anti-aphrodisiac compounds of male butterflies increase the risk of egg parasitoid attack by inducing plant synomone production. J Chem Ecol 35:1373–1381

Fattorusso E, Taglialatela-Scafati O (2007) Modern alkaloids: structure, isolation, synthesis and biology. Wiley, Hoboken

Feeny PP (1968) Effect of oak leaf tannins on larval growth of the winter moth *Operophtera brumata*. J Insect Physiol 14:805–817. https://doi.org/10.1016/0022-1910(68)90191-1

Felton GW (2008) Caterpillar secretions and induced plant responses. In: Schaller A (ed) Induced plant resistance to herbivory. Springer, New York, pp 369–387

Felton GW, Tumlinson JH (2008) Plant-insect dialogs: complex interactions at the plant-insect interface. Curr Opin Plant Biol 11:457–463

Felton GW, Bi JL, Summers CB, Mueller AJ, Duffey SS (1994) Potential role of lipoxygenases in defense against insect herbivory. J Chem Ecol 20:651–666

Ferrari S, Savatin DV, Sicilia F, Gramegna G, Cervone F, Lorenzo GD (2013) Oligogalacturonides: plant damage associated molecular patterns and regulators of growth and development. Front Plant Sci 4:49

Ferry N, Edwards MG, Gatehouse JA, Gatehouse AMR (2004) Plant-insect interactions: molecular approaches to insect resistance. Curr Opin Biotechnol 15:155–161

Forslund K, Morant M, Jørgensen B, Olsen CE, Asamizu E, Sato S, Tabata S, Bak S (2004) Biosynthesis of the nitrile glucosides rhodiocyanoside A and D and the cyanogenic glucosides lotaustralin and linamarin in *Lotus japonicus*. Plant Physiol 135:71–84

Frey M, Stettner C, Pare PW, Schmelz EA, Tumlinson JH, Gierl A (2000) An herbivore elicitor activates the gene for indole emission in maize. Proc Natl Acad Sci U S A 97:14801–14806

Fürstenberg-Hägg J, Zagrobelny M, Bak S (2013) Plant defense against insect herbivores. Int J Mol Sci 14:10242–10297. https://doi.org/10.3390/ijms140510242

García-Lara S, Bergvinson DJ, Burt AJ, Ramputh AI, Díaz-Pontones DM, Arnason JT (2004) The role of pericarp cell wall components in maize weevil resistance. Crop Sci 44:1560–1567. https://doi.org/10.2135/cropsci2004.1546

Gardner H (1980) Lipid Enzymes: lipases, lipoxygenases, and "hydroperoxidases". https://doi.org/10.1007/978-1-4757-9351-2_25

Gatehouse AMR, Gatehouse JA (1998) Identifying proteins with insecticidal activity: use of encoding genes to produce insect-resistant transgenic crops. Pestic Sci 52(2):165–175

Gill RS, Gupta K, Taggar GK, Taggar MS (2010) Role of oxidative enzymes in plant defenses against herbivory. Acta Phytopathol Entomol Hung 45:277–290. https://doi.org/10.1556/APhyt.45.2010.2.4

Giovanini MP, Saltzmann KD, Puthoff DP, Gonzalo M, Ohm HW, Williams CE (2007) A novel wheat gene encoding a putative chitin-binding lectin is associated with resistance against Hessian fly. Mol Plant Pathol 8:69–82; PMID: 20507479; https://doi.org/10.1111/j.1364-3703.2006.00371.x

Glauser G, Marti G, Villard N, Doyen GA, Wolfender JL, Turlings TCJ, Erb M (2011) Induction and detoxification of maize 1,4-benzoxazin-3-ones by insect herbivores. Plant J 68:901–911

Grayer RJ, Kimmins FM, Padgham DE, Harborne JB, Ranga Rao DV (1992) Condensed tannin levels and resistance in groundnuts (*Arachis hypogaea* (L.)) against *Aphis craccivora* (Koch). Phytochemistry 31:3795–3799. https://doi.org/10.1016/S0031-9422(00)97530-7

Hagenbucher S, Olson DM, Ruberson JR, Wackers FL, Romeis J (2013) Resistance mechanisms against arthropod herbivores in cotton and their interactions with natural enemies. Crit Rev Plant Sci 32:458–482

Handley R, Ekbom B, Agren J (2005) Variation in trichome density and resistance against a specialist insect herbivore in natural populations of Arabidopsis thaliana. Ecol Entomol 30:284–292. https://doi.org/10.1111/j.0307-6946.2005.00699.x

Hanley ME, Lamont BB, Fairbanks MM, Rafferty CM (2007) Plant structural traits and their role in antiherbivore defense. Perspect Plant Ecol Evol Syst 8:157–178. https://doi.org/10.1016/j.ppees.2007.01.001

Heil M (2004) Induction of two indirect defences benefits lima bean in nature. J Ecol 92:527–536

Herbert L, Fraenkel GS, Liener IE (1954) Growth inhibitors. Effect of soybean inhibitors on growth of Tribolium confusum. J Agric Food Chem 2(8):410–414

Hettenhausen C, Baldwin IT, Wu J (2013) *Nicotiana attenuata* MPK4 suppresses a novel jasmonic acid (JA) signaling-independent defense pathway against the specialist insect *Manduca sexta*, but is not required for the resistance to the generalist *Spodoptera littoralis*. New Phytol 199:787–799

Hiraga S, Sasaki K, Ito H, Ohashi Y, Matsui H (2001) A large family of class III plant peroxidases. Plant Cell Physiol 42:462–468

Hopkins RJ, van Dam NM, van Loon JJ (2009) Role of glucosinolates in insect-plant relationships and multitrophic interactions. Annu Rev Entomol 54:57–83

Howe GA, Jander G (2008) Plant immunity to insect herbivores. Annu Rev Plant Biol 59:41–66

Hummelbrunner LA, Isman MB (2001) Acute, sublethal, antifeedant, and synergistic effects of monoterpenoid essential oil compounds on the tobacco cutworm, Spodoptera litura (Lep., Noctuidae). J Agric Food Chem 49:715–720

Irmisch S, Clavijo McCormick A, Günther J, Schmidt A, Boeckler GA, Gershenzon J, Unsicker SB, Köllner TG (2014) Herbivore-induced poplar cytochrome P450 enzymes of the CYP71 family convert aldoximes to nitriles which repel a generalist caterpillar. Plant J 80:1095–1107

James DG (2005) Further field evaluation of synthetic herbivore-induced plan volatiles as attractants for beneficial insects. J Chem Ecol 31(3):481–495

John M, Röhrig H, Schmidt J, Walden R, Schell J (1997) Cell signalling by oligosaccharides. Trends Plant Sci 2:111–115

Johnson ET, Dowd PF (2004) Differentially enhanced insect resistance, at a cost, in *Arabidopsis thaliana* constitutively expressing a transcription factor of defensive metabolites. J Agric Food Chem 52:5135–5138; PMID: 15291486; https://doi.org/10.1021/jf0308049

Kessler A, Baldwin IT (2002) Plant responses to insect herbivory. Annu Rev Plant Biol 53:299–328
Koiwa H, Bressan RA, Hasegawa PM (1997) Regulation of protease inhibitors and plant defense. Trends Plant Sci 2:379–384
Koptur S (1992) Extrafloral nectary-mediated interactions between insects and plants. In: Bernays E (ed) Insect–plant interactions, vol 4. CRC Press, Boca Raton, FL, pp 81–129
Lane GA, Sutherland ORW, Skipp RA (1987) Isoflavonoids as insect feeding deterrents and antifungal components from root of *Lupinus angustifolius*. J Chem Ecol 13:771–783. https://doi.org/10.1007/BF01020159
Herbert Lipke, G. S. Fraenkel, Irvin E. Liener, (1954) . Journal of Agricultural and Food Chemistry 2 (8):410–414
Lorenzo O, Piqueras R, Sánchez-Serrano JJ, Solano R (2003) Ethylene response factor1 integrates signals from ethylene and jasmonate pathways in plant defence. Plant Cell 15:165–178
Louis J, Luthe DS, Felton GW (2014) Salivary signals of European corn borer induce indirect defenses in tomato. Plant Signal Behav 8(11):e27318
Luczynski A, Isman MB, Rawirth DA (1990) Strawberry foliar phenolics and their relationship to development of the two-spotted spider mite. J Econ Entomol 83:557–563
Macedo MLR, Oliveira CFR, Costa PM, Castelhano EC, Silva-Filho MC (2015) Adaptive mechanisms of insect pests against plant protease inhibitors and future prospects related to crop protection: a review. Protein Pept Lett 22:149–163
Maffei ME (2010) Sites of synthesis, biochemistry and functional role of plant volatiles. S Afr J Bot 76(4):612–631
Mahanil S, Attajarusit J, Stout MJ, Thipyapong P (2008) Overexpression of tomato polyphenol oxidase increases resistance to common cutworm. Plant Sci 174:456–466. Int. J. Mol. Sci. 2013, 14 10291
Malagón J, Garrido A (1990) Relación entre el contenido de glicósidos cianogénicos y la resistencia a *Capnodis tenebrionis* l. En frutales de hueso. Bol Sanid Veg Plagas 16:499–503
Mao YB, Cai WJ, Wang JW, Hong GJ, Tao XY, Wang LJ, et al (2007) Silencing a cotton bollworm P450 monooxygenase gene by plant-mediated RNAi impairs larval tolerance of gossypol. Nat Biotechnol 25:1307–1313; PMID: 17982444; https://doi.org/10.1038/nbt1352
Maxwell FG, Lafever HN, Jenkins JN (1965) Blister beetles on glandless cotton. J Econ Entomol 58:792–798
McMahon JM, White WLB, Sayre RT (1995) Cyanogenesis in cassava (*Manihot esculenta* crantz). J Exp Bot 46:731–741
Melan MA, Dong X, Endara ME, Davis KR, Ausubel FM, Peterman TK (1993) An *Arabidopsis thaliana* lipoxygenase gene can be induced by pathogens, abscisic acid, and methyl jasmonate. Plant Physiol 101:441–450
Mellway RD, Tran LT, Prouse MB, Campbell MM, Constabel CP (2009) The wound-, pathogen-, and ultraviolet B-responsive MYB134 gene encodes an R2R3 MYB transcription factor that regulates proanthocyanidins synthesis in poplar. Plant Physiol 150:924–941; PMID: 19395405; https://doi.org/10.1104/pp.109.139071
Møller BL, Seigler DS (1991) Biosynthesis of cyanogenic glycosides, cyanolipids and related compounds. In: Singh BK (ed) Plant amino acids, biochemistry and biotechnology. Marcel Dekker, New York, pp 563–609
Nuessly GS, Scully BT, Hentz MG, Beiriger R, Snook ME, Widstrom NW (2007) Resistance to *Spodoptera frugiperda* (lepidoptera: Noctuidae) and *Euxesta stigmatias* (diptera: Ulidiidae) in sweet corn derived from exogenous and endogenous genetic systems. J Econ Entomol 100:1887–1895
O'Connell DM, Monks A, Dickinson KJ, Lee WG (2015) Is domatia production in Coprosma rotundifolia (Rubiaceae) induced by mites or foliar pathogens? N Z J Ecol 39:214–220
O'Doherty I, Yim JJ, Schmelz EA, Schroeder FC (2011) Synthesis of caeliferins, elicitors of plant immune responses: accessing lipophilic natural products via cross metathesis. Org Lett 13:5900–5903

Oliver JE, Doss RP, Williamson RT, Carney JR, DeVilbiss ED (2000) Bruchins–mitogenic 3-(hydroxypropanoyl) esters of long chain diols from weevils of the Bruchidae. Tetrahedron 56:7633–7641

Osborn TC, Alexander DC, Sun SSM, Cardona C, Bliss FA (1988) Insecticidal activity and lectin homology of arcelin seed protein. Science 240:207–210

Ozawa R, Arimura G, Takabayashi J, Shimoda T, Nishioka T (2000) Involvement of jasmonate- and salicylate related signaling pathways for the production of specific herbivore-induced volatiles in plants. Plant Cell Physiol 41:391–398

Pare PW, Tumlinson JH (1999) Plant volatiles as a defense against insect herbivores. Plant Physiol 121:325–332

Peng C, He DP, Li S, Yang ZD (2016) A facile and efficient synthesis of caeliferin a 16:0. Chem Nat Compd 52:199–201

Petterson DS, Harris DJ, Allen DG (1991) Alkaloids. In: D'Mello JPF, Duffus CM, Duffus JH (eds) Toxic substances in crop plants. The Royal Society of Chemistry, Cambridge, pp 148–179

Phillips MA, Croteau RB (1999) Resin-based defenses in conifers. Trends Plant Sci 4:184–190

Pieterse CMJ, Van Loon LC (2004) NPR1: the spider in the web of induced resistance signaling pathways. Curr Opin Plant Biol 7:456–464; PMID: 15231270; https://doi.org/10.1016/j.pbi.2004.05.006

Puthoff DP, Sardesai N, Subramanyam S, Nemacheck JA, Williams CE (2005) Hfr-2, a wheat cytolytic toxin-like gene, is up-regulated by virulent Hessian fly larval feeding. Mol Plant Biol 6:411–423

Renwick JAA, Zhang W, Haribal M, Attygalle AB, Lopez KD (2001) Dual chemical barriers protect a plant against different larval stages of an insect. J Chem Ecol 27:157583; PMID:11521397; https://doi.org/10.1023/A:1010402107427

Rosenthal GA (1991) Nonprotein amino acids as protective allelochemicals. In: Rosenthal GA, Berenbaum MR (eds) Herbivores: their interactions with secondary plant metabolites, vol 1. Academic Press, San Diego, pp 1–34

Ruuhola T, Tikkanen O, Tahvanainen O (2001) Differences in host use efficiency of larvae of a generalist moth, *Operophtera brumata* on three chemically divergent salix species. J Chem Ecol 27:1595–1615

Schäfer M, Fischer C, Meldau S, Seebald E, Oelmuller R, Baldwin IT (2011a) Lipase activity in insect oral secretions mediates defense responses in Arabidopsis. Plant Physiol 156:1520–1534

Schäfer M, Fischer C, Baldwin IT, Meldau S (2011b) Grasshopper oral secretions increase salicylic acid and abscisic acid levels in wounded leaves of Arabidopsis thaliana. Plant Signal Behav 6:1256–1258

Schmelz EA, Carroll MJ, LeClere S, Phipps SM, Meredith J, Chourey PS, et al (2006) Fragments of ATP synthase mediate plant perception of insect attack. Proc Natl Acad Sci U S A 103:8894–8899; PMID:16720701; https://doi.org/10.1073/pnas.0602328103

Schmelz EA, Carroll MJ, LeClere S, Phipps SM, Meredith J, Chourey PS, Alborn HT, Teal PEA (2007a) Fragments of ATP synthase mediate plant perception of insect attack. Proc Natl Acad Sci U S A 103:8894–8899

Schmelz EA, LeClere S, Carroll MJ, Alborn HT, Teal PE (2007b) Cowpea chloroplastic ATP synthase is the source of multiple plant defense elicitors during insect herbivory. Plant Physiol 144:793–805

Schmelz EA, Engelberth J, Alborn HT, Tumlinson JH III, Teal PEA (2009) Phytohormone-based activity mapping of insect herbivore-produced elicitors. Proc Natl Acad Sci U S A 106:653–657

Schoonhoven LM, Van Loon JJA, Dicke M (2005) Insect plant biology. Oxford University Press, Oxford

Sharma HC, Agarwal RA (1983) Role of some chemical components and leaf hairs in varietal resistance in cotton to jassid, *Amrasca biguttula biguttula* Ishida. J Entomol Res 7:145–149

Sharma P, Narayanan A, El-Shall HE, Moudjil BM (2015) Engineered particulate systems for controlled release of pesticides and repellants. US Patent US20150056259 A1

Simmonds MSJ, Stevenson PC (2001) Effects of isoflavonoids from Cicer on larvae of *Helicoverpa armigera*. J Chem Ecol 27:965–977; PMID: 11471948; https://doi.org/10.1023/A:1010339104206

Steppuhn A, Baldwin IT (2007) Resistance management in a native plant: nicotine prevents herbivores from compensating for plant protease inhibitors. Ecol Lett 10:499–511; PMID: 17498149; https://doi.org/10.1111/j.1461-0248.2007.01045.x

Telang MA, Giri AP, Pyati PS, Gupta VS, Tegeder M, Franceschi VR (2009) Winged bean chymotrypsin inhibitors retard growth of *Helicoverpa armigera*. Gene 431:80–85. https://doi.org/10.1016/j.gene.2008.10.026

Tempfli B, Pénzes B, Fail J, Szabó Á (2015) The occurrence of tydeoid mites (Acari: Tydeoidea) in Hungarian vineyards. Syst Appl Acarol 20:937–954

Tiku RA (2018) Antimicrobial compounds and their role in plant defense. In: Molecular aspects of plant-pathogen interaction. University of Geneva, Geneva, pp 283–307

Ton J, D'Alessandro M, Jourdie V, Jakab G, Karlen D, Held M, Mauch-Mani B, Turlings TC (2007) Priming by airborne signals boosts direct and indirect resistance in maize. Plant J 49:16–26

Trapp S, Croteau R (2001) Defensive resin biosynthesis in conifers. Annu Rev Plant Physiol Plant Mol Biol 52:689–724

Treutter D (2006) Significance of flavonoids in plant resistance: a review. Environ Chem Lett 4:147–157. https://doi.org/10.1007/s10311-006-0068-8

Truitt CL, Wei H-X, Pare PW (2004) A plasma membrane protein from *Zea mays* binds with the herbivore elicitor volicitin. Plant Cell 16:523–532

Turlings TCJ, Benrey B (2016) Effects of plant metabolites on the behavior and development of parasitic wasps. Écoscience 5(3):321–333

Vandenborre G, Miersch O, Hause B, Smagghe G, Wasternack C, Van Damme EJM (2009) *Spodoptera littoralis* induced lectin expression in tobacco. Plant Cell Physiol 50:1142–1155; PMID:19416954; https://doi.org/10.1093/pcp/pcp065

Vandenborre G, Smagghe G, Van Damme EJM (2011) Plant lectins as defense proteins against phytophagous insects. Phytochemistry 72:1538–1550; PMID: 21429537; https://doi.org/10.1016/j.phytochem.2011.02.024

Verma V, Ravindran P, Kumar PP (2016) Plant hormone mediated regulation of stress responses. BMC Plant Biol 16:86. https://doi.org/10.1186/s12870-016-0771-y

Von Dahl CC, Winz RA, Halitschke R, Kühnemann F, Gase K, Baldwin IT (2007) Tuning the herbivore-induced ethylene burst: the role of transcript accumulation and ethylene perception in *Nicotiana attenuata*. Plant J 51:293–307; PMID: 17559506; https://doi.org/10.1111/j.1365-313X.2007.03142.x

Vos IV, Moritz L, Pieterse CMJ, van Wees SCM (2015) Impact of hormonal crosstalk on plant resistance and fitness under multi-attacker conditions. Front Plant Sci 6:639

Wang KL, Li H, Ecker JR (2002) Ethylene biosynthesis and signaling networks. Plant Cell 14 (Suppl):S131–S151

War AR, Paulraj MG, Ahmad T, Buhroo AA, Hussain B, Ignacimuthu S, Sharma HC (2012) Mechanisms of plant defense against insect herbivores. Plant Signal Behav 7(10):1306–1320. https://doi.org/10.4161/psb.21663

Wasternack C, Hause B (2013) Jasmonates: biosynthesis, perception, signal transduction and action in plant stress response, growth and development. An update to the 2007 review in Annals of Botany. Ann Bot 111(6):1021–1058

Wei JN, van Loon JJA, Gols R, Menzel TR, Li N, Kang L, Dicke M (2014) Reciprocal crosstalk between jasmonate and salicylate defence-signalling pathways modulates plant volatile emission and herbivore hostselection behaviour. J Exp Bot 65:3289–3298

Welinder KG (1986) The plant peroxidase superfamily. In: Lobarzewski J, Greppin H, Penel C, Gaspar TH (eds) Biochemical, molecular and physiological aspects of plant peroxidases. University of Geneva, Geneva, pp 3–13

Woldemariam MG, Onkokesung N, Baldwin IT, Galis I (2012) Jasmonoyl-l-isoleucine hydrolase 1 (JIH1) regulates jasmonoyl-l-isoleucine levels and attenuates plant defenses against herbivores. Plant J 72:758–767

Xu S, Zhou W, Pottinger S, Baldwin IT (2015) Herbivore associated elicitor-induced defences are highly specific among closely related nicotiana species. BMC Plant Biol 15:2. https://doi.org/10.1186/s12870-014-0406-0

Yu H, Zhang Y, Wu K, Gao XW, Guo YY (2008) Field-testing of synthetic herbivore-induced plant volatiles as attractants for beneficial insects. Environ Entomol 37:1410–1415

Zhang SZ, Hau BZ, Zhang F (2008) Induction of the activities of antioxidative enzymes and the levels of malondialdehyde in cucumber seedlings as a consequence of *Bemisia tabaci* (Hemiptera: Aleyrodidae) infestation. Arthropod-Plant Interact 2:209–213. https://doi.org/10.1007/s11829-008-9044-5

Zhang P, Broekgaarden C, Zheng S, Snoeren TAL, van Loon JJA, Gols R, Dicke M (2013) Jasmonate and ethylene signaling mediate whitefly-induced interference with indirect plant defense in *Arabidopsis thaliana*. New Phytol 197:1291–1299

Zhao LY, Chen JL, Cheng DF, Sun JR, Liu Y, Tian Z (2009) Biochemical and molecular characterizations of *Sitobion avenae*- induced wheat defense responses. Crop Prot 28:435–442. https://doi.org/10.1016/j.cropro.2009.01.005

Zhuang X, Fiesselmann A, Zhao N, Chen H, Frey M, Chen F (2012) Biosynthesis and emission of insect Herbivory-induced volatile indole in rice. Phytochemistry 73:15–22

Ziegler J, Facchini PJ (2008) Alkaloid biosynthesis: metabolism and trafficking. Annu Rev Plant Biol 59:736–769

Signalling During Insect Plant Interaction

Vibha Gulyani Checker and Meenakshi Sharma

Abstract

Insects and plants share refined interactions as plants recognize insects via mechanical and chemical hints. The initial response begins at plant cell membrane. Insects interact physically with the membrane and triggers production of signalling molecules in the plant. Herbivore contact causes charge distribution differences across the membrane which eventually leads to calcium signalling cascade. Generation of reactive oxygen and nitrogen species also follows membrane depolarization. Multiple hormone response pathways leads to appropriate responses, but the primary signalling cassette mediating the information received at the plant-insect interface and starting defence responses in plants is the jasmonate (JA) pathway. Defence responses are initiated by the accumulation of several secondary metabolites and defence proteins. Plants differently prioritize defence response at different developmental stages. Plants release volatile cues to mediate ecological interactions, and these metabolites play pivotal role in plant-herbivore interactions.

Keywords

Ion flux · Calcium · ROS · RNS · JA · SA · Volatiles

V. G. Checker (✉)
Department of Botany, Kirori Mal College, University of Delhi, Delhi, India

M. Sharma
Department of Botany, Daulat Ram College, University of Delhi, Delhi, India

I. K. Singh, A. Singh (eds.), *Plant-Pest Interactions: From Molecular Mechanisms to Chemical Ecology*, https://doi.org/10.1007/978-981-15-2467-7_9

1 Introduction

Insects represent the abundant group of organisms on Earth and occur in virtually all habitats. Most of the insects have specific feeding habits in having narrow host range of one or two related species of plants. Apart from providing food to the insects, plants provide multitude of ecological opportunities to the insects. Many herbivorous insects spend an important part of their life cycle on their host, and plants provide them with oviposition site, the shelter (from predators and for hibernation) and even a place to associate in the process of sexual reproduction (Kergoat et al. 2017). The intimate association of herbivorous insects with plants has led to species formation in insects. This hypothesis of species diversification of insects because of plants was formulated 150 years ago (Walsh 1867). Recent advances in molecular biology have led to identification of genes important in selection of feed by herbivorous insects. Significant correlation has been observed between plant chemical signatures and insect molecular adaptation coevolving diversification rates within the plants and their insect predators. Therefore it is important to figure out role of signalling events to gain insights in plant-insect interactions.

Plants recognize insects via mechanical and chemical cues. Plants have evolved many strategies against herbivorous insects which protect the plant against herbivory. The direct physical response of plants to cope with these herbivorous insects includes thick cuticle, thorns, glandular trichomes and suberin coating. Plants also defend themselves by releasing some chemical defences upon insect attack, which include alkaloids, glucosinolates, phenylpropanoids, terpenoids and flavonoids. The indirect plant responses that protect the plant employ molecular interactions of the host plant with usual enemies of herbivores (Sabelis et al. 1998) (Fig. 1).

Initially plants sense insects by mechanical stimulation as insects walk on plant surfaces and by the contact of salivary components during feeding. The perception of insects causes increase in concentration of cytosolic calcium, depolarization of plasma membrane potential, ion efflux/influx, protein phosphorylation, activation of MAPK and NADPH oxidase, release of reactive oxygen (ROS) and nitrogen species (RNS). These cascades trigger regulatory responses in the plants that include multiple hormone response pathways, and the pathway playing a core role in increasing plant adaptations to stay fit and survive in the presence of aggressors is the JA-mediated signalling pathway. This in turn leads to gene expression changes and production of defence response proteins producing volatile and toxic compounds (Zebelo and Maffei 2015).

2 Molecular Recognition of Insects by the Plants

Plants possess a strong immune system with multiple layers to sense, detect and restrict pathogen entry and expansion. The primary response is initiated by the pattern recognition receptors (PRRs) which detect and analyse pathogen-/microbe-associated molecular patterns (PAMPs/MAMPs) and initiate pattern-triggered immunity (PTI). PRRs are present on the surface of the plant cell; therefore PTI

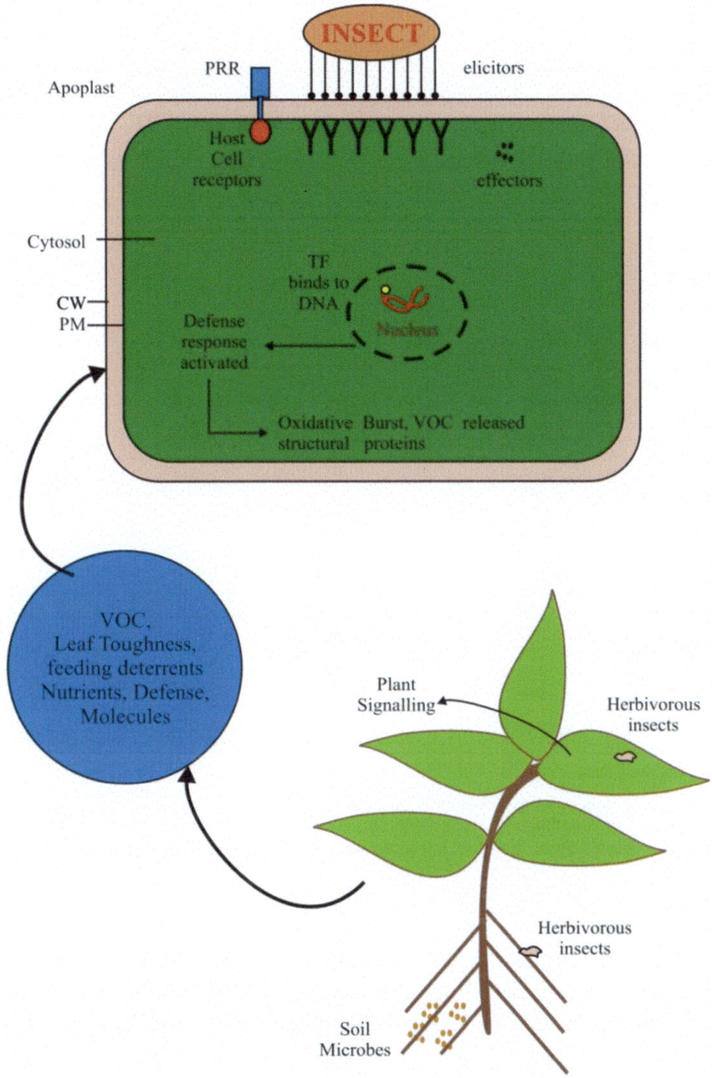

Fig. 1 Model of early events in insect-plant communication

limits pathogen inception. Damage-associated molecular patterns (DAMPs) are produced upon infection by insects which are also recognized by PRRs. However, a few pathogens have developed the ability to fight PTI by developing effector proteins. These effector proteins increase the ability of the pathogen to combat basal defence of the plant. Hence, these pathogens can easily grow and reproduce inside the host plant. Plants supress these pathogens through production of intracellular resistance (R) proteins. These proteins trigger effector-triggered immunity (ETI), a defence response of the plant to suppress pathogen and recover resistance by starting

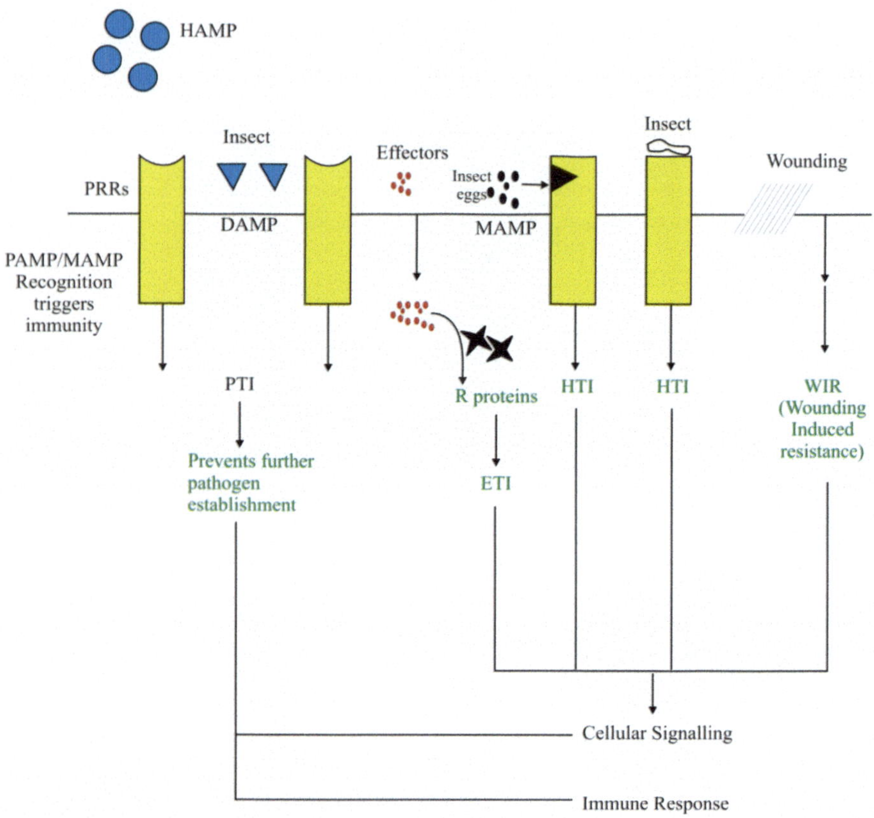

Fig. 2 Molecular recognition of insect by the plant

defence against pathogens (Jones and Dangl 2006). Plant-insect interaction differs from plant-pathogen interaction, but clues from PTI/ETI model research can be used to recognize the threats of herbivory (Fig. 2).

The earliest interaction of the plant with the insect happens when the insect's tarsi comes in contact with the leaf surface. The movement of the insect tarsi on the surface of the leaf will break the trichomes due to pressure. During this time, certain chemicals are released from tarsal pads and are deposited on the leaf surface (Hilker and Meiners 2010). Plants can perceive this pressure. The Venus fly trap (*Dionaea muscipula*) shows high receptivity to touch as it was observed to close immediately when its trichomes are touched by insects (Forterre et al. 2005). Studies in *Medicago truncatula* have shown that mechanical stimulation by repeated touching can stimulate the accumulation of phytohormone jasmonic acid (Tretner et al. 2008). Glandular trichomes also act as initial sensors, thus warning the plants. Induction of defence responses was observed in tomato (*Solanum lycopersicum*) plant. The early signals induced after the leaf trichomes are broken by moths or caterpillars include release of hydrogen peroxide (H_2O_2) and production of proteinase inhibitors (Peiffer et al.

2009). Plants provide herbivorous insects with the oviposition site. The oviposition-associated changes in plants act as MAMP to recognize and predict herbivore attack. Oviposition-associated compounds are also identified by unknown receptors on the plants, thereby activating defence reactions. The constitution of necrotic zones or undifferentiated cells was observed following egg deposition in *Brassica nigra* (Shapiro and DeVay 1987) and *Pisum sativum* plants (Doss et al. 2000). Bruchins, the insect-induced plant regulators, incite neoplasm formation following oviposition (Doss et al. 2000). The neoplasm helps to boost plant resistance by preventing the larvae entry into the pod (Doss et al. 2000). Insects also produce oviposition effectors which manipulate host defence response to suppress the plant immunity (Bruessow et al. 2010; Peñaflor et al. 2011).

Insects interfere with integrity of the plant tissue, and mechanical wounding can activate many plant defence responses. This is due to the fact that mechanical wounding disrupts intracellular compartmentalization of the plant leading to mechanisms for the generation of molecules that stimulate general plant immune responses. Wounding causes the release of self-derived molecules of the cell in response to stress and initiate wound-induced resistance (WIR) in the cell. At the same time, WIR can be supressed by effector-like molecules from insects. Plants also recognize compounds released by insects during feeding, and wound-induced resistance mechanisms are activated. In addition to WIR, there are mechanisms by which plants can sense insects. The recognition pathway for Hemipteran and Dipteran insects follows the PTI/ETI theory (Hogenhout and Bos 2011). Many of the insect herbivore species include chewing insects like beetles and caterpillars. Recent reports have demonstrated that the application of oral secretions of the insect to the plant injuries results in amplification of wound response of the plant (Erb et al. 2009). Numerous insect-derived compounds have been identified as elicitors, namely, fatty acid-amino acid conjugates (FACs) (Alborn et al. 2007), sulphur-containing fatty acids like inceptins (Schmelz et al. 2006), peptides like glycolate oxidase (Eichenseer et al. 1999) and lipases (Schäfer et al. 2011. PRRs can also induce herbivore-triggered immunity (HTI) by recognizing putative herbivore-associated molecular patterns (HAMPs). These chemical signatures initiate plant response upon insect attack, for example, inceptins control the production of ethylene, increase the levels of jasmonic acid and salicylic acid and further regulate the expression of various defence reactions to modulate plant perception of herbivory.

3 Induction of Wound-Activated Surface Potential

Leaf damage by the insect initiates a cascade of events involving production of molecules which bind specific receptors at plant cell membrane or direct delivery of elicitors on the cell membrane. This generates difference in electrochemical gradient between the exterior and interior of plant cell (V_m). The differences between the two sides of plant cell lead to depolarization or hyperpolarization resulting in either more positive or more negative V_m values. These differences have been demonstrated across taxa (Zebelo and Maffei 2015; Bricchi et al. 2013; Mohanta et al. 2012).

Chewing by the insect is sufficient to induce an electrophysiological response which can even be recorded in phloem sieve elements and contributes to wound-activated surface potentials (Salvador-Recatala et al. 2014). Oral secretions (OS) from certain groups of insect herbivores hold effectors that have the ability to overcome basal defence of the plant. Oral secretions also contain various microbes that can modify plant-insect interaction and are responsible for defence suppression (Chung et al. 2013). Furthermore, oral secretions of insect herbivores are known to transmit electrical signals including action potentials (APs) and variation potentials (VPs) on nearby receiver plants. Action potential comprises of a momentary long-distance signalling system which can perceive environmental stimuli and start intercellular and intracellular communication. These signals propagate and travel throughout the plant from the point of origin (Volkov 2012). A number of herbivore-induced plant volatiles trigger electrical potential across the membrane, thereby depolarizing the membrane and activating voltage-gated ion channels. Activation of these channels accelerates depolarization of membrane (Zebelo et al. 2012). Long-distance communication between wounded leaves and other parts of plant was demonstrated by Mousavi et al. (2013). The group demonstrated reduced expression of JA responsive gene in the leaves farthest to the wound, thus demonstrating existence of genes functioning in wound signalling. Further, it has been suggested that plants can differentiate between vibrations caused because of different reasons like chewing, wind or insect song. This discovery of vibration-induced signalling cascade is the focus of several research groups (Gagliano et al. 2012; Monshausen and Haswell 2013).

4 Calcium Sensors in Plant Immune Response

Insect feeding or insect-derived elicitors cause variations in the V_m because of increase in ion species including protons (H^+), calcium (Ca^{2+}), potassium (K^+) and chloride (Cl^-). Feeding by herbivores is known to drive changes in calcium concentrations, and calcium sensor proteins detect calcium signals and manage downstream targets to begin signalling pathway. Proteins functioning as calcium sensor in plant immune response are calmodulin (CaM), CaM-like (CML) proteins, calcineurin B-like (CBL) proteins and calcium-dependent protein kinases (CPKs). Figure 3 illustrates ion channels in plant immune response.

CaM: Calmodulin is a key calcium sensor upon herbivory and plays important role in wound signalling. Seven genes encoding four CaM isoforms are encoded by *Arabidopsis* genome (Batistic and Kudla 2012). It has also been shown that Ca^{2+}/CaM binding plays critical role in herbivore-induced wound response mediated by *AtSr1* (the SIGNAL RESPONSIVE1) which encodes a calmodulin-binding transcription factor in *Arabidopsis* (Galon et al. 2008).

CMLs: In *Arabidopsis* there are 50 CaM-like (CML) proteins which function as calcium sensors. These proteins are present in cytoplasm and undergo lipid modifications to bind membrane (Batistic and Kudla 2012). *CML43* and *CML42* are the key CMLs conferring plant immune response against herbivory. Oral

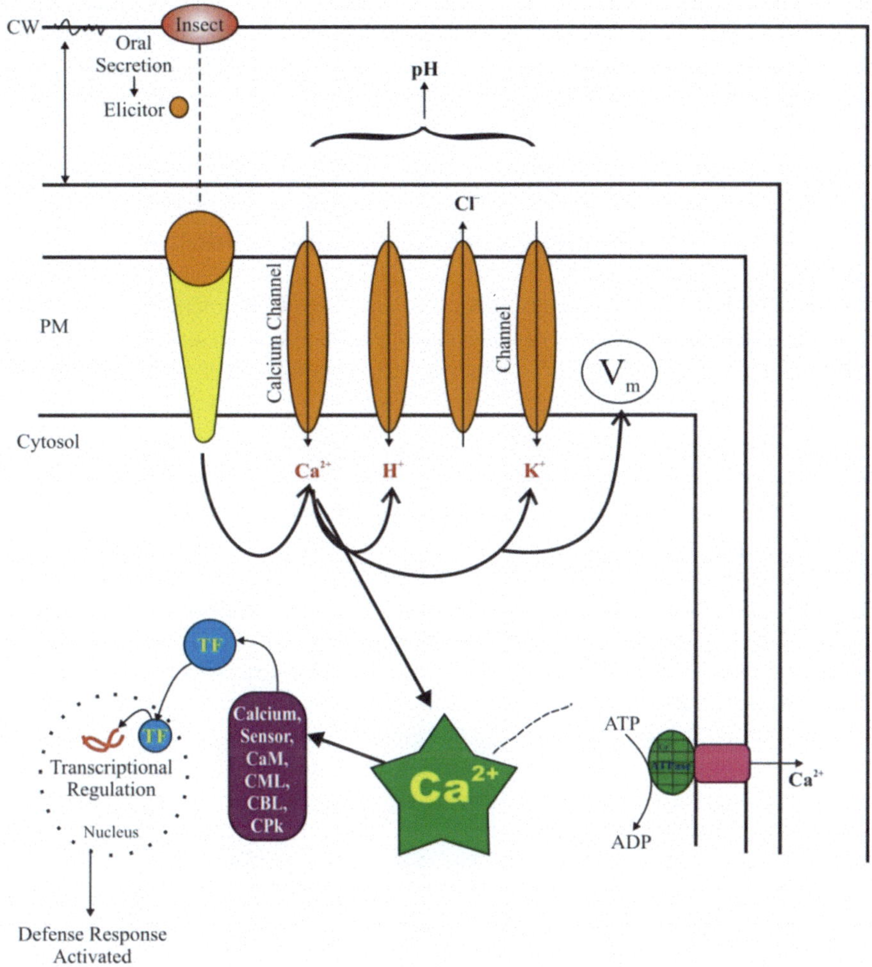

Fig. 3 Ion channels in plant immune response

secretion upregulates expression of *CML42*, and this upregulation is adversely regulated by JA receptor CORONATINE INSENSITIVE 1. CML42 is a major signalling component linking Ca^{2+} and JA signalling (Vadassery et al. 2012).

CBLs and CPKs: These calcium-sensing proteins interact specifically with CBL-interacting protein kinases and form CBL-CIPK complexes. Apart from decoding calcium signals, these complexes were shown to coordinate the activity of K^+ channel in grapevine (Cuellar et al. 2013). Moreover, herbivory elicits genes involved in CBL-CIPK pathway (Yang et al. 2011). CPKs serve as calcium-binding proteins and modulate expression of other genes by translating calcium signals into phosphorylation events, for example, CPK3 and CPK13 regulate transcription of *defensin* gene, independent of ethylene, JA and ABA (Kanchiswamy et al. 2014).

Subsequent to herbivory there is burst of cytosolic calcium. Therefore, to struggle with negative effects of calcium, plants have evolved Ca^{2+}-ATPases, which belong to P-type ATPase superfamily and regulate calcium concentration by maintaining homeostasis. It acts by actively transporting ions across membranes (Bose et al. 2011).

5 Potassium Channels in Plant Immune Response

Potassium performs fundamental functions in controlling membrane potential. Herbivory-induced increased cytosolic calcium concentration causes opening of inward potassium channels. These channels are responsible for herbivore-induced V_m depolarization.

6 ROS and RNS in Cell Signalling upon Herbivory

Reactive oxygen species (ROS) plays pivotal role in defence response of plants, and accumulating evidences suggest involvement of ROS in plant-insect interactions (Kerchev et al. 2012). As elicitors originating from herbivory bind putative receptors on the membrane, there is increase in cytosolic calcium. This in turn triggers CBL-CIPK signalling pathway which further activates plant cell membrane NADP oxidase – the ROS-generating enzyme. Superoxide radical is generated and is soon converted into H_2O_2 by the enzyme superoxide dismutase. H_2O_2 plays a pivotal role in plant defence via oxidative crosslinking of cell wall proteins, modulating host defence-related genes and killing herbivorous insects. H_2O_2 also enters the cytosol, thereby increasing cytosolic concentration of H_2O_2. The cytosolic levels of H_2O_2 are reduced by the enzymatic actions (Zebelo and Maffei 2015). The antioxidant enzymes were found to be more active upon herbivore wounding, indicating their role in conferring resistance against insects (An et al. 2010).

Nitric oxide (NO) has a noticeable role in plant-insect interaction (Wuensche et al. 2011). There is increase in nitric oxide synthase (NOS) activity because of increased cytosolic calcium. NOS further increase cytosolic NO levels. Evidences indicate herbivory induced transient and rapid increase in NO levels (Liu et al. 2011; Bricchi et al. 2010). Various *nitroso* species are produced having possible involvement in JA signalling pathway and inducing expression of defence genes. NO acts as a central player in activation of the phenylpropanoid pathway by promoting cell death and activating the enzyme phenylalanine ammonia lyase (PAL) (Fig. 4). The phenylpropanoid metabolism is involved in common lignin or flavonoid biosynthesis and also synthesizes variety of other aromatic metabolites (coumarins, phenolic volatiles or hydrolyzable tannins). These aromatic metabolites contribute to stability and robustness towards mechanical or environmental damage. Phenylpropanoids are also key players in mediating the resistance of plants from pests (La Camera et al. 2004).

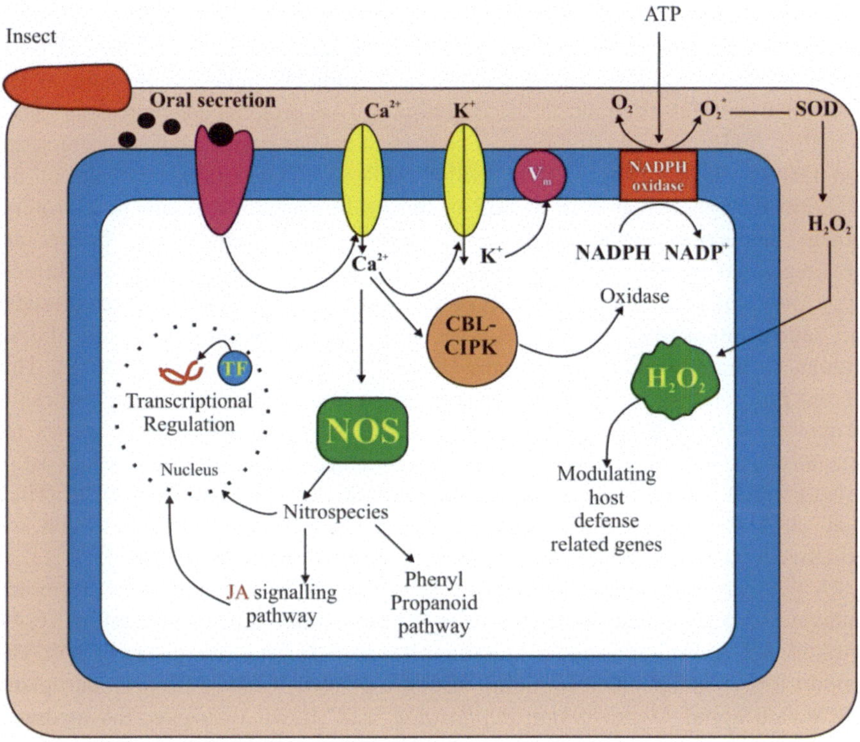

Fig. 4 ROS and RNS signalling in insect-plant interaction

7 Plant Hormones Mediate Insect-Specific Plant Reactions

Plants produce various chemical signatures to hinder attackers. These chemicals are synthesized only upon need as constitutive generation of chemicals and volatiles that are hazardous to insects can be detrimental to plants as well. Following insect attack, an intricate web of signalling molecules is synthesized in plants to stimulate a resistance response and to produce defensive molecules (Wu and Baldwin 2010). Certain pathogens induce production of jasmonic acid and ethylene which in turn trigger defence reactions that deter these tissue damaging insects. These responses include transcriptional activation of defence-responsive genes like protease inhibitors and secondary metabolites. Salicylic acid-regulated responses and JA/ ET-mediated responses are antagonistic to each other. Though the plant uses diverse and specific signalling mechanisms for the recognition of the insect, SA and JA play a pivotal role in mediating plant defence responses against a few sap-feeding insects (hemipterans like aphids and whiteflies) and biotrophic pathogens (Zarate et al. 2007).

Chemical signatures like HAMPs, DAMPs and wounding induce JA signalling pathway leading to increased accumulation of jasmonoyl-L-isoleucine (JA-Ile) which regulates most of the responses. JAZ proteins have Jas domain at the C-terminal and ZIM domain at the N-terminal. JAZ proteins are repressors of JA signalling under normal conditions. The C-terminal Jas domain binds to transcription factors like CORONATINE INSENSITIVE1 (COI1) (a component of the ubiquitin E3 ligase SCF^{COI1}), MYC2 or the JA-Ile receptor and acts as a protein-protein interaction surface (Yan et al. 2009). These transcription factors are downregulated by accumulating levels of JAZ proteins under normal conditions. Upon mechanical wounding or insect attack, jasmonoyl-L-isoleucine accumulates and causes interaction of COI1-JAZ and ubiquitin-dependent destruction of JAZ proteins mediated by 26S proteasome (Koo et al. 2009; Sheard et al. 2010). This process promotes release of other transcription factors like MYC2, which pushes the formation of proteins that fosters defence and inhibits expression of genes for vegetative growth. Plant defence against insect attack occurs at a metabolic cost of growth which can be considered as a balancing act to optimize plant health (Huot et al. 2014). Herbivory also causes changes in intracellular ion concentration and oxidative burst and regulates the JA pathway either directly or indirectly.

JA pathway is a central pathway in providing resistance, but other hormonal pathways also play their respective roles. These include auxin, gibberellins (GA), cytokinins (CKs), ethylene (ET), brassinosteroids (BRs) and salicylic acid (SA). The support for JA-independent hormone response pathways comes from tomato plant. The potato aphid *Macrosiphum euphorbiae was* shown to trigger SA-mediated signalling. SA acts as a dominant player communicating plant immunity by antagonizing various pathogen-produced hormones. In general, SA and JA act antagonistically, whereas ET doesn't have clear effects on JA-mediated resistance response as it can act positively or negatively.

A rapid signal also runs through the plant as higher levels of defence related genes were found in distal leaves of the plant along with the affected leaf. This indicates the involvement of a mobile signal from wounded tissues to distal organs. Insect oral secretions containing elicitors are required to start the response of this signal. Little is known about the nature of this molecule, but xylem and phloem are good candidates for passing the signal.

8 Jasmonic Acid (JA)

Jasmonic acid (JA) is found naturally in plants. Several genes are expressed in multiple defensive mechanisms highly regulated by JAs and their derivatives. JA has significant physiological roles during development and in response to biotic and abiotic stress (osmotic and drought stress) in plants. There are various forms of JA like jasmonates (JAs), jasmonic acid (JA) and methyl jasmonates (MeJAs). These forms regulate various physiological processes such as fertility, sex determination, differentiation, elongation of root system and response to stress, along with other phytohormones; JAs concentration increases in wounded leaves where herbivore attacks because of exposure to chitins, oligosaccharides and oligogalacturonides. It is also reported that concentration of various plant secondary metabolites (PSMs) such as terpenes, terpenoids, alkaloids and phenolic compounds increased in response to herbivores which is dynamically controlled by JAs. The jasmonate (JA) pathway is crucial for plant-insect interaction to develop broad-spectrum defence responses. Various microbial pathogens and insects not only manipulate the JA biosynthesis and signalling mechanisms but also modulate the concentration of several other hormones.

9 Ethylene (ET)

Ethylene (ET) plays important roles in different processes of plant development in the life cycle of plant like organ abscission, cell death, seedling emergence, leaf and flower senescence and ripening. ET also plays central role in responsive mechanisms of the plant during various insect or other pathogenic microorganisms. The rapid production and accumulation of ethylene ET in different concentration is one of the first indigenous signalling events after the attack of insect on plant species. Various studies showed that high level of ethylene ET biosynthesis and accumulation increased the resistance line in plant to various infections. ET regulates the plant defence mechanisms dynamically as ET helps in making networks with different signalling pathways of hormones.

10 Abscisic Acid (ABA)

Stress signals are known to increase the levels of endogenous ABA, the stress hormone. Abscisic acid belongs to sesquiterpenes and is an isoprenoid kind of metabolite. ABA is derived from a (C5) precursor isopentenyl (IDP). In 1960, ABA was isolated and identified from cotton balls and is found to be the most important stress signal universal to all kingdoms of life. ABA plays dynamic role at each level of cell growth from the biosynthesis of enzyme to defensive response against biological or chemical stresses such temperature, low moisture, high salts and high radiations. The involvement of ABA in the process of stomatal closure is well documented. ABA dynamically works on fine-tuning of signalling networks in both abiotic and biotic stress. Abscisic acid (ABA) regulates biotic stress pathway during plant interaction with pathogens. Signal transduction pathways responsive to ABA vary in different organisms. Several studies reported that tissue concentration of ABA increased in corn (*Zea mays*) during pathogen infection. The level of ABA increased in plant root system when western corn rootworm (*Diabrotica virgifera*) attaches to the corn plant. The goldenrod species (*Solidago altissima*) synthesized more ABA after the infection of tobacco budworm (*Heliothis virescens*) caterpillar. In contrast goldenrod species (*Solidago altissima*) synthesized ABA in low level during infection with gall-inducing caterpillar *Gnorimoschema gallaesolidaginis* (Tooker and Moraes 2008).

11 Brassinosteroids (BRs)

Brassinosteroids (BRs) are a universal class of plant steroids. Structurally brassinolide has a highly unique ring structure and consists of a lactone moiety in the B ring and a pair of vicinal diols in A ring and side chain. This ubiquitous hormone (BR) was isolated from *Brassica napus* pollen grain in 1979. Several physiological and phenotypic anomalies are associated with defects in BRs biosynthesis. BR signaling and its regulating proteins play role in diverse physiological processes in plant growth and development including photomorphogenesis, stem elongation, stomatal development, flowering and pollen tube growth. BRs play prominent role in conferring tolerance against abiotic stresses (high temperature, chilling, drought, salinity) and in increasing plant immunity to pathogens and insects. BRs also increase the activation mechanism of proton pumps and modulate the cellulose microtubules and xylogenesis. Plant developed different levels of herbivore defence system which is triggered by the wounded or infected tissues. Peptide hormone system acts as the spreading signal in infected plants. Various studies documented that BRs interact with other growth regulators such as auxin, cytokinin, ethylene, gibberellin, jasmonic acid, abscisic acid, salicylic acid and polyamine in mediating array of physiological processes and in synthesizing stress defensive processes in plants. BRs regulate glucosinolate (GS) biosynthesis to function in plant defence against herbivory (Belkhadir and Jaillais 2015).

12 Gibberellins (GAs)

Gibberellins belong to tetracyclic diterpenoid family and are synthesized from geranylgeranyl diphosphate. They regulate processes like seed germination, stem elongation, leaf expansion, induction of flowering, pollen maturation and fruit senescence and control growth and development of plant at various stages. Gibberellic acid (GA) and its signalling components are the critical regulators against various types of necrotrophic and biotrophic pathogens. However, the significance of GA is less understood, but various scientific studies suggest that GA interfere with the immune system to develop plant resistance. GA are known to control plant disease resistance by inducing the degradation of DELLA proteins. This interaction between GA and DELLA increased the resistance response of plants towards pathogenic fungus. GAs fine-tune interactions between plant and insect.

13 Cytokinins

Cytokinin is an essential class of diverse phytohormones involved in controlling signalling pathway during different growth phases of plant. Cytokinins are formed from N (6)-substituted adenine derivatives. Cytokinin was first reported as crucial inducer molecule during cell division. Cytokinins are plant hormones that are not only key regulator hormone of cell cycle but are also important for several mechanisms like morphological changes, defence, senescence, ion uptakes and biotic and abiotic interactions. Plant adaptations to different stresses are negatively regulated by cytokinins. Identification of various cytokinin receptors helped to understand the cytokinin signalling pathway. Cytokinin signalling is coupled to environmental stimuli by a multistep phosphorelay. The two components of cytokinin signalling cascade include hybrid histidine kinases (HKs) (membrane-bound) and a cytokinin-binding input domain (CHKs). The concentration of cytokinin increases and starts accumulating in plant tissues after insect or pathogen attacks the plant to develop resistance. CK plays role as a pivotal signalling compound for modulation of primary and secondary metabolism. Interestingly, some insects such as leaf miners use cytokinin to modify the tissue surrounding their mines, resulting in the well-described phenomenon of "green islands" to increase their fitness. The green islands are characterized by photosynthetically active green patches on senescing leaves and correspond to regions with an increased concentration in cytokinins and an enriched nutritional environment. CKs are integral components of wounding, and HAMP (herbivore-associated molecular pattern) triggered resistance responses in most of the stressed plant to protect them from pathogens (Walters et al. 2008).

14 Interplay Between Ethylene (ET), Jasmonic Acid (JA) and Salicylic Acid (SA) Network of Regulated Interactions

As we have discussed earlier, induction of defence mechanisms in plant is regulated by phytohormones like ethylene (ET), jasmonic acid (JA), abscisic acid (ABA), auxin and gibberellic acid (GA). Ethylene, salicylic acid, cytokinins and jasmonic acid play multiple roles in plant defence and infection mechanisms in response to insects and pathogenic microorganisms. Interestingly, ethylene and cytokinins modulate defence signalling pathways which are regulated mainly by jasmonic and salicylic acid. The JA acts synergistically with ET to activate defensive response against necrotrophic pathogens and insects including chewing-biting herbivores, cell-content feeders and necrotrophic pathogens. SA exerts effect against biotrophic pathogens (piercing-sucking insects). SA downregulates JA biosynthesis genes to target JA signalling. The antagonistic action of SA on JA varies with type of pathogens and timing of infection. Plant-insect interaction decides the action of ethylene and SA as ethylene can have both positive and negative effect on JA. ET plays a crucial role in fine-tuning as a traffic controller on the hormonal crossroads to defence responses (Broekgarden et al. 2015).

15 Interplay of Gibberellic Acid (GA) and Jasmonic Acid in (JA) Mediated Defence Response

As discussed before about highly regulated crosstalk between SA, JA and ET phytohormones, there exists strong interaction between GA and JA mediated response of plants against insects. JA prioritizes on defence networks during infection and interferes with gibberellin signalling cascade to suppress the plant growth. One of the important signalling regulator proteins, DELLA represses transcription of GA-responsive genes and antagonistically interacts with JA signalling. GA basically disables JA-mediated pathogen resistance by degrading DELLAs (Yanga et al. 2013).

16 Interplay of ABA, Auxin and JA in Defence Response

ABA and JA hormones have synergistic action in response to different stresses. Auxin is well-known for plant growth and development and also plays an important role in plant defence against pathogen attack. Gene expression involved in GA biosynthesis is downregulated by ABA (Alberico et al. 2018). Concentration of nicotine and jasmonates decreases due to IAA during stress. IAA signalling may activate different kind of plant responses and modulate the other hormonal pathways during pathogen attack.

17 Interplay of Cytokinins and JA in Plant Defence

Cytokinin plays a crucial role in regulation of cell cycle and develops synergistic and antagonistic interactions with other hormones for defensive mechanisms to several biotic and abiotic stresses. Plant tissues produced different concentration of JA and CK due to crosstalk. Enough reports are not available to explain hormonal interaction between JA and CK. In several studies, the JA and CK levels increase significantly which accelerate the defence response in plant during pathogen attack. JA treatments in infected plant can induce the accumulation of CK ribosides. These observations show that there is specific interaction between CK and JA and also interplay with other hormones during infection or based on type of stresses (Brien and Benková 2013).

18 Age-Dependent Dynamics of Plant Insect Resistance

Plant age has important implications on insect resistance as aged plants display higher resistance against insect herbivores. A recent study indicates that JA response decreases steadily with plant age but the defence compounds increase progressively, thereby conferring insect resistance to the plant throughout its growth (Mao et al. 2017). The group demonstrated that this reduced JA response is primarily modulated by the *miR156* which targets a group of transcription factors called Squamosa promoter binding protein-like (SPL) and functions as an important regulator in age-dependent development. The abundance of miR156 is high during the young stage and decreases during later plant growth and development. This leads to steady increase in the level of SPLs which controls a wide range of processes including shoot regeneration, flowering, vernalization, trichome initiation and secondary metabolite production. SPL9 interacts with JAZ proteins, promotes JAZ accumulation by preventing its degradation and therefore negatively regulates JA response and insect resistance (Mao et al. 2017). GA interferes with JA signalling. DELLAs, the master growth repressors, also work as repressors of GA analogous to JAZ proteins of the JA signalling pathway. DELLA proteins are degraded when the concentration of GAs increases. Recent investigations revealed that DELLAs directly bind to SPL proteins and negatively regulate their transcriptional activity, thereby positively regulating insect resistance. DELLAs also interact with MYC2 (Hong et al. 2012) and JAZ proteins to control suppression of JA-responsive genes (Hou et al. 2010).

Though JA pathway plays a dominant role in providing high resistance against herbivores, it sacrifices plant growth by interfering with other hormone signalling pathways like auxin and GA (Yang et al. 2012). Plants maintain the balance between defence and growth by decreasing the JA signal with age. Alleviated levels of nutritional, structural and mechanical components as well defence and secondary metabolites provide a higher level of resistance to aging plants and, therefore, compensate for the delayed JA response. The decline of JA signal could be a trade-off strategy of plants to safeguard successful development, in which the SPL proteins form the regulatory loop to form the balance between defence and the growth (Huot et al. 2014).

19 Effect of Soil Microbes on Plant Heath Improvement

A three-way interaction exists between host plant, insects and microbes as microbial communities also exert influence on insect-plant relationship. It has been demonstrated that aphid-barley interactions are reliant on the genotypes of the interacting species and rhizosphere bacteria (Tetard-Jones et al. 2012). Beneficial soil microorganisms can modulate plant defence against insect herbivores by interfering with hormone signalling pathways. Plants continuously interact with soil microorganisms which can help plant fitness in different ways (Fig. 5). The beneficial genera can help plants with important abilities and confer resistance against insect herbivores. Plant growth-promoting fungi and rhizobacteria help the plant by fixing atmospheric nitrogen and making unobtainable foods available. Beneficial microorganisms improve nutrient composition of the plant and enable growth of tissues subsequent to herbivory stimulating plant tolerance (Kempel et al. 2009). Many beneficial microbes can also synthesize plant hormones including IAA, cytokinin, auxins, JA and GA which together form important network during plant-insect interaction (Contreas-Cornejo et al. 2009). The plant hormones ET, JA and

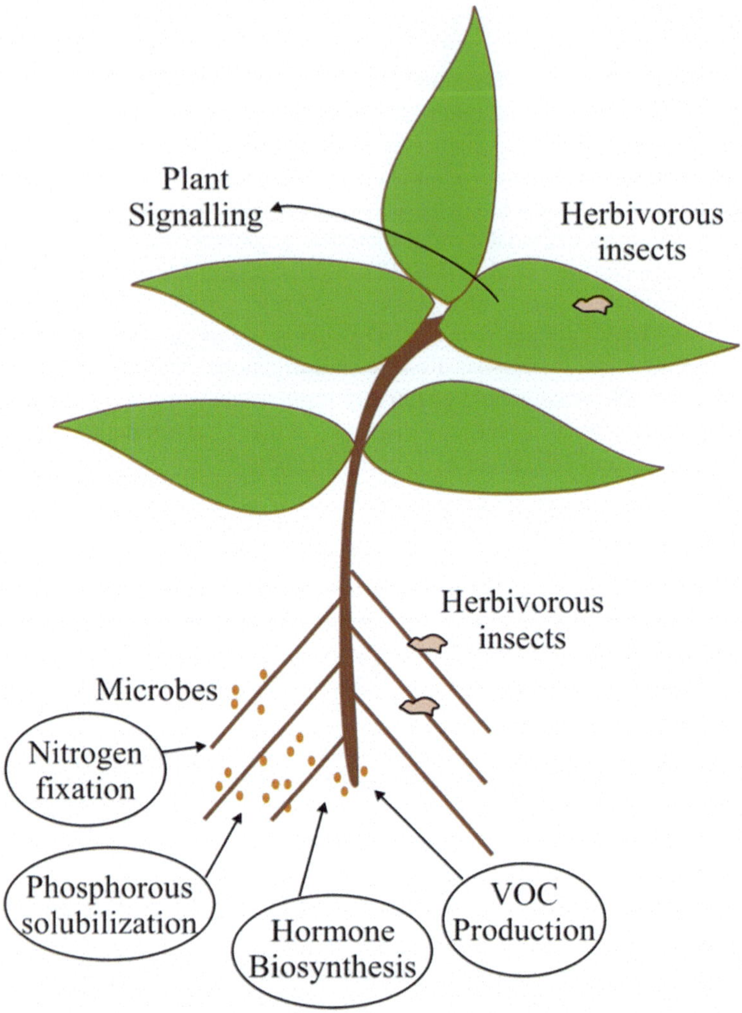

Fig. 5 Beneficial effects on plant health post interaction with insects

SA can modulate resistance by beneficial microbes. In a recent study, Pangesti et al. (2016) have shown that when *Arabidopsis* roots were treated with rhizobacterium *P. simiae*, the plant showed resistance to chewing insects and there was higher expression of JA/ET-responsive genes. Similarly *Bacillus subtilis* promotes resistance of tomato plants against phloem insect whitefly by regulating expression of JA-dependent and JA-independent genes (Valenzuela-Soto et al. 2010). Resistance mechanisms are alleviated through recognition of microbial elicitors. The biosynthetic pathways for secondary metabolites, defence compounds and plant volatiles are activated by root colonization of soil microorganisms. Among these compounds, the biosynthesis of camalexin and glucosinolates is increased during rhizobacterial colonization. These compounds play important role in plant defence against leaf chewers and sap-sucking insect herbivores (Clay et al. 2009; Pangesti et al. 2016). Flavonoids are also found in root exudates and are important insect feeding inhibitors. Herbivory can activate cascade of JA signalling pathway which positively regulates the biosynthesis of flavonoids and anthocyanins (Dombrecht et al. 2007). The flavonoid tricin has insecticidal activity against mosquito larva and inhibits infestation of brown plant hopper in resistant rice cultivar (Bing et al. 2007). Phenolics are the most common group of defensive compounds against herbivorous insects. These compounds are accumulated by plant growth-promoting rhizobacteria. Phenolics produce hypersensitive response and confer resistance against attack by herbivorous insects. Lignin, an important phenolic compound, enhances leaf hardiness and physically limits entrance and feeding of insect herbivores (Johnson et al. 2009; Barakat et al. 2010). Another phenolic sesquiterpenoid gossypol confers resistance against many chewing and sucking insects. Inoculation of cotton plants with *Bacillus* species induces the expression of JA related genes which initiates transcription of gossypol genes to reduce herbivory (Wu and Baldwin 2010).

20 Chemical Signals in Insect-Plant Interaction

Signal can be defined as a specific mixture of molecules which are generated with specific ratio and released in minute quantities (10^{-8}, 10^{-9} gr). Once these signals are perceived and analysed in the antennal lobes of the brain, several behavioural or physiological responses are observed in insects. Leaves, stem, roots and flowers release diverse volatile organic compounds (VOC). These have biological origin and are known as biogenic VOCs. The released metabolites vary between different plant species and different genotypes within a single species. These metabolites play many ecological functions for the benefit of plants. They act as pollinator attractants and herbivore and pathogen repellents. These metabolites indirectly help the plant against insects by attracting natural enemies of herbivores (Tholl 2015). Chemical signals released by plants are species and variety specific. These signals vary in nature due to change in plant physiology, periodicity and environmental factors (climate and pollution) (Frerot et al. 2016). Chemically most of plant volatiles are alkanes, alkenes, aldehydes, ketones, esters and alcohols. Plants evolved these secondary metabolites as their defence, but insects have coevolved and use them against plants. Three major biosynthetic pathways terpenes, oxylipins and shikimate and benzoic acid are responsible for producing majority of plant volatiles (Maffei et al. 2010).

Terpenes: The basic structure of terpenes is isoprene unit (C5). These isoprene units are produced in plant chloroplasts. Their synthesis is dependent on light and temperature. Terpenes have been extensively studied and represent highly variable group of volatile compounds in plants. Terpenoids vary with each other according to the number of comprising isoprene units. Monoterpenes (C10) and sesquiterpenes (C15) are usually volatile in nature. Diterpenes can be either volatile or non-volatile.

Some of the homoterpenes like E-4,8-dimethyl-1,3,7-nonatriene (C11, DMNT) and 4,8,12-trimethyltrideca-1,3,7,11-tetraene (C16, TMTT) are herbivore stimulated. These are important compounds in signal transduction during plant-herbivore interaction. Dutta et al. (2014) have demonstrated that plants release volatile terpene-based compounds in long time to participate in ecological communications.

Oxylipins: These are known as green leaf volatiles (GLV) and are fatty acid derived. They are emitted by plants suffering from mechanical damage (Brilli et al. 2011). They can easily be smelled by humans after cutting of grass. They are immediately released after an insect feeds on a plant. Within 2–3 min from the beginning of feeding activity, emission peaks of leaf aldehydes (E)-3-hexenal and (E)-2-hexenal were recorded. GLV emissions store chemical signals for insects to identify spots occupied by other herbivores in plants.

Benzenoid Compounds: These are aromatic compounds produced in high rates by plants and have pivotal role in plant to plant communication.

21 Conclusion

Regardless of the nature of insect, depolarization is the initial response of the plant cell membrane. Depolarization of membrane results from ion fluxes, opening of K^+ channels and increase in cytosolic calcium. The cytosolic calcium activates calcium sensors including calmodulin, CPKs, CML and calcineurin. These sensors in turn activate cascade of signals modulating various transcription factors. Calcium also triggers ROS and RNS production. ROS and RNS are potent molecules that unequivocally affect plant response to herbivory. Molecular pathways connect plant-insect interactions. Together, multiple hormone response pathways lead to appropriate responses, but majority of these can be related to the JA pathway, which forms the key network in coordinating the plant responses to insect attack. SCF ubiquitin ligase SCF^{COI1}/JAZ pathway is activated in response to increased accumulation of jasmonoyl-L-isoleucine (JA-Ile). The crucial step for defence process in plants is the activation of interconnected hormonal network for development of local or systemic resistance. Synthesis of secondary metabolites is one of the prominent changes observed in plant response to insect attack.

References

Alberico B, Louis M, Carolin S, Philipp F, Eva LM (2018) Unraveling the initial plant hormone signaling, metabolic mechanisms and plant defence triggering the endomycorrhizal symbiosis behavior. Front Plant Sci 9:1–28

Alborn HT, Hansen TV, Jones TH, Bennett DC, James H, Schmelz EA, Teal PEA (2007) Disulfooxy fatty acids from the American bird grasshopper. *Schistocerca americana*, elicitors of plant volatiles. PNAS 104(32):12976–12981

An Y, Shen Y, Zhang Z (2010) Effects of damage stress on membrane lipid peroxidation and antioxidant enzymes in poplar leaves. Res Soil Water Conserv 17:241–244

Barakat A, Bagniewska-Zadworna A, Frost CJ, Carlson JE (2010) Phylogeny and expression profiling of *CAD* and *CAD*-like genes in hybrid *Populus* (*P. deltoides* × *P. nigra)*: evidence from herbivore damage for subfunctionalization and functional divergence. BMC Plant Biol 10:100

Batistic O, Kudla J (2012) Analysis of calcium signaling pathways in plants. Biochim Biophys Acta 1820:1283–1293

Belkhadir Y, Jaillais Y (2015) The molecular circuitry of brassinosteroid Signalling. New Phytol 206:522–540

Bing L, Hongxia D, Maoxin Z, Di X, Jingshu W (2007) Potential resistance of tricin in rice against brown planthopper *Nilaparvata lugens* (Stål). Acta Ecol Sin 27:1300–1306

Bose J, Pottosin II, Shabala SS, Palmgren MG, Shabala S (2011) Calcium efflux systems in stress signaling and adaptation in plants. Front Plant Sci 2:85

Bricchi I, Leitner M, Foti M, Mithöfer A, Boland W, Maffei ME (2010) Robotic mechanical wounding (MecWorm) versus herbivore-induced responses: early signaling and volatile emission in Lima bean (*Phaseolus lunatus* L). Planta 232:719–729

Bricchi I, Occhipinti A, Bertea CM, Zebelo SA, Brillada C, Verrillo F, De Castro C, Molinaro A, Faulkner C, Maule AJ, Maffei ME (2013) Separation of early and late responses to herbivory in Arabidopsis by changing plasmodesmal function. Plant J 73:14–25

Brien JA, Benková E (2013) Cytokinin cross-talking during biotic and abiotic stress responses. Front Plant Sci 4:451

Brilli F, Ruuskanen TM, Schnitzhofer R, Muller M, Breitenlechner M, Bittner V, Wohlfahrt G, Loreto F, Hansel A (2011) Detection of plant volatiles after leaf wounding and darkening by proton transfer reaction "time-of-flight" mass spectrometry (PTR-TOF). PLoS One 6:e20419

Broekgaarden C, Caarls L, Vos IA, Pieterse CMJ, Saskia CM, Wees V (2015) Ethylene: traffic controller on hormonal crossroads to defence. Plant Physiol 169:2371–2379

Bruessow F, Gouhier-Darimont C, Buchala A, Metraux JP, Reymond P (2010) Insect eggs suppress plant defence against chewing herbivores. Plant J 62:876–885

Chung SH, Rosa C, Scully ED, Peiffer M, Tooker JF, Hoover K, Luthe DS, Felton GW (2013) Defence suppression by insect-associated bacteria. PNAS 110(39):15728–15733

Clay NK, Adio AM, Denoux C, Jander G, Ausubel FM (2009) Glucosinolate metabolites required for an *Arabidopsis* innate immune response. Science 323:95–101

Contreras-Cornejo HA, Macías-Rodríguez LI, Cortés-Penagos C, López-Bucio J (2009) Trichoderma virens, a plant beneficial fungus, enhances biomass production and promotes lateral root growth through an auxin-dependent mechanism in Arabidopsis. Plant Physiol 149:1579–1592

Cuellar T, Azeem F, Andrianteranagna M, Pascaud F, Verdeil JL, Sentenac H, Zimmermann S, Gaillard I (2013) Potassium transport in developing fleshy fruits: the grapevine inward K^+channel VvK1.2 is activated by CIPK–CBL complexes and induced in ripening berry flesh cells. Plant J 73:1006–1018

Dombrecht B, Xue GP, Sprague SJ, Kirkegaard JA, Ross JJ, Reid JB et al (2007) MYC2 differentially modulates diverse jasmonate-dependent functions in *Arabidopsis*. Plant Cell 19:2225–2224

Doss RP, Oliver JE, Proebsting WM, Potter SW, Kuy SR, Clement SL, Williamson RT, Carney JR, DeVilbiss ED (2000) Bruchins: insect-derived plant regulators that stimulate neoplasm formation. Proc Natl Acad Sci U S A 97:6218–6223

Dutta S, Saxena R, Singh H (2014) Exceptional preservation of angiosperm markers in Miocene and Eocene ambers. Geology 42:155–159

Eichenseer H, Mathews MC, Bi JL, Murphy JB, Felton GW (1999) Salivary glucose oxidase: multifunctional roles for helicoverpa zea? Arch Insect Biochem Physiol 42(1):99–109

Erb M, Flors V, Karlen D, De Lange E, Planchamp C, D'Alessandro M (2009) Signal signature of aboveground-induced resistance upon belowground herbivory in maize. Plant J 59(2):292–302

Forterre Y, Skotheim JM, Dumais J, Mahadevan L (2005) How the Venus flytrap snaps. Nature 433:421–425

Frérot B, Leppik E, Groot A, Unbehend M, Jarmo H (2016) Chemical signatures in plant–insect interactions. Adv Bot Res. https://doi.org/10.1016/bs.abr.2016.10.003
Gagliano M, Mancuso S, Robert D (2012) Towards understanding plant bioacoustics. Trends Plant Sci 17:323–325
Galon Y, Nave R, Boyce JM, Nachmias D, Knight MR, Fromm H (2008) Calmodulin-binding transcription activator (CAMTA) 3 mediates biotic defence responses in Arabidopsis. FEBS Lett 582:943–948
Hilker M, Meiners T (2010) How do plants 'notice' attack by herbivorous arthropods? Biol Rev 85:267–280
Hogenhout SA, Bos JIB (2011) Effector proteins that modulate plant–insect interactions. Curr Opin Plant Biol 14:422–428
Hong GJ, Xue XY, Mao YB, Wang LJ, Chen XY (2012) *Arabidopsis* MYC2 interacts with DELLA proteins in regulating sesquiterpene synthase gene expression. Plant Cell 24:2635–2648
Hou X, Lee LY, Xia K, Yan Y, Yu H (2010) DELLAs modulate jasmonate signaling via competitive binding to JAZs. Dev Cell 19:884–894
Huot B, Yao J, Montgomery BL, He SY (2014) Growth-defence tradeoffs in plants: a balancing act to optimize fitness. Mol Plant 7(8):1267–1287
Johnson MT, Smith SD, Rausher MD (2009) Plant sex and the evolution of plant defences against herbivores. Proc Natl Acad Sci U S A 106:18079–18084
Jones JDG, Dangl JL (2006) The plant immune system. Nature 444:323–329
Kanchiswamy CN, Malnoy M, Occhipinti A, Maffei ME (2014) Calcium imaging perspective in plants. Int J Mol Sci 15:3842–3859
Kempel A, Brandl R, Schädler M (2009) Symbiotic soil microorganisms as players in aboveground plant-herbivore interactions—the role of rhizobia. Oikos 118:634–640
Kerchev PI, Fenton B, Foyer CH, Hancock RD (2012) Plant responses to insect herbivory: interactions between photosynthesis, reactive oxygen species and hormonal signalling pathways. Plant Cell Environ 35:441–453
Kergoat GJ, Meseguer S, Emmanuelle AJ (2017) Evolution of plant–insect interactions: insights from macroevolutionary approaches in plants and herbivorous insects. Adv Bot Res 81:25–53
Koo AJ, Gao X, Jones AD, Howe GA (2009) A rapid wound signal activates the systemic synthesis of bioactive jasmonates in Arabidopsis. Plant J 59:974–986
LaCamera S, Gouzerh G, Dhondt S, Hoffmann L, Frittig B, Legrand M, Heitz T (2004) Metabolic reprogramming in plant innate immunity: the contributions of phenylpropanoid and oxylipin pathways. Immunol Rev 198:267–284
Liu Y, He J, Jiang L, Wu H, Xiao Y, Liu Y, Li G, Du Y, Liu C, Wan J (2011) Nitric oxide production is associated with response to brown planthopper infestation in rice. J Plant Physiol 168:739–745
Maffei A, Lambo ME, Turrigiano GG (2010) Developmental regulation of experience-dependent inhibitory plasticity. J Neurosci 30:3304–3309
Mao YB, Liu YQ, Chen DY et al (2017) Jasmonate response decay and defence metabolite accumulation contributes to age-regulated dynamics of plant insect resistance. Nat Commun 8:13925
Mohanta TK, Occhipinti A, Atsbaha Zebelo S, Foti M, Fliegmann J, Bossi S et al (2012) *Ginkgo biloba* responds to herbivory by activating early signaling and direct defences. PLoS One 7(3): e32822
Monshausen GB, Haswell ES (2013) A force of nature: molecular mechanisms of mechanoperception in plants. J Exp Bot 64(15):4663–4680
Mousavi SA, Chauvin A, Pascaud F, Kellenberger S, Farmer EE (2013) Glutamate receptor-like genes mediate leaf-to-leaf wound signalling. Nature 500(7463):422–426
Pangesti N, Reichelt M, van de Mortel JE et al (2016) Jasmonic acid and ethylene signaling pathways regulate glucosinolate levels in plants during rhizobacteria-induced systemic resistance against a leaf-chewing herbivore. J Chem Ecol 42(12):1212–1225

Peiffer M, Tooker JF, Luthe DS, Felton GW (2009) Plants on early alert: glandular trichomes as sensors for insect herbivores. New Phytol 184(10):644–656

Peñaflor MF, Erb M, Robert C, Miranda LA, Dossi F, Turlings T, Bento JM (2011) Oviposition by a moth suppresses constitutive and herbivore-induced plant volatiles in maize. Planta 234:207–215

Sabelis MW, Baalen M, Bakker F, Bruin J, Drukker B, Egas M, Janssen A, Lesna I, Pels B, Rijn P, Scutareanu P (1998) The evolution of direct and indirect plant defence against herbivorous arthropods. In: Olff H, Brown LK, Drent RH (eds) Herbivores: between plants and predators. Blackwell Science, Oxford, pp 109–166

Salvador-Recatalà V, Tjallingii WF, Farmer EE (2014) Real-time, in vivo intracellular recordings of caterpillar-induced depolarization waves in sieve elements using aphid electrodes. New Phytol 203(2):674–684

Schäfer M, Fischer C, Meldau S, Seebald E, Oelmüller R, Baldwin IT (2011) Lipase activity in insect oral secretions mediates defence responses in *Arabidopsis*. Plant Physiol 156:1520–1534

Schmelz EA, Carroll MJ, LeClere S, Phipps SM, Meredith J, Chourey PS, Alborn HT, Teal PEA (2006) Fragments of ATP synthase mediate plant perception of insect attack. PNAS 103 (23):8894–8899

Shapiro AM, DeVay JE (1987) Hypersensitivity reaction of *Brassica nigra* L. (Cruciferae) kills eggs of *Pieris* butterflies (Lepidoptera: Pieridae). Oecologia 71:631–632

Sheard LB, Tan X, Mao H et al (2010) Jasmonate perception by inositol-phosphate-potentiated COI1-JAZ co-receptor. Nature 468:400–405

Tetard-Jones C, Kertesz MA, Preziosi RF (2012) Identification of plant quantitative trait loci modulating a rhizobacteria-aphid indirect effect. PLoS ONE 7:e41524

Tholl D (2015) Biosynthesis and biological functions of terpenoids in plants. In: Schrader J, Bohlmann J (eds) Biotechnology of isoprenoids. Springer, Cham, pp 63–106

Tooker J, De Moraes C (2008) Gall insects and indirect plant defenses: a case of active manipulation? Plant Signal Behav 3:503–504. https://doi.org/10.4161/psb.3.7.6184

Tretner C, Huth U, Hause B (2008) Mechanostimulation of *Medicago truncatula* leads to enhanced levels of jasmonic acid. J Exp Bot 59(10):2847–2856

Vadassery J, Scholz SS, Mithofer A (2012) Multiple calmodulin-like proteins in Arabidopsis are induced by insect-derived (*Spodoptera littoralis*) oral secretion. Signal Behav 7:1277–1280

Valenzuela Soto J, Estrada G, Ibarra-Laclette E, Délano-Frier JP (2010) Inoculation of tomato plants (Solanum lycopersicum) with growth-promoting Bacillus subtilis retards whitefly Bemisia tabaci development. Planta 231:397–410

Volkov A (2012) Plant electrophysiology: methods and cell electrophysiology. Heildelberg Springer, New York

Walsh BD (1867) The apple-worm and the apple maggot. J Hortic Sci Biotechnol 2:338–343

Walters DR, McRoberts N, Fitt BD (2008) Are green-islands red herrings? significance of green-islands in plant interactions with pathogens and pests. Biol Rev Camb Philos Soc 83:79–102

Wu JQ, Baldwin IT (2010) New insights into plant responses to the attack from insect herbivores. Annu Rev Genet 44:1–24

Wuensche H, Baldwin IT, Wu J (2011) Silencing NOA1 elevates herbivory-induced jasmonic acid accumulation and compromises most of the carbon-based defence metabolites in *Nicotiana attenuata*. J Integr Plant Biol 53:619–631

Yan J, Zhang C, Gu M et al (2009) The Arabidopsis CORONATINE INSENSITIVE1 protein is a jasmonate receptor. Plant Cell 21:2220–2236

Yang HM, Xie SX, Wang L, Jing SL, Zhu XP, Li XW, Zeng W, Yuan HY (2011) Identification of up-regulated genes in tea leaves under mild infestation of green leafhopper. Sci Hortic 130:476–481

Yang DH, Hettenhausen C, Baldwin IT, Wu J (2012) Silencing *Nicotiana attenuata* calcium-dependent protein kinases, CDPK4 and CDPK5, strongly up-regulates wound- and herbivory-induced jasmonic acid accumulations. Plant Physiol 159:1591–1607

Yanga DL, Yangc Y, Hea Z (2013) Roles of plant hormones and their interplay in rice immunity. Mol Plant 6(3):675–685

Zarate SI, Kempema LA, Walling LL (2007) Silverleaf whitefly induces salicylic acid defences and suppresses effectual jasmonic acid defences. Plant Physiol 143:866–875

Zebelo SA, Maffei ME (2015) Role of early signalling events in plant-insect interaction. J Exp Bot 66(2):435–448

Zebelo S, Matsui K, Ozawa R, Maffei M (2012) Plasma membrane potential depolarization and cytosolic calcium flux are early events involved in tomato *(Solanum lycopersicon)* plant-to-plant communication. Plant Sci Int J Exp Plant Biol 196:93–100

Role of Phytohormones in Plant Defence Against Insects: Signalling and Crosstalk

A. R. Sakthi, C. Selvi, and R. Poorniammal

Abstract

Plants possess their own defence mechanisms to combat infestations of herbivorous insects and pathogens. It has few preexisting defence mechanisms, whereas most of the defence response is activated only after insect or pathogen invasion. Crosstalk between defence signalling pathways induced in response to herbivore/pathogen attack offer plants to regulate defence mechanism in an effective manner. Crosstalk and interaction of signalling pathways can function synergistically, or either one is suppressed under stress condition. Crosstalk enables the plant to utilize less energy for defence response and allocate energy for its growth and development. Host plant resistance against herbivores can be manipulated using chemical elicitors, which induces defence signalling mechanism in plants. Unravelling of defence mechanisms and phytohormonal signalling pathway under stress condition, the induced responses of the herbivores can be predicted, that can be further utilized for pest management and reduction in crop losses.

1 Introduction

Insects cause direct losses to the agricultural production by reducing the yield and quality of the harvested produce either directly or indirectly, and some of these insects carry plant pathogen-causing diseases (Kumar et al. 2006). Also, the cost of cultivation is increased by additional usage of pesticides and their application. Plants respond to environmental stresses by initiating a number of physiological and

A. R. Sakthi (✉) · C. Selvi
Tamil Nadu Agricultural University, Coimbatore, India

R. Poorniammal
Horticultural College and Research Institute, Periyakulam, Theni, India

I. K. Singh, A. Singh (eds.), *Plant-Pest Interactions: From Molecular Mechanisms to Chemical Ecology*, https://doi.org/10.1007/978-981-15-2467-7_10

developmental changes by producing different secondary metabolic compounds that are generally involved in plant-insect interactions.

The feeding strategies of insects vary depending on intake of nutrients from all the plant parts. All insects feeding on plants leads to mechanical damage, and the extent of injury differs on the nature of insect feeding on plants. Commonly known phytophagous insects are consumers of leaves belonging to the order Coleoptera or Lepidoptera, and the damage is caused by using their chewing and snipping mouthparts (Schoonhoven et al. 1998). Piercing and sucking herbivores including thrips and mites have mouthparts with tube-like structures that consume liquids from lacerated cells. The leaf miners feed on tender portion of leaf tissues (mesophyll cell layers). The hemipteran insects, viz. aphids, whiteflies and other insects belonging to this group, feed on phloem cells by inserting specialized stylets in between the cells. In the above-mentioned relationship between the plants and insects, the outcome of interaction is determined by both partners through exchange of chemical cues. To counter the insect attack, the plants have developed a broad range of resistance mechanisms by continuous interaction and evolution of plants and insects (Schoonhoven et al. 2005). The insect defence mechanisms are grouped as constitutive and inducible defences. Constitutive defences include physical characters like presence of hairs, thorns and phytotoxins in plants, whereas inducible defences include the inherent plant resistance mechanisms activated after the infestation. Also, some of the already existing constitutive defences may increase to higher levels after insect infestation (Howe and Jander 2008).

Host plants counter the attack of herbivores through their morphological, biochemical and three-dimensional architecture. Similarly, plants may interact with beneficial insects such as pollinators and predators, and it alters interactions with their herbivores (Strauss 1997; Dicke and Van Loon 2000). The host plants' characteristics influence inhabitation of beneficial insects. Herbivores may encounter an enemy-free environment or presence of natural enemies in plants, and the plants may develop direct or indirect defence response against herbivores. Plants defend herbivores directly through the plant morphological characteristics such as thorns, nature of leaf and stem or phytotoxins like lignin (Franceschi et al. 2005) that affect the herbivore's biology. These phytotoxins have anti-nutritive properties or it possesses repellent effects on herbivores. For example, protease inhibitors present in wild species of pigeon pea shows resistance against *Helicoverpa armigera* (Parde et al. 2012), and threonine deaminase in tomato provides resistance against insects by degrading threonine in insects' gut (Gonzales-Vigil et al. 2011). In contrast, the indirect defence provides favourable environment for carnivorous insects, by providing shelter and alternative food for survival of these insects (Dicke and Vet 1999). Plants also utilize chemicals or secondary metabolites to promote the survival of natural enemies to fight against herbivores. Immediately after infestation of spider mite, plants emit secondary metabolites like terpenoids and methyl salicylate for attracting the predatory mite *Phytoseiulus persimilis*, which reduces the damage by insect populations. Beneficial insects attracted by herbivory-induced plant volatiles include predators, parasitoids, and it lays eggs on the plant (Turlings et al. 1995). Laying of egg by the pest induces both types of defence mechanism which serves as an early alert to induce defence responses (Hilker and Meiners 2008; Hilker and Fatouros 2016). For example, in *Vicia faba*, the induced defence response was shown by Rondoni et al. (2018) by challenging the invasive stinkbug (*Halyomorpha halys)* females which impair the development of its offspring. Nymphs feeding on these plants weighed less, and shorter dimensions of tibia length were observed in third instar larvae. Also, feeding of these nymphs on plants induces expression of cysteine proteinase inhibitor gene and *NAI1* (jasmonic acid-dependent genes).

A study in *Arabidopsis thaliana* revealed plants exposed to aphids require jasmonate signalling pathway to release herbivore-induced plant volatiles (HIPVs) (De Vos and Jander 2009), and similarly feeding of hawk moth larvae induces HIPVs in tomato (Degenhardt et al. 2010). The HIPVs can increase plant stability (Schuman et al. 2012), which was also reported by Tholl et al. (2011) in *Arabidopsis*. Also these volatile compounds attract predators and reduce loads of herbivores by 50%. The genetic lines showed varied responsiveness to insects.

A field study in HIPV-emitting *Nicotiana attenuata* plants showed HIPVs increase plant fitness by producing twice the buds and flowers compared to HIPV-silenced plants. The herbivore loads were reduced up to 50% on HIPV-emitting plants by predators (*Geocoris* spp.), and the response varied among genetic lines (Degen et al. 2012). The herbivores also develop adaptive strategy to counter the attack of natural enemies on its offspring by offering undamaged plants for oviposition (Signoretti et al. 2012).

2 Elicitors and plant hormones

Elicitors are diverse molecules that trigger hypersensitivity/defence response in plants. These molecules were present in insect oral secretions, and these molecules have more significance in studying plant-herbivore interaction. These elicitors enter into the plant tissues while insect feeding, and these molecules activate defence response specific to herbivores. Volicitin (OS: N-(17-hydroxylinolenoyl)-L-glutamine) was the first identified elicitor in beet armyworm, and it induces volatile compounds in infected plants to attract parasitic wasps (Alborn et al. 1997). Many other compounds were identified in lepidopteran insects, *Manduca sexta* (L.) and other *Spodoptera* spp. (Halitschke et al. 2001; Spiteller and Boland 2003).

Signals may be produced at the site of insect attack, and it is further transmitted systemically to other parts of the plant, inducing defence response to counter future herbivore attacks (Howe and Jander 2008).

Parasitoids find their prey using HIPVs as cues, and it is also an adaptive strategy of the emitting plants to combat insect attack. These insect plant interactions are specific in nature. For example, terpene synthase (TPS10) secreted in maize elicits indirect plant defence against *Spodoptera litura* in *Arabidopsis thaliana (*Schnee et al. 2006) by attracting a parasitoid, Cotesia marginiventris. HIPVs are also emitted by roots to attract natural enemies against belowground insect pests. *Heterorhabditis megidi, an entomopathogenic nematode, was attracted towards* (*E*)-β-caryophyllene, a sesquiterpene produced by maize roots on infestation by the larvae of western corn rootworm *(*Rashmann et al. 2005*)*.

2.1 Plant Hormones

The phytohormonal crosstalk plays a major role to optimize plant responses against various biotic and abiotic stresses occurring simultaneously in the environment. In natural environment, plants were prone to infestation by herbivorous insects, including leaf-eating beetles belonging to Coleoptera or caterpillars; piercing-sucking type of insects like thrips, spider mites and other types of sucking pests; and phloem-sucking aphids or whiteflies (Schaller 2008).

The common defence response induced in affected plants under both biotic and abiotic stresses is the activation of complex phytohormone signalling networks (Schenk et al. 2000). The primary phytohormones involved in plant defence are jasmonic acid (JA), salicylic acid (SA), and ethylene (ET), while other plant hormones like abscisic acid (ABA), auxins (IAA), brassinosteroids (BR), cytokinins (CK), gibberellins (GA) and strigolactones (STR) involve directly in plant defence response by interacting with other primary plant hormones. The three main signalling molecules involved in different signalling pathways for regulating biotic stress-related defence responses are jasmonic acid, salicylic acid and ethylene (Kunkel and Brooks 2002). These signalling molecules are involved in the three main signal transduction pathways which are octadecanoid pathway, ethylene pathway and shikimate pathway. Jasmonic acid pathway is involved in defence against chewing and necrotrophic insects, whereas salicylic acid pathway functions in response to pathogen-derived infections on plants. Thus, activation of specific signalling pathways depends on pathogen or insect attack on plants. For instance, the mirid bug (*Nesidiocoris tenuis*) induces both jasmonic acid and abscisic acid pathways in tomato plants (Pérez-Hedo et al. 2015b; Naselli et al. 2016), while only JA pathway was activated by infection of *Macrolophus pygmaeus* (Pérez-Hedo et al. 2015a).

Herbivores feed on plants by chewing and sucking their plant parts. *Spodoptera exigua* causes massive leaf damage by chewing, and the defence response to wounding was initiated through volicitin, an elicitor released from oral secretions of insects. Sucking pests like whitefly activate SA-dependent pathway since they are perceived as pathogens and rarely JA-/ET-dependent pathways (Peng et al. 2004; Li et al. 2012).

Plant and herbivore interaction starts when the insect lands on the leaf surface for feeding. Insect's presence on the leaf surface will exert pressure, trichomes are damaged and it releases elicitors from tarsal pad (Hilker and Meiners 2010). Plants are highly sensitive to external stimuli, and they have developed internal sensory mechanisms to sense pressure exerted by insect or other organisms (Braam 2005). Tretner et al. (2008) reported that repeated touching of leaves induces defence response by accumulation of jasmonic acid.

Herbivore-associated molecular patterns (HAMPs) and endogenous damage-associated molecular patterns (DAMPs) are released after feeding of plant parts by insects (Acevedo et al. 2015). After recognition of signal, plants initiate accumulation of defence-responsive phytohormones, which activates signalling cascades for regulating downstream transcriptional responses. Among the phytohormones, JA and its isoleucine conjugate (JA-Ile) which is the most active form of JA are generally accepted by many herbivores as the core inducers (Howe and Jander 2008; Wasternack and Hause 2013; Tytgat et al. 2013). The plants which are deficient in synthesis of JA do not showed defence response against herbivores belonging to different insect orders (Thaler and Bostock 2004; Bodenhausen and Reymond 2007; Schweizer et al. 2013).

After recognition of insect attack, not only jasmonic acid but also other signalling phytohormones are also induced to combat the defences against the attacker (Acevedo et al. 2015; Xu et al. 2015), and due to differences in mouthparts of insects and its feeding strategies, the inducing signal was in specific plant parts. Sucking insects secrete specific enzymes on the salivary sheet along with their mandibles which interact with plant cells during infestation (Foyer et al. 2016). Thus, sucking and chewing types of insects induce different types of defence signalling pathways (DeVos et al. 2005; Bidart-Bouzat and Kliebenstein 2011). The defence signal induced by the feeding of herbivores depends on insect species (Nguyen et al. 2016). Plants produce JA and ET on infestation by *M. Sexta* larvae, whereas SA is induced in addition to JA by *S. exigua* in *Nicotiana attenuata* (Diezel et al. 2009). In maize and *Arabidopsis*, feeding of S. *exigua* induces JA and ET (Schmelz et al. 2003; Rehrig et al. 2014), whereas ABA was induced by *Pieris rapae* in addition to JA (Vos et al. 2013). In tomato, infestation of Colorado potato beetle and the *Solenopsis* mealy bug induces accumulation of both important phytohormones, JA and SA (Chung et al. 2013; Zhang et al. 2015). Plants fine-tune their defence responses against various insect attackers through interaction of phytohormones (Pieterse et al. 2012; Erb et al. 2012).

2.2 Jasmonic Acid (JA)

JA is one of the phytohormones synthesized by plants on receiving signal of herbivore attack and also activates both direct and indirect defences (Usha Rani and Jyothsna 2010; Shivaji et al. 2010; War et al. 2011). The main precursor of JA is linolenic acid, which was derived through octadecanoid pathway, and it accumulates upon wounding by herbivorous insects in plant tissues (Zhang et al. 2008). Chewing of different parts of the plant by insects causes dioxygenation of linoleic acid by specific lipoxygenases (LOX) at C9 or C13 to form (9S)- or (13S)-hydroperoxy-octadecadi(tri)enoic acids, which are converted into 12-oxophytodienoic acid (12-OPDA) by allene oxide synthase and allene oxide cyclase. OPDA is reduced by OPDA reductase 3 (OPR3) forming JA in peroxisome.

To combat abiotic and biotic stresses, plants possess three main JA-signalling components which include JAR1 (jasmonate resistant 1), COI1 (coronatine insensitive 1) and JIN1/MYC2 (jasmonate insensitive 1/MYC2). The most required component for JA signalling is COI1, and it codes for an F-box protein involved in the SCF-mediated protein degradation by the 26S proteasome (Xie et al. 1998). The enzyme JA-amino acid synthetase is synthesized from gene JAR1, which converts JA to its bioactive form by conjugation of isoleucine to JA. This conjugated form is perceived by the plants (Staswick and Tiryaki 2004; Thines et al. 2007). JIN1/

MYC2 codes for a transcription factor which regulates transcription of some JA-responsive genes (Lorenzo et al. 2004).

OPDA is also involved in the plant defence signalling pathways by regulating the transcription of COI1 gene (Ribot et al. 2008), alteration of the calcium level in the cells and cellular redox status (Walter et al. 2007). Jasmonates promote binding of the COI1-unit to JAZ (jasmonate ZIM-domain) proteins on interacting with the COI1 unit of an E3 ubiquitin ligase complex, termed SCFCOI1 (Skip/Cullin/F-box–COI1), which results in degradation of JAZ proteins (Sheard et al. 2010). The transcripts of calcium-dependent protein kinases (CDPK) have been affected by JA (Ulloa et al. 2002), and they play an important role in combating various biotic and abiotic stresses through signal transduction (Ludwig et al. 2004). JA induces signalling molecules, thereby contributing to both direct and indirect insect resistance (Barbehenn et al. 2009). Plants produce extrafloral nectar induced by JA which is consumed as alternate food by natural enemies of herbivorous insects (Kost and Heil 2005). JA also induces enzymes involved in defence such as peroxidase (POD), (Usha Rani and Jyothsna 2010; War et al. 2011) and polyphenoloxidase (PPO) (Usha Rani and Jyothsna 2010).

Jasmonate-responsive genes were suppressed by few JAZ family members by acting as repressors of jasmonate-responsive genes. In response to JA treatment, JAZ proteins which are repressors of JA-responsive genes were degraded in a COI1- and 26S proteasome-dependent manner. The responsiveness of plants to jasmonic acid is reduced by dominant mutations in the conserved C-terminal domain of JAZ proteins by stabilizing them against SCFCOI1-mediated degradation (Ulloa et al. 2002; Chini et al. 2007; Yan et al. 2007). The interaction of COI1 and JAZ1 is stimulated by JA-Ile based on dose-dependent manner. COI1-dependent biological activity of JA, MeJA and OPDA requires its bioactive form (e.g. JA-Ile) to promote this interaction. The stimulation of COI1-JAZ1 interaction by JA-Ile in the yeast two-hybrid system by non-inclusion of other plant proteins defines that the COI1-JAZ complex is a receptor for JA-Ile (Ulloa et al. 2002).

2.3 Ethylene (ET)

ET signalling in response to herbivory is also common among plants, and mostly ET acts as a modulator of herbivore-induced defence responses and has variable effects on defence regulation (Von Dahl and Baldwin 2007). Ethylene modulates JA-mediated insect defences synergistically for regulating defensive genes like PDF1.2 and PR1, 4 and 5 which are induced against infestation by necrotrophic pathogens through the co-regulation of the AP2/ERF TFs, ERF1 and ORA59 (Lorenzo et al. 2003; Pre et al. 2008). ET signalling mediated by JA also contributes to the emission of volatile compounds on infestation by *S. exigua* in maize or infestation by *Bemisia tabaci* on *Arabidopsis* (Schmelz et al. 2003; Zhang et al. 2013). Two main steps are involved in ethylene biosynthesis mediated by two enzymes, viz. ACC synthase and ACC oxidase: ACC synthase is responsible for conversion of *S*-adenosyl-L-Met to 1-aminocyclopropane-1-carboxylic acid (ACC) and oxidation of ACC to form ethylene by ACC oxidases. In *Arabidopsis,* Liu and Zhang (2004) reported that direct phosphorylation of ACS2 and 6 by MPK6 (orthologue of tobacco SIPK) enhances their stability and increases release of ethylene. The lack of phosphorylation of ACS2 and 6 resulted in degradation by 26S proteasome pathway (Joo et al. 2008).

ETR1, ETR2, ERS1, ERS2 and EIN4 were five receptors through which ethylene is perceived in *Arabidopsis* (Chen et al. 2005).

CTR1 is a negative regulator of ethylene signalling by interacting with ETR1 receptor (Hua and Meyerowitz 1998; Huang et al. 2003), whereas EIN2 and EIN3 are positive regulators of ethylene signalling (Chao et al. 1997; Alonso et al. 1999). EIN3 and other related proteins mediate transcription of various ethylene-responsive genes by acting as transcription factors (Ohme-Takagi and Shinshi 1995).

3 Molecular Players Involved in JA Signal Transduction

Mitogen-activated protein kinase (MAPK) is one of the important signalling elements, and it is responsible for transfer of signalling information from sensors in eukaryotes (Menke et al. 2004; Nakagami et al. 2005). Among MAPKs, MAPK4 is involved in negative crosstalk between two important phytohormones, SA and JA, in *Arabidopsis*. MAPK4 acts as a negative regulator of SA signalling, and it regulates JA signalling in a positive manner. The mapk4 mutant was generated by inactivation of MAPK gene, and these mutants showed higher expression of SA and SA-responsive genes in *Arabidopsis* and tomato (Petersen et al. 2000; Brodersen et al. 2006).

MAPK4 phosphorylates Map kinase 4 substrate 1 (MKS1) which is one of the targets to repress SA signalling. The expression of MAPKs increases rapidly after infestation/wounding by herbivores, for example, wound-induced protein kinase (WIPK, member of MAPK subfamily A) expression increases rapidly after wounding in tobacco (Seo et al. 1995), and Hettenhausen et al. (2015) observed reduced levels of JA when the WIPK gene was subjected to antisense expression. The above observed results revealed that MAPK4 regulates JA signalling positively and represses genes involved in SA signalling.

Glutaredoxin (GRX480) is a disulphide reductase catalysing thiol-disulphide reductions, and it is an important regulating enzyme involved in crosstalk among SA- and JA-mediated signalling. Glutaredoxins are involved in various cellular processes and in the redox regulation of protein activities (Meyer et al. 2008). GRX480 regulates the SA-responsive PR genes by interacting with TGA transcription factors (Ndamukong et al. 2007). The SA induces the expression of GRX480, and it needs TGA transcription factors and NPR1 for induction. GRX480 along with these components represses the activity of JA-responsive genes.

WRKY53 is a senescence-specific transcription factor, and in addition to other factors, it interacts with JA and SA signalling. WRKY53 interacts with the epithiospecifying senescence regulator (ESR) which is a jasmonic acid-inducible protein. Both these factors are involved in negative crosstalk between resistance to pathogen and senescence in *Arabidopsis* (Miao and Zentgraf 2007). Another transcription factor, JIN1/MYC2, is a JA-responsive element which suppresses the expression SA-responsive genes. The study reported by Laurie-Berry et al. (2006) developed *jin1* mutant plants which showed increased expression of SA-responsive genes and gained resistance against pathogen infection.

The MYC2 is a basic helix-loop-helix transcription factor which functions to combat resistance against biotic and abiotic stresses. The role of MYC2 as transcriptional factor in regulating abiotic stress was first identified in *Arabidopsis* (Abe et al. 2003). MYC2 upregulates the expression of JA-responsive genes and negatively regulates the expression of JA-/ET-mediated PR genes (Anderson et al. 2004; Lorenzo et al. 2004). It is involved in ABA, JA, ET and SA signalling and also functions in combating abiotic stress response (Fujita et al. 2009). In drought-responsive genes, MYC2 binds to a CACNTG core that responds not only to drought condition but also to ABA-, light- and JA-regulated genes (Abe et al. 2003; Yadav et al. 2005; Dombrecht et al. 2007).

The transcription factors involved in crosstalk between major phytohormones (SA, JA and ABA) were MYB2, RD26, ATAF1 and 2, WRKY82 and ZAT7. These transcription factors regulate the crosstalk among phytohormonal signalling pathways when the plants are prone to wounding by herbivores and pathogen attack and also under abiotic stress condition (Wu et al. 2009; Atkinson and Urwin 2012). The transcription factors belonging to NAC family are involved in crosstalk between the ABA and JA signalling. The transcription factors included under NAC family were ATAF, NAM and CUC (Mauch-Mani and Flors 2009). The transcription factor ATAF2 responds to SA and JA treatments and wounding, whereas its homologue, ATAF1, responds to ABA treatment, wounding and dehydration. ATAF1 also responds to salinity treatments, and it downregulates responses to soil pathogens (Wu et al. 2009). The function of RD26 is stimulated by JA and ABA treatment, H_2O_2 and pathogen infections (Zimmermann et al. 2004; Fujita et al. 2009).

ERFs are a group of transcription factors involved in JA signalling by interacting and binding to GCCGCC boxes present in the ET-inducible genes. Park et al. (2001) reported overexpression of TSI1 enhances expression of PR proteins which confers osmotic stress tolerance and resistance against biotrophic pathogen infection. ERFLP1 is another ethylene response transcription factor reported to be induced under both pathogen infection and wounding by insects; thus, it contributes to both biotic stresses in pepper (Lee et al. 2004).

4 JA/ET and SA Signalling and Crosstalk with Respect to Insect Defence

Exploration of interaction and crosstalk among phytohormones and defence signalling pathways provides information for the development of host resistance mechanisms (Grant and Jones 2009). Phytohormonal pathways exchange signals among themselves for regulation of growth and plant development under varied environmental conditions (Gfeller et al. 2010). Interaction and crosstalk among

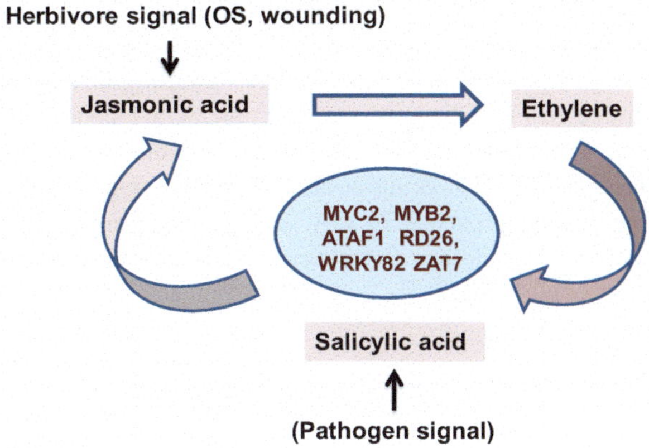

Fig. 1 The relationship among phytohormone signalling pathways in plants in response to herbivory attack

phytohormone-mediated signalling pathways depend on the nature of stresses (Fig. 1). The main biochemical response mechanisms induced under various stresses were JA and SA signalling pathway.

In response to external stimuli, the activation of plant defence mechanism on hormonal signalling implies allocation of plant resources (Pieterse et al. 2012). The utilization of plant resources under one type of stress condition can affect the plant's ability to fight against other types of biotic or abiotic stresses. Thus, for allocation of resources against a particular type of stress, the crosstalk among phytohormonal signalling pathways is essential. Plants infected by pathogens induce SA signalling pathway which suppresses JA-dependent defence signalling pathway, prioritizing the utilization of resources for SA signalling mechanism and associated transcription factors (Spoel et al. 2007; Uppalapati et al. 2007). SA reduces accumulation of the AP2/ERF-type transcriptional activator ORA59 (octadecanoid-responsive arabidopsis59) by affecting GCC-box motifs in JA-responsive promoters and downregulates JA signalling. The GCC-box motif present in JA-responsive promoters is sufficient for the suppression genes induced by JA (Van der Does et al. 2013). The suppression of genes involved in JA signalling mechanism is controlled by regulating the accumulation of GCC-box binding TF ORA59. SA cannot affect JA signalling mechanism at reduced level of ORA59.

The MAT1 gene accumulates reducing agents to control plant redox homeostasis at the time of herbivore attack, thereby avoiding oxidative damage and death of plant cell. The expression of genes involved in jasmonic acid and salicylic acid biosynthesis was affected by MATI gene (Santamaría et al. 2017).

Non-expressor of PR1 (NPR1) genes is an important component involved in the activation of genes induced by salicylic acid, and it functions by interacting with TGA transcription factors (Dong 2004). Npr1 gene is involved in the interaction of SA- and JA-responsive genes, and it suppresses genes responsive to JA (Spoel et al. 2007). The transcription factor WRKY70 reacts positively upon infection by pathogen to initiate genes induced by salicylic acid. It plays a pivotal role in balancing the expression of genes involved in SA and JA pathways. The SA inducible PR genes are constitutively expressed by increased expression of WRKY70 (Li et al. 2006).

GA signalling interacts with JA signalling through the DELLA protein. DELLAs induce resistance against necrotrophs via JA signalling, and it is known as plant growth repressors. DELLAs and JAZs deactivate each other by binding directly (Hou et al. 2013; Song et al. 2014). The degradation of DELLAs is promoted by GA. In the presence of GA, 26S proteasome releases JAZs to degrade DELLA protein by binding with repressor RGA to suppress MYC2 activity (Hou et al. 2010; Wild et al. 2012).

4.1 Plant and Herbivore Interaction

Feeding of plants by some beneficial insects activates defence mechanisms against biotic stresses. For example, feeding of the predatory bug *Orius laevigatus* increased resistance against sucking pests (thrips and whiteflies) in tomato (De Puysseleyr et al. 2011). Naselli et al. (2016) reported ABA and JA signalling pathways are activated by feeding of mirid bug on tomato plants, whereas feeding of the insect *Macrolophus pygmaeus* induces genes involved in JA pathway (Pérez-Hedo et al. 2015a).

Sun et al. (2017) investigated the tritrophic interactions among insect vectors, viruses and plants. Three tomato varieties with varying levels of JA-inducible resistance were used to study involvement of plant resistance mechanism in mediating the interactions between whitefly and begomovirus. The plants expressing JA-responsive genes at high level showed minimum infestation by whitefly, whereas the plants lacking resistance induced by JA are subjected to more infestation.

Plants showed age-related resistance against pathogens, as older plants are more resistant to pathogens than younger plants (Kus et al. 2002). The plant vigour hypothesis is based on observations that old and mature plants are less frequently attacked than young and vigorous plants which were prone to herbivore attack (Price 1991). In plants, miR156 is an important regulator for age-dependent development which functions through SQUAMOSA promoter binding protein-like (SPL) transcription factors (Wu and Poethig 2006). Expression of JA-responsive genes was highly induced in young plants than old plants after application of methyl JA and also showed increased JA response than old plants. In yeast two-hybrid assays, it was observed that SPL9 interacts directly with JAZ proteins including JAZ1, JAZ3, JAZ4, JAZ6, JAZ10 and JAZ11, but it does not display interaction with COI1 and MYC2. SPL2, a homologue of SPL9, displays interaction with JAZ1, JAZ4 and JAZ9 (Mao et al. 2017).

The undergoing biological processes in response to the pest Asian corn borer, *Ostrinia furnacalis* and JA were profiled by Wang et al. (2017) using RNA sequencing and clustering of differentially expressed transcripts. ZmNAC60 was identified as a novel positive regulator of JA-responsive genes.

MYC2 is a transcriptional factor involved in coordination of JA-mediated activation of wounding and pathogen-responsive genes. Du et al. (2017) identified

655 JA-responsive genes targeted by MYC2 using chromatin immune precipitation sequencing coupled with RNA sequencing assays. Further, it was revealed that MYC2 and MTF JA2-Like (JA2L) transcription factors regulate herbivore-associated wound-responsive genes, whereas MYC2 and MTF ETHYLENE RESPONSE FACTOR.C3 (ERF.C3) regulate pathogen-responsive genes by forming transcription module. MYC2 forms hierarchical transcriptional cascade with its associated transcription factors during JA-mediated defence response.

The natural indirect interactions between plants and herbivores depend on herbivore infesting the plant and plant species involved, induction of species-insect specific signalling pathways and sequence of attack by insects. The role of phytohormones in response to two invasive herbivores [elongate hemlock scale (EHS, *Fiorinia externa*) and hemlock woolly adelgid (HWA, *Adelges tsugae*)] in Eastern hemlock (*Tsuga canadensis*) was reported by Schaeffer et al. (2018). SA and ABA content seems to be elevated in local tissues fed by HWA, while feeding by EHS showed no significant effect on the phytohormonal levels. The young ones of herbivore (HWA) preferred ASM-treated foliage for its settlement and avoided feeding on methyl jasmonate-treated foliage.

5 Conclusion

Survival strategies of plants under adverse environmental conditions induced after crosstalk among various hormonal signalling pathways may lead to the development of new hypotheses. Several studies on plant hormone and herbivore interaction revealed plants effectively modulate availability of salicylic acid, jasmonic acid and ethylene levels after infection/damage by various types of pathogens and pests. After receiving signal, it triggers the expression of pathogen-/herbivore-specific resistant genes accordingly and coordinates interactions between phytohormonal defence signalling pathways. Identification of genes/transcription factors and signalling mechanism involved in the interactions of phytohormones and understanding the crosstalk between different signalling pathways will provide information on phytohormone-mediated defence signalling network in plants. The scientific evidence on phytohormone-induced defence responses of plants against herbivores is useful in designing effective strategies for designing crop management strategies through genetic engineering or genomics approach for pest resistance.

References

Abe H, Urao T, Ito T et al (2003) *Arabidopsis* AtMYC2 (bHLH) and AtMYB2 (MYB) function as transcriptional activators in abscisic acid signaling. Plant Cell 15:63–78

Acevedo FE, Rivera-Vega LJ, Chung SH et al (2015) Cues from chewing insects-the intersection of DAMPs, HAMPs, MAMPs and effectors. Curr Opin Plant Biol 26:80–86

Alborn T, Turlings TCJ, Jones TH et al (1997) An elicitor of plant volatiles from beet armyworm oral secretion. Science 276:945–949

Alonso JM, Hirayama T, Roman G et al (1999) EIN2, a bifunctional transducer of ethylene and stress responses in *Arabidopsis*. Science 284:2148–2152

Anderson JP, Badruzsaufari E, Schenk PM et al (2004) Antagonistic interaction between abscisic acid and jasmonate-ethylene signaling pathways modulates defense gene expression and disease resistance in *Arabidopsis*. Plant Cell 16:3460–3479

Atkinson NJ, Urwin E (2012) The integration of plant biotic and abiotic stresses: from genes to the field. J Exp Bot 63(10):3523–3543

Barbehenn RV, Jaros A, Lee G et al (2009) Hydrolyzable tannins as "quantitative defenses": limited impact against *Lymantria dispar* caterpillars on hybrid poplar. J Insect Physiol 55:297–304

Bidart-Bouzat MG, Kliebenstein D (2011) An ecological genomic approach challenging the paradigm of differential plant responses to specialist versus generalist insect herbivores. Oecologia 167:677–689

Bodenhausen N, Reymond P (2007) Signaling pathways controlling induced resistance to insect herbivores in Arabidopsis. Mol Plant-Microbe Interact 20:1406–1420

Braam J (2005) In touch: plant responses to mechanical stimuli. New Phytol 165:373–389

Brodersen P, Petersen M, Bjorn Nielsen H et al (2006) *Arabidopsis* MAP kinase 4 regulates salicylic acid- and jasmonic acid/ethylene-dependent responses *via* EDS1 and PAD4. Plant J 47:532–546

Chao Q, Rothenberg M, Solano R et al (1997) Activation of the ethylene gas response pathway in *Arabidopsis* by the nuclear protein ETHYLENE INSENSITIVE3 and related proteins. Cell 89:1133–1144

Chen YF, Etheridge N, Schaller GE (2005) Ethylene signal transduction. Ann Bot 95:901–915

Chini A', Fonseca S, Fernandez G (2007) The JAZ family of repressors is the missing link in jasmonate signalling. Nature 448:666–671

Chung SH, Rosa C, Scully ED et al (2013) Herbivore exploits orally secreted bacteria to suppress plant defenses. Proc Natl Acad Sci U S A 110:15728–15733

De Puysseleyr V, Höfte M, De Clercq P (2011) Ovipositing *Orius laevigatus* increase tomato resistance against *Frankliniella occidentalis* feeding by inducing the wound response. Arthropod Plant Interact 5:71–80

De Vos M, Jander G (2009) *Myzus persicae* (green peach aphid) salivary components induce defence responses in *Arabidopsis thaliana*. Plant Cell Environ 32:1548–1560

Degen T, Bakalovic N, Bergvinson D, Turlings TCJ (2012) Differential performance and parasitism of caterpillars on maize inbred lines with distinctly different herbivore-induced volatile emissions. PLoS One:7

Degenhardt DC, Refi-Hind S, Stratmann JW, Lincoln DE (2010) Systemin and jasmonic acid regulate constitutive and herbivore induced systemic volatile emissions in tomato, *Solanum lycopersicum*. Phytochemistry 71:2024–2037

DeVos M, Van Oosten VR, Van Poecke RMP et al (2005) Signal signature and transcriptome changes of *Arabidopsis* during pathogen and insect attack. Mol Plant-Microbe Interact 18:923–937

Dicke M, Van Loon JJA (2000) Multitrophic effects of herbivore-induced plant volatiles in an evolutionary context. Entomol Exp Appl 97:237–249

Dicke M, Vet LEM (1999) Plant–carnivore interactions: evolutionary and ecological consequences for plant, herbivore and carnivore. In: Olff H, Brown VK, Drent RH (eds) Herbivores: between plants and predators. Blackwell Science, Oxford, pp 483–520

Diezel C, von Dahl CC, Gaquerel E et al (2009) Different lepidopteran elicitors account for cross-talk in herbivory-induced phytohormone signaling. Plant Physiol 150:1576–1586

Dombrecht B, Xue GP, Sprague SJ et al (2007) MYC2 differentially modulates diverse jasmonate-dependent functions in *Arabidopsis*. Plant Cell 19:2225–2245

Dong X (2004) NPR1, all things considered. Curr Opin Plant Biol 7:547–552

Du M, Zhao J, Tzeng TW, Liu Y, Deng, Yang T et al (2017) MYC2 orchestrates a hierarchical transcriptional cascade that regulates jasmonate mediated plant immunity in tomato. Plant Cell 29(8):1883–1906

Erb M, Meldau S, Howe GA (2012) Role of phytohormones in insect specific plant reactions. Trends Plant Sci 17:250–259
Foyer CH, Rasool B, Davey JW et al (2016) Cross-tolerance to biotic and abiotic stresses in plants: a focus on resistance to aphid infestation. J Exp Bot 67:2025–2037
Franceschi VR, Krokene P, Christiansen E, Krekling T (2005) Anatomical and chemical defenses of conifer bark against bark beetles and other pests. New Phytol 167:353–375
Fujita M, Shinozaki Y, Shinozaki K (2009) Transcription factors involved in the crosstalk between abiotic and biotic stress signaling-networks. In: Yoshioka K, Shinozaki K (eds) Signal crosstalk in plant stress responses. Whiley-Blackwell, New York, pp 43–59
Gfeller A, Liechti R, Farmer EE (2010) *Arabidopsis* Jasmonate signaling pathway. Sci Signal 3 (109):cm4
Gonzales-Vigil E, Bianchetti CM, Phillips GN Jr, Howe GA (2011) Adaptive evolution of threonine deaminase in plant defense against insect herbivores. Proc Natl Acad Sci U S A 108:5897–5902
Grant MR, Jones JDG (2009) Hormone (dis)harmony moulds plant health and disease
Halitschke R, Schittko U, Pohnert G et al (2001) Molecular interactions between the specialist herbivore *Manduca sexta* (Lepidoptera, Sphingidae) and its natural host *Nicotiana attenuata*. III. Fatty acid-amino acid conjugates in herbivore oral secretions are necessary and sufficient for herbivore-specific plant responses. Plant Physiol 125:711–717
Hettenhausen C, Schuman MC, Wu J (2015) MAPK signaling – a key element in plant defense response to insects. Insect Sci 22(2):157–164
Hilker M, Fatouros NE (2016) Resisting the onset of herbivore attack: plants perceive and respond to insect eggs. Curr Opin Plant Biol 32:9–16
Hilker M, Meiners T (2008) Chemoecology of insect eggs and egg deposition. Blackwell, Berlin
Hilker M, Meiners T (2010) How do plants "notice" attack by herbivorous arthropods? Biol Rev 85:267–280
Hou X, Lee LYC, Xia K et al (2010) DELLAs modulate jasmonate signaling via competitive binding to JAZs. Dev Cell 19:884–894
Hou X, Ding L, Yu H (2013) Crosstalk between GA and JA signalling mediates plant growth and defense. Plant Cell Rep 32:1067–1074
Howe GA, Jander G (2008) Plant immunity to insect herbivores. Annu Rev Plant Biol 59:41–66
Hua J, Meyerowitz EM (1998) Ethylene responses are negatively regulated by a receptor gene family in *Arabidopsis thaliana*. Cell 94:261–271
Huang Y, Li H, Hutchison CE, Laskey J, Kieber JJ (2003) Biochemical and functional analysis of CTR1, a protein kinase that negatively regulates ethylene signalling in Arabidopsis. Plant J 33 (2):221–233
Joo S, Liu Y, Lueth A et al (2008) MAPK phosphorylation induced stabilization of ACS6 protein is mediated by the non-catalytic C-terminal domain, which also contains the cis determinant for rapid degradation by the 26S proteasome pathway. Plant J 54:129–140
Kost C, Heil M (2005) Increased availability of extra floral nectar reduces herbivory in Lima bean plants (*Phaseolus lunatus*, Fabaceae). Basic Appl Ecol 6:237–248
Kumar S, Chandra A, Pandey KC (2006) Genetic transformation of lucerne (*Medicago sativa* L.) for weevil (Hypera postica) resistance. In: Extended summaries. National Seminar Transgenic Crops Indian Agriculture: Status, Risks and Acceptance, Hisar, pp 35–37
Kunkel BN, Brooks DM (2002) Cross talk between signaling pathways in pathogen defense. Curr Opin Plant Biol 5:25–331
Kus JV, Zaton K, Sarkar R et al (2002) Age-related resistance in *Arabidopsis* is a developmentally regulated defense response to *Pseudomonas syringae*. Plant Cell 14:479–490
Laurie-Berry N, Joardar V, Street IH et al (2006) The *Arabidopsis thaliana* JASMONATE INSENSITIVE 1 gene is required for suppression of salicylic acid-dependent defences during infection by *Pseudomonas syringae*. Mol Plant-Microbe Interact 19:789–800
Lee JH, Hong JP, Oh SK et al (2004) The ethylene-responsive factor like protein 1 (CaERFLP1) of hot pepper (*Capsicum annuum* L.) interacts in vitro with both GCC and DRE/CRT sequences

with different binding affinities: possible biological roles of CaERFLP1 in response to pathogen infection and high salinity conditions in transgenic tobacco plants. Plant Mol Biol 55:61–81

Li J, Brader G, Kariola T, Palva ET (2006) WRKY70 modulates the selection of signaling pathways in plant defence. Plant J 46:477–491

Li YF, Zhang ZH, Nie YF et al (2012) Proteomic analysis of salicylic acid-induced resistance to *Magnaporthe oryzae* in susceptible and resistant rice. Proteomics:12

Liu Y, Zhang S (2004) Phosphorylation of 1-aminocyclopropane1-carboxylic acid synthase by MPK6, a stress-responsive mitogen activated protein kinase, induces ethylene biosynthesis in *Arabidopsis*. Plant Cell 16:3386–3399

Lorenzo O, Piqueras R, Sanchez-Serrano JJ et al (2003) Ethylene response Factor1 integrates signals from ethylene and jasmonate pathways in plant defense. Plant Cell 15:165–178

Lorenzo O, Chico JM, Sanchez-Serrano JJ et al (2004) JASMONATE-INSENSITIVE1 encodes a MYC transcription factor essential to discriminate between different jasmonate regulated defense responses in *Arabidopsis*. Plant Cell 16:1938–1950

Ludwig AA, Romeis T, Jones JD (2004) CDPK-mediated signalling pathways: specificity and cross-talk. J Exp Bot 55:181–188

Mani M, Flors V (2009) The ATAF1 transcription factor: at the convergence point of ABA-dependent plant defense against biotic and abiotic stresses. Cell Res 19(12):1322–1323

Mao YB, Liu YQ, Chen DY et al (2017) Jasmonate response decay and defense metabolite accumulation contributes to age-regulated dynamics of plant insect resistance. https://doi.org/10.1038/ncomms13925

Menke FLH, van Pelt JA, Pieterse CMJ, Klessig DF (2004) Silencing of the mitogen-activated protein kinase MPK6 compromises disease resistance in *Arabidopsis*. Plant Cell 16:897–907

Meyer Y, Siala W, Bashandy T (2008) Glutaredoxins and thioredoxins in plants. Biochim Biophys Acta 1783:589–600

Miao Y, Zentgraf U (2007) The antagonist function of Arabidopsis WRKY53 and ESR/ESP in leaf senescence is modulated by the jasmonic and salicylic acid equilibrium. Plant Cell 19:819–830

Nakagami H, Pitzschke A, Hirt H (2005) Emerging MAP kinase pathways in plant stress signalling. Trends Plant Sci 10:339–346

Naselli M, Urbaneja A, Siscaro G, Jaques JA, Zappalà L, Flors V, Pérez-Hedo M (2016) Stage-related defense response induction in tomato plants by *Nesidiocoris tenuis*. Int J Mol Sci 17:1210

Ndamukong I, Abdallat AA, Thurow C et al (2007) SA-inducible Arabidopsis glutaredoxin interacts with TGA factors and suppresses JA-responsive PDF1.2 transcription. Plant J 50:128–139

Nguyen D, Rieu I, Mariani C et al (2016) How plants handle multiple stresses: hormonal interactions underlying responses to abiotic stress and insect herbivory. Plant Mol Biol 91:727. https://doi.org/10.1007/s11103-016-0481-8

Ohme-Takagi M, Shinshi H (1995) Ethylene-inducible DNA binding proteins that interact with an ethylene-responsive element. Plant Cell 7:173–182

Parde VD, Sharma HC, Kachole MS (2012) Protease inhibitors in wild relatives of pigeonpea against the cotton bollworm/legume pod borer, *Helicoverpa armigera*. Am J Plant Sci 3:627–635

Park JM, Park CJ, Lee SB et al (2001) Overexpression of the tobacco *Tsi1* gene encoding an EREBP/ AP2-type transcription factor enhances resistance against pathogen attack and osmotic stress in tobacco. Plant Cell 13:1035–1046

Peng JY, Deng XJ, Huang JH (2004) Role of salicylic acid in tomato (*Lycopersicon esculentum*) plant defense against cotton bollworm, *Helicoverpa armigera* Hubner. Z Naturforsch C 59:856–862

Pérez-Hedo M, Bouagga S, Jaques JA et al (2015a) Tomato plant responses to feeding behavior of three zoophytophagous predators (Hemiptera: Miridae). Biol Control 86:46–51

Pérez-Hedo M, Urbaneja-Bernat P, Jaques JA et al (2015b) Defensive plant responses induced by *Nesidiocoris tenuis* (Hemiptera: Miridae) on tomato plants. J Pest Sci 88:543–554

Petersen M, Brodersen P, Naested H et al (2000) *Arabidopsis* map kinase 4 negatively regulates systemic acquired resistance. Cell 103:1111–1120
Pieterse CM, Van Der Does D, Zamioudis C et al (2012) Hormonal modulation of plant immunity. Annu Rev Cell Dev Biol 28:489–521
Pre M, Atallah M, Champion A et al (2008) The AP2/ERF domain transcription factor ORA59 integrates jasmonic acid and ethylene signals in plant defense. Plant Physiol 147:1347–1357
Price PW (1991) The plant vigor hypothesis and herbivore attack. Oikos 62:244–251
Rashmann S, Kollner T, Deenhardt J, Hiltpold I, Toepfer S, Kuhlmann U et al (2005) Recruitment of entomopathogenic nematodes by insect damaged maize roots. Nature 434:732–737
Rehrig EM, Appel HM, Jones D, Schultz JC (2014) Roles for jasmonate and ethylene induced transcription factors in the ability of Arabidopsis to respond differentially to damage caused by two insect herbivores. Front Plant Sci 5:1–14
Ribot C, Zimmerli C, Farmer EE et al (2008) Induction of the *Arabidopsis PHO1;H10* gene by 12-oxo-phytodienoic acid but not jasmonic acid via a CORONATINE INSENSITIVE1-dependent pathway. Plant Physiol 147:696–706
Rondoni G, Bertoldi V, Malek R, Djelouah K, Moretti C, Buonaurio R, Conti E (2018) *Vicia faba* plants respond to oviposition by invasive *Halyomorpha halys* activating direct defences against offspring
Santamaría ME, Martinez M, Arnaiz A et al (2017) MATI, a novel protein involved in the regulation of herbivore-associated signaling pathways. Front Plant Sci 8:975
Schaeffer RN, Wang Z, Thornber CS, Preisser EL, Orians CM (2018) Two invasive herbivores on a shared host: patterns and consequences of phytohormone induction. Oecologia. https://doi.org/10.1007/s00442-018-4063-0
Schaller A (2008) Induced plant resistance to herbivory. Springer, Dordrecht, pp 2340–2354
Schenk PM, Kazan K, Wilson I et al (2000) Coordinate plant defense responses in *Arabidopsis* revealed by microarray analysis. Proc Natl Acad Sci U S A 97:11655–11660
Schmelz EA, Alborn HT, Banchio E (2003) Quantitative relationships between induced jasmonic acid levels and volatile emission in *Zea mays* during *Spodoptera exigua* herbivory. Planta 216:665–673
Schnee C, Kollner TG, Held M, Turlings TCJ, Gershenzon J, Degenhardt J (2006) The products of a single maize sesquiterpene synthase form a volatile defense signal that attracts natural enemies of maize herbivores. Proc Natl Acad Sci U S A 103:1129–1134
Schoonhoven LM, Jermey T, van Loon JJA (1998) Insect-plant biology: from physiology to evolution. Chapman & Hall, London
Schoonhoven LM, van Loon JJA, Dicke M (2005) Insect plant biology. Oxford University Press Inc, New York
Schuman MC, Barthel K, Baldwin IT (2012) Herbivory-induced volatiles function as defenses increasing fitness of the native plant *Nicotiana attenuata* in nature. elife 1:e00007
Schweizer F, Fernandez-Calvo P, Zander M et al (2013) *Arabidopsis* basic helix-loop-helix transcription factors MYC2, MYC3, and MYC4 regulate glucosinolate biosynthesis, insect performance, and feeding behavior. Plant Cell 25:3117–3132
Seo S, Okamoto M, Seto H et al (1995) Tobacco MAP kinase: a possible mediator in wound signal transduction pathways. Science 270:1988–1992
Sheard LB, Tan X, Mao H et al (2010) Jasmonate perception by inositol phosphate-potentiated COI1-JAZ co-receptor. Nature 468:400–405
Shivaji R, Camas A, Ankala A et al (2010) Plants on constant alert: elevated levels of jasmonic acid and jasmonate-induced transcripts in caterpillar-resistant maize. J Chem Ecol 36:179–191
Signoretti AGC, Peñaflor MFGV, Bento JMS (2012) Fall armyworm, *Spodoptera frugiperda* (JE smith) (Lepidoptera: Noctuidae), female moths respond to herbivore-induced corn volatiles. Neotrop Entomol 41:22–26
Song S, Qi T, Wasternack C et al (2014) Jasmonate signaling and crosstalk with gibberellin and ethylene. Curr Opin Plant Biol 21:112–119

Spiteller D, Boland W (2003) N-(15, 16-Dihydroxylinoleoyl)-glutamine and N-(15, 16-epoxylinoleoyl)-glutamine isolated from oral secretions of lepidopteran larvae. Tetrahedron 59:135–139

Spoel SH, Johnson JS, Dong X (2007) Regulation of tradeoffs between plant defenses against pathogens with different lifestyles. Proc Natl Acad Sci U S A 104:18842–18847

Staswick PE, Tiryaki I (2004) The oxylipin signal jasmonic acid is activated by an enzyme that conjugates it to isoleucine in *Arabidopsis*. Plant Cell 16:2117–2127

Strauss SY (1997) Floral characters link herbivores, pollinators, and plant fitness. Ecology 78:1640–1645

Sun YC, Pan LL, Ying FZ, Li P, Wang XW, Sheng S (2017) Jasmonic acid-related resistance in tomato mediates interactions between whitefly and whitefly-transmitted virus. Li Sci Rep 7:566

Thaler JS, Bostock RM (2004) Interactions between abscisic-acid mediated responses and plant resistance to pathogens and insects. Ecology 85:48–58

Thines B, Katsir L, Melotto M et al (2007) JAZ repressor proteins are targets of the SCF(COI1) complex during jasmonate signaling. Nature 448:661–665

Tholl D, Sohrabi R, Hyun Huh J, Lee S (2011) The biochemistry of homoterpenes- common constituents of floral and herbivore induced plant volatile bouquets. Phytochemistry 72 (13):1635–1646

Tretner C, Huth U, Hause B (2008) Mechano stimulation of *Medicago truncatula* leads to enhanced levels of jasmonic acid. J Exp Bot 59:2847–2856

Turlings TCJ, Loughrin JH, McCall PJ, Rose USR, Lewis WJ, Tumlinson JH (1995) How caterpillar-damaged plants protect themselves by attracting parasitic wasps. Proc Natl Acad Sci U S A 92:4169–4174

Tytgat TOG, Verhoeven KJF, Jansen JJ et al (2013) Plants know where it hurts: root and shoot jasmonic acid induction elicit differential responses in *Brassica oleracea*. PLoS One 8:e65502

Ulloa RM, Raices M, MacIntosh GC et al (2002) Jasmonic acid affects plant morphology and calcium-dependent protein kinase expression and activity in *Solanum tuberosum*. Physiol Plant 115:417–427

Uppalapati SR, Ishiga Y, Wangdi T et al (2007) The phytotoxin coronatine contributes to pathogen fitness and is required for suppression of salicylic acid accumulation in tomato inoculated with *Pseudomonas syringe* epv. Tomato DC3000. Mol Plant-Microbe Interact 20:955–965

Usha Rani P, Jyothsna Y (2010) Biochemical and enzymatic changes in rice as a mechanism of defense. Acta Physiol Plant 32:695–701

Van der Does D, Leon-Reyes A, Koornneef A et al (2013) Salicylic acid suppresses jasmonic acid signaling downstream of SCFCOI1-JAZ by targeting GCC promoter motifs via transcription factor ORA59. Plant Cell 25(2):744–761

Von Dahl CC, Baldwin IT (2007) Deciphering the role of ethylene in plant–herbivore interactions. J Plant Growth Regul 26:201–209

Vos IA, Verhage A, Schuurink RC et al (2013) Onset of herbivore-induced resistance in systemic tissue primed for jasmonate-dependent defenses is activated by abscisic acid. Front Plant Sci 4:539

Walter A, Mazars C, Maitrejean M et al (2007) 2007. Structural requirements of jasmonates and synthetic analogues as inducers of Ca2+ signals in the nucleus and the cytosol of plant cells. Angew Chem Int Ed Eng 46:4783–4785

Wang H, Li S, Teng S, Liang H et al (2017) Transcriptome profiling revealed novel transcriptional regulators in maize responses to *Ostrinia furnacalis* and jasmonic acid. PLoS One 12(5): e0177739. https://doi.org/10.1371/journal.pone.0177739

War AR, Paulraj MG, War MY et al (2011) Jasmonic acid- mediated induced resistance in groundnut (*Arachis hypogaea* L.) against *Helicoverpa armigera* (Hubner) (Lepidoptera: Noctuidae). J Plant Growth Regul 30:512–523

Wasternack C, Hause B (2013) Jasmonates: biosynthesis, perception, signal transduction and action in plant stress response, growth and development. An update to the 2007 review in annals of botany. Ann Bot 111:1021–1058

Wild M, Daviére JM, Cheminant S et al (2012) The *Arabidopsis* DELLA RGA-LIKE3 is a direct target of MYC2 and modulates jasmonate signaling responses. Plant Cell 24:3307–3319

Wu G, Poethig RS (2006) Temporal regulation of shoot development in Arabidopsis thaliana by miR156 and its target SPL3. Development 133:3539–3547

Wu Y, Deng Z, Lai J et al (2009) Dual function of *Arabidopsis* ATAF1 in abiotic and biotic stress responses. Cell Res 19:1279–1290

Xie DX, Feys BF, James S, Nieto-Rostro M, Turner JG (1998) COI1: an Arabidopsis gene required for jasmonate-regulated defense and fertility. Science 280:1091–1094

Xu S, Zhou W, Pottinger S et al (2015) Herbivore associated elicitor-induced defences are highly specific among closely related *Nicotiana* species. BMC Plant Biol 15:2

Yadav V, Mallappa C, Grangappa SN et al (2005) A basic helix-loop-helix transcription factor in *Arabidopsis*, MYC2, acts as a repressor of blue light-mediated photomorphogenic growth. Plant Cell 17:1953–1966

Yan Y, Stolz S, Chetelat A (2007) A downstream mediator in the growth repression limb of the jasmonate pathway. Plant Cell 19:2470–2483

Zhang SZ, Hau BZ, Zhang F (2008) Induction of the activities of antioxidative enzymes and the levels of malondialdehyde in cucumber seedlings as a consequence of *Bemisia tabaci* (Hemiptera: Aleyrodidae) infestation. Arthropod-Plant Interact 2:209–213

Zhang PJ, Broekgaarden C, Zheng SJ et al (2013) Jasmonate and ethylene signaling mediate whitefly-induced interference with indirect plant defense in *Arabidopsis thaliana*. New Phytol 197:1291–1299

Zhang P-J, Huang F, Zhang JM et al (2015) The mealybug *Phenacoccus solenopsis* suppresses plant defense responses by manipulating JA–SA crosstalk. Sci Rep 5:9354

Zimmermann P, Hirsch-Hoffmann M, Hennig L et al (2004) GENEVESTIGATOR: *Arabidopsis* microarray database and analysis toolbox. Plant Physiol 136:2621–2632

Plant Proteinase Inhibitor and Protease Interaction During Insect-Plant Communication

Mahaswetta Saikia and Rakesh Kumar

Abstract

The interaction of proteinases with plant proteinase inhibitors (PPIs) has always been important in insect and plant interaction studies. Available literature suggests that proteinase inhibitors have the ability to defend plants against herbivorous insects. There are three modes of action of PPIs with proteinases: insects adapt to the presence of PPI (i) by overexpressing native proteinases, (ii) by expressing new proteinases that are resistant to the inhibitors, and (iii) by proteolytic inactivation of PIs by insect's own midgut proteinases. Midgut proteinases have been widely studied in lepidopteran pests. In the case of *Antheraea assamensis*, a silk-producing lepidopteran, midgut proteinase interaction with PIs of host plants shows resistance to proteolysis by larval midgut proteases. It suggests that PPIs are stable in the alkaline environment of the gut and can potentially interact with gut proteinases. Moreover, the expression levels of PPIs differed in *A. assamensis* larvae when reared on different species of host plants. *Pieris brassicae* (euryphagous pest of crucifers) midgut proteases studies reveal that those larvae which fed on *Brassica oleracea* var. *botrytis* but not those which fed on *Tropaeolum majus* (garden nasturtium) adapted to ingested PPIs. These results suggest that midgut physiology of this versatile pest show considerable plasticity enabling it to feed on wide range of host plants.

Keywords

Digestive physiology · Host plants · Insect pests · Proteinase · Plant proteinase inhibitors

M. Saikia (✉) · R. Kumar
Lab 15/18, Insect-Plant Interactions Group. Department of Botany, University of Delhi, Delhi, India

I. K. Singh, A. Singh (eds.), *Plant-Pest Interactions: From Molecular Mechanisms to Chemical Ecology*, https://doi.org/10.1007/978-981-15-2467-7_11

1 Introduction

Insects encounter various compounds in the ingested food material to be used as a food source to derive nutrition. These include plant defence compounds like proteinase inhibitors. The proteinase inhibitors **(PIs)** in plants are ubiquitous in nature, and their significance in plant's natural defence system against proteinases of pests and pathogens has been well established (Ryan 1978, 1979, 1990; Laskowski and Kato 1980; Richardson 1991; Valueva and Mosolov 1999; Lawrence and Koundal 2002; Mello et al. 2002; Christeller and Liang 2005; Fan and Wu 2005; Telang et al. 2009; Upadhyay and Chandrashekhar 2012; Rufino et al. 2013; Tamaki and Terra 2015; Souza et al. 2016; Pandey et al. 2016; Bezerra et al. 2017; Ramalho et al. 2018). Proteinase inhibitors found in plant kingdom possess some general properties. These polypeptides are entirely made up of L-amino acids which are connected through peptide bonds. The methionine, histidine and tryptophan content are typically low or absent but aspartic acid, glutamic acid, serine, and lysine content are often rich in plant inhibitors. At neutral pH, they show strong association with their respective proteinases. Most of the plant proteinase inhibitors exhibit the size range of 4000–60,000 MW but are generally in the range of 8000–20,000 MW. It is well established that the active sites of plant proteinase inhibitors are hyper-variable and show positive Darwinian selection (Odani and Ikenaka 1976; Creighton and Darby 1989; Christeller and Liang 2005).

Another intriguing aspect of their interactions with insects is the differential regulation and transcription of sequence divergent digestive serine proteinase in Lepidoptera in response to ingestion of plant proteinase inhibitors (Bown et al. 1997; Broadway 1997; Mazumdar-Leighton and Broadway 2001a, b). Digestive serine proteinases from Lepidoptera are likely to have evolved in response to the selection pressures imposed by the plant proteinase inhibitors, other defence compounds and substrates encountered in the ingested host plant tissues (Cipollini et al. 2003; Howe and Schaller 2008). Plant proteinase inhibitors inhibit serine, cysteine and aspartyl proteinases from insects, fungi, bacteria and nematodes (Laskowski and Kato 1980; Ryan 1989, 1990; Richardson 1991; McPherson and Harrison 2001; Liang and McManus 2002; Lawrence and Koundal 2002; Fan and Wu 2005; Zavala et al. 2008; Dantzer et al. 2015; Jamal et al. 2015; Swathi et al. 2016). A plant proteinase inhibitor database (PLANT-PIs) developed by De Leo et al. (2002) is available online (http://bighost.area.ba.cnr.it/PLANT-PIs/), where all information regarding sequence and functional properties of all known inhibitors are easily obtainable. There is some debate regarding the susceptibility of PIs to gut proteinases. Several theories have been put forward to explain the status of PIs within the insect gut. One prevailing view supported by proteomic studies on tomato-reared *Manduca sexta* is that PIs are hyper-stable proteins resistant to gut proteases and are eventually egested almost intact by peristalsis as frass (Chen et al. 1997). Stability of PIs to alkaline gut proteases has also been observed in several other Lepidoptera (Richardson 1991; Macedo et al. 2002, 2003). Another opinion is that ingested PIs are digested and hydrolysed by gut proteases. These observations have mostly been reported from Coleopterans, and they use digestive cysteine proteinases or a dual digestive system of serine and cysteine proteinases (Michaud 1997; Girard et al. 1998; Vinokurov et al. 2006). In Lepidoptera, breakdown by gut proteases of *H. armigera* of various leguminous seed PIs has been reported (Giri et al. 1998; Telang et al. 2005; Jamal et al. 2015). The later scenario argues against PIs affecting digestive enzyme profile of these herbivorous larvae. The response of larval gut proteinases to PIs has been widely studied at the physiological, biochemical as well as molecular levels in Lepidoptera (Broadway and Duffey 1986a, b; Christeller et al. 1992; Broadway 1995; Jongsma et al. 1995; Broadway 1997; Jongsma and Bolter 1997; Bown et al. 1997; Wu et al. 1997; Jouanin et al. 1998; De Leo et al. 2001a, b; Volpicella et al. 2003; Diaz-Mendoza et al. 2005). Most responses to ingested PI in the insect involve differential regulation of midgut transcripts encoding diverse serine proteinases with varied susceptibility to the PPI (Broadway 1996a, b; Gatehouse et al. 1997; Mazumdar-Leighton and Broadway 2001a, b; Diaz-Mendoza et al. 2005; Saikia et al. 2011; Kumar et al. 2015).

In this review an attempt has been made to comprehend the interaction of plant proteinase inhibitors with the digestive midgut proteases of two lepidopteran insects, viz. *Pieris brassicae*, which is known as a pest of crucifers, and *A. assamensis*, a valuable beneficial lepidopteran exploited for production of commercial silk in the north-eastern region of India.

1.1 Plant Proteinase Inhibitors

Plant proteinase inhibitors (PPIs) are those polypeptides or proteins that are found naturally in a large number of plants and are considered to play an important role in plant's natural defence system against herbivores (Ryan 1990; Jongsma and Bolter 1997; Zavala et al. 2004). The classification of PIs usually follows the proteinase catalytic type (Laskowski and Kato 1980). Many families of PPIs exist, each family specific for each of the four mechanistic classes of proteolytic enzymes; these are classified on the basis of the active site amino acid (Koiwa et al. 1997). PPIs are ubiquitous in plants and appear to play a significant role in protection of plants against proteinases of pests and pathogens (Ryan 1990; McManus et al. 1994, 1999; McManus and Burgess 1995). The hypothesis that they are an integral component of natural plant defence mechanism was first proposed by Applebaum (1964). This has been substantiated by the work of Green and Ryan (1972) where they observed that PIs are actively induced in the leaves of tomato and potato upon attack by insect herbivores. Table 1 shows a list of inhibitor clans according to Liang and McManus (2002).

Serine proteinase inhibitors are found universally all over the plant kingdom with trypsin inhibitors being the most common one. Families of all serine inhibitor from plants are competitive inhibitors and having an almost similar standard mechanism to inhibit proteinases (Laskowski and Kato 1980). Serine proteinase inhibitors show antinutritional effects against many lepidopteran species (Schukle and Murdock 1983; Applebaum 1985). The first well-characterized plant inhibitor was soybean trypsin inhibitor (Kunitz) (SKTI). One of the classical achievements of inhibitor

Table 1 Classification of PIs

Proteinase inhibited	# Name	Active unit molecular weight range	Known distribution and tissue
Serine	1S Bowman-Birk 2S Cereal type 3S Kunitz 4S Mustard seed inhibitors 5S Proteinase inhibitor 1 6S Proteinase inhibitor 2 7S Serpins 8S Squash	6000–9500 Da (except rice, wheat and barley which are 14,000–15,000 Da) 11,500–14,000 Da 19,000–24,000 6600–7100 7200–9100, synthesized as a large precursor Base unit is 5000–6000 Da 42,000–44,000 Da (*Arabidopsis*) more variable 3000–3500 Da	Legume seeds, *Zea mays*, rice, wheat, barley Cereals: Barley, wheat, sorghum, *Eleusine coracana*, rice, rye Legumes. Also reported in potato, cereals and *Arabidopsis* and in other species. Usually found in seeds Brassicaceae including mustard, rape and *Arabidopsis* Solanaceae species, cereals, squash, *Arabidopsis*, legumes Solanaceous species including tomato, potato, tobacco, aubergine and capsicum in tubers, fruits, seeds, leaves and flowers Cereals wheat, rice and barley, squash, *Arabidopsis* Cucurbit seeds
Cysteine	9C Cystatins Muticystatin 3C Kunitz 1C Pineapple bromelain[a]	10,000–16,000 Da, occasionally even smaller or greater than this 6000–87,000 Da 20,089 5800–5900	Rice, maize, wheat, potatoes, soybean and *Arabidopsis* A multiple active site multicystatin has been reported from potato (8 active sites), sunflower (3) and other species Potato Pineapple
Aspartic	3A Kunitz 11A SQAPI 12A Wheat inhibitor[a]	20,000–21,000 Da 10,500 Da 58,000 Da	Potato Squash Wheat
Metallo-	13M Carboxypeptidase inhibitor	4100–4300 Da	*Solanum*

S, A, M, C refer to the catalytic type of proteinases inhibited by the inhibitors (Barrett et al. 1998)
[a]Denotes not cloned or sequenced at the DNA level. Source: Liang and McManus (2002)

chemistry is the isolation and crystallization of SKTI and that of its complex with trypsin by M. Kunitz (Kunitz 1945). Kunitz's STI is known to play a key role in the early study of proteases, through extensive use in biochemical and kinetic studies that led to the definition of the standard mechanism of protease inhibitor action. A total of 24 residues of Porcine Pancreatic Trypsin (PPT) make contact with 9 residues of STI, in the tetragonal STI: PPT complex (PDB ID 1AVX). Similar crystal structures are available for other serine proteinase inhibitors like Bowman-Birk inhibitor (Chen et al. 1997; Suzuki et al. 1993; Tsunogae et al. 1986; Werner and

Wemmer 1992; Voss et al. 1996; Song and Suh 1998; Song et al. 1999; Park et al. 2004; Koepke et al. 2000; Barbosa et al. 2007; Rao and Suresh 2007).

The second most studied class of inhibitors belongs to the plant cystatins or phytocystatins which have been identified and characterized from several different plants, viz. cowpea, potato, cabbage, carrot, etc. Cystatins from the seeds of a wide range of crop plants have also been isolated, of which the rice cysteine proteinase inhibitors are most widely studied (Abe and Arai 1985). The inhibition of cysteine proteinases by cystatin is reversible and competitive. E-64 (L-*trans*-epoxy-succinyl-l-leucylamido (4-guanido) butane) is one of the most commonly used diagnostic proteinase inhibitors of cysteine proteinases (Beynon and Salvesen 1989; Barrett 1994; Novillo et al. 1997) and was isolated from cultures of *Aspergillus japonicus* (Hanada et al. 1978). It is a non-competitive inhibitor as well as an irreversible inhibitor of thiol proteinases of the papain and calpain family (Hanada et al. 1978; Hashida et al. 1980; Barrett 1986; Parkes et al. 1985).

Two families of metalloproteinase inhibitors, viz. metallocarboxypeptidase inhibitor family in potato (Rancour and Ryan 1968) and tomato (Graham and Ryan 1981) and cathepsin D inhibitor family from potatoes (Keilova and Tomasek 1976) have been reported in plants. Isolation of aspartic PIs has been reported from sunflower (Park et al. 2000), barley (Kervinen et al. 1999) and cardoon *Cynara cardunculus* (Frazao et al. 1999).

An understanding of different classes of midgut proteinases based on their catalytic mechanism will help to better understand their interaction with PPIs.

1.2 Midgut Proteinases in Phytophagous Insects

Phytophagous insects have evolved or adapted in their feeding habits, digestive physiology and gene expression in relation to their host plants to meet their nutritional requirements. Generally, digestion of food coming from plant parts occurs in the midgut by the action of digestive proteases. The term 'protease' as described by Barrett (1986) encompasses exopeptidases as well as endopeptidases, while the term 'proteinases' describes only endopeptidases (enzymes which cleave internal peptide bonds of polypeptides). The Classification of proteinases is based on their catalytic mechanism. IUBMB recognized four mechanistic classes of proteinases based on the essential amino acid residue present at the active site, the optimum pH range, amino acid sequence similarity and similarity in response to inhibitors (Cygler and Morb 1997; Matsumoto et al. 1997). These include serine proteinases, cysteine proteinases, aspartic proteinases and metalloproteinases (Table 2). The classification of the proteinases based on the catalytic mechanism is extended by another classification on the basis of evolutionary relationship of proteases (Rawlings and Barrett 1993). Food digestion in silkworm occurs in the midgut (Horie et al. 1963) where pH is highly alkaline and serine proteinases like trypsin predominate (Eguchi and Iwamoto 1976). Lepidopteran digestive proteases have been extensively studied (Christeller et al. 1992; Houseman and Chin 1995; Novillo et al. 1999; Pereira et al. 2005; Chougule et al. 2007; George et al. 2008). The midgut proteinases of

Table 2 Different classes of proteinases

Proteinases	Amino acid in active site	pH Optima (range)	Proteins
Serine proteinases	Ser, His, Asp	7–9	Trypsin, chymotrypsin, elastase, cathepsin (+) G
Cysteine proteinases	Cys	4–7	Papain, ficin, bromelain, ananain, cathepsins (+) B,C,H,K,L,O,S,W
Aspartic proteinases	Asp, Try	Below 5	Renin, pepsin
Metalloproteinase	Metal ion	7–9	Carboxypeptidases A and B, aminopeptidases

Note: (+) the term cathepsin is also referred for lysosomal cysteine protease (Cygler and Morb 1997; Matsumoto et al. 1997)

the Lepidoptera *B. mori* are the most widely studied among sericigenous insects (Eguchi and Iwamoto 1976; Eguchi et al. 1982; Eguchi and Kuriyama 1983; Sasaki et al. 1993; Sasaki and Suzuki 1982; Shinbo et al. 1996). The midgut of *B. mori* fifth instar larvae has been used to construct an extensive EST database (Mita et al. 2002). Compared to *B. mori* the study of midgut proteinases of *A. assamensis* is at its infantile stage. Saikia et al. (2011) reported for the first time differential gene expression of midgut proteinases (trypsin and chymotrypsin) feeding on diverse host plant species.

1.3 Serine Proteinases

Serine proteinases comprise one of the largest and best known families of proteolytic hydrolases. The digestive serine proteases (also known as the S1 clan of serine proteinases) include trypsin, chymotrypsin and elastases (Barrett and Rawlings 1995). All digestive serine proteinases have a highly conserved catalytic triad of mainly three residues His57, Asp102 and Ser195 (numbering after the sequence of bovine chymotrypsinogen), where His57 acts as a general acid and base, Asp102 functions to orient His57, while Ser195 forms a covalent bond with the peptide which is to be cleaved (Kiel 1965). Trypsin, chymotrypsin and elastase have the same active site residues but differ dramatically in their substrate specificity and in the amino acids that confer this specificity. The substrate binding site in trypsin is occupied by Asp189, Gly 216 and Gly226. Asp189 gives negative charge to the substrate binding pocket and is perfectly placed to form an ion pair with the positive charge of lysine or arginine residues present in the scissile bond of a substrate or inhibitor (Evnin et al. 1990; Perona et al. 1994). Asp189 is replaced by Ser189 in chymotrypsins which cleave the peptide bonds at carboxyl termini of amino acids tryptophan, phenylalanine and methionine (Boyer 1971). In the elastase specificity pocket, Gly216 and Gly226 are replaced by Val216 and Thr226 which block the entrance of large bulky amino acids into the substrate binding cavity (Shotton and

Hartley 1970). One of the best understood insect digestive proteinases are serine proteinase with specificities resembling mammalian trypsin and chymotrypsin (Applebaum 1985). Digestive serine proteinases from various economically important phytophagous lepidopterans have been reported. A large number of encoding genes, cDNAs and (in a few cases) purified digestive serine proteinases have been isolated from these insects (Johnston et al. 1991; Peterson et al. 1994, 1995; Noriega et al. 1996; Jiang et al. 1997; Zhu et al. 1997; Mazumdar-Leighton et al. 2000). The presence of at least six serine proteinases in the midgut of *B. mori* is suggested, of which two have been isolated, purified and characterized (Sasaki and Suzuki 1982; Eguchi et al. 1982; Eguchi and Kuriyama 1983; Kotani et al. 1999). Their cDNAs have also been isolated and resemble other lepidopteran enzymes. In fact the first insect serine proteinase to be characterized was 'cocoonase' responsible for the dissolution of the *B. mori* cocoon before emergence of the metamorphosed adult (Kramer et al. 1973).

Trypsins (EC.3.4.21.4) are enzymes which preferentially catalyse the hydrolysis of ester and peptide bonds at the carboxyl group of basic amino acids such as arginine and lysine. The three-dimensional structure of trypsin and its mode of action have been well studied with the help of crystallography (Stroud 1974; Kraut 1977; Huber and Bode 1978). Genes encoding trypsin protein have been sequenced from a wide variety of organisms including sequences from both vertebrates and invertebrates. Trypsin from beef pancreas was one of the first proteolytic enzymes which was isolated in pure form (Northrop et al. 1948). Bovine trypsin was one of the first proteins to be sequenced (Walsh et al. 1964), while the genes for rat trypsin were isolated and sequenced in the 1980s (Craik et al. 1984). Most trypsin genes are found as members of gene families in eukaryotes. The number of trypsin genes within a gene family varies widely, from species to species. In mammals the trypsin is encoded by multiple genes. A family of at least ten trypsin genes is presented in rat (Craik et al. 1984). In the case of invertebrates, much work has centred on the trypsins in the blood-sucking insects like *Aedes aegypti* (Barillas-Mury et al. 1991; Noriega et al. 1996; Jiang et al. 1997). Trypsin sequences have also been reported from crayfish (Zwilling et al. 1975), dogfish (Titani et al. 1983), hornet (Jany and Haug 1983) and *Drosophila melanogaster* (Davis et al. 1985). All Dipteran insects have been found to contain trypsin gene families, for example, eight trypsin members were found in *D. melanogaster* (Wang et al. 1999) and four were found in *Lucilia cuprina* (Casu et al. 1997). In the case of lepidopteran trypsin genes, the first reports came in the 1990s from *Manduca sexta* (Peterson et al. 1994, 1995), *Choristoneura fumiferana* (Wang et al. 1995) and *Plodia interpunctella* (Zhu et al. 1997). They occur as gene families consisting of multiple members in each family (Bown et al. 1997; Gatehouse et al. 1997; Mazumdar-Leighton et al. 2000; Diaz-Mendoza et al. 2005).

Chymotrypsin (EC 3.4.21.1) is another prominent group of enzyme belonging to the digestive serine proteinase family. Chymotrypsin is synthesized as a catalytically inert precursor chymotrypsinogen. Trypsin cleavage between Arg15 and Ile16 of bovine chymotrypsinogen results in the formation of a fully active chymotrypsin (Kraut 1977). Trypsins and chymotrypsins share the same catalytic residues

His57-Asp102-Ser195 and also three disulphide bridges known as His loop (42–58), Met loop (168–182) and Ser loop (191–220) (Hartley 1964). Chymotrypsins have been isolated from a number of invertebrates in the 1970s and 1980s (Hermodson et al. 1971; Grant et al. 1980; Jany and Haug 1983; Jany et al. 1983). Since then a large number of chymotrypsins have been identified in Lepidoptera. The first report of pancreatic elastase activity dates back to 1878 when Walchli found that ox pancreas digested ligamentum nuchae elastin, and Kuhne reported that impure trypsin preparations dissolved elastin (Boyer 1971). Elastase (E.C 3.4.4.7) is an endopeptidase that can digest elastin, the elastic fibrous protein of connective tissue (Boyer 1971). It is present in all mammals investigated and also has been reported in pancreatic extracts of many other vertebrates and invertebrates. The ability of elastase to degrade elastin to soluble peptides is not shared by other pancreatic endopeptidases. In 1970, Shotton and Hartley determined the complete sequence of porcine elastase. The elastase molecule consists of a single polypeptide chain of 240 amino acid residues which contain 4 disulphide bridges. Like other serine proteinases, elastase contains the highly conserved catalytic triad of His57-Asp102-Ser195 (Vered et al. 1986). The specificity pocket of elastases is small, lipophilic and occupied by Val216-Thr226 (http:/www.kubinyi.de/dd-21.pdf). The first elastase gene to be cloned was from pig (Shotton and Hartley 1970). In later years, elastase genes in rat (Swift et al. 1984) and mouse were also cloned and sequenced (Stevenson et al. 1986). In Lepidoptera, the first elastase sequence was reported from *Manduca sexta* (Peterson et al. 1994).

1.4 Cysteine Proteinases

Cysteine proteinases (3.4.22) are endopeptidases with a cysteine residue involved in catalysis (Cristofoletti et al. 2005). Rawlings and Barrett (1993) reported that there are 14 structural families of cysteine proteinases. Of these, the C1 (including cathepsin B, L, H) group of cysteine proteinases prefers substrates with a bulky hydrophobic side chain occupying enzyme subsite (Ser 205) (Barrett et al. 1998). Cathepsin H has aminopeptidase activities; cathepsin B has peptidyl dipeptidase activities, while cathepsin L exhibits only endopeptidase activity (Barrett et al. 1998; Cristofoletti et al. 2005). Thie and Houseman (1990) reported the presence of cysteine proteinases in the more acidic anterior midgut, while in the posterior midgut of *T. molitor*, serine proteinases dominated where the region has a more alkaline pH. The occurrence of such a condition in the midgut of *A. assamensis* cannot be ruled out.

Cysteine proteinase found in insects may be digestive (Rawlings and Barrett 1993; Terra and Ferreira 1994; Oliveira et al. 2003) but are also found in several other tissues, ovary, fat body, hemocyte and hemolymph (Yamamoto et al. 1994; Matsumoto et al. 1995, 1997; Oliveira et al. 2003). The pH-dependent activity of cysteine proteinase in crude midgut extract of insect larvae has indicated the presence of alkaline range (Bode and Huber 1992; Oliveira et al. 2003). Recent research has indicated that the cathepsin L-like enzymes are the only insect cysteine proteinases quantitatively important (Cristofoletti et al. 2005). The molecular characterization of insect digestive CALs having peculiar secretory routes has not progressed very far (Terra and Ferreira 2005; Cristofoletti et al. 2005) despite their importance in insect digestion. A thiol proteinase has been reported from eggs of *B. mori* (Kageyama et al. 1981). Genes encoding cysteine proteinases have been reported from various Lepidoptera including *Helicoverpa armigera*, *Helicoverpa assulta*, *Spodoptera exigua*, *Manduca sexta*, etc. Cysteine proteinases (cathepsin B) have also been reported from larval midguts of *A. assamensis* and *Samia cynthia ricini* (EU126819, EU126818).

1.5 Aspartic Proteinases and Metalloproteinases

Most of aspartic proteinases (EC 3.4.23.6) belong to the pepsin family and consist of digestive enzymes such as pepsin, chymosin, lysosomal cathepsins D, renin and fungal proteases (penicillopepsin, rhizopuspepsin, endothiapepsin). The catalytic mechanism in aspartic proteinases does not involve a covalent intermediate; however, a tetrahedral intermediate exists (Fan and Wu 2005).

Metalloproteases (EC 3.4.24) bind a metal ion such as Zn^{2+} in the active site. Carboxypeptidases A and B, aminopeptidases and thermolysin are a few most widely studied digestive metalloproteases. One of the important physiological roles of midgut aminopeptidases is in dietary protein digestion (Terra and Ferreira 1994). Aminopeptidases have been reported from *B. mori* and *Plutella xylostella* (Nakanishi et al. 2002) as well as *Helicoverpa armigera* (Rajagopal et al. 2003). Four midgut aminopeptidases and five carboxypeptidases have been known to be reported from *Trichoplusia ni* (Wang et al. 2004, 2005).

1.6 Interaction Between Proteinases and Proteinase Inhibitors

The binding mechanism of a plant protease inhibitor with the insect proteases appears similar among all four classes of inhibitors. The interaction mechanism of protein inhibitors with mammalian serine proteinases is known. Most of these studies are available as diffraction data and NMR (nuclear magnetic resonance) spectral data that can be easily retrieved from the Brookhaven Protein Database (USA) as PDB files and displayed as molecular models on computer. Here the action of serine proteases with PIs is described in detail. Based on the studies of many serine proteinases using amino acid sequence analysis and X-ray crystallographic, it has been found that the catalytic site lies in the cleft on the surface of the enzyme molecule. The placement of substrate chain is along the active site cleft. On either side of active site cleft, the subsites present are adapted to bind amino acid chains of a substrate (Fig. 1). The naming of individual subsites of an enzyme (.....S3,S2,S1, S1′,S2′,S3′...) and the complementary part of a substrate (.....P3,P2,P2,P1′,P2′, P3′...) is according to Berger and Schechter (1970). The catalytic triad consisting of His57, Asp102 and Ser195 (chymotrypsinogen numbering) is in the centre of the

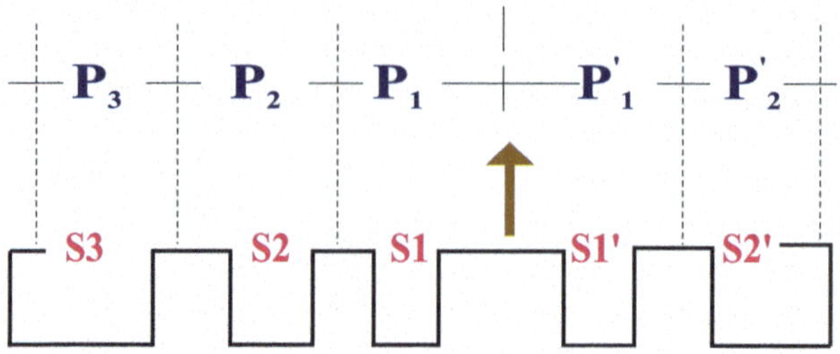

Fig. 1 Specificity subsites of proteases and their corresponding complementary substrate sites. The subsites of proteases, S_1, S_2, S_3. . ., are away from the active site towards N-terminus while S_1', S_2' and S_3' are towards C-terminus. The cleavage site is indicated by an arrow. Similarly, the substrate sites corresponding to subsites of enzymes P_1, P_2, P_3. . . . are oriented towards N-terminus, while P_1', P_2', P_3'. . . . . . are oriented towards C-terminus. (Source: Ashoka 2005)

cleft. During the initial catalysis step, an acyl-enzyme intermediate is formed between the substrate and serine residue at the active site of the enzyme. A tetrahedral intermediate is formed through this covalent complex (via negatively charged transition state). The acyl-enzyme intermediate is hydrolysed by a water molecule during the process of diacylation and releases a cleaved peptide and thus restores the Ser hydroxyl of the enzyme (Fastrez and Fersht 1973).

A protein is called a 'protease inhibitor' if it can interact with the active site of the protease and act in a way like a substrate to either block or reduce the protease activity. The standard mechanism of inhibition proposed by Laskowski and Kato (1980) refers to the mechanism by which the inhibitors suppress protease activity by acting as substrates of their target enzymes. Inhibitors with similar reactive site loop configuration follow the standard mechanism. The reactive site loop is composed of a scissile bond between two amino acid residues, i.e. P1 and P1′ within the substrate (Fig. 1). In trypsin inhibitors, the P1 residue at the N-terminal end can either be a lysine or arginine, resembling a substrate. The P1 residue could be either Tyr, Trp, Phe, Leu, or Met in chymotrypsin inhibitors. In the case of elastase, the P1 residue is either alanine or serine. At least one disulphide bond at the reactive site peptide bond is involved in inhibitors that follow the standard mechanism (Fig. 2). Such mechanism ensures close proximity without dissociation of two peptide chains during the conversion of virgin inhibitor to cleaved inhibitor.

Figure 3 shows enzyme-inhibitor interaction mechanism of Laskowski and Kato (1980). In this mechanism, the enzyme (E) and inhibitor (I) interact to make a stable complex (C) after having small conformational change. Subsequently, the stable complex (C) dissociates very slowly to produce the free enzyme and virgin inhibitor (I) or cleaved inhibitor (I*). The scissile bond at P1-P1′ dissociates in the cleaved inhibitor (I*). The enzyme and inhibitor interact largely in a canonical fashion. The reactive site loop of the inhibitor associates with the catalytic residues in like manner as that of bound substrates in complexes with their cognate enzymes (Fig. 4). In trypsin-like proteinases, the amino terminal segment of the scissile bond of inhibitor fits into the enzyme in an antiparallel β-strand. It is assisted via main chain hydrogen

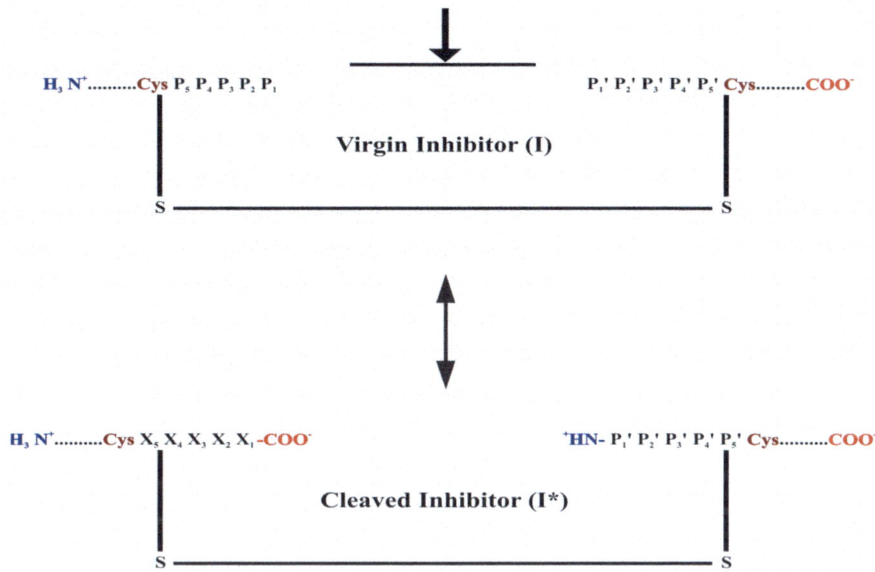

Fig. 2 Depiction of inhibitor reactive site. I refers to the virgin inhibitor and I* represents cleaved inhibitor. The peptide bond cleavage site between P_1 and P_1' residues in the virgin inhibitor is shown by an arrow. The disulphide bond encompasses the reactive centre site in the virgin inhibitor. In the case of cleaved inhibitor (I*), the two peptide chains are protected by disulphide bond. (Source: Ashoka 2005)

bond formation between P3 and P1. The adjoining carboxy-terminal residue at P2′ position interacts to make one more hydrogen bond. During such condition, the inhibitor reactive site is in close vicinity with the catalytic residues of the protease. Through Van der Waals interaction, contact is formed between P1 carbonyl carbon and Ser195 Oγ. By projecting into the oxyanion hole, the carbonyl group of P1 forms two hydrogen bonds with Gly193N and Ser N (Bode and Huber 1992). The scissile peptide bond remains intact. The exposed P1 residue of the inhibitor gets buried in the hydrophobic S1 specificity pocket. The interactions of the P1 side chain with S1 specificity pocket are energetically most important that determine the specificity of a given inhibitor for a particular protease (Ashoka 2005).

1.7 Plant Protease Inhibitors in Crucifers

The crucifer family contains more than one type of proteinase inhibitors. Developmentally regulated serine proteinase inhibitors have been reported from cabbage (Broadway 1989a, b; Broadway and Missurelli 1990; Colvin and Broadway 1992; Broadway 1993). The napin family trypsin inhibitor isolated from *Brassica napus* var. *rapifera* inhibited subtilisin DY but not chymotrypsin (Svendsen et al. 1989). The first plant-derived inhibitor of thrombin showing inhibition of several proteases

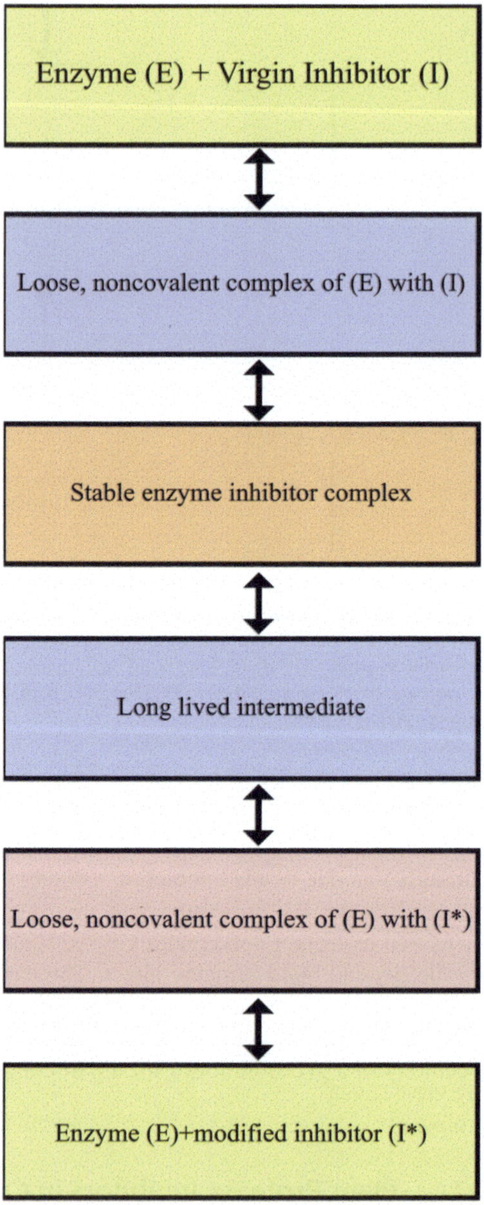

Fig. 3 Mechanism of enzyme-inhibitor interaction. Figure adapted from Laskowski and Kato (1980). The alphabets in the figure denote the following: E, enzyme; I and I^*, virgin and modified inhibitors; L and L^*, loose, noncovalent complexes of E with I and I^*, respectively; C, stable enzyme inhibitor complex; X, long-lived intermediate in the E + I^* reaction. (Source: Ashoka 2005)

playing central role in haemostasis was reported from cabbage seeds (Carter et al. 1990). Serine proteinase inhibitors from *Brassica napus* var. *oleifera* (rapeseed) are reported to inhibit the β-trypsin and bovine α-chymotrypsin (Ceciliani et al. 1994). A 15.5 kDa napin family storage protein having inhibitory activity against trypsin was isolated and characterized from *Sinapis arvensis* (Svendsen et al. 1994). The

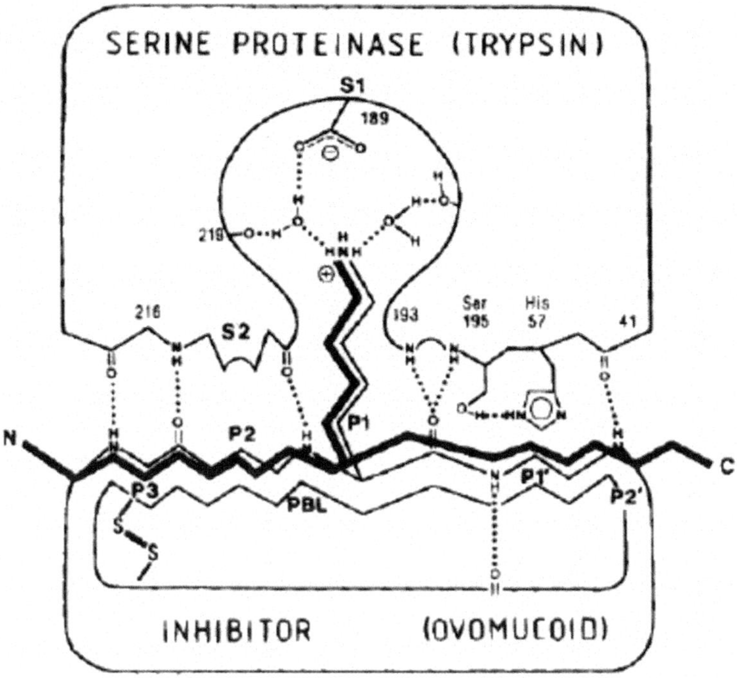

Fig. 4 Depiction of complex formation between enzyme and inhibitor (trypsin and ovomucoid inhibitor). Adapted from Bode and Huber (1992). The reactive site loop extends between scaffold-anchored pillars. The connection of inhibitor to its molecular core is through intra-disulphide bond and other spacer elements. The bold lines show the binding of the inhibitor similar to that of substrate through hydrogen bonds forming an antiparallel β-structure between P_1 and P_3 of inhibitor and amino acids 216 to 218 of protease. The catalytic Ser195 O^{γ} is linked to P_1 carbonyl carbon of inhibitor through van der Waals forces. (Source: Ashoka 2005)

distinctive features of some of the inhibitory proteins from crucifers enable themselves to be classified into a new family of inhibitors (Ceci et al. 1995). Mustard trypsin inhibitor (MTI-2) reported from seeds of white mustard (*Sinapis alba*) (Menegatti et al. 1992) of mustard inhibitor family (MSI) is a trypsin inhibitor effective against lepidopteran and other insect pests (De Leo et al. 1998, 2001b; Volpicella et al. 2000; De Leo and Gallerani 2002; Ceci et al. 2003; Yang et al. 2009). The MTI-2 is a small (63 residues) low molecular weight (7 kDa) inhibitor having eight cystine residues and four disulphide bonds and expressed during seed germination and in leaves due to wounding and jasmonic acid treatment in mustard and *Arabidopsis* (Ceci et al. 1995; De Leo et al. 2001a, 2001b; Clauss and Mitchell-Olds 2003). A low molecular weight serine proteinase isoinhibitor designated as RTI-III was obtained from oil rape (*Brassica napus* L. var. *oleifera*) and white mustard (*Sinapis alba* L. MTI-2) seeds showing homology with MTI-2 (Ascenzi et al. 1999; Trovato et al. 2000; Ruoppolo et al. 2000). About 29 serpin genes were reported from *A. thaliana* genome (Silverman et al. 2001). Genes homologous to mustard and rapeseed inhibitor were also reported from the genome of *A. thaliana* (Zhao et al. 2002). Three proteinase inhibitor genes (rti-1, rti-2 and rti-3), with high homology to genes of mustard inhibitor family (MSI), have been found in *B. napus* genome (De Leo et al. 2006). A Kunitz type of protease inhibitor induced in response to environmental stresses from cauliflower and *Arabidopsis* has also been reported (Halls et al. 2006). The host plant protease inhibitor from crucifers such as MTI-2 appears to be non-effective for crucifer specialist *Plutella xylostella*, which showed its insensitivity towards MTI by specific degradation using its gut proteases (Yang et al. 2009). Recently, protease inhibitors from two plants, viz. *Diplotaxis muralis* and *Diplotaxis tenuifolia*, belonging to mustard trypsin inhibitor 2 were characterized (Volpicella et al. 2009). They showed similar inhibitory activity as MTI-2 against trypsin-like gut proteases of *Helicoverpa zea* larvae.

The trypsin inhibitory activity was investigated in the foliage of cultivated and wild crucifers and in the storage organs of cultivated crucifers. The highest tryptic inhibitory activity was found in the foliage of cultivated crucifers (Broadway 1989a, b). Low tryptic inhibitory activity was recorded in the succulent foliage during the seedling stage in cabbage, but the highest tryptic inhibitory activity was found in young foliage, when plants matured with solid cabbage heads. The tryptic inhibitory activity increased significantly due to acute wounding in the succulent leaves, but factors such as environmental stresses had shown no significant effect on the foliar tryptic inhibitory activity (Broadway and Missurelli 1990). The purified cabbage proteinase inhibitors from the foliage showed both trypsin and chymotrypsin inhibitory activity and ranged from 9000 to 25,000 kDa (Broadway 1993). The unpublished data of Broadway suggested that cabbage PIs fall in the Kunitz soybean trypsin inhibitor family (Broadway 1996a, b). Transgenic plants containing serine PIs from *B. oleracea* performed well when compared with transgenic plants having different insect resistance transgenes against *Heliothis virescens* (Pulliam et al. 2001). The MTI-2 expressed in transgenic tobacco, *Arabidopsis* and oilseed rape affects growth and development of *Plutella xylostella*, *Mamestra brassicae* and *Spodoptera littoralis* larvae at varying concentrations (De Leo et al. 2001a). Trypsin inhibitor levels in crucifers were reported to be increased due to mechanical wounding in *Alliaria petiolata* (Cipollini 2002) and induction by *Trichoplusia ni* in *Brassica rapa* (Cipollini et al. 2003). The expression of defence-related genes due to early season herbivory by *P. rapae* larvae in *B. oleracea* showed enhanced or similar expression levels compared to primary herbivory by specialist *P. rapae* and *Plutella xylostella*, whereas compared to primary herbivory, low levels of gene expression were observed by secondary herbivory due to generalist *Mamestra brassicae* (Poelman et al. 2008). Trypsin inhibitors from non-host plant such as *Cassia obtusifolia* (COTI) showed significant inhibitory activity towards *P. rapae* larval midgut proteases, both in vitro and in vivo (Liao et al. 2007; Hai et al. 2008). The trypsin-like proteases from *P. rapae* are also shown to be inhibited by *Albizia kalkora* inhibitors (Zhou et al. 2008).

1.8 Plant Proteinase Inhibitor in Lauraceae

There are no published reports of PI from leaves of Lauraceae family. Unpublished data from the work of Saikia (2008) (Ph. D dissertation, 2008) reports for the first time detection of PIs from leaves and seeds of host plants of *A. assamensis*. Further work is required to isolate, characterize and better understand the PPIs in *P. bombycina* and *L. monopetala*.

1.9 *P. brassicae* and Its Host Plant Interaction

Insect pests cause about 40% yield losses in various vegetable crops (Ali and Rizvi 2007). The importance of *P. brassicae* as a pest of crucifers in tropical agro-ecosystems has been well documented on many host and non-host plants (Vats et al. 1977). The distribution of *P. brassicae* as an important pest of crucifers in many regions of world and in India in states of Punjab, Himachal Pradesh, Bihar, Bengal and Assam causing appreciable damage has been recorded. Among a number of insect pests (about 38) of cole crops (Pajmon 1999), *P. brassicae* is a destructive polyphagous pest of crucifers in India (Sachan and Gangwar 1980; Mohan and Gujar 2003; Lal and Bhajan 2004; Younas et al. 2004; Ashfaq et al. 2006; Khalid 2006; Saljoqi et al. 2006; Ali and Rizvi 2007; Blatt et al. 2008). In China and other parts of Asia, it is known to feed on *B. oleracea, B. rapa* and *B. napus* (Hai et al. 2008) damaging almost all stages of plant development. The larvae are also known to feed on other related families like Resedaceae (Guan-Soon and Yuan 1990; Zafar et al. 2002) and Capparidaceae (Hwang et al. 2008). Marketable losses of about 68% have been reported in cabbage due to *P. brassicae* in Meghalaya (Thakur 1996; Dhandapani et al. 2003). The occurrence of cabbage white on *Brassica* oilseeds crop is becoming a serious problem in India especially in the eastern part of Uttar Pradesh (Lal and Bhajan 2004).

Pieris brassicae is a gregarious, multivoltine, lepidopteran pest of crucifers around the world. It was observed as a recurrent pest on cauliflower (*Brassica oleracea* var. *botrytis*) fields in Haryana. Other cultivated cruciferous host plants were cabbage (*Brassica oleracea* var. *capitata*), radish (*Raphanus sativus*) and mustard (*Brassica juncea*) (Feltwell 1982). The insect was also observed feeding prolifically on garden nasturtium (*Tropaeolum majus*). Other host plants of this family include *Tropaeolum brasiliense* Casaretto (Terofal 1961) and *Tropaeolum peregrinum* L. (Nicholson 1939). The polyphagous nature of *Pieris* enables them to utilize a broad range of plants as their food material. The preference of larvae for a particular host plant as their food source can depend upon a number of factors such as larval age, previously eaten food and air movement (Hovanitz and Chang 1962). The *Pieris* larvae had shown higher viability, higher growth rate and increased size as a result of feeding on kale (*B. oleracea* var. *acephala*) compared to feeding on mustard (*B. nigra*) (Hovanitz and Chang 1963b). But the preference for nasturtium had also been shown if larvae previously fed on nasturtium (Hovanitz and Chang 1963a). The previous exposure of *P. brassicae* to a host plant correlates with the selection of that plant for factors such as ovipositional response by adults (Hovanitz and Chang 1963c, 1965). Oviposition of *P. brassicae* on clumped vegetation is benefited for host resource exploitation (Davies and Gilbert 1985). Compared to other host plant species, *P. brassicae* larvae showed preference for *B. oleracea capitata* but showed no significant difference between two varieties, i.e. *B. oleracea capitata* and *B. oleracea botrytis* (Rather and Azim 2009). For *P. brassicae* the highest assimilation rates and food consumption were recorded on *Brassica oleracea* var. *capitata* and lowest on *Nasturtium montanum* (Kaushal and Vats 1983). Ali and Rizvi (2007) demonstrated that *P. brassicae* developed more rapidly on cabbage in comparison to other cole crops both in laboratory and field conditions and lowest generation mortality was observed on cabbage under both conditions. The suitability of *Arabidopsis thaliana* as a food source for *P. rapae* under laboratory was also well demonstrated (Harvey et al. 2007).

Induced defences in cultivated as well as wild species of crucifers such as *B. oleracea* (Mattiacci et al. 2001; van Dam et al. 2004), *B. nigra* (Traw and Dawson 2002; van Dam et al. 2004), *B. rapa* (Siemens et al. 2002), *B. vulgaris* (Agerbirk et al. 2001, 2003), *Sinapis alba* (Bodnaryk 1992; McCaffrey et al. 2004), *Raphanus raphanistrum* (Agrawal 2000) and *Alliaria petiolata* (Cipollini 2002) are adequately documented. Both specialist (*P. brassicae*) and generalist (*Spodoptera littoralis*) herbivores show no effect on their growth and mobility as a result of wounding of cabbage foliage as a consequence of direct or indirect resistance (Coleman et al. 1996, 1997), and in *P. brassicae*-damaged wild radish, no fitness cost due to tolerance is detected (Agrawal 2000). Herbivory in *B. nigra* by *P. brassicae* does not affect plant fitness and is tolerated during vegetative phase through compensation (Blatt et al. 2008). Higher plant nutrient availability in crucifers influences the performance of *Pieris* butterflies in terms of shorter larval development duration, reduced consumption rate and enhanced growth rate and higher efficiency of food processing (Chen et al. 2004; Hwang et al. 2008). But in *P. rapae* the slow growth and high mortality (SG-HM) hypothesis depends upon the interactive effects of plant variation on both herbivore development and upon the direct defences of the plant on enemy search and access to herbivores (Benrey and Denno 1997). The herbivore density of crucifer pests, viz. *P. brassicae, Plutella xylostella* and *Trichoplusia ni*, also depended upon plant patch shape (Muriel and Grez 2002). Habituation to a particular deterrent by larvae of *Pieris rapae* enables them to feed and grow on that plant containing deterrent (Huang and Renwick 1995). The metabolic profiles of *P. brassicae* that fed on host plants such as kale and *B. rapa* var. *rapa* revealed their ability to sequester, metabolize and excrete phenolic compounds (Pereira et al. 2009; Ferreres et al. 2009). The change in food quality had major consequences on the next generation of *P. rapae* helping them to cope with seasonal changes in food quality (Rotem et al. 2003).

1.10 *A. assamensis* and Its Host Plant Interaction

The desired human perspective defines the concept of '*beneficial*' insects, i.e. such insects produce economically viable product for human consumption. *Antheraea assamensis* is a phytophagous, semi-domesticated, multivoltine Lepidoptera from North-East India. Its cocoon silk is used to make expensive fabric sold locally and exported. Farmers prefer to rear larvae mainly on two host plants, namely, *Persea bombycina* and *Litsea monopetala* of Lauraceae family. Cocoon silks of larvae reared on *P. bombycina* differ significantly from cocoon silks of larvae reared on *L. monopetala* in their physico-chemical properties and total amino acid content. The cocoon silks of *P. bombycina*-reared larvae have marketable features like a golden-yellow hue, high denier length and good reeling quality. In contrast cocoons from *L. monopetala*-reared larvae are used as 'seeds' of healthy, fecund moths that can be used for breeding in the next generation. Improved understanding of the interactions of this insect with its host plants may be crucial in designing effective strategies to enhance silk production as the commercial production of this silk from the region has remained low and stagnant in the past decade.

An important prerequisite of insect improvement programmes is to study the insect and host plant interaction patterns. The variable effect of host plant species on the relative survival of herbivore is mediated by differentially influencing its food intake, digestion and assimilation, thus directly affecting larval growth and development (Waldbauer 1968; Scriber and Slansky 1981). The role of nutrition in improving the growth and development of the silkworm *B. mori* L is well established. Synthesis of silk is a vital function of the silkworms, which is primarily determined by the nutritional quality of the host plants (Horie and Watanabe 1980. Legay (1958) affirmed that silk production is governed by the larval nutrition and production of good-quality cocoons is very much influenced by nutritive value of the mulberry leaves. The host plant provides energy and nutrition to the developing larvae for silk biosynthesis. The protein and amino acid content of mulberry (*Morus alba*) leaves affects silk yield in *B. mori* (Gopinathan 1992). Although the extensively studied mulberry leaf-consuming *B. mori* produces the finest natural silks, there are several other Lepidoptera that spin large cocoons. Host plant improvement is crucial in increasing the yield and quality of silk produced by non-mulberry silkworms, *A. assamensis* (Gopinathan 1992). According to Bharali (1974) adequate information was lacking on the nutritive value of *P. bombycina,* the primary host plant of *A. assamensis*. *Litsea monopetala* (synonym – *L. polyantha*) induces fecundity, while *P. bombycina* (synonym – *M. bombycina*) improves the quantity of silk production (Choudhury 1992). The nutritional composition of *P. bombycina* and *L. monopetala* leaves influenced the growth and development of the *A. assamensis* and also the economic characteristics of its cocoons to a great extent (Saikia et al. 1993). Hazarika et al. (1998) evaluated the physico-chemical characteristics of silk fibres of *A. assamensis* that fed on *P. bombycina* and *L. monopetala.* It was established that larvae reared on *P. bombycina* produced cocoons that could retain the golden colour (commercially important) of the fibre much better than those reared on *L. monopetala*. The quality and quantity of the silk produced mainly have been attributed to the quality of leaves eaten by the larvae (Pant and Unni 1980; Unni 1996). Abiotic factors, seasonal variations and food plants have been shown to greatly influence the growth and development of *A. assamensis*, its shell weight, yield of cocoons and quality of the silk produced (Thangavelu 1986; Barah et al. 1988; Sharma and Devi 1997; Chaudhuri et al. 1999; Brahma et al. 2000). Varieties of *P. bombycina* collected from different parts of Assam differentially influenced the growth and life cycle of *A. assamensis* (Choudhury et al. 2000).

1.11 Utilization of Proteinase Inhibitors for Insect Pest Management

Several transgenic plants expressing proteinase inhibitors have been developed with varied success rates. The different approaches that have been tried successfully to improve plant defence against pests by different groups of researchers are as follows.

1.12 Phage Display

Phage display technology is a promising field for selection of heterologous protein variants (PIS) with novel specificities. Several workers have reported its use in selecting inhibitors with greater insecticidal activity as is evident in the published work of Ceci et al. (2003). But efforts are still on to develop proteinase inhibitors with improved activities. Volpicella et al. (2001) demonstrated the efficient selection on trypsin of phage particle displaying active trypsin inhibitor (MTI-2) from a large population of phages exhibiting an inactive mutant. The mustard trypsin inhibitor (MTI-2) is toxic for lepidopteran pests but has low activity against aphids. In order to improve its activity against aphids, Ceci et al. (2003) constructed a large library of MTI-2 variants to be displayed on phage particles and used for identification of novel MTI-2 derived antitrypsin and antichymotrypsin inhibitors. The inhibitor variant Chy8 identified has high affinity for bovine chymotrypsin and is the strongest known recombinant chymotrypsin inhibitor of the MTI-2 family. Thus, with this method, a new class of genes are available for increasing plant defence against aphids. In this way phage display selection and identification of more active inhibitors can assist in improving the natural plant defence mechanisms.

1.13 Single Chain Antibody

Single chain antibody uses genetic engineering techniques to specifically design antibodies or antibody fragments specific to the pest's essential protein and express them in the crop plant. This process ensures that only pests and not beneficial insects can be targeted. The potential of this technique has already been demonstrated in the works of Atkinson (1993), Rosso et al. (1996), Tavladoraki et al. (1993) and van Engelen et al. (1994).

1.14 RNAi

RNAi, a gene silencing mechanism in insect and pests, is a promising area in pest control strategy. It relies on transgenic plants stably expressing dsRNA that target essential genes in insect pests. In 2007 studies on two economically important pests demonstrated the concept of plants expressing dsRNA that directed dsRNAs to target gene regions of cotton bollworm (*H. armigera*; Lepidoptera) and the western corn rootworm (Coleoptera, *Diabrotica virgifera virgifera*) (Baum et al. 2007; Mao et al. 2007). This successful demonstration prompted vigorous use of RNAi technology to control important agricultural pests belonging to large number of species and orders (Rodrigues and Figueira 2016). Although there is widespread use of RNAi in crop protection strategies, it has been found that the proportion of silencing varied across the insect populations, more so in the case of lepidopterans (Zhu-Salzman and Zeng 2015). Therefore, understanding the potential of this mechanism, detailed study should be done on the limiting factors affecting the efficiency of RNAi. Accurate information on the uptake and silencing process of RNAi in pests would aid in making this technology more suitable for pest control in the future.

1.15 Gene Stacking

The practice of producing transgenic plants expressing combination of transgenes, viz. inhibitors or Cry toxins, is known as gene stacking. Stacking PIs with different protease target sites may help to broaden pest susceptibility while delaying acquisition of resistance to PI (Kushwaha et al. 2013). Different transformation techniques have been utilized successfully by several workers to introduce ‘stacked genes’ in crop plants as evident from the review works of Halpin (2005) and Macedo et al. (2015). Although, this technique has few limitations, nevertheless, gene stacking technology has the potential to overcome the limiting factors that hinder crop yield. It is an expanding and exciting area of research for developing transgenic plants with improved defence traits against pests (Douglas and Halpin 2009).

2 Conclusion

The study of proteinase interaction with PPIs is an exciting avenue that has always attracted continued investigation since decades. Till date hundreds of PPIs have been isolated, but research has demonstrated that insects are able to develop strategies to overcome the detrimental effects of inhibitors. One of their adaptive responses can be changed in gene expression level which was evident in the case of the lepidopteran discussed in this review. Various studies on *Pieris brassicae* demonstrated the ability of its larvae to develop at different rates on various host plants, suggesting that alternate host plant utilization incurred only metabolic costs. It can be compensated by increasing food consumption and more efficient food utilization. This suggests that simply developing new PI-expressing transgenic crop as a means of pest management strategies is not very useful. The focus should be on prudent management of cropping practices. Transgenic approaches can also be attempted, but would require judicious use of promoters and choice of transgenes. As compared to *Pieris brassicae,* less is known about the midgut proteinase gene expression and its regulation in the beneficial silk-producing insect *A. assamensis*. Trypsin and chymotrypsin inhibitors were detected in herbivore-induced leaves of *L. monopetala* and *P. bombycina* that could inhibit midgut proteinases of *A. assamensis.* Such interactions may affect proteolytic digestion in larvae reared on different host plant species. Understanding mechanisms of digestion and host plant proteins that affect digestion may aid in developing an artificial diet for rearing *A. assamensis*, a very useful development for Indian sericulture. The significance of this work may be manifested in quality of silk produced by this economically important insect, with potential implication for silkworm rearers.

Acknowledgement The technical support received from Lab no. 15, Department of Botany, University of Delhi, and financial assistance from the Department of Biotechnology, Government of India, for carrying out the work on the digestive physiology of two lepidopteran insects is greatly acknowledged.

References

Abe K, Arai S (1985) Purification of a cysteine proteinase inhibitor from rice, *Oryza sativa* L. *japonica*. Agric Biol Chem 49:3349–3350

Agerbirk N, Mulle C, Olsen CE, Halkier BA, Nielsen JK (2001) 1,4-Dimethoxyglucobrassicin in Barbarea and 4-hydroxyglucobrassicin in *Arabidopsis* and Brassica. J Agric Food Chem 49:1502–1507

Agerbirk N, Orgaard M, Nielsen JK (2003) Glucosinolates, flea beetle resistance, and leaf pubescence as taxonomic characters in the genus *Barbarea* (Brassicaceae). Phytochemistry 63 (1):69–80

Agrawal AA (2000) Specificity of induced resistance in wild radish: causes and consequences for the two specialist and two generalist caterpillars. Oikos 89:493–500

Ali A, Rizvi PQ (2007) Developmental response of cabbage butterfly *Pieris brassicae* L. (Lepidoptera: Pieridae) on different cole crops under laboratory and field condition. Asian J Plant Sci 6(8):1241–1245

Applebaum SW (1964) Physiological aspect of host specificity in the Bruchidae-I. General considerations of developmental compatibility. J Insect Physiol 10:783–788

Applebaum SW (1985) Biochemistry of digestion. In: Kerkut GA, Gilbert LI (eds) Comparative physiology, biochemistry and pharmacology of insects, vol 4. Pergamon Press, Oxford, pp 279–312

Ascenzi P, Ruoppolo M, Amoresano A, Pucci PM, Consonni R, Zetta L, Pascarella S, Bortolotti F, Menegatti E (1999) Characterization of low-molecular-mass trypsin isoinhibiors from oil-rape (*Brassica napus* var. *oleifera*) seed. Eur J Biochem 261:275–284

Ashfaq M, Gill TI, Ali A (2006) Consumption and utilization of various plants by parasitized and unparasitized larvae of cabbage butterfly (*Pieris brassicae* L.). Pakistan Entomol 28(1):1–4

Ashoka MA (2005) Molecular characterization of protease inhibitors from the hessian fly, *Mayetiola destructor* (Say.). PhD thesis, Kansas State University, Manhattan, Kansas

Atkinson H (1993) Opportunities for improved control of plant parasitic nematodes via plant biotechnology. In: Beadle DJ, Copping DHL, Dixon GK, Holloman DW (eds) Opportunities for molecular biology in crop production. BCPC, Farnham, pp 257–266

Barah A, Phukan SN, Samson MV (1988) Variation in cocoon characters of *Antheraea assama* Ww. during different seasons. Sericol 28(2):215–218

Barbosa JRG, Silva LP, Teles RCL, Esteves GF, Azevedo RB, Ventura MM, Freitas SM (2007) Crystal structure of the Bowman-Birk inhibitor from *Vigna unguiculata* seeds in complex with β-trypsin at 1.55 Å resolution and its structural properties in association with proteinases. J Biophys 92:1638–1650

Barillas-Mury C, Graf R, Hagedorn HH, Wells MA (1991) cDNA and deduced amino acid sequence of a blood meal-induced trypsin from the mosquito *Aedes aegypti*. Insect Biochem. 21:825–831

Barrett AJ (1986) Peptidases, the classes of proteolytic enzyme. In: DaUing MJ (ed) Plant proteolytic enzymes. CRC Press, Boca Raton, pp 1–16

Barrett AJ (1994) Classification of peptidases. In: Barrett AJ (ed) Methods in enzymology, Vol, vol 244. Academic, New York, pp 1–15

Barrett AJ, Rawlings RD (1995) Families and clans of serine peptidases. Arch Biochem Biophys 318:247–250

Barrett AJ, Rawlings ND, Woessner JF (1998) Handbook of proteolytic enzymes. Academic, London

Baum JA, Bogaert T, Clinton W, Heck GR, Feldmann P, Ilagan O, Johnson S, Plaetinck G, Munyikwa T, Pleau M, Vaughn T, Roberts J (2007) Control of coleopteran insect pests through RNA interference. Nat Biotechnol 25:1322–1326

Benrey B, Denno RF (1997) The slow-growth-high-mortality hypothesis: a test using the cabbage butterfly. Ecology 78(4):987–999

Berger A, Schechter I (1970) Mapping the active site of papain with the aid of peptide substrates and inhibitors. Philosophical transaction of the royal society B. Biol Sci B257:249–264

Beynon RJ, Salvesen G (1989) Commercially available protease inhibitor. In: Beynon RJ, Bond JS (eds) Proteolytic enzymes. A practical approach. IRL Press, New York, pp 149–241

Bezerra CDS, Oleviera CT, Macedo MLR (2017) Inga vera trypsin inhibitor interferes in the proteolytic activity and nutritional physiology of Ephestia kuehniella larvae. Entomol Exp Appl 165(2–3):109–119

Bharali N (1974) Some studies on the ecology of muga silkworm. In: Proceedings of the International Seminar on Non-mulberry Silks. Central Silk Board, Bombay, pp 254–256

Blatt SE, Smallegange RC, Hess L, Harvey JA, Dicke M, van Loon JJA (2008) Tolerance of *Brassica nigra* to *Pieris brassicae* herbivory. Botany 86(6):641–648

Bode W, Huber R (1992) Natural protein proteinase inhibitors and their interaction with proteinases. Eur J Biochem 204:433–451

Bodnaryk RP (1992) Effects of wounding on glucosinolates in the cotyledons of oilseed rape and mustard. Phytochemistry 31:2671–2677

Bown DP, Wilkinson H, Gatehouse JA (1997) Differentially regulated inhibitor-sensitive and insensitive protease genes from the phytophagous insect pest, *Helicoverpa armigera*, are members of complex multigene families. Insect Biochem Mol Biol 27:625–638

Boyer PD (1971) The enzymes–hydrolysis: peptide bonds, vol III, 3rd edn. Academic, New York

Brahma KC, Benchamin KV, Senapati MD (2000) Studies on rearing of muga silkworm based on host plant and season. In: Abstract: seminar on sericulture R & D in Muga and Eri, CMER & TI, Ladoigarh, 8th & 9th November 2000

Broadway RM (1989a) Characterization and ecological implications of midgut proteolytic activity in larval *Pieris rapae* and *Trichoplusia ni*. J Chem Ecol 15(7):2101–2113

Broadway RM (1989b) Tryptic inhibitory activity in wild and cultivated crucifers. Phytochemistry 28(3):755–758

Broadway RM (1993) Purification and partial characterization of trypsin/chymotrypsin inhibitors from cabbage foliage. Phytochemistry 33(1):21–27

Broadway RM (1995) Are insects resistant to plant proteinase inhibitors? J Insect Physiol 41:107–116

Broadway RM (1996a) Dietary proteinase inhibitors alter complement of midgut proteases. Arch Insect Biochem Physiol 32:39–53

Broadway RM (1996b) Resistance of plants to herbivorous insects: can this resistance fail? Can J Plant Pathol 18:476–481

Broadway RM (1997) Dietary regulation of serine proteinase that are resistant to serine proteinase inhibitors. J Insect Physiol 43(9):855–874

Broadway RM, Duffey SS (1986a) The effect of dietary protein on the growth and digestive physiology of larval *Heliothis zea* and *Spodoptera exigua*. J Insect Physiol 32(8):673–680

Broadway RM, Duffey SS (1986b) Plant proteinase inhibitors: mechanism of action and effect on the growth and digestive physiology of larval *Heliothis zea* and *Spodoptera exigua*. J Insect Physiol 32:827–833

Broadway RM, Missurelli DL (1990) Regulatory mechanisms of tryptic inhibitory activity in cabbage plants. Phytochemistry 29(12):3721–3725

Carter TH, Everson BA, Ratnoff OD (1990) Cabbage seed protease inhibitor: a slow, tight-binding inhibitor of trypsin with activity toward thrombin, activated Stuart factor (factor Xa), activated Hageman factor (factor XIIa), and plasmin. Blood 75(1):108–115

Casu RE, Pearson RD, Jarmey JM, Cadogan LC, Riding GA, Tellam RL (1997) Excretory/ secretory chymotrypsin from *Lucilia cuprina*: purification, enzymatic specificity, and amino acid sequence deduced from mRNA. Insect Mol Biolo 3:201–211

Ceci LR, Spoto N, de Virgilio M, Gallerani R (1995) The gene coding for the mustard trypsin inhibitor-2 is discontinuous and wound-inducible. FEBS Lett 364:179–181

Ceci LR, Volpicella M, Rahbe Y, Gallerani R, Beekwilder J, Jongsma MA (2003) Selection by phage display of a variant mustard trypsin inhibitor toxic against aphids. Plant J 33:557–566

Ceciliani F, Bortolotti F, Menegatti E, Ronchi S, Ascenzi P, Palmieri S (1994) Purification, inhibitory properties, amino acid sequence and identification of the reactive site of a new serine proteinase inhibitor from oil rape (*Brassica napus*) seed. FEBS Lett 342:221–224

Chaudhuri M, Singh SS, Das B, Dhar NJ, Basumatary B, Goswami D, Das K, Barah A, Sahu M, Kakoty LN, Mandal T, Chatterjee SN (1999) Climatic variability in nine locations of North East India and their effect on cocoon productivity of muga silkworm (*Antheraea assama* Westwood). Sericolo 39(4):577–591

Chen PW, Chow SH, Chen LJ (1997) Nucleotide sequence of a cDNA encoding rice Bowman-Birk proteinase inhibitor. Plant Physiol 113:663–665

Chen Y-Z, Lin L, Wang C-W, Yeh C-C, Hwang Y-S (2004) Response of two *Pieris* (Lepidoptera: Pieridae) species to fertilization of a host plant. Zool Stud 43(4):778–786

Choudhury SN (1992) Silk and sericulture, 1st edn. Directorate of Sericulture, Government of Assam

Choudhury SN, Ahmed RZ, Bhattacharya PR, Dutta S, Das AM, Rajkhowa A (2000) Performance of muga silkworm (*Antheraea assama* Westwood) on different som (*Persea bombycina* King Ex, Hook (F) Kost.) plant collections from Assam, India. In: Proceedings: The 3rd international conference on wild silkmoths, Bhubaneswar, 11–14th November, 1998, 17. International Journal of Wild Silkmoths and Silk 5:66–69

Chougule NP, Doyle E, Fitches E, Gatehouse JA (2007) Biochemical characterization of midgut digestive proteases from *Mamestra brassicae* (cabbage moth; Lepidoptera: Noctuidae) and effect of soybean Kunitz inhibitor (SKTI) in feeding assays. J Insect Physiol 54:563–572

Christeller JT, Liang W (2005) Plant serine protease inhibitors. Protein Pept Lett 12:439–447

Christeller JT, Liang WA, Markwick NP, Burgess EPJ (1992) Midgut protease activities in 12 phytophagous lepidopteran larvae: dietary and proteases inhibitory interactions. Insect Biochem Mol Biol 22:248–254

Cipollini D (2002) Variation in the expression of chemical defenses in *Alliaria petiolata* (Brassicaceae) in the field and common garden. Am J Bot 89(2):1422–1430

Cipollini DF, Busch JW, Stowe AK, Simms EL, Bergelson J (2003) Genetic variation and relationships of constitutive and herbivore-induced glucosinolates, trypsin inhibitors, and herbivore resistance in *Brassica rapa*. J Chem Ecol 29(2):285–302

Clauss MJ, Mitchell-Olds T (2003) Population genetics of tandem trypsin inhibitor genes in *Arabidopsis* species with contrasting ecology and life history. Mol Ecol 12:1287–1299

Coleman RA, Barker AM, Fenner M (1996) Cabbage (*Brassica oleracea* var. capitata) fails to show wound induced defense against a specialist and a generalist herbivore? Oecologia 108:105–112

Coleman RA, Barker AM, Fenner M (1997) A test of possible indirect mediation of wound-induced resistance in cabbage against *Pieris brassicae*. Oikos 80:43–50

Colvin AA, Broadway RM (1992) Influence of cabbage proteinase inhibitors in situ on the growth of larval *Trichoplusia ni* and *Pieris rapae*. J Chem Ecol 18(7):1009–1024

Craik CS, Choo QL, Swift GH, Quinto C, MacDonald RJ, Rutter WJ (1984) Structure of two related rat pancreatic trypsin genes. J Biol Chem 259:14255–14264

Creighton TE, Darby NJ (1989) Functional evolutionary divergence of proteolytic enzymes and their inhibitors. Trends Biochem Sci 14:319–324

Cristofoletti PT, Ribeiro AF, Terra WR (2005) The cathepsin L-like proteinases from the midgut of *Tenebrio molitor* larvae: sequence, properties, immunocytochemical localization and function. Insect Biochem Mol Biol 35(8):883–901

Cygler M, Morb JS (1997) Proregion structure of members of the papain superfamily mode of inhibition of enzymatic activity. Biochemist 79:645–652

Dantzer M, Vasconcelos IM, Scorsato V, Aparicio R, Marangoni S, Macedo MLR (2015) Bowman-Birk proteinase inhibitor from *Clitoria fairchildiana* seed: isolation, biochemical properties and insecticidal potential. Phytochemistry 118:224–235

Davies CR, Gilbert N (1985) A comparative study of the egg-laying behavior and larval development of *Pieris rapae* L. and *P. brassicae* L. on the same host plants. Oecologia 10:278–281

Davis CA, Riddell DC, Higgins MJ, Holden JJ, White BN (1985) A gene family in *Drosophila melanogaster* coding for trypsin-like enzymes. Nucleic Acids Res 13:6605–6619

De Leo F, Gallerani R (2002) The mustard trypsin inhibitor 2 affects the fertility of *Spodoptera littoralis* larvae fed on transgenic plants. Insect Biochem Mol Biol 33:489–496

De Leo F, Bonadé-Bottino MA, Ceci LR, Gallerani R, Jouanin L (1998) Opposite effects on *Spodoptera littoralis* larvae of high and low level of a trypsin proteinase inhibitor in transgenic plants. Plant Physiol 118:997–1004

De Leo F, Ceci LR, Jouanin L, Gallerani R (2001a) Analysis of mustard trypsin inhibitor-2 gene expression in response to developmental or environmental induction. Planta 212:710–717

De Leo F, Bonadé-Bottino M, Ceci LR, Gallerani R, Jouanin L (2001b) Effects of a mustard trypsin inhibitor expressed in different plants on three lepidopteran pests. Insect Biochem Mol Biol 31:593–602

De Leo FM, Volpicella F, Licciulli S, Liuni R, Gallerani CLR (2002) PLANT-PIs: a database for plant protease inhibitors and their genes. Nucleic Acids Res 30:347–348

De Leo F, Volpicella M, Sciancalepore M, Gallerani R, Ceci LR (2006) One of the three proteinase inhibitor genes newly identified in the *Brassica napus* genome codes for an inhibitor of glutamyl endopeptidase. FEBS Lett 580:948–954

Dhandapani N, Shelkar UR, Murugan M (2003) Bio-intensive pest management (BIPM) in major vegetable crops: an Indian perspective. J Food Agric Environ 1(2):333–339

Diaz-Mendoza M, Ortego F, Garcia deLecabo M, Magana C, Poza M, Farinos GP, Castanera P, Hernandez-Crespo P (2005) Diversity of trypsins in the Mediterranean corn borer *Sesamia nonagrioides* (Lepidoptera: Noctuidae), revealed by nucleic acid sequences and enzyme purification. Insect Biochem Mol Biol 35(9):1005–1020

Douglas E, Halpin C (2009) Gene Stacking. In: Jain SM, Brar DS (eds) Molecular techniques in crop improvement. Springer, Dordrecht, pp 613–629

Eguchi M, Iwamoto A (1976) Alkaline proteases in the midgut tissue and digestive fluid of the silkworm. Insect Biochem 6:491–496

Eguchi M, Kuriyama K (1983) Relationship between alkaline proteases from the midgut lumen and epithelia of the silkworm: partial purification and comparison of properties of both proteases-6B3. Comp Biochem Physiol-B 76(1):29–34

Eguchi M, Iwamoto A, Yamauchi M (1982) Interrelation of proteases from the midgut lumen, epithelia and peritropic membrane of the silkworm *Bombyx mori* L. Comp Biochem Physiol-A 72(2):359–363

Evnin LB, Vasquez JR, Craik CS (1990) Substrate specificity of trypsin investigated by using a genetic selection. Proc Natl Acad Sci USA 87:6659–6663

Fan S-G, Wu G-J (2005) Characteristics of plant proteinase inhibitors and their application in combating phytophagous insects. Bot Bull Acad Sin 46:273–292

Fastrez J, Fersht AR (1973) Demonstration of the acyl-enzyme mechanism for the hydrolysis of peptides and anilides by chymotrypsin. Biochemist 12:2025–2034

Feltwell J (1982) Large white butterfly, the biology, biochemistry and physiology of *Pieris brassicae* (Linnaeus). Dr W. Junk, The Hague

Ferreres F, Fernandes F, Oliveira JM, Valentão P, Pereira JA, Andrade PB (2009) Metabolic profiling and biological capacity of *Pieris brassicae* fed with kale (*Brassica oleracea* L. var. *acephala*). Food Chem Toxicol 47(6):1209–1220

Frazao C, Bento I, Costa J, Soares CM, Verissimo P, Faro C, Pires E, Cooper J, Carrondo MA (1999) Crystal structure of cardosin a glycosylated and Arg-Gly-Asp- containing aspartic proteinase from the flowers of *Cyanara cardunculus* L. J Biol Chem 274(39):27694–27701

Gatehouse AMR, Davison GM, Newell CA, Merryweather A, Hamilton WDO, Burgess EPJ, Gilbert RJC, Gatehouse JA (1997) Transgenic potato plants with enhanced resistance to the tomato moth *Lacanobia oleracea*: growth room trials. Mol Breed 3:49–63

George D, Ferry N, Back EJ, Gatehouse AMR (2008) Characterisation of midgut digestive proteases from the maize stem borer *Busseola fusca*. Pest Manag Sci 64:1151–1158

Girard C, Metayer L, Bonade-Bottino M, Pham-Delegue MH, Jouanin L (1998) High level of resistance to proteinase inhibitors may be conferred by proteolytic cleavage in beetle larvae. Insect Biochem Mol Biol 28:229–237

Giri AP, Harsulkar AM, Deshpande VV, Sainani MN, Gupta VS, Ranjekar PK (1998) Chickpea defensive proteinase inhibitors can be inactivated by podborer gut proteinases. Plant Physiol 116:393–401

Gopinathan KP (1992) Biotechnology in sericulture. Curr Sci 62(3):282–287

Graham JS, Ryan CA (1981) Accumulation of a metallo-carboxypeptidase inhibitor in leaves of wounded potato plants. Biochem Biophys Res Commun 101(4):1164–1170

Grant GA, Henderson KO, Eisen AZ, Bradshaw RA (1980) Amino acid sequence of a collagenolytic protease from the hepatopancreas of the fiddler crab *Uca pugilator*. Biochemist 19:4653–4659

Green TR, Ryan CA (1972) Wound-induced proteinase inhibitor in plant leaves: a possible defense mechanism against insects. Science 175:776–777

Guan-Soon L, Yuan BD (1990) Status and management of major vegetable pests in the Asia-Pacific region with specific focus towards integrated pest management. RARA Publication (3) Bangkok, Thailand

Hai L, Wei R, Xiao-jun Z, Lin-fang D (2008) Identification of kunitz inhibitor from *Cassia obtusifolia* L. and its inhibitory effect against *Pieris rapae* proteases. J Zhejiang Univ 34 (1):29–37

Halls CE, Rogers SW, Oufattole M, Ostergard O, Sevensson B, Rogers JC (2006) A Kunitz-type cysteine protease inhibitor from cauliflower and Arabidopsis. Plant Sci 170:1102–1110

Halpin C (2005) Gene stacking in transgenic plants – the challenge for 21st century plant biotechnology. Plant Biotechnol J 3:141–155

Hanada K, Tamai M, Yamagishi M, Ohmura S, Sawada J, Tanaka I (1978) Isolation and characterization of E-64, a new thiol protease inhibitor. Agric Biol Chem 42:523–528

Hartley BS (1964) Amino acid sequence of bovine chymotrypsinogen-A. Nature 201:1284–1287

Harvey JA, Witjes LMA, Benkirane M, Duytes H, Wagenaar R (2007) Nutritional suitability and ecological relevance of *Arabidopsis thaliana* and *Brassica oleracea* as food plants for the cabbage butterfly *Pieris rapae*. Plant Ecol 189:117–126

Hashida S, Towatari T, Kominami E, Katunuma N (1980) Inhibitions by E-64 derivatives of rat liver cathepsin-B and cathepsin-L *in vitro* and *in vivo*. J Biochem 88:1805–1811

Hazarika LK, Saikia CN, Kataky A, Bordoloi S, Hazarika J (1998) Evaluation of physico chemical characteristics of silk fibres of *Antheraea assama* reared on different host plants. Bioresour Technol 64:67–70

Hermodson MA, Tye RA, Reeck GR, Neurath H, Walsh KA (1971) Comparison of amino acid terminal sequences of bovine, dogfish and lungfish Trypsinogens. FEBS Lett 14:222–224

Horie Y, Watanabe H (1980) Recent advances in sericulture. Annu Rev Entomol 25:49–71

Horie Y, Tanaka M, Ito T (1963) Proteolytic enzymes of digestive juice and midgut of the silkworm, *Bombyx mori*. J Sericultural Sci Japan 32:8–15

Houseman JG, Chin P-S (1995) Distribution of digestive proteinases in the alimentary tract of the European corn borer *Ostrinia nubilalis* (Lepidoptera: Pyralidae). Arch Insect Biochem Physiol 28(2):103–111

Hovanitz W, Chang VCS (1962) Three factors affecting larval choice of food plants. J Res Lepid. 1 (1):51–61

Hovanitz W, Chang VCS (1963a) Changes of food plant preference by larvae of *Pieris rapae* controlled by strain selection, and the inheritance of this trait. J Res Lepid 1(2):163–168

Hovanitz W, Chang VCS (1963b) The effect of hybridization of host plant strains on growth rate and mortality of *Pieris rapae*. J Res Lepid 1(2):157–162

Hovanitz W, Chang VCS (1963c) Ovipositional preference tests with *Pieris*. J Res Lepid 2 (3):185–200

Hovanitz W, Chang VCS (1965) The alteration of host plant specificity in larvae of *Pieris rapae* by induction. J Res Lepid 4(1):13–21

Howe GA, Schaller A (2008) Direct defenses in plants and their induction by wounding and insect herbivores. In: Induced plant resistance to herbivory, pp.7–29. Springer, Dordrecht

Huang XP, Renwick JAA (1995) Chemical and experimental basis for rejection of *Tropaeolum majus* by *Pieris rapae* larvae. J Chem Ecol 21(10):1601–1617

Huber R, Bode W (1978) Structural basis of the activation and action of trypsin. Acc Chem Res 11:114–122

Hwang SW, Liu CH, Shen TC (2008) Effects of plant nutrient availability and host plant species on the performance of two *Pieris* butterflies (Lepidoptera: Pieridae). Biochem Syst Ecol 36:505–513

Jamal F, Pandey PK, Singh D, Ahmed W (2015) A kunitz-type serine protease inhibitor from *Butea monosperma* seed and its influences on developmental physiology of *Helicoverpa armigera*. Process Biochem 50(2):211–316

Jany KD, Haug H (1983) Amino acid sequence of the chymotryptic protease II from the larvae of hornet, (*Vespa crabro*). FEBS Lett 158:98–102

Jany KD, Bekelar K, Pfleiderer G, Ishay J (1983) Amino acid sequence of an insect chymotrypsin from larvae of hornet (*Vespa orientalis*). Biochem Biophys Res Commun 110:1–7

Jiang Q, Hall M, Norienga FG, Wells M (1997) cDNA cloning and pattern of expression of an adult, female specific chymotrypsin from *Aedes aegypti* midgut. Insect Biochem Mol Biol 27 (4):283–289

Johnston KA, Lee MJ, Gatehouse JA, Anstee JH (1991) The partial purification and characterization of serine proteases activity in the midgut of the larval *Helicoverpa armigera*. Insect Biochem 21:389–397

Jongsma MA, Bolter C (1997) The adaptation of insects to plant protease inhibitors. J Insect Physiol 143(10):885–895

Jongsma MA, Bakker PL, Peters J, Boss D, Stiekema WJ (1995) Adaptation of *Spodoptera exigua* larvae to plant proteinase inhibitor by induction of gut proteinase activity insensitive to inhibition. Proc Natl Acad Sci USA 92:8041–8045

Jouanin L, Bonadi-Bottino M, Girard C, Morrot G, Giband M (1998) Transgenic plants for insect resistance. Plant Sci 131:1–11

Kageyama T, Takahashi SY, Takahashi K (1981) Occurrences of thiol proteinases in the eggs of the silkworm *B. mori*. J Biochem 90:665–671

Kaushal BR, Vats LK (1983) Energy budget of *Pieris brassicae* L. larvae (Lepidoptera: Pieridae) fed on four host plant species. Agric Ecosyst Environ 10:385–398

Keilova H, Tomasek V (1976) Isolation and properties of cathepsin-D inhibitor from potatoes. Collect Czechoslov Chem Commun 41:489–497

Kervinen J, Tobin GJ, Costa J, Waugh DS, Wlodawer A, Zdanov A (1999) Crystal structure of plant aspartic proteinase prophytepsin: inactivation and vacuolar targeting. EMBO J 18 (14):3947–3955

Khalid S (2006) Bio-ecological studies of *Pieris brassicae* (Lepidoptera: Pieridae) on different hosts. MSc thesis. AMU, Aligarh, India

Kiel B (1965) The chemistry and structure of peptides and proteins. Annu Rev Biochem 34:175

Koepke J, Ermler U, Warkentin E, Wenzl G, Flecker P (2000) Crystal structure of cancer chemopreventive Bowman-Birk inhibitor in ternary complex with bovine trypsin at 2.3 Å resolution. Structural basis of Janus-faced serine protease inhibitor specificity. J Mol Biol 298 (3):477–491

Koiwa H, Bressan RA, Hasegawa PM (1997) Regulation of protease inhibitors and plant defense. Trends Plant Sci 2:379–383

Kotani E, Niwa T, Suga K, Sugimura Y, Oda K, Mori H, Furusawa T (1999) Cloning and sequencing of a cDNA for a highly basic protease from the digestive juice of the silkworm, *Bombyx mori*. Insect Mol Biol 8(2):299–304

Kramer KJ, Felsted RL, Law JL (1973) Structural studies on an insect serine protease. J Biol Chem 248(9):3021–3028

Kraut J (1977) Serine proteases: structure and mechanism of catalysis. Annu Rev Biochem 46:331–358

Kumar R, Bhardwaj U, Kumar P, Mazumdar-Leighton S (2015) Midgut serine proteases and alternative host plant utilization in *Pieris brassicae* L. Front Physiol 31(6):95. https://doi.org/10.3389/fphys.2015.00095

Kunitz M (1945) Crystallization of a trypsin inhibitor from soybean. Science 101:668–669

Kushwaha R, Payne CM, Downie AB (2013) Uses of phage display in agriculture: a review of food related protein-protein interactions discovered by biopanning over diverse baits. Comput Math Methods Med:653759. https://doi.org/10.1155/2013/653759

Lal MN, Bhajan R (2004) Cabbage butterfly, *Pieris brassicae* L.-an upcoming menace for *Brassica* oilseed crops in Northern India. Cruciferae Newsl 25:83–86

Laskowski M, Kato I (1980) Protease inhibitors of proteinases. Annu Rev Biochem 49:593–626

Lawrence PK, Koundal KR (2002) Plant protease inhibitors in control of phytophagous insects. EJB 5(1):93–109

Legay JM (1958) Recent advances in silkworm nutrition. Ann Rev Entomol 3:75–86

Liang W, McManus MT (2002) Proteinase inhibitors. In: McManus MT, Liang WA, Allan AC (eds) Protein-Protein Interaction in Plant Biology. CRC Press, Boca Raton, pp 77–119

Liao H, Ren W, Kang Z, Jiang J, Zhao X, Du L (2007) A trypsin inhibitor from *Cassia obtusifolia* seeds: isolation, characterization and activity against *Pieris rapae*. Biotechnol Lett 29:653–658

Macedo MLR, Mello GC, Freire MGM, Novello JC, Marangoni S, Matos DGG (2002) Effect of a trypsin inhibitor from *Dimorphandra mollis* seeds on the development of *Callosobruchus maculatus*. Plant Physiol Biochem 40:891–898

Macedo MLR, Freire MGM, Cabrini EC, Toyama MH, Novello JC, Marangoni S (2003) A trypsin inhibitor from *Peltophorum dubium* seeds active against pest proteases and its effect on the survival of *Anagasta kuehniella* (Lepidoptera: Pyralidae). Biochimica et Biphysica Acta 1621:170–182

Macedo L, Oliviera C, Silva-Filho FC (2015) Adaptive mechanisms of insect pests against plant protease inhibitors and future prospects related to crop protection: a review. Protein Pept Lett 22 (2):149–163

Mao YB, Cai WJ, Wang JW, Hong GJ, Tao XY, Wang LJ, Huang YP, Chen XY (2007) Silencing a cotton bollworm P450 monoxygenase gene by plant-mediated RNAi impairs larval tolerance of gossypol. Nat Biotechnol 25:1307–1313

Matsumoto I, Watanabe H, Abe K, Arai S, Emori Y (1995) A putative digestive cysteine proteinases from *Drosophila melanogaster* is predominantly expressed in the embryonic and larval midgut. Eur J Biochem 227:582–587

Matsumoto I, Emori Y, Abe K, Arai S (1997) Characterization of a gene family encoding cysteine proteinases of *Sitophilus zeamais* (maize weevil), and analysis of the protein distribution in various tissues including alimentary tract and germ cells. J Biochem 121:464–476

Mattiacci L, Rudell S, Roca BA, Genini S, Dorn S (2001) Systemically-induced response of cabbage plants against a specialist herbivore *Pieris brassicae*. Chemoecol 11:167–173

Mazumdar-Leighton S, Broadway RM (2001a) Transcriptional induction of diverse midgut trypsins in larval *Agrotis ipsilon* and *Helicoverpa zea* feeding on the soybean trypsin inhibitor. Insect Biochem Mol Biol 31:645–657

Mazumdar-Leighton S, Broadway RM (2001b) Identification of six chymotrypsin cDNA from larval midguts of *Helicoverpa zea* and *Agrotis ipsilon* feeding on the soybean (Kunitz) trypsin inhibitor. Insect Biochem Mol Biol 31:633–644

Mazumdar-Leighton S, Babu CR, Bennett J (2000) Identification of novel serine proteinase gene transcripts in the midgets of two tropical insect pests *Scirpophaga incertulas* (Wk) and *Helicoverpa armigera* (Hb). Insect Biochem Mol Biol 30:57–68

McCaffrey JP, Harmon BL, Brown J, Davis JB (2004) Resistance of canola-quality cultivars of yellow mustard, *Sinapis alba* L., to the cabbage seedpod weevil, *Ceutorhynchus obstrictus* (Marsham). Can J Plant Sci 84:397–399

McManus MT, Burgess EPJ (1995) Effects of the soybean (Kunitz) trypsin inhibitor on growth and digestive proteases of larvae of *Spodoptera litura*. J Insect Physiol 41:731–738

McManus MT, White DWR, McGregor PG (1994) Accumulation of a chymotrypsin inhibitor in transgenic tobacco can affect the growth of insect pests. Transgenic Res 3:50–58

McManus MT, Burgess EPJ, Philip B, Watson LM, Laing WA, Voisey CR, White DWR (1999) Expression of the soybean (Kunitz) trypsin inhibitor in transgenic tobacco: effects on larval development of *Spodoptera litura*. Transgenic Res 8:383–395

McPherson MJ, Harrison DJ (2001) Protease inhibitors and directed evolution: enhancing plant resistance to nematodes. Biochem Soc Symp 68:125–142

Mello GC, Oliva MLV, Sumikawa JT, Machado OLT, Marangoni S, Novello JC, Macedo MLR (2002) Purification and characterization of a new trypsin inhibitor from *Dimorphandra mollis* seeds. J Protein Chem 20:625–632

Menegatti E, Tedeschi G, Ronchi S, Bortolotti F, Ascenzi P, Thomas RM, Bolognesi M, Palmier S (1992) Purification, inhibitory properties and amino acid sequence of a new serine proteinase inhibitor from white mustard (*Sinapis alba*) seed. FEBS Lett 301:10–14

Michaud D (1997) Avoiding protease-mediated resistance in herbivorous pests. Trends Biotechnol 15:4–6

Mita K, Morimyo M, Okano K, Koike Y, Nohata J, Kawasaki H, Kadono-Okuda K, Yamamoto K, Suzuki MG, Shimada T (2002) Construction of an EST database for *Bombyx mori* and its application. Recent Adv Silkworm Biol 83(4):426–431

Mohan M, Gujar GT (2003) Characterization and comparison of midgut proteases of *Bacillus thuringiensis* susceptible and resistant diamondback moth (Plutellidae: Lepidoptera). J Invertebr Pathol 82(1):1–11

Muriel SB, Grez AA (2002) Effect of plant patch shape on the distribution and abundance of three lepidopteran species associated with *Brassica oleracea*. Agric For Entomol 4:179–185

Nakanishi K, Yaoi K, Nagino Y, Hara H, Kitami M, Atsumi S, Muira N, Sato R (2002) Aminopeptidase N isoforms from the midgut of *Bombyx mori* and *Plutella xylostella*-their classification and the factors that determine their binding specificity of *Bacillus thuringiensis* Cry1Ac toxin. FEBS Lett 519:215–220

Nicholson C (1939) Larvae of *Pieris brassicae* (?) in December. Entomology 72:111

Noriega FG, Wang XY, Pennington JE, Barillas-Mury CV, Wells MA (1996) Early trypsin, a female specific midgut protease in *Aedes aegypti*: isolation, amino terminal sequence determination and cloning and sequence of gene. Insect Biochem Mol Biol 26(2):119–126

Northrop JH, Kunitz M, Herriott R (1948) Crystalline Enzymes, 2nd edn. Columbia University Press, New York

Novillo C, Castañera P, Ortego F (1997) Inhibition of digestive trypsin like proteases from larvae of several lepidopteran species by the diagnostic cysteine protease inhibitor E64. Insect Biochem Mol Biol 27:247–254

Novillo C, Castanera P, Ortego F (1999) Isolation and characterization of two digestive trypsin-like proteinases from larvae of the stalk corn borer, *Sesamia nonagrioides*. Insect Biochem Mol Biol 29(2):177–184

Odani S, Ikenaka T (1976) The amino acid sequences of two soybean double headed proteinase inhibitors and evolutionary consideration on the legume proteinase inhibitors. J Biochem 80:641–643

Oliveira AS, Xavier-Filho J, Sales MP (2003) Cysteine proteinases and cystatins. Braz Arch Biol Technol 46(1):91–104

Pajmon A (1999) Pest of cabbage. SodobnaKemetijstvo 32:537–540

Pandey PK, Singh BD, Singh R, Sinha MK, Jamal F (2016) *Cassia fistula* seed trypsin inhibitor (s) as antibiotic agent in *Helicoverpa armigera* pest management. ISBAB 6:202–208

Pant R, Unni BG (1980) Free amino acid of haemolymph and silk gland in the developing fifth instar and spinning larvae of *Philosamia ricini*. Curr Sci 49:538–541

Park H, Yamanaka N, Mikkonen A, Kusakabe I, Kobayashi H (2000) Purification and characterization of aspartic proteinase from sunflower seeds. Biosci Biotechnol Biochem 64:931–939

Park EY, Kim JA, Kim HW, Kim YS, Song HK (2004) Crystal structure of the Bowman-Birk inhibitor from barley seeds in ternary complex with porcine trypsin. J Mol Biol 343(1):173–186

Parkes C, Kembhavi AA, Barrett AJ (1985) Calpain inhibition by peptide epoxides. Biochem J 230:509–516

Pereira ME, Dorr FA, Peixoto NC, Lima-Garcia JF, Dorr F, Brito GG (2005) Perspectives of digestive pest control with proteinase inhibitors that mainly affect the trypsin-like activity of *Anticarsia gemmatalis* Hübner (Lepidoptera: Noctuidae). Braz J Med Biol Res 38 (11):1633–1641

Pereira DM, Noites A, Valentao P, Ferreres F, Pereira JA, Vale-Silva L, Pinto E, Andrade PB (2009) Targeted metabolite analysis and biological activity of *Pieris brassicae* fed with *Brassica rapa* var. *rapa*. J Agric Food Chem 57:483–489

Perona JJ, Hedstorm L, Wagner RL, Rutter WJ, Craik CS, Fletterick RJ (1994) Exogenous acetate reconstitutes the enzymatic activity of trypsin Asp189Ser. Biochemist 33(11):3252–3259

Peterson AM, Barillas-Mury CV, Wells MA (1994) Sequence of three cDNAs encoding an alkaline midgut trypsin from *Manduca sexta*. Insect Biochem Mol Biol 24:463–471

Peterson AM, Fernando GJP, Wells MA (1995) Purification, characterization and cDNAs sequence of an alkaline chymotrypsin from the midgut of *Manduca sexta*. Insect Biochem Mol Biol 25 (7):765–774

Poelman EH, Broekgaarden C, van Loon JJA, Dicke M (2008) Early season herbivore differentially affects plant defence responses to subsequently colonizing herbivores and their abundance in the field. Mol Ecol 17:3352–3365

Pulliam DA, Williams DL, Broadway RM, Stewart CN (2001) Isolation and characterization of a serine proteinase inhibitor cDNA from cabbage and its antibiosis in transgenic tobacco plants. PCBMB 2(1&2):19–32

Rajagopal R, Agrawal N, Selvapandiyan A, Sivakumar S, Ahmad S, Bhatnagar RK (2003) Recombinantly expressed isoenzymic aminopeptidases from *Helicoverpa armigera* (American cotton bollworm) midgut display differential interaction with closely related *Bacillus thuringiensis* insecticidal proteins. Biochem J 370:971–978

Ramalho SR, Bezerra CDS, De ODG L, Souza LL, Mario NS, De OCF R, Valerio VN, Rodrigues MML (2018) J Agric Food Chem 66(6):1349–1358

Rancour JM, Ryan CA (1968) Isolation of a carboxypeptidase-B inhibitor from potatoes. Arch Biochem Biophys 125:380–382

Rao KN, Suresh CG (2007) Bowman-Birk protease inhibitor from the seeds of *Vigna unguiculata* forms a highly stable dimeric structure. Biochim Biophys Acta 1774(10):1264–1273

Rather AH, Azim MN (2009) Feeding response in *Pieris brassicae* larvae to host/non-host plants. WJAS 5(2):143–145

Rawlings ND, Barrett AJ (1993) Evolutionary families of peptidases. Biochem J 290:205–218

Richardson M (1991) Seed storage proteins: the enzyme inhibitors. In: Rogers LJ (ed) Methods in plant biochemistry, vol 5. Academic, New York, pp 259–305

Rodrigues TB, Figueira A (2016) Management of insect pest by RNAi – a new tool for crop protection in RNA interference. In: Abdurakhmonov IY (ed) Chapter 18, RNA interference. InTech

Rosso MN, Schooten A, Roosien J, Borst Vrenssen T, Hussey RS, Gommers FJ, Bakker J, Schots A, Abad P (1996) Expression and functional characterization of a single chain FV antibody directed against secretions involved in plant nematode infection process. Biochem Biophys Res Commun 220:255–263

Rotem K, Agarwal AA, Kott L (2003) Parental effects in *Pieris rapae* in response to variation in food quality: adaptive plasticity across generations? Ecol Entomol 28:211–218

Rufino FPS, Pedroso VMA, Araujo JN, Franca AFJ, Rabelo LMA, Migliolo L, Kiyota S, Santos EA, Franco OL, Oliviera AS (2013) Plant Physiol Biochem 63:70–76

Ruoppolo M, Amoresano A, Pucci P, Pascarella S, Polticelli F, Trovato M, Menegatti E, Ascenzi P (2000) Characterization of five new low- molecular-mass trypsin inhibitors from white mustard (*Sinapis alba* L.) seed. Eur J Biochem 267:6486–6492

Ryan CA (1978) Proteinase inhibitors in plant leaves: a biochemical model for pest-induced natural plant protection. TIBS 5:148–150

Ryan CA (1979) Proteinase inhibitors. In: Rosenthal GA, Janzen DH (eds) Herbivores: their interactions with secondary plant metabolites. Academic, New York, pp 599–618

Ryan CA (1989) Insect-induced chemical signals regulating natural plant protection responses. In: Denno RF, Mcclure MS (eds) Variable plants and herbivores in natural and managed systems. Academic, New York, pp 43–60

Ryan CA (1990) Proteinase inhibitors in plants: genes for improving defenses against insects and pathogens. Annu Rev Phytopathol 28:425–449

Sachan JN, Gangwar SK (1980) Vertical distribution of important pest of cole crop in Meghalaya as influenced by the environment factors. Indian J Entomol 42:414–421

Saikia PC, Khanikor D, Unni BG (1993) Impact of muga host plants on quality and quantity of the muga silk fibres of *Antheraea assama*. In: Proceedings of 62nd Society of Biological Chemists (India) annual meet, Madurai Kamaraj University, Madurai, p 32

Saikia M, Singh YT, Bhattacharya A, Mazumdar-Leighton S (2011) Expression of diverse midgut serine proteinases in the sericigenous Lepidoptera *Antheraea assamensis* (Helfer) is influenced by choice of host plant species. Insect Mol Biol 20:1–13

Saljoqi AUR, Rehman S, Hussain N, Khan SA (2006) Insect pest of canola crop (other than aphid). J Agric Biol Sci 1(4):19–21

Sasaki T, Suzuki Y (1982) Alkaline proteases in digestive juice of the silkworm *B. mori*. Biochim Biophys Acta 703:1–10

Sasaki T, Hishida T, Ichikawa K, Asari S (1993) Amino acid sequence of alkaliphilic serine protease from silkworm, *Bombyx mori*, larval digestive juice. FEBS Lett 320:35–37

Schukle RH, Murdock LL (1983) Lipoxygenase, trypsin inhibitor, and lectins from soybean: effects on larval growth of *Manduca sexta* (Lepidoptera: Sphigidae). Environ Entomol 12:787–791

Scriber JM, Slansky F (1981) The nutritional ecology of immature insects. Annu Rev Entomol 26:183–211

Sharma DK, Devi D (1997) Seasonal variation of the foliar constituents of the primary food plants of the muga silkworm (*Antheraea assama* Ww.). Sericologia 37(2):251–258

Shinbo H, Konno K, Hirayama C, Watanabe K (1996) Digestive sites of dietary proteins and absorptive sites of amino acids along the midgut of the silkworm *Bombyx mori*. J Insect Physiol 42:1129–1138

Shotton DM, Hartley BS (1970) Amino acid sequence of porcine pancreatic elastase and its homologies with other serine proteinases. Nature 225:802–806

Siemens DH, Garner SH, Mitchell-Olds T, Callaway RM (2002) Cost of defense in the context of plant competition: *Brassica rapa* may grow and defend. Ecology 83:505–517

Silverman GA, Bird PI, Carrell RW, Church FC, Coughlin PB, Gettins PWG, Irving JA, Lomas DA, Luke CJ, Moyer RW, Pemberton PA, Remold-O'Donnell E, Salvesen GS, Travis J, Whisstock JC (2001) The serpins are an expanding superfamily of structurally similar but functionally diverse proteins. Evolution, mechanism of inhibition, novel function, and a revised nomenclature. J Biol Chem 276(36):33293–33296

Song HK, Suh SW (1998) Kunitz-type soybean trypsin inhibitor revisited: refined structure of its complex with porcine trypsin reveals an insight into the interaction between a homologous inhibitor from *Erythrina caffra* and tissue-type plasminogen activator. J Mol Biol 275:347–363

Song HK, Kim YS, Yang JK, Moon J, Lee JY, Suh SW (1999) Crystal structure of a 16 kDa double-headed Bowman-Birk trypsin inhibitor from barley seeds at 1.9 Å resolution. J Mol Biol 293 (5):1133–1144

Souza TP, Dias RO, Castelhano EC, Brandao MM, Moura DS, Silva-Filho MC (2016) Comp Biochem Physiol Part B 196–197:67–73

Stevenson BJ, Hagenbuchle O, Wellaue PK (1986) Sequence organization and transcriptional regulation of the mouse elastase II and trypsin genes. Nucleic Acids Res 14:8307–8331

Stroud R (1974) A family of protein-cutting proteins. Sci Am 231(1):74–88

Suzuki A, Yamane T, Ashida T, Norioka S, Hara S, Ikenazka T (1993) Crystallographic refinement of Bowman-Birk type protease inhibitor A-II from peanut (*Arachis hypogaea*) at 2.3 Å resolution. J Mol Biol 234(3):722–734

Svendsen I, Nicolova D, Goshev I, Genov N (1989) Isolation and characterization of a trypsin inhibitor from the seeds of kohlrabi (*Brassica napus* var. *rapifera*) belonging to the napin family of storage proteins. Carlsb Res Commun 54:231–239

Svendsen I, Nicolova D, Goshev I, Genov N (1994) Primary structure, spectroscopic and inhibitory properties of a two-chain trypsin inhibitor from the seeds of charlock (*Sinapis arvensis* L.), a member of the napin protein family. Int J Pept Protein Res 43:425–430

Swathi M, Mishra PK, Lokya V, Swaroop V, Mallikarjuna N, Dutta-Gupta A, Padmashree K (2016) Purification and partial characterization of trypsin-specific proteinase inhibitors from pigeon pea wild relative *Cajanus platycarpus* L. (Fabaceae) active against gut proteases of lepidopteran pest *Helicoverpa armigera*. Front Physiol 7:388

Swift GH, Craik CS, Stary SJ, Quinto C, Lahaie RG, Rutter WJ, MacDonald RJ (1984) Structure of the two related elastase genes expressed in the rat pancreas. J Biol Chem 259(22):14271–14278

Tamaki FK, Terra WR (2015) Molecular insights into mechanisms of lepidopteran serine proteinase resistance to natural plant defense. Biochem Biophys Res Commun 467(4):885–891

Tavladoraki P, Benvenuto E, Trinca S, De Martinis D, Cattaneo A, Galeffi P (1993) Transgenic plants expressing a functional single-chain Fv antibody are specifically protected from virus attack. Nature 366(6454):469–472

Telang MA, Giri AP, Sainani MN, Gupta VS (2005) Characterization of two midgut proteinases of *Helicoverpa armigera* and their interaction with proteinase inhibitors. J Insect Physiol 51:513–522

Telang MA, Pyati P, Sainani M, Gupta VS, Giri AP (2009) *Momordica charantia* trypsin inhibitor II inhibits growth and development of *Helicoverpa armigera*. Insect Sci 16(5):371–379

Terofal F (1961) Zum Problem der Wirtsspezifitat bei Pieriden (Lepidoptera), Unter besonderer Berucksichtigung der einheimischen Arten Pieris brassicae L., Pieris rapae L, und Pieris napi L. Dissertation, University of Munich

Terra WR, Ferreira C (1994) Insect digestive enzymes: properties, compartmentalization and function. Comp Biochem Physiol-B 109:1–62

Terra WR, Ferreira C (2005) Biochemistry of digestion. In: Gilbert LI, Iatrou K, Gill SS (eds) Comprehensive molecular insect science, vol 4. Elsevier, Oxford, pp 171–224

Thakur NSA (1996) Relationship of cabbage butterfly larval (*Pieris brassicae* Linn.) population on the marketable yield of cabbage. J Hill Res 9(2):356–358

Thangavelu K (1986) Muga silk production. Lecture on sericulture. Suramya Publishers, Bangalore, pp 166–174

Thie NMR, Houseman JG (1990) Cysteine and serine proteolytic activities in larval midgut of yellow mealworm, *Tenebrio molitor* L. (Coleoptera: Tenebrionidae). Insect Biochem 20:741–744

Titani K, Sasagawa T, Woodbury RG, Ericsson LH, Dorsam H, Kraemer M, Neurath H, Zwilling R (1983) Amino acid sequence of crayfish (*Astacus fluviatilis*) trypsin I. Biochemist 22:1459–1465

Traw MB, Dawson TE (2002) Reduced performance of two specialist herbivores (Lepidoptera: Pieridae, Coleoptera: Chrysomelidae) on new leaves of damaged black mustard plants. Environ Entomol 31:714–722

Trovato M, Maras B, Polticelli F, Costantino P, Ascenzi P (2000) A chimeric mini-trypsin inhibitor derived from the oil rape proteinase inhibitor type III. Biochem Biophys Res Commun 275:817–820

Tsunogae Y, Suzuki A, Sone T, Takahashi K, Tanaka I, Yamane T, Ashida T, Norioka S, Hara S, Ikenaka T (1986) Crystallization of Bowman-Birk type protease inhibitor (peanut) and its complex with trypsin. J Biochem 100(1):243–246

Unni BG (1996) Amino acid analysis of *Cricula trifnestrata* cocoons-a pest of *Antheraea assama* host plants *Machilus bombycina*. Sericologia 36:173–175

Upadhyay SK, Chandrashekhar K (2012) Interaction of salivary and midgut proteins of *Helicoverpa armigera* with soybean trypsin inhibitor. Protein J 31(3):259–264

Valueva TA, Mosolov VV (1999) Protein inhibitors of proteinases in seeds: 2. Physiological functions. Russ J Plant Physiol 46:379–387

Van Dam NM, Witjes L, Svatos A (2004) Interactions between above ground and belowground induction of glucosinolates in two wild *Brassica* species. New Phytol 161:801–810

van Engelen FA, Schouen A, Moltho JW, Roosien J, Salinas J, Dirkse WG, Schots A, Bakker J, Gommers F, Jongsma MA, Bosch D, Stiekma WJ (1994) Coordinate expression of antibody subunit genes yields high levels of functional antibodies in roots of transgenic tobacco. Plant Mol Biol 26(6):1701–1710

Vats LK, Singh JS, Yadava PS (1977) Food energy budget of *Pieris brassicae* larvae-a pest of cruciferous agro-ecosystems. Agro-Ecosystems 3:303–312

Vered M, Gertler A, Burstein Y (1986) Partial amino acid sequence of porcine elastase II. Active site and activation peptide regions. Int J Pept Protein Res 27(2):183–190

Vinokurov KS, Elpidina EN, Oppert B, Prabhakar S, Zhuzhikov DP, Dunaevsky YE, Belozersky MA (2006) Diversity of digestive proteinases in *Tenebrio molitor* (Coleoptera: Tenebrionidae) larvae. Comp Biochem Physiol-B 145:126–137

Volpicella M, Shipper A, Jongsma MA, Spoto N, Gallerani R, Ceci LR (2000) Characterisation of recombinant mustard trypsin inhibitor 2 (MTI-2) expressed in *Pichia pastoris*. FEBS Lett 468:137–141

Volpicella M, Ceci LR, Gallerani R, Jongsma MA, Beekwilder J (2001) Functional expression on bacteriophage of the mustard trypsin inhibitor MTI-2. Biochem Biophys Res Commun 280 (3):813–817

Volpicella M, Ceci LR, Cordewener J, America T, Gallerani R, Bode W, Jongsma MA, Beekwilder J (2003) Properties of purified gut trypsin from *Helicoverpa zea*, adapted to proteinase inhibitors. Eur J Biochem 270:10–19

Volpicella M, Leo FD, Sciancalepore M, Sonnante G, Pignone D, Gallerani R, Ceci LR (2009) Identification and characterization of protease inhibitors in *Diplotaxis* species. Plant Physiol Biochem 47:175–180

Voss RH, Ermler U, Essen LO, Wenzl G, Kim YM, Flecker P (1996) Crystal structure of the bifunctional soybean Bowman-Birk inhibitor at 0.28-nm resolution. Structural peculiarities in a folded protein conformation. Eur J Biochem 242(1):122–131

Waldbauer GP (1968) The consumption and utilization of food by insects. Adv Insect Phys 5:229–288

Walsh KA, Kauffman DL, Sampath Kumar KSV, Neurath H (1964) On the structure and function of bovine trypsinogen and trypsin. Proc Natl Aca Sci USA 51:301–308

Wang S, Young F, Hickey DA (1995) Genomic organization and expression of a trypsin gene from the spruce budworm, *Choristoneura fumiferana*. Insect Biochem Mol Biol 25(8):899–908

Wang S, Magoulas C, Hickey D (1999) Concerted evolution within a trypsin gene cluster in *Drosophila*. Mol Biol 16(9):1117–1124

Wang P, Li G, Kain W (2004) Characterization and cDNA cloning of midgut carboxypeptidases from *Trichoplusiani*. Insect Biochem Mol Biol 34:831–843

Wang P, Zhang X, Zhang J (2005) Molecular characterization of four midgut aminopeptidase N isozymes from the cabbage looper, *Trichoplusia ni*. Insect Biochem Mol Biol 35:611–620

Werner MH, Wemmer DE (1992) Three dimensional structure of soybean trypsin/chymotrypsin Bowman-Birk inhibitor in solution. Biochemist 31(4):999–1010

Wu Y, Llewellyn D, Mathews A, Dennis ES (1997) Adaptation of *Helicoverpa armigera* (Lepidoptera: Noctuidae) to a proteinase inhibitor expressed in transgenic tobacco. Mol Breed 3:371–380

Yamamoto Y, Takimoto K, Izumi S, Toriyama-Sakurai M, Kageyama T, Takahashi SY (1994) Molecular cloning and sequencing of cDNA that encode cysteine proteinases in the eggs of the silkmoth, *Bombyx mori*. J Biochem 116:1330–1335

Yang L, Fang Z, Dicke M, vanLoon JJA, Jongsma MA (2009) The diamondback moth *Plutella xylostella*, specifically inactivates mustard trypsin inhibitor 2 (MTI-2) to overcome host plant defense. Insect Biochem Mol Biol 39:55–61

Younas M, Naeem M, Raquib A, Masud S (2004) Population dynamics of *Pieris brassica* on five cultivar of cauliflower at Peshawar. Asian J Plant Sci 3:391–393

Zafar AU, Nasir IA, Shahid AA (2002) Performance evaluation of camb biopesticides to control cabbage butterfly (*Pieris brassicae*) in cauliflower crop. Pakistan J Biol Sci 5(10):1041–1043

Zavala JA, Patankar AG, Dequan Hui KG, Baldwin IT (2004) Manipulation of endogenous trypsin proteinase inhibitor production in *Nicotiana attenuata* demonstrates their function as antiherbivore defenses. Plant Physiol 134:1181–1190

Zavala JA, Giri AP, Jongsma MA, Baldwin IT (2008) Digestive duet: Midgut digestive proteinases of *Manduca sexta* ingesting *Nicotiana attenuata* with manipulated trypsin protease inhibitor expression. PLoS One 3(4):e2008

Zhao Q, Chae Y, Markley J (2002) NMR solution structure of ATTp, an *Arabidopsis thaliana* trypsin inhibitor. Biochemist 15:12284–12296

Zhou JY, Liao H, Zhang NH, Tang L, Xu Y, Chen F (2008) Identification of a kunitz inhibitor from *Albizia kalkora* and its inhibitory effects against pest midgut proteases. Biotechnol Lett 30:1495–1499

Zhu YC, Oppert B, Kramer KJ, Mc Gaughey WH, Dowdy AK (1997) cDNAs for a chymotrypsin like protein from two strains of *Plodia interpunctella*. Insect Biochem Mol Biol 27:1027–1037

Zhu-Salzman K, Zeng R (2015) Insect response to plant defensive protease inhibitors. Annu Rev Entomol 60:233–252

Zwilling R, Neurath H, Ericsson LH, Enfield DL (1975) The amino terminal sequences of an invertebrate trypsin crayfish (*Astacus leptodactylus*): homology with other serine proteases. FEBS Lett 60:247–224

Natural Insecticidal Proteins and Their Potential in Future IPM

Amit Roy and Amrita Chakraborty

Abstract

Increasing population and global food security is the foremost challenge for this century. Insect pests cause substantial damage to our crops by direct as well as indirect means such as vectoring plant viruses. Introduction of *Bacillus thuringiensis* originated toxins, namely, cry toxins, in the crop plants that showed significant resistance to insect damage during the early years (1990s). However, its societal unacceptability, nontarget effects, and the frequent development of resistance in target insects jeopardize Cry-toxin-mediated pest resistance. Alternatively, plant proteins with insecticidal activity hold great potential for future insect pest management strategies (IPM). Present chapter mainly deals with the ongoing advances in research on plant lectins. However, the entomotoxic potential of other plant proteins such as digestive inhibitors and plant peptides is also stated briefly. Further, future challenges and possibilities for developing sustainable pest management strategies are also discussed.

Keywords

Chitin · Entomotoxic protein · GM plants · IPM · Lectin · Protease inhibitor · Plant peptide · RNAi

A. Roy (✉)
Faculty of Forestry and Wood Sciences, EXTEMIT-K, Czech University of Life Sciences, Prague, Czech Republic
e-mail: Roy@fld.czu.cz

A. Chakraborty
Faculty of Forestry and Wood Sciences, EVA 4.0, Czech University of Life Sciences, Prague, Czech Republic

Department of Plant Protection Biology, Swedish University of Agricultutral Sciences, Alnarp, Sweden

I. K. Singh, A. Singh (eds.), *Plant-Pest Interactions: From Molecular Mechanisms to Chemical Ecology*, https://doi.org/10.1007/978-981-15-2467-7_12

1 Introduction

Rapid technological advancements and mass globalization are fundamental contributors to the modernization of society. Such development though embraced as "boon" carries the inevitable "bane" in loss of productive agricultural land, thus fueling the fire of malnourishment and poverty. Besides, an outbreak of pest and pathogens add on to the problem. The amount of preharvest crop loss, which is around 35% by insect pest infestation (Oerke 2006), has kept the crisis of "food sufficiency" inflamed. Pests are not only jeopardizing crop productivity and sinking the farmer's net income but may also upset the supply of food and feed as well as the economies of rural areas from all over the world (Zadoks and Schein 1979). Alternatively, the global population is expected to rise at an alarming rate (70 million per annum) leading to 9.2 billion by 2050. Such an increase in population demands the rise in food production by 70% (Ray et al. 2013). Thus, reduction of crop loss by pest insect is a significant challenge for global agriculture (Popp 2011; Popp et al. 2013). Although sincere efforts have been made to develop pest management strategies over the last three decades, we are still incapable of controlling several insect pests in an environmentally safe manner. Additionally, indiscriminate application of pesticides has not only resulted in the demolition of natural enemies, components of food and environment, and expansion of resistance in a large number of insects, but it also poses a constant threat to public health and therefore failed to receive social acceptability (Swamy et al. 2009).

The journey from advent to the advancement of plant biotechnology has placed high hopes on the development of eco-friendly strategies for integrated pest management (IPM) (Waage 1997). Notably, incorporation of insecticidal genes in crop plants offers tremendous improvement in pest control management through the fulfillment of the demand for better, useful, eco-friendly, broad-spectrum biocontrol elements. Thus, the need for such bio-control components that would effectively add to the pest management arsenal is still at a rise. Research for developing active resistance against this devastating group of insects dates back to the late 1980s. The expression of insecticidal crystal protein (ICP) of *Bacillus thuringiensis* (Bt) in tobacco and cotton plants was considered a milestone achievement in applied biotechnology against Lepidopteran and Coleopteran class of pest insects (Andrews et al. 1987; Grimaldi et al. 2005). Unfortunately, Bt toxins seem to be ineffective against sap-sucking pests due to the absence of appropriate receptors on the brush border membrane vesicle of this specially adapted insect class. Bt toxins may not evolve to kill hemipterans as *B. thuringiensis* bacteria are not naturally selected for toxicity against hemipterans (Chougule and Bonning 2012). Moreover, there is uncertainty regarding the ecological impacts of the newly introduced insecticidal proteins from the bacterial origin as well as biosafety concerns related to the expressed Bt proteins in edible plants for nontarget animals (Vazquez-Padron et al. 1999, 2000). Consumers often misunderstood these problems and denied to accept the so-called "Frankenstein" materials considering them anti-natural due to the crossing of the species barrier (i.e., bacterial origin). A promising alternative could be taking advantage of our growing knowledge of plant defense mechanisms and boost them using state-of-the-art plant biotechnological methodologies. Transgenic expression of natural insecticidal genes (insect control gene) *in planta* is also a favorable option.

Herbivore insects have interacted with plants for millions of years. Chemical cues released during herbivore insect feeding allow the plant to detect the attack and mount sophisticated defense responses. Plant defenses are either constitutive or inducible. Constitutive defense responses are more general, nonspecific, and continuous such as physical barriers created by a thick cell wall and/or waxy epidermal cuticles. Inducible defenses primarily include chemical defenses, for instance, the release of volatiles to attract natural enemies or production of secondary metabolites that are entomotoxic or induce apoptosis (programmed cell death) (Grossi-de-Sá et al. 2017). Based on the nature of the attack, plants execute a battery of responses that often include preexisting constitutive defenses complemented with direct, induced defenses (i.e., secretion of secondary metabolites, proteins, microRNAs) and indirect defenses (i.e., emission of volatiles to attract predators and parasites) (Stahl et al. 2018). Thus, plants produce an impressive diversity of natural insecticide proteins to cope up with the improper influences by the pest insects (Maag et al. 2014). Some of the well-known families of entomotoxic proteins of plant origin include lectins, ribosome-inactivating proteins (RIPs), ureases, chitinases, proteases, digestive inhibitors, and small peptides such as cyclotides and defensins. It is worth mentioning that some of these entomotoxic proteins are from edible plant origin (i.e., lectins from tomato, potato, garlic, taro, and banana; RIPs from pumpkin, beet, cucumber, etc.) (Barbieri et al. 2006; Van Damme et al. 1998) and thus presume safe for plant biotechnical applications (i.e., transgenic expression). However, the biosafety assessment for each of the transgenic expressed proteins needs to be experimentally validated before marked as safe for mammals (FAO 2001). Despite the accessibility of a massive amount of information on insecticidal proteins from plant origin, a recent comprehensive outline of natural insecticidal proteins and their potential applications in agriculture is lacking. Present chapter mostly focuses on giving a brief overview of some common entomotoxic proteins from plant origin, their substrate specificity, and mode of action keeping the primary focus on plant lectins. Potential applications and future challenges regarding the use of these natural insecticidal proteins in agriculture are also discussed.

2 Lectins: Plant Storage Proteins with Entomotoxicity

In 1888, a toxic protein with hemagglutinating activity (i.e., agglutinate red blood cells from different animals) named as ricin was first reported by Stillmark (1992) from castor beans (*Ricinus communis L.).* With the advancement of protein purification methodologies, another lectin, namely, concanavalin A (Con A), was isolated from *Canavalia ensiformis* seeds in highly purified crystalline form (Sumner 1919). A few years later, Landsteiner (1990) documented variability in hemagglutinating activity of various seed extracts after testing with erythrocytes from different animals. He also compared such sugar-binding specificity with that of antibodies of animal blood serum. The specificity of lectins toward specific erythrocytes was further investigated by Boyd and Shapleigh who first coined the term "lectin," that came from the "Latin" word *Legere*, to pick, select, or choose (Boyd and Shapleigh 1954). Lectins are also denoted as agglutinins or phytohemagglutinins (Brown and Hunt 1978). Later, it was discovered that some lectins did not possess hemagglutinating activity. Hence, hemagglutination is no longer a benchmark for proteins to be designated as lectins. By definition, lectins are proteins of non-immune origin with one or more non-catalytic domains that can reversibly bind to mono- or oligosaccharides (Peumans and Van Damme 1995). Studies on lectins also revealed additional non-sugar-binding domains with other biological activity (Van Damme et al. 1998, 2003).

The contributions incorporated by Stillmark marked the beginning of the centennial on lectin biology. Until the 1970s, little was known about lectins, as only a few of them were isolated (mostly from plants and few invertebrates) (Moreira et al. 1991). Since 1970, several hundreds of lectins are reported, and their biological activity is fully or partly characterized (Van Damme et al. 1998). Although lectins are ubiquitous in plants, their presence is also stated in fungi, bacteria, virus, insects, and animals (Van Damme 2014). Lectins are most abundant in developing seeds and other plant storage parts indicating its nature to be of primary storage type (Vandenborre et al. 2011). However, their role as plant defense protein is also experimentally demonstrated (Murdock and Shade 2002). However, the selection pressure underlying such functional switch from storage proteins to defensive molecule remains unclear (Gupta and Das 2012).

2.1 Classification, Expression, Structure, and Sugar-Binding Property

Lectins can accurately recognize and bind to a particular sugar moiety through its carbohydrate-binding domain (CBD). However, the sugar-binding specificity is highly diverse in lectins. Different lectins recognize distinct carbohydrate structures. Some of them recognize and bind to simple monosaccharides such as mannose, galactose, glucose, fructose, and so on, whereas most other plant lectins interact with further complex oligosaccharides such as O and N -linked glycans (Ghazarian et al. 2011). Lectins bind to complex sugar or glycan moiety with much higher affinity compared to simple sugar, i.e., affinity for complex and simple sugars are in the range of 10^{-6}–10^{-8} and 10^{-3}–10^{-4} Kd value, respectively (Garcia-Pino et al. 2007). Recently, glycan array analysis revealed that the carbohydrate binding profile rather than binding to a particular sugar structure could better characterize the carbohydrate-binding domain (CBD) of different lectins (Taylor and Drickamer 2009). The CBD typically consists of five to six amino acid residues, i.e., monocot mannose-binding lectins have highly conserved mannose-binding motif composed of five amino acid residues: glutamine, aspartic acid, asparagine, valine, and tyrosine (QDNVY) comprising the polar surface of the sugar-binding pocket (Ramachandraiah and Chandra 2000). The lectin-carbohydrate interaction is made

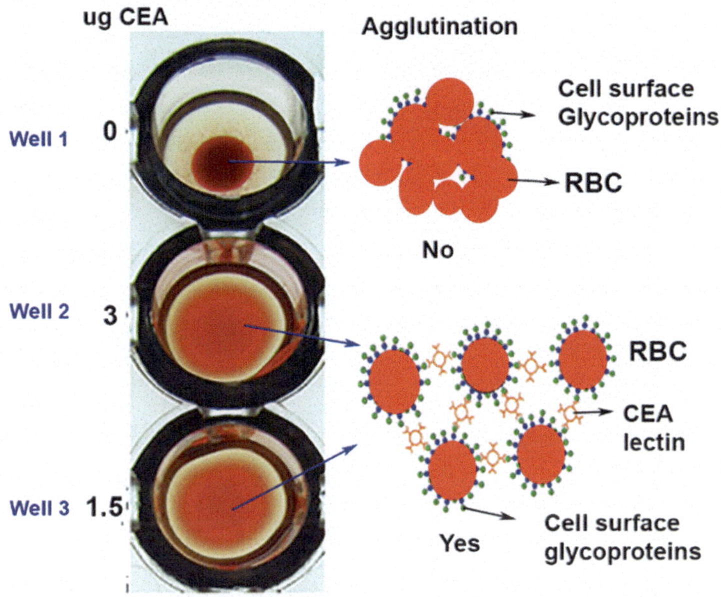

Fig. 1 Agglutination assay and pictorial illustration. Different doses of *Colocasia esculenta* tuber agglutinin (well 2, 3) cause agglutination of rabbit blood cells by forming a carpet over the wells whereas in control (well 1) without lectin shows a tight button of red cells revealing a negative reaction. (*RBC* red blood cells)

by a series of H-bonds that is often sustained by piling of the hydrophobic pyranose ring of the sugar moiety to the aromatic ring of the aromatic amino acid residues (i.e., phenylalanine, tryptophan, tyrosine) situated in the nearby neighborhood of the CBD (del Carmen et al. 2012; Schwefel et al. 2010). Presence of such extended binding sites endorses the complex sugar structures (i.e., N-linked glycan) as the natural binding site for the majority of the plant lectins (Van Damme et al. 2008).

Based on CBD architecture, plant agglutinins or lectins are classified into four major groups: merolectins, chimerolectins, hololectins, and superlectins (Van Damme et al. 1998). Merolectins have only one CBD and thus lost the capacity to agglutinate the red blood cells whereas hololectins are made up of two or more homologous CBDs and display agglutination activity (Fig. 1). Most of the plant lectins, isolated and characterized until now, are belonging to this class. The superlectins are the lectins with two or multivalent CBDs recognizing structurally unrelated carbohydrate structures. Finally, chimerolectins with one or more CBDs fused to another domain with independent biological activity.

Recent advancement of next-generation sequencing-based studies revealed that many plant lectins are chimerolectins, reinforcing the notion of their multifunctional role (Van Damme et al. 2008). Moreover, identification of plant lectins with additional domains, i.e., kinase domain in rice and soybean lectin besides CBD, implies to the evolutionary acquisition of additional functionalities among plant lectins.

Table 1 Brief classification of lectins

Type	Lectin domain	Carbohydrate specificity	Example
Entomotoxic	*Galanthus nivalis* agglutinin (GNA) domain	High Man N-glycans, Man, N-glycans	GNA, ASAL, CEA, AMTL
	Hevein domain	Chitin, Man, High-Man, N-glycans	WGA, Hevein
	Legume lectin domain	Complex N glycans, Man/Glc, Gal/GalNAc	PSA, PHA, ConA
	Jacalins	Gal, Man, T-antigen	Jacalin
	Nicotiana tabacum agglutinin domain	High Man N glycans, GlcNAc	NICTABA
	Ricin-B domain	Gal.GalNac, Sia	Ricin, SNAI
	Amaranthins	T-antigen	Amaranthin
Non-entomotoxic	*Agaricus bisporus* agglutinin domain	T-antigen	ABA
	LsyM	Chitin-oligosaccharide	LySM, CEBiP
	Class V chitinase homologs	High-Man N-glycans, blood gr B	RobpsCRA
	Cyanovirin domain	High-Man N-glycans	CV-N
	Euonymus europaeus agglutinin domain	High-Man N-glycans, blood gr B	EEA

Adapted from Vandenborre et al. (2011)
Abbreviations: *Man* mannose, *ASAL Allium sativum* leaf agglutinin, *WGA* wheat germ agglutinin, *Gal* galactose, *GalNAc* N-acetylglucosamine, *Glc* glucose *ConA Canavalia ensiformis* agglutinin, *PSA Pisum sativum* agglutinin, *PHA Phaseolus vulgaris* agglutinin, *NICTABA Nicotiana tabacum* agglutinin, *Sia* salicylic acid, *SNAI Sambucus nigra* agglutinin I, *ABA Agaricus bisporus* agglutinin, *LsyM* lysin domain, *CEBiP* chitin elicitor-binding protein, *RobpsCRA* chitinase-related agglutinin from *Robinia pseudoacacia*, *CV-N* cyanovirin-N, *EEA Euonymus europaeus* agglutinin, *CEA Colocasia esculenta* tuber agglutinin, *AMTL Amorphophallus paeoniifolius* tuber agglutinin

Based on all the diversities and similarities within CBDs, currently plant lectins are clustered into 12 distinct families (Table 1). So far, all discovered plant lectins are belonging to one of these families except maltose-binding lectin extracted from *Dioscorea batatas* (DB3L) because of its unique structure and sugar specificity (Gaidamashvili et al. 2004). Among these 12 plant lectin families, seven families showed different degrees of entomotoxic potential. They are *Galanthus nivalis* agglutinin (GNA) domain, hevein domain, legume lectin domain, jacalins, *Nicotiana tabacum* agglutinin domain, ricin-B domain, and amaranthins.

Since the plants have limited nutrient resources, they need to synchronize their energy investment to produce defense compounds for survival. Continuous expression of defense-related compounds is a costly affair (Zavala and Baldwin 2004). Consequently, induced expression of defense compounds had evolved as a cost-effective alternative (Chen 2008). Lectin gene expression is also regulated by plants. Depending on the temporal and developmental regulations on lectin gene expression, they are grouped into two types: constitutively expressed lectins and inducible lectins. Constitutively expressed lectins always occur in the plant regardless of abiotic or biotic stress levels. Most of the common insecticidal lectins are constitutively expressed in plants such as GNA, *Allium sativum* leaf agglutinin (ASAL), *Colocasia esculenta* tuber agglutinin (CEA), *Allium cepa* agglutinin (ACA), *Arum maculatum* lectin (ATL), wheat germ agglutinin (WGA), *Pisum sativum* agglutinin (PSA), ricin-B lectins, etc. to name a few. In plants, such lectins are manufactured in the endoplasmic reticulum (ER) and accumulated inside the vacuole or other extracellular spaces. These lectins are copious in plant storage tissues such as seeds, tubers, rhizomes, bulbs, and so on.

In contrast, inducible lectins are only expressed during biotic (i.e., insect herbivory) and abiotic stresses (i.e., wounding, drought, etc.). Inducible lectins are mostly found in the non-storage tissues or tissues usually affected by biotic and abiotic stresses such as leaves, root, or flowers. Inducible lectins are produced in the cytoplasm and resided in the cytoplasmic or nuclear component; their expression level is also relatively low compared to constitutive lectins (Van Damme et al. 2008). ORYSATA, a mannose-specific jacalin-related lectin, was the first inducible lectin from the salt-treated rice seedling (Zhang et al. 2000). Under the normal physical condition, rice roots and sheaths did not express ORYSATA, but upon salt or drought stress, it started producing ORYSATA. Another lectin from *N. tabacum*, called NICTABA, was documented to express only after insect herbivory or after application of the plant hormone, methyl jasmonate (MeJA) (Chen et al. 2002; Vandenborre et al. 2009a, b).

Generally, the 3D structure of lectins is made up of β-sheets joined by loops forming antiparallel chains with little contribution from α-helixes. Structural stability of dimeric or tetrameric forms is primarily attended by H-bond, hydrophobic interactions, and salt bridges (Sharon and Lis 1990). Precisely, the lectin CBD is composed of three overlapping regions: the central conserved core region with residues interacting with metal ions essential for carbohydrate binding; surrounding the core region, aromatic residues are located giving monosaccharide specificity to lectins; lastly, nonspecific residues with higher diversity are found in the outer region involving the interactions with complex oligosaccharide molecules (Sharon and Lis 1990; Young and Oomen 1992). The structure of GNA and other GNA-related lectins (monocot mannose-binding lectins, MMBL) were investigated in depth due to their entomotoxic potential. Crystal structure of GNA showed a three fold symmetry β-sheet polypeptide fold composed of three subunits (I, II, III); each one of which contains four stranded β-sheets forming a 12-stranded β-barrel (Hester et al. 1995). The tetrameric structure of GNA has 12 mannose-binding pockets (QXDXNXVXY). The "subunit I" have the highest mannose-binding specificity.

Interestingly, the tetrameric structure is formed and stabilized by two dimers interacted by C-terminal exchange, a signature characteristic of the β prism II fold structure. The hydrophobic residues present at the center of the prism stabilize the complex quaternary structure through robust Van der Waals interaction with three residues at the top along the centers of an equilateral triangle. The side chains of the hydrophobic residues also enhance the stability (Sinha et al. 2007). Notably, the biological role of lectins is significantly dependent on its oligomerization features (Chandra et al. 1999). Commonly, monomeric lectin shows strong fungicidal activity; dimeric lectin exhibits insecticidal activity; whereas tetrameric lectin demonstrates insecticidal and antiretroviral activity. For instance, tetrameric lectin GNA has both insecticidal and antiretroviral activity (i.e., can bind to gp120, a major glycoprotein from HIV) (Balzarini et al. 1991), while dimeric ASAL has only insecticidal activity (Bandyopadhyay et al. 2001). Fascinatingly, conversion of dimeric ASAL to monomeric mASAL by introducing a site-specific mutation (altering fine amino acid residues near C-terminal of ASAL) resulted in the loss of anti-insecticidal activity and simultaneously gain on significant antifungal activity (Banerjee et al. 2011).

2.2 Entomotoxicity: Mode of Action

Lectins are multifunctional proteins serving various biological functions in an organism. Among several physiological functions of plant lectins such as induction of antifeedant activity or alteration of oviposition behavior in insects (Michiels et al. 2010), insecticidal activity received highest attention and has been tested on different orders of insects (Bandyopadhyay et al. 2001; Fitches et al. 2008; George et al. 2018; Mondal et al. 2012; Oliveira et al. 2017; Powell et al. 1998; Roy and Das 2015; Van Damme et al. 1998; Vandenborre et al. 2010). The snowdrop lectin (GNA) had received particular attention due to its insecticidal activity against a number of hemipteran groups of pest insects (Foissac et al. 2000; Hilder et al. 1995; Nagadhara et al. 2004; K. S. Powell et al. 1998; Rao et al. 1998; Wang et al. 2005). GNA has a specific binding affinity for α-1-3-linked mannose residues, which are scarce in brush border membrane vesicles (BBMVs) of the mammalian small intestine; thus, GNA is presumed to be safe for mammals. Moreover, studies on rats fed on GNA add experimental evidence to that (Pusztai et al. 1995, 1996). Furthermore, GNA-related lectins with insecticidal activities from other sources such as leek, garlic, and onion are reported to have no toxicity to mammals (Peumans and Van Damme 1995); thus, GNA-related lectins are considered as promising candidates for transgenic expression. However, the search for new lectins with entomotoxic potential is still ongoing. Recently, *S. terebinthifolius* leaf lectin (SteLL) is found to affect the survival and nutrition parameters of *S. zeamais* (maize weevil) (Camaroti et al. 2018); lectin from *W. somnifera* (ashwagandha) leaves (WsMBP1, mannose-binding) showed insecticidal activity in *H. puera* (Lepidoptera) and *P. sanguinolens* (Hemiptera) through interruption of digestive process and nutrient assimilation (George et al. 2018); a chitin-binding lectin from *Moringa oleifera* seeds (WSMoL) diminishes *Anagasta kuehniella* (Mediterranean fruit fly) larval weight gain by 50% by altering normal digestive enzyme activities (Oliveira et al. 2017).

In general, the negative influence of plant lectins on insect physiology, for instance, growth, fecundity, and development are already well-documented (Vandenborre et al. 2011). Binding to insect gut epithelium is one of the fundamental prerequisites for lectin toxicity. Lectins interact with diverse glycan or glycoprotein structures on the insect BBMV and cause toxicity by altering normal physiological processes. For example, GNA and *concanavalin A* (ConA) can cause toxicity to pea

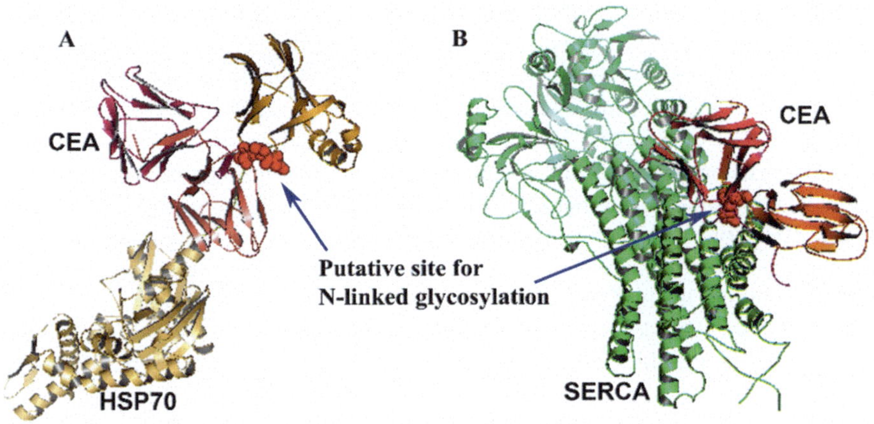

Fig. 2 (**a**) CEA-HSP70 interactive region and (**b**) CEA-SERCA interactive region show putative N-linked glycosylation sites in the near vicinity. (Adapted from Roy et al. 2014)

aphid through binding to gut membrane-anchored digestive enzyme with glycosylated mannose residues, aminopeptidase N (APN) (Cristofoletti et al. 2006). GNA alters iron metabolism after binding to ferritin subunit in various insects such as *Nilaparvata lugens* (brown planthopper, BPH) and *S. littoralis* (Du et al. 2000; Sadeghi et al. 2008). Moreover, immunohistochemical studies revealed the interaction between GNA and cell surface carbohydrate residues from insect gut epithelium (Powell et al. 1998). Likewise, *Colocasia esculenta* lectin (CEA, a GNA-related lectin) from taro was observed to bind midgut epithelial cells of red cotton bug (RCB) (Roy and Das 2015) indicating the primary mechanism of entomotoxicity among GNA-related lectins. Several studies have reported midgut receptors or interactive partners of various entomotoxic lectins from different insect orders. Recently Roy et al. (Roy et al. 2014) identified midgut interaction partners for CEA from *Bemisia tabaci* (whitefly), namely, vacuolar ATP synthase, sarcoplasmic reticulum-type Ca^{++} ATPase (SERCA), and from *Lipaphis erysimi* (mustard aphid) heat shock protein 70 (HSP70), ATP synthase β subunit, clathrin heavy chain protein. Proteins such as ATP synthase α-subunit, cytochrome P450, and actin from red cotton bug were also documented as binding partners of CEA (Roy and Das 2015). In silico docking analysis further supports the interaction of CEA with its binding partners, i.e., SERCA, vacuolar ATP synthase, HSP70, and clathrin heavy chain protein. Sequence analysis also showed conserved glycosylation sites around the close vicinity to the interactive residues of CEA (Fig. 2) (Roy et al. 2014). Other investigations have identified ferritin (Du et al. 2000) and NADH-quinone oxidoreductase (Bala et al. 2013a, b) from BPH as binding partners for GNA and *Allium sativum* leaf agglutinin (ASAL), respectively. The entomotoxicity resulted due to binding of lectins to its binding partners probably entails an altered metabolic functioning in target insects leading to delayed development, abnormal reproductive physiology (i.e., loss of fecundity), and death.

Hemipteran insects lack PM; hence, lectins can freely interact with the gut epithelium for exerting toxicity. GNA can also cross the insect gut epithelium and get access to other target tissues such as ovaries and fat body (Fitches et al. 2001b; Powell et al. 1998). Recently, Caccia et al. (2012) showed that FITC-leveled lectin from *Amaryllis* bulb (FITC-HHA) was not only bind to the cell membrane of columnar cells but also internalized into the cells suggesting its ability to cross insect epithelial cells and reach to hemolymph. The authors further proved the internalization of HHA through clathrin-mediated endocytosis. Interestingly, clathrin heavy chain protein from mustard aphid was also documented as the interaction partner for mannose-binding lectin, CEA (Roy et al. 2014).

Furthermore, Walski et al. (2014) demonstrated that the capacity to penetrate through the peritrophic membrane (PM) is the key to lectin toxicity against *T. castaneum* (Tc, red flour beetle, Coleopteran). However, such penetration capabilities of lectins solely depend on its molecular dimensions, charge, and PM pore size and can be valid for toxicity against Lepidopterans. Higher oligomeric forms of lectins that are larger than the PM pore size will be retained in the endoperitrophic space. For instance, *Rhizoctonia solani* agglutinin (RSA, ~32 kDa, dimer) showed the highest activity against Tc larvae as compared to *Sambucus nigra* agglutinin II (SNAII, 64 kDa, dimer) that showed higher toxicity to Tc-cell line where there is no requirement for penetration to PM. Being smaller in size, RSA can penetrate the PM in vivo more efficiently than SNAII, thus showing higher insecticidal potential. The partial permeability of SNAII through PM attributed to its monomeric form (32 kDa) in the solution. Interestingly, in the same study, SNAI did not show any insecticidal activity in vivo due to proteolytic degradation in the gut of the beetle (Walski et al. 2014). Hence, it is important to note that resistance to proteolytic degradation in insect gut lumen is also one of the primary prerequisites for lectin toxicity (Felton 2005). Generally, insecticidal lectins showed a diverse amount of stability inside the gut lumen to induce toxicity, i.e., *Moringa oleifera* lectin (cMOL) is resistant to proteases from Mediterranean fruit fly for up to 12 h (De Oliveira et al. 2011); *Olneya tesota* lectin (PF2) can resist digestive enzymes from *Zabrotes subfasciatus* for a day (Lagarda-Diaz et al. 2008); *N. tabacum* agglutinin (NICTABA) remains undigested after 3 days of incubation with *S. littoralis* midgut extracts (Vandenborre et al. 2011).

Lepidopteran PM is mainly composed of glycoproteins and chitin-microfibrils (Hegedus et al. 2009), which serves as easy targets for lectins. Interaction of PM with lectins can alter PM morphology leading abnormal digestion and absorption process. Wheat germ agglutinin (WGA) feeding causes disintegrations of gut microvilli in *Ostrinia nubilalis* (European corn borer) indicating the negative impact of WGA binding to chitin microfibrils and other glycosylated peritrophic matrix proteins. The voids in the PM caused due to WGA interaction allowed direct interaction of the food particles with the gut microvilli brush border (Harper et al. 1998). WGA feeding also resulted in an alteration in the microvilli of *D. melanogaster* (H. M. Li et al. 2009). Interestingly, WGA did not show such changes in another lepidopteran pest insect *Manduca sexta* suggesting that the difference in PM formation and structure can make selected insects more resistant to WGA than the others.

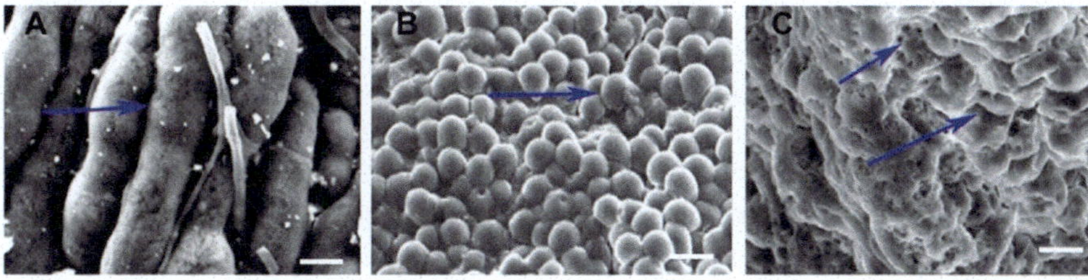

Fig. 3 Scanning electron microscopic (SEM) analysis displays morphological alterations in *Colocasia esculenta* lectin amended diet-fed insect midgut epithelial cells of red cotton bug. (**a**) The normal midgut cells of well-nourished insect gut completely covered with a perimicrovillar membrane (PMM); (**b**) the morphology of famished insect midgut with no PMM cover; (**c**) the unusual morphology of CEA amended diet-fed insect midgut with substantial degradation of PMM. (Bar- 800× magnification). (Reproduced from Roy and Das 2015)

In some mannose-binding lectins such as *Dioscorea batatas* agglutinin (DB1), CEA can also affect the normal gut morphology upon binding and induce toxicity. DB1 strongly binds to the PM and brush border membrane of *H. armigera* (Ohizumi et al. 2009). Scanning electron microscopic analysis with the gut of CEA supplemented diet-fed *Dysdercus cingulatus* (red cotton bug, RCB) displayed substantial degradation of the perimicrovillar membrane (PMM), a typical lipoprotein membrane covering the microvilli (Fig. 3) (Roy and Das 2015). Similar membrane damage in the midgut of *S. littoralis* by plant lectins was also documented (Chougule and Bonning 2012). Precisely, such membrane damage potential and permeability of plant lectins to insect epithelial cells can aid its interaction with diverse intracellular binding partners to induce toxicity. Perhaps, alteration of PM can compromise the insect immunity by allowing the entry of the harmful microbes or pathogens associated with plant material to midgut epithelial cells.

Li et al. (2009) conducted *Drosophila* midgut gene expression analysis to study the altered midgut morphology in detail. Microarray study showed differential expression of 61 transcripts in WGA supplemented-fed *D. melanogaster* midgut cells (Li et al. 2009). These transcripts primarily associated with chitin metabolism, cytoskeletal organization, digestion, detoxification, and energy metabolism in *D. melanogaster*. Differential expression of genes related to the cytoskeletal organization can be directly linked with the alteration of midgut morphology in insects. Similarly, induction of chitin metabolism can be attributed to the repair of the damaged PM. Reduced weight gain and antifeedant activity in lectin-fed insects may result from the differences in digestive enzyme expression and energy metabolism. Such alteration in insect physiology (i.e., disruption of PM) by plant lectins leads to oxidative stress and release of detoxifying enzymes such as glutathione-S-transferase (GST).

2.2.1 Source of Variability in Entomotoxicity

Many plant lectins are lethal to insects; however, the target insect and the level of toxicity vary lectin to lectin. Some lectins are active against a specific group of pest insects, whereas others showed broad toxicity against many distantly related insect groups. Some of them are even toxic to beneficial insects (Hogervorst et al. 2006). Such a diverse range of toxic activity of lectins resulted primarily due to variability in insect gut physiology and carbohydrate specificity of the lectins. Insect midgut environment can be acidic or alkaline. Plant lectins need to tolerate such a hostile environment to induce toxicity. For instance, GNA is stable in a wide pH range (pH 2 to pH 12), thus effective against a wide range of target pests, whereas NICTABA is active against insects with the alkaline gut environment such as lepidopterans due to its stability only in alkaline pH (Chen et al. 2007; Vandenborre et al. 2009a, b). Another source of variability is the resistance against proteolysis inside the insect gut. Every insect has its specific set of proteolytic enzymes (i.e., metalloproteinase, serine, or cysteine proteinase) with specialized proteolytic degradation capability (Felton 2005). Hence, given plant lectin will only induce toxicity if it can resist proteolysis inside the target insect gut.

The glycan profile of insect midgut tissues will solely depend on the developmental stage of the insect, hence causes variability in lectin toxicity. Therefore, an insect can be susceptible for toxicity to a given plant lectin at the larval stage while the adults may remain unaffected. The most excellent example of such a scenario is the selective toxicity of PSA on larvae and adult beetles. Larvae of pollen beetle showed significant toxicity after ingestion of PSA, while the adult pollen beetles showed less profound effect (Lehrman 2007; Melander et al. 2003). Furthermore, plant lectins also have carbohydrate-binding specificity; therefore, the presence of particular carbohydrate structures determines the success or failure of a given lectin. For instance, GNA and ConA showed a toxic effect on pea aphid (lacking PM), but WGA was nontoxic as it binds to the chitin microfibrils found in the PM of target insects (i.e., *D. melanogaster*) (Li et al. 2009; Rahbé et al. 1995).

2.3 Biotechnological Application: Insect Resistance GM Plants

Genetically engineered plants with foreign insecticidal proteins offer excellent potential for pest management and enhance the crop productivity. Over the last two decades, there is rapid progress in the direction of using this technology for generating insect resistance GM plants. Due to its insecticidal property and public concern for using Bt (*Bacillus thuringiensis*) genes for transgenic application, lectins (i.e., mainly those that are from edible plant origin) become a superior candidate for the transgenic application. Many lectins with proved entomotoxic potential are used for transgenic expression and subsequently tested for insecticidal activity. GNA was used for transgenic expression in *Nicotiana tabacum, Solanum tuberosum, Triticum aestivum, Zea mays* L., *Saccharum officinarum*, and *Oryza sativa* resulting generation of insect-resistant plants against various insect pests such as *Nilaparvata lugens, Myzus persicae, Nephotettix cincticeps, Helicoverpa zea, Nephotettix virescens, Sitobion avenae, Lacanobia oleracea, Rhopalosiphum maidis*, etc. to name a few (Down et al. 2003; Fitches et al. 1997; Gatehouse et al. 1997; Hilder et al. 1995; Maqbool et al. 2001; Powell et al. 1993; Sétamou et al. 2002; Wang and Guo 1999; Wang et al. 2005).

Fig. 4 Insect bioassay setups (A, B) on transgenic mustard plants expressing ASAL. The inset shows *L. erysimi* on control mustard plant. (Redrafted from Bala et al. 2013a, b)

Besides GNA, *N. tabacum* plant was also transformed to express different lectins targeting various pest insect. Some of the expressed insecticidal lectins are soybean lectin (SBL, resistant to *Spodoptera exigua*) (Guo et al. 2013), *Helianthus tuberosus* agglutinin (HTA, resistant against *Myzus persicae*) (Chang et al. 2003), ORYSATA (resistant against *A. pisum*, *S. exigua*, *M. persicae*) (Al Atalah et al. 2014), PSA (resistance against *H. virescens*) (Boulter et al. 1990), *Zephyranthes grandiflora* agglutinin (ZGA, resistance against *M. nicotianae*) (Ye et al. 2009), ASAL (resistant against *M. persicae*) (Dutta et al. 2005), and ASA II (resistance against *M. nicotianae; S. littoralis*) (Sadeghi et al. 2007, 2008). *In planta* bioassays on these transgenic plants showed a noteworthy effect on the fecundity and survival of the target pests mentioned above. Furthermore, transgenic rice plants expressing ASAL caused significant mortality and reduction in fecundity of *N. lugens, L. erysimi*, and *N. virescens* (green planthopper) (Bala et al. 2013a; b; Saha et al. 2006; Sengupta et al. 2010) (Fig. 4). Transgenic *Cicer arietinum* (chickpea) transformed with ASAL showed resistance against *Aphis craccivora*; survival and fecundity were decreased up to 26% and 42%, respectively, during *in planta* bioassay with T1 transgenic plants compared to control plants (untransformed) that supported 85% aphid survival (Chakraborti et al. 2009). Chandrasekhar et al. (2014) conducted *in planta* bioassay on rice lines (T2 homozygous) expressing ASAL under phloem-specific promoter and observed up to 80% of the reduction in the fecundity, development, and survival of *N. lugens* compared to control untransformed plant. Similar experiments with ASA I and ASA II expressing transgenic tobacco plant resulted in a delay in development and metamorphosis in *S. littoralis* (Sadeghi et al. 2007, 2008).

Selectable marker-free transgenic *B. juncea* (mustard plant) expressing ASAL showed significant toxicity to *L. erysimi* (70%), which is comparable to the earlier results (89% mortality) where ASAL gene is constitutively expressed under a phloem-specific promoter in mustard plant (Bala et al. 2013a, b; Dutta et al. 2005). GM cotton plant with a phloem-specific expression of *Amaranthus caudatus* agglutinin (ACA) showed resistance against cotton aphid, *Aphis gossypii* (Wu et al. 2006). Furthermore, transgenic expression of WGA (in *Zea mays and B. juncea*), *P. vulgaris* agglutinin (PHA, in *A. thaliana*), and ConA (in *S. tuberosum*) also demonstrated significant resistance against different target pests (Fitches et al. 2001a; Gatehouse et al. 1999; Kanrar et al. 2002; Maddock 1991). Very recently, *in planta* bioassay on CEA expressing mustard lines displayed higher insect mortality of up to 81%, whereas fecundity of mustard aphid (*L. erysimi*) was diminished by 49–62% compared to the untransformed control plants (Das et al. 2018).

2.4 Biosafety: Are Lectin-Expressing GM Plants Safe for Mammals?

There is a rising concern in the society regarding the potential risk accompanying the use of genetically modified food crops on the health of the mammals and other nontarget organisms. It is now obligatory to perform the biosafety assessment studies on the gene of interest for transgenic expression. To date, no GM plant expressing lectin is commercially available. Biosafety measures also need to be performed rigorously before releasing GM plants with expressed lectin. The toxicity of few insecticidal lectins are tested on mammals, and most of the cases, where no adverse effect was observed (Macedo et al. 2015), may be due to dissimilarity in the glycosylation profile in mammals compared to insects. The biosafety assessment study strongly depends on monitoring the allergenicity potential of the lectins, which is primarily evaluated concentrating on the source of the gene, searching for sequence hormology with the known allergens, the expressed protein level in GM crops, the reactivity of the protein with IgE, heat, and digestive stability of the introduced protein (Taylor and Hefle 2001).

Interestingly, Mondal et al. (2011) monitored allergenicity potential of ASAL by conducting in vivo and in vitro experiments following the prescribed guidelines of FWO/WHO (2001) and found ASAL as safe for transgenic application. Later on, Ghosh et al. (2013) proved monomeric ASAL (mASAL) with antifungal activity is also safe for mammals and thus for transgenic application, too. Histochemical studies with mASAL fed Balb/c mice gut did not show any symptoms for hypersensitive reactions, whereas mouse treated with ovalbumin (allergenic to mice) as a positive control showed signature hypersensitive reactions in the lung (Fig. 5) and

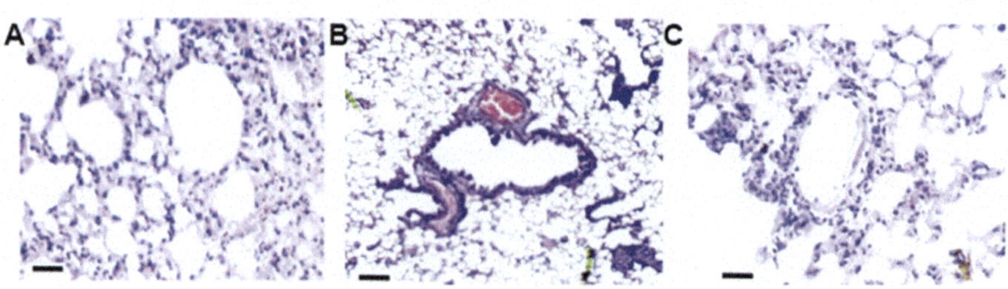

Fig. 5 Histopathological images of the lung of the sensitized Balb/c mice. (**a**) Lung section of mice sensitized with PBS control, (**b**) lung section of mice sensitized with ovalbumin (a known allergen) showing pathological symptoms such as congested lung structure with peribronchial and perivascular inflammatory cell infiltrate, (**c**) monomeric *Allium sativum* leaf agglutinin (mASAL) sensitized lung section of mice with no allergenic symptoms. (Bar = 1 μm). (Redrafted from Ghosh et al. 2013)

causes loss of normal gut morphology. Very recently, insecticidal CEA expressed in mustard plants showed no significant risk of allergenicity for mammalian consumption (Das et al. 2018); thus, CEA is evidenced safe for future transgenic application. Notably, all the lectins do not show similar results when tested against mammals. Mice fed on GM potato expressing GNA demonstrated proliferation of gastric mucosa (Ewen and Pusztai 1999). PHA from red kidney bean induced prominent hypersensitivity reactions in the intestine, lung, and spleen of mice (Kumar et al. 2013). Therefore, biosafety assessments need to be conducted for each of the expressed lectins on a case to case basis even before thinking of its commercial use.

2.5 Impact on Nontarget Organisms: Concern for Biotechnological Application

In addition to evaluating GM for food safety, it is essential to evaluate the environmental safety of expressed (transgenic) lectins toward nontarget organisms such as insect predators and parasitoids, beneficial fungus, and insects. One of the key features underlying the entomotoxicity of lectins is their ability to sustain the hostile gut environment of the target pests. Moreover, lectins are incredibly resistant to gut proteolysis and can be retained inside the gut after binding with appropriate binding partners. Excess lectin can be released with feces or honeydew and became accessible to other nontarget organisms. Unfortunately, these features often make some lectins less suitable for transgenic application due to its nontarget effects, i.e., GNA was undigested and accumulated in the gut of three aphid predator species (Hogervorst et al. 2006). Many studies have already investigated the effect of lectins on natural enemies (predators and parasitoids) in details (Bell et al. 2001, 2004; Birch et al. 1999; Couty and Poppy 2001; Down et al. 2000).

Sucrose or honey supplemented with GNA was used to stimulate the lectin feeding by parasitoids (Bell et al. 2004; Hogervorst et al. 2006; Romeis et al. 2003). Sucrose solution supplemented with 1% GNA (w/v) caused declined fecundity and longevity on diverse parasitoid species. GNA was also spotted in the gut and hemolymph of some parasitoid. It is worth to mention here that the concentration of the exposed lectin to parasitoids predator in the field condition is rather low as compared to its expression level in the transgenic crops. Often the concentration was below the detection level (Nagadhara et al. 2004).

Nonetheless, an indirect toxic effect of transgenic potato plant expressing GNA was also documented on aphid predator, two-spotted ladybird (*Adalia bipunctata*). Precisely, the fecundity, longevity, and egg viability of ladybird was affected when they fed on *M. persicae* that reared on GNA-expressing transgenic potato plants (Birch et al. 1999). It was not clear if the observed effect was due to the reduced quality of food (GNA fed aphids) or GNA, although GNA was located in BBMV (brush border membrane villi) and gut epithelial cells of ladybird. The same aphid species when fed on an artificial diet supplemented with GNA at a particular concentration showed delayed growth ascribing aphids as a suboptimal food source (Down et al. 2000).

Auspiciously, often there is no evidence for lectins being toxic to the nontarget organisms. *A. abdominalis* (an endoparasitoid of aphids) did not show any symptoms of toxicity after feeding on potato aphids (*Macrosiphon euphorbiae*) that fed on 0.1% GNA-supplemented artificial diet. Since there was no GNA detected in the hemolymph of potato aphid, it could be assumed that the parasitoid was not exposed to threshold amount GNA to show any toxic effect (Couty and Poppy 2001). Similarly, the ectoparasitoid *E. pennicornis* feeding on *L. oleracea* that reared on transgenic potato leaves with GNA did not show any adverse effect (Bell et al. 2001). However, when the ectoparasitoid was fed on honey with 0.1–1% GNA, adverse effects were observed on longevity and fecundity of the insects. Effects on fecundity were rather evident after high doses (1%) of GNA feeding. Hence it is understandable from the above observations that environmental impact of the GM crops with expressed lectin is quite complicated and depends on various factors such as level of lectin expression in the transgenic plant, level of exposure to the nontarget pests, mechanism of toxicity on the target pest (i.e., lectin reaches to hemolymph or not), and many more.

3 Other Classes of Insecticidal Proteins: A Brief Overview

Plants have evolved ample mechanisms to defend themselves from enemies such as insects. These mechanisms differ from species to species depending on the exposed biotic challenges. In addition to lectins, several other insecticidal proteins exist in nature originated from different plants targeting specific physiology of the insects. In the coming section, some of them are discussed briefly.

3.1 Plant Enzymes with Entomotoxicity

Plants produce a plethora of enzymes to maintain their physiological well-being. Interestingly, some of the enzymes displayed entomotoxic activity by upsetting the normal physiology of the attacked insects. For instances, ribosome-inactivating proteins, proteases, chitinases, ureases, etc. are few representatives of this category.

3.1.1 Ribosome-Inactivating Proteins (RIPs)

Ribosome-inactivating proteins represent a particular set of cytotoxic proteins with specific ribosomal RNA N-glycosidase activity. RIPs can activate prokaryote and eukaryote large ribosomal subunits; thus, protein biosynthesis is directly affected in the target cells (Peumans et al. 2001; Stirpe 2013). Precisely, RIPs can recognize highly conserved GAGA sequence and eliminate the adenine residue from the exposed loop (sarcin/ricin) of 28 s rRNA of animals and 23 s rRNA from prokaryotic ribosomes via their N-glycosidase activity. The elimination of adenosine residue interrupts the interaction of large subunit of rRNA with elongation factor 2 (eIF-2) causing arrest of the protein synthesis. Moreover, RIPs are also reported to possess superoxide dismutase, DNase, chitinase, or phospholipase activity (Virgilio et al.

2010), thus showing its capability to interact with diverse substrates. RIP domain is quite common within plant kingdom but is not copious, i.e., recent genome sequencing studies indicated absence of RIP domains from model plant *Arabidopsis thaliana* and other 23 plants (Shang et al. 2014). RIPs are grouped into two major clusters: type I RIPs (RIP-I) and type II RIP (RIP-II). RIP-I proteins composed of a single peptide chain with polynucleotide adenine glycosylase (PAG) domain, whereas RIP-II is a chimeric protein with PAG domain (A chain) connected to a C-type lectin domain (B chain) (Stirpe 2013).

RIPs showed insecticidal activity against different insects due to their capacity to alter protein production in the cells. Presence of lectin domain in RIP-II aids the access of the protein inside the cell via binding to glycoconjugate receptors on cell surface followed by the endocytic pathway. Once entered in the cell, RIP arrives at the ER lumen via a retrograde transport. Finally, upon reaching to cytoplasm, RIPs show enzymatic activity on rRNA, block protein synthesis, and cause target cell death by apoptosis. Ricin purified from the castor bean (*Ricinus communis*) is a well-documented example for RIP-II type protein with high entomotoxic activity (Carlini and Grossi-de-Sá 2002). Biosynthesis of ricin is already studied in details (Lord and Spooner 2011). Mature ricin protein is composed of two chains: 32 kDa A chain and 34 kDa B chain linked by a disulfide bond. Preproricin, the precursor of ricin, consists of a 26-residue signal peptide, 9-residue propeptide in front of A chain, and a linker (12-residue) in between A and B chain. The signal peptide aids in translocation of ricin to ER inside the cell. During this process, the signal peptide is cleaved. At this stage, ricin protein, with propeptide and linker residues, is inactive. Further modifications in Golgi complex and eventually in vacuole by vacuolar enzymes resulted in the removal of the propeptide and linker to generate fully active ricin. Unlike RIP-II, very little is documented about the synthesis of RIP-I. However, multiple RIP-I proteins with insecticidal activity are already documented, i.e., momordin (from *Momordica charantia*), gelonin (from *Gelonium multiflorum*), pokeweed antiviral protein (PAP from *Phytolacca americana*), saporin (from *Saponaria officinalis*), etc. with insecticidal activity against fall armyworm (*S. frugiperda*) and velvetbean moth (*Anticarsia gemmatalis*) (Carlini and Grossi-de-Sá 2002).

Due to its entomotoxic potential, RIP protein is used for transgenic expression. For instance, transgenic tobacco plant (*N. tabacum*) expressing maize RIP (MRIP) showed resistance against *M. sexta and H. zea* (Dowd et al. 2003). Similarly, tobacco plant with expressed SNAI (*Sambucus nigra* agglutinin I) was resistant against tobacco aphid (*M. nicotianae*) and beet armyworm (*S. exigua*). Nevertheless, RIPs often shows toxicity to mammals and other nontarget organisms, which limit its potential use in transgenic plants (Virgilio et al. 2010).

3.1.2 Proteases

Proteases or peptidases are a particular class of enzymes with the ability to hydrolyze peptide bonds among the amino acids inside a polypeptide chain. These enzymes are ubiquitous and found in plants, animals, bacteria, virus, and archaea. Some plant proteases display entomotoxic potential against different target insect herbivores.

Moreover, ectopic administration of proteases within insects showed toxic effects (Harrison and Bonning 2010). Peritrophic membranes (PM) in insects are composed of chitin fibrils connected to proteoglycans and glycoproteins. Proteases targeted these proteoglycans and glycoproteins resulting disruption of the integrity of the PM that affects normal midgut digestive physiology and increases midgut susceptibility to entomotoxic molecules associated to food in target insects (Harrison and Bonning 2010). For instance, *S. frugiperda* feeding resulted in the production of a papain-like cysteine protease (Mir1-CP, 33-kDa) in maize inbred herbivore-resistant lines (Lopez et al. 2007). *S. frugiperda* larvae displayed growth retardation after rearing on calluses transformed with the Mir1-CP construct and exhibited altered PM morphology as observed under scanning electron microscopy (Jiang et al. 1995; Pechan et al. 2000, 2002).

Other cysteine proteases from the latex of papaya (*Carica papaya*) and a wild fig (*Ficus virgata*) also demonstrated insecticidal activity against various lepidopterans, namely, cabbage moth (*M. brassicae*) and tobacco cutworm (*S. litura*) (Konno et al. 2004). Microarray studies also revealed broad spectrum arthropod-inducible candidate proteases with insecticidal potential (Zhu-Salzman et al. 2008). Often plant proteases do not only cause toxicity but also enhance the toxicity of other insecticides (i.e., Bt toxins) by facilitating higher exposure to insect midgut epithelial cells through disruption of PM (Mohan et al. 2008). Hence, plant proteases are promising as well as unexplored candidates for future IPM applications (Fig. 6).

3.1.3 Chitinases

Chitinases are enzymes around 25 to 35 kDa molecular mass and cause hydrolysis of molecules with β-1, 4-linked N-acetylglucosamine residues (Cohen 1993; Nagpure et al. 2014). Based on the site of cleavage on the chitin molecule, plant chitinases can be of two types, exochitinase and endochitinase. However, structurally chitinases are classified into four groups (Collinge et al. 1993):

- Class I chitinases composed of the highly conserved structure with an N-terminal cysteine-rich domain.
- Class II chitinases have highly conserved structures close to class I chitinases but lacking the N-terminal cysteine-rich region.
- Class III chitinases shared low sequence homology with the class I and II chitinases but showed similar biochemical properties.
- Class IV chitinases are smaller proteins with fewer amino acids (45 to 60) than other classes of chitinases. It contains an N-terminal cysteine-rich region with some deletions.

Interestingly, some of the chitinases from plant origin showed insecticidal potential due to their competence of hydrolyzing the chitin molecules present at the extracellular layer of insect exoskeleton and peritrophic membrane (PM) in the gut. Chitinases termed as WIN6 extracted from poplar plants (*P. trichocarpa*) showed toxicity against Colorado potato beetle (*L. decemlineata*) when expressed in transgenic tomato plants (Lawrence and Novak 2006). Kitajima et al. (2010) found two chitinases (LA-a, LA-b) from mulberry (*Morus* sp.) showing toxicity

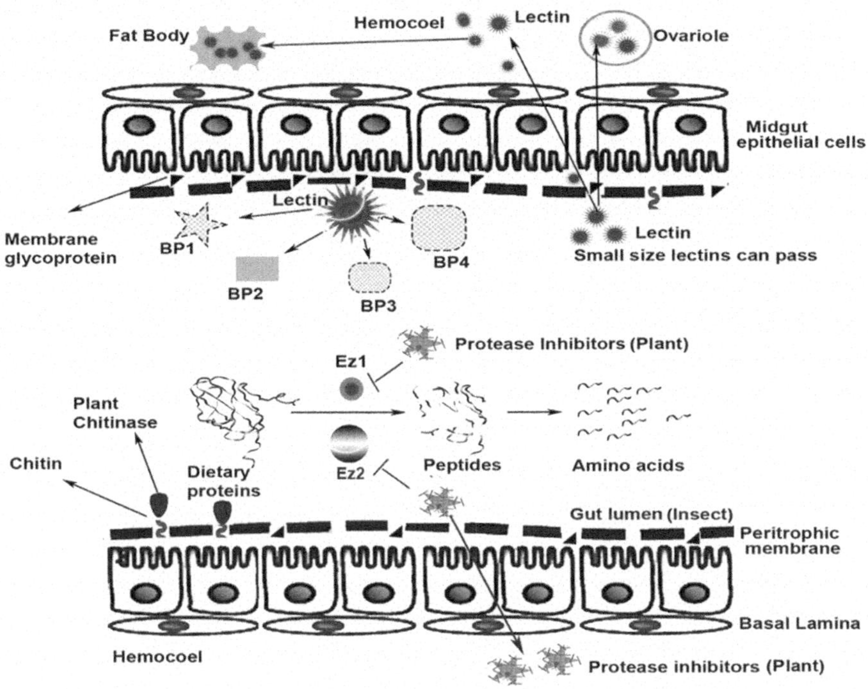

Fig. 6 Diagram of the insect gut (generic) showing the mechanism of action of lectins, plant protease inhibitors, and plant chitinases. Lectins can interact with different BPs to induce toxicity. Smaller size lectins can cross the midgut epithelial cells and reach to other target sites. Plant PIs inhibit different Ezs and thus affect insect digestion. They often reach to insect hemocoel, but their targets are unknown. Plant chitinases degrade chitins resided on the PM causing membrane damage. (*BP* represents binding partners of lectins such as sucrose, ferritin, etc. *Ez* represents insect midgut digestive enzymes such as aminopeptidase, cathepsin, or chymotrypsin)

against *D. melanogaster*. Such observations endorse the insecticidal potential of plant chitinases.

3.1.4 Ureases

Ureases are enzymes with metal ions (metalloenzymes), which breaks urea ultimately into ammonia and carbonic acid via formation of carbamate. Nearly 100 years ago, urease from jack bean seeds was the first protein to be crystallized and reported to have six identical 90.7 kDa chains (homohexamers) (Summer 1926). Ureases primarily found in fungi, bacteria, and plants. Ureases aid plants to exploit urea as nitrogen (N_2) source. It promotes seed germination via hydrolysis of the N_2 stored in the seeds (Fernanda Stanisçuaski and Carlini 2012). Interestingly, some plant ureases also aid in the defense against insect herbivores (Becker-Ritt et al. 2017). Other biological activities of urease include membrane permeabilization and fungicidal activities (Becker-Ritt and Carlini 2012).

Urease is not toxic until cathepsin-like enzymes hydrolyze it and form peptides. These urease-derived peptides are entomotoxic (Stanisçuaski and Carlini 2012). Hence, insects such as *R. prolixus* (kissing bug) and *C. maculatus* (cowpea weevil) who produce cathepsin-like enzymes in their gut are susceptible to ureases, whereas other insects (i.e., fruit fly, yellow fever mosquito, etc.) lacking cathepsin-like enzymes are not. The toxicity of JBURE-I (jack bean urease) solely relies on the discharge of the entomotoxic peptide, pepcanatox, by enzymatic action (Ferreira-DaSilva et al. 2000). Precisely, upon ingestion and eventually reaching to the midgut of the target insects, JBURE-I releases pepcanatox after enzymatic digestion. Entomotoxic pepcanatox is then transported via an ion channel-based mechanism to hemolymph and affect diuresis in Malpighian tubule (Piovesan et al. 2014; Stanisçuaski and Carlini 2012).

Based on the pepcanatox sequence, recombinant peptides such as jaburetox (JBTX) and jaburetox2Ec with entomotoxic and fungicidal activities were synthesized (Carlini and Ligabue-Braun 2016; Mulinari et al. 2007; Postal et al. 2012). The insecticidal activity of these recombinant peptides is high even in very low-dose application (0.01 or less w/w) comparing to other plant-derived entomotoxic proteins (Carlini and Ligabue-Braun 2016). Further studies revealed that JBTX peptide could adopt a β-hairpin structure at its C-terminal end similar to antimicrobial peptides with membrane pore-forming competence (F Mulinari et al. 2007). JBTX also reported to disrupt lipid membranes and form cation-selective ion channels (Barros et al. 2009; Piovesan et al. 2014). Further mutagenesis studies by Martinelli et al. (2014) demonstrated that the N-terminal region of the JBTX is vital for its entomotoxicity. Recently, *E.coli* cells expressing JBTX tested against *A. aegypti* showed induction of mortality in larvae and adults (Kappaun 2011). Similarly, injection of 0.1 μg of JBTX/mg of body weight into adult *T. infestans* (main vector of Chagas disease) resulted in the death of the insect within 24 h. Interestingly, insect started showing neurotoxic behavior such as uncoordinated leg movements, the irregular behavior of antennae, etc. just after 3 h of injection. It is found that JBTX binds insect neuronal cells and impedes the activity of the enzyme called nitric oxide synthase causing the decline of nitric oxide neurotransmitter inside the target insects. JBTX also interacts with UDP-N-acetylglucosamine pyrophosphorylase (UDP-GlcNAcP, an enzyme involved in chitin biosynthesis) and induces its activity in *T. infestans* and *D. peruvianus* (Galvani et al. 2015; Stanisçuaski et al. 2005). In kissing bug, JBTX interferes with the ability of the insect to raise an immune response against bacterial infection (Fruttero et al. 2016). Unlike the JBTX and JBURE-I, canatoxin jack bean urease isoform (homodimer of 95 kDa subunit) showed entomotoxicity against Hemipteran and Coleopteran insects (Carlini and Grossi-de-Sá 2002). Interestingly, canatoxin displayed toxicity against two major pests such as *D. peruvianus* and *N. viridula* who became resistant to some cry proteins and other applied chemical insecticides (Carlini and Grossi-de-Sá 2002; Carlini et al. 1997; Ferreira-DaSilva et al. 2000; Stanisçuaski and Carlini 2012). Soybean embryo-specific ureases also reported SBU to be more toxic to *D. peruvianus* (Follmer et al. 2004).

The high entomotoxic potential of JBTX automatically makes it a potential candidate for transgenic expression. Such options become more lucrative when biosafety studies indicate that the higher doses of JBURE-I and SBU do not induce lethality in mice and rats after intraperitoneal administration (Follmer et al. 2004). Mulinari (2008) made the first attempt at generating transgenic plants. Transgenic tobacco expressed different levels of JBTX that was generated and subsequently challenged against fall armyworm (*S. frugiperda*) resulting in higher resistance to the caterpillar attack. Similarly, transgenic sugarcane expressing JBTX when challenged with stem borer (*D. saccharalis*) displayed high resistance to caterpillar attack and induced mortality even up to 100% (Becker-Ritt et al. 2017). Undoubtedly, a plant expressing urease-derived entomotoxic peptides or JBTX holds enormous potential in the expansion of transgenic plants with pest resistance and aid to the formulation of effective biopesticides that are safe for other mammals.

3.2 Plant Protein Acts as Digestive Inhibitors

Insect digestion is localized in the midgut region where different enzymes are acting on food materials. Understandably, many enzymes such as amylases, proteases, etc. are identified in high concentration in the insect midgut. Often plant defense mechanisms are based on targeting such digestive enzymes from insect midgut. Blocking of digestive enzymes can induce a different degree of toxicity in insects and often can lead to death. Plant alpha-amylase inhibitors and protease inhibitors are the examples of entomotoxic enzymes that can disrupt normal digestive physiology of the target insects. In the subsequent section, we briefly discuss these two enzymes and their entomotoxic potential.

3.2.1 Alpha-Amylase Inhibitors

Alpha-amylase (α-A) catalyzes the hydrolysis of α-D-(1, 4) glucan linkages from complex carbohydrate molecule such as starch or glycogen (Franco et al. 2002). In insects, α-amylase aids breakdown of oligosaccharides, which then further digested by α-glucosidases leading the formation of glucose. Breakdown of glucose after glycolysis and TCA cycle produces energy, essential for all the activities of a living organism. Hence, inhibition of insect α-amylases by plant α-amylase inhibitor (α-AI) leads to insect growth retardation and ultimately death due to starvation (Kaur et al. 2014). Commonly α-AIs are reported in beans (white, red, black beans). Two isoforms of α-AI, namely, α-AI-1 and α-AI-2, with entomotoxicity to different target insects have been identified and characterized from common beans (Franco et al. 2002). Precisely, α-AI-1 inhibits the α-A from *C. chinensis* (pulse beetle) and *C. maculatus* (cowpea weevil) but cannot do so in *Z. subfasciatus* (Mexican bean weevil), whereas α-AI-2 shows entomotoxicity to only Mexican bean weevil (Feng et al. 1996; Silva et al. 2001). It can be an example of a plant's fine-tuning of defense response for specific enemies. Over the years, transgenic plants expressing only α-AI-1 or α-AI-1 plus Cry1Ac/b or Cry2Aa showed entomotoxicity against *H. armigera*, *B. pisorum* (pea weevil), and *H. hampei* (coffee borer beetle) (Acharjee and Sarmah 2013; Barbosa et al. 2010; Morton et al. 2000). Interestingly, transgenic chickpea plants expressing the other isoform (α-AI-2) also showed entomotoxicity against *B. pisorum* indicating that the target insects for both isoforms are often overlapping. Transgenic cowpea (*Vigna unguiculata*) expressing α-AI from kidney bean conferred resistance to bruchid beetles (*A. obtectus*) (Solleti et al. 2008). Recently, Luthi et al. (2018) documented that α-AI-1 from common bean expressed in chickpea seeds does not show any toxicity to Hymenopteran bruchid parasitoids, which are important natural enemies for cosmopolitan bruchids.

There are few other reports demonstrating toxic effects of α-AIs from cereal plant origin such as α-AI BIII from rye (*S. cereal*), PvCAI from kidney bean (*Phaseolus vulgaris*), α-AI from amaranth (*Amaranthus hypochondriacus*), α-AI from scarlet runner bean (*Phaseolus coccineus*), and α-AIs (i.e., wheat α-AI 0.19, wheat α-AI 0.28, WRP25, WRP 26,WRP27) against various pests (Dayler et al. 2005; Feng et al. 1996; Franco et al. 2002; Oliveira-Neto et al. 2003; Titarenko and Chrispeels 2000). Recently, the entomotoxic effect of α-AI from kidney bean was reported against an important grain storage pest, namely, the rice moth (*Corcyra cephalonica* Stainton) (Rani et al. 2018). These plant α-AIs are resistant to insect gut proteases and have high α-amylase inhibitory activity in low concentrations. Hence, transgenic plants expressing α-AIs hold potential for the future IPM. However, the inhibitory activity of α-AI against plant's endogenous α-amylases essential for germination and mammalian α-amylases needs to be tested before transgenic application of the entomotoxic α-AIs. Encouragingly, some reports already showed that α-AI BIII from rye and α-AI from amaranth were not inhibiting mammalian α-amylases (Chagolla-Lopez et al. 1994; Dias et al. 2005). However, the major challenge is the target pest resistance against α-AI. Insects showed the capacity to alter their digestive enzymes or detoxify the toxic plant components. Therefore, identification of new α-amylase inhibitors with the novel mechanism of action is crucial (Kaur et al. 2014).

3.2.2 Proteinase Inhibitors

Plant protease inhibitors (PIs) are recognized as plants own defense against insect herbivores. PIs are targeting insects' digestive system. They compete with the substrates for the binding to the active site of the insect proteases, thus alter insect digestive physiology and often cause insect death. Details of plants PI is already described in this book or elsewhere and in plant PIs database (Consiglio et al. 2011). Insect proteases such as cysteine (Cys), serine (Ser), aspartyl (Asp), and metalloproteinases (MP) are the prime target for PIs of plant origin. However, the bulk of plant PIs reported to date are serine PIs. Kunitz-type and Bowman-Birk inhibitors (BBIs) are two best-studied serine PIs with one and two active sites, respectively. Kuntiz PIs are 20 kDa proteins with low Cys content, whereas the molecular weight (MW) of BBIs is 9 kDa with high Cys content (Dang and Van Damme 2015).

Transgenic expression of PIs increases the GM plant resistance against target pests on numerous occasions. Transgenic GM plants with PIs such as *Oryza sativa* expressing Kunitz-type PI from *G. max*, PI-II from *S. tuberosum*, CpT1 from

V. unguiculata, and other PIs showed resistance against *N. lugens* (brown planthopper), *S. littoralis* (Egyptian cotton leafworm), *S. inferens* (Asiatic pink stem borer), *C. suppressalis* (striped rice stem borer), and so on (Alfonso-Rubí et al. 2003; Duan et al. 1996; Lee et al. 1999; Mochizuki et al. 1999; Xu et al. 1996). Transgenic tobacco, mustard, wheat, cotton, tomato, etc. expressing PIs were reported to show higher resistance against several notorious agricultural pests (Altpeter et al. 1999; Chen et al. 2005; Gatehouse et al. 1997; Kang et al. 2006; Li et al. 1998; Marchetti et al. 2000; Zhang et al. 2012). Nevertheless, rapid resistance development in pests jeopardizes the PI-mediated plant resistance to insect herbivores (Macedo et al. 2015; Zhu-Salzman and Zeng 2015).

Recently, transgene pyramiding has been applied to reduce the chances of resistance development and increase the efficiency of existing PIs. For instances, GM cotton expressing PI NaPI from tobacco and PI StPin1A from potato together delivered resistance to *H. armigera* (Dunse et al. 2010). Similarly, rice expressing PCI from tomato and MPI from maize showed higher resistance against striped rice stem borer (Quilis et al. 2014). Hence, better strategic use of PIs can have the potential for future IPM.

3.3 Plant Peptides: Small Molecules with Insecticidal Activity

Identification of peptides with entomotoxicity from plants has increased the possibilities for IPM (Da Silva et al. 2010). These peptides can serve as a valuable alternative to chemical pesticides through transgenic expression. However, understanding the structure and mechanism of entomotoxic action of the plant-derived peptides is one of the fundamental prerequisites for their transgenic application. In the following section, some of the well-known plant peptides with entomotoxicity are discussed, and their mode of action will also be explained briefly.

3.3.1 Cyclotides

Cyclotides, first reported from kalata-kalata (*O. affinis*, African medicinal plant) in the 1970s, are a small circular polypeptide with 28 to 37 amino acids. N and C terminal of cyclotide backbones are linked by peptide bond (Isaacs 1995). It also contains six conserved cysteine residues that together with the cyclical backbone form a stable structural motif called cyclic cysteine knot (Isaacs 1995). Based on the structural features (i.e., twist formation) and presence of cis-Pro-motif, cyclotides are classified into two subfamilies: Möbius and bracelet (Pelegrini et al. 2007). In plants, precursor proteins undergo cleavage and cyclization to form functional cyclotides. Most often cyclotides contain 1–3 domains (Jennings et al. 2001). More than 200 different types of cyclotides reported to date with functional activity such as antibacterial, antiviral, and insecticidal (Pelegrini et al. 2007). Cyclotides, namely, kalata B1, extracted from *O. affinis*, showed insecticidal activity against *H. punctigera* leading to growth retardation to the Kalata B1 supplemented diet-fed larvae. The mortality rate was almost 50% (Jennings et al. 2001). Interestingly, a diet amended with Kalata B1 or Kalata B2 (also extracted from *O. affinis*) inhibited the growth of *H. armigera* (Jennings et al. 2005). Furthermore, a diet supplemented with finotin, extracted from *Clitoria ternatea* (blue pea), caused 100% mortality to *A. obtectus* and *Z. subfasciatus* larvae (Kelemu et al. 2004). Paragidin-BR-1, cyclotide from *Palicourea rigida* (Brazilian Savannah Rubiaceae flower plant) caused up to 60% mortality *against Diatraea saccharalis* and showed significant efficacy against SF-9 cell line at a micromolar concentration (Pinto et al. 2012).

Interestingly, cyclotides are not affecting the digestive physiology of their target insects. They induce toxicity by causing physical damage to the midgut membrane leading disruption of normal absorption of nutrient through insect midgut. Kalata B1 was reported to form pores of 41–47 Å with ion channel activities (Huang et al. 2009). Light and electron microscopic studies with midgut tissues from kalata B1 supplemented diet-fed *H. armigera* larvae demonstrated the membrane disruption of midgut epithelial cells and pore formation that ultimately caused swelling and subsequent lysis of the cells (Barbeta et al. 2008).

3.3.2 Defensins

Plant defensins are made up of 40–55 amino acids with a molecular mass around 5 kDa. The 3D structure of defensins is composed of α-helix and three antiparallel β-sheets stabilized by 3–4 disulfide bridges. Plant defensins are isolated from root, stem, leaf, and endosperm tissue and displayed various biological activities including entomotoxicity (Lacerda et al. 2014). The basis underlying entomotoxicity solely relies on their capability of inhibiting the insect α-amylases and proteases similar to plant α-amylase inhibitor proteins. Plant defensins with entomotoxicity were first identified in *S. bicolor* (sorghum) displaying toxicity to *P. Americana* (cockroach) and *S. americana* (American grasshopper) via inhibition of gut α-amylases (Bloch and Richardson 1991). However, VrD1 from bruchid-resistant mung bean was the first reported defensins showing in vivo and in vitro entomotoxicity (Thevissen et al. 1999). *N. alata* defensin exhibited toxicity toward *H. armigera* and *H. punctigera* (Lay et al. 2003). Similarly, defensin extracted from *C. papaya* (papaya) inhibited α-amylase from *C. maculatus* (cowpea weevil). Interestingly, VrD1 defensin from cowpea exhibited toxicity against *Z. subfasciatus* and *A. obtectus* rather than *C. maculatus* (Pelegrini et al. 2008). Molecular modeling analysis revealed the occurrence of high-density surface anionic residues to facilitate the interaction with α-amylases and proteases (Liu et al. 2000).

Attempts were made to modify the insecticidal efficacy of defensins and to remove their interaction with mammalian digestive enzymes. The defensins TVD1, extracted from weedy legume *T. villosa,* were modified around β2-β3 loop region via in vitro mutagenesis leading the formation of α-TVD1. Such modification displayed higher insecticidal efficacy of α-TVD1 compared to wild-type against *T. molitor* (mealworm) (Vijayan et al. 2012). Recombinant VuD1 was entomotoxic to *C. maculatus* and no effect on mammalian enzymes similar to native VuD1 (dos Santos et al. 2010; Pelegrini et al. 2008).

Few studies have already demonstrated that transgenic expression plant defensins hold promise for future pest management. For instances, GM rice expressing BrD1 from *B. rapa* (turnip) exhibited increased resistance against *N. lugens* (Choi et al.

2009). Similarly, GM tobacco (*N. tabacum*) expressing TvD1 and NaD1 was resistant against *S. litura* and *H. armigera* attack, respectively (Lay et al. 2003; Vijayan et al. 2013). Hence, transgenic expression of plant defensins can also serve as a promising alternative to environmentally aggressive chemical pesticides. However, the nontarget effects of potential enterotoxins need to be evaluated meticulously.

4 Next-Generation GM Plants: Lessons from Lectin

Agricultural pests cause an estimation of 20–40% crop yield loss through various means such as direct damage during feeding and indirect damage by transmitting pathogenic organisms responsible for plant disease (FAO 2009). Diminishing the burden of insect pests from our agroecosystem is of an utmost priority now considering the rapidly increasing human population and the growing demand of food (FAO 2009; Ray et al. 2013). Frequent incidences of resistance development in target pest jeopardize the chemical pesticide-based pest management methods. Considering the growing consensus on the harmful impact of environmentally aggressive chemical pesticides, crop management strategies with higher potential for increased crop productivity without any ecological degradation consequences are on high demand. Biotechnological innovations can achieve the above-mentioned sustainable escalations (King 2017).

Reduction of crop resistance is one of the unfavorable outcomes of modern agriculture practices. Cultivated crop varieties are supplied with quality fertilizers, which, in turn, enhance their nutrient content and reduces their defensive abilities compared to their wild relative, thus favoring good source of foods for the insects (Olsen and Wendel 2013). Therefore, increase resistance to cultivated crop against pest insect is an ongoing challenge. Conventional breeding with marker-assisted selection held promise but limited within sexually compatible related plant species (Collard and Mackill 2008). Another considerable limitation of the conventional breeding program is that the trait of interest needs to be present in the same or within the related plant species. Nevertheless, a transgenic technology that deals with foreign entomotoxic protein expression in crops showed higher resistance against target pests until the pests became resistant. Therefore, novel strategies with reduced risk of resistance development need to be formulated for sustainable pest management. Targeted mutagenesis of entomotoxins can increase the efficacy against previously unaffected or resistant pests. For instance, transgenic cotton expressing the variant protein of Cry51Aa2, namely, Cry51Aa2.834_16 (generated by mutation), reduced the *Lygus* spp. population up to 30-fold in whole plant caged field trials (Gowda et al. 2016).

Alternatively, transgenic plants expressing more than one entomotoxins with diverse origin and receptor specificity displayed great potential. For instance, transgenic *Arabidopsis*, rice, and tobacco expressing a fusion of GNA (active against Hemiptera) lectin domain and scorpion neurotoxin domain (As1T, active against Lepidoptera) exhibited resistance against several hemipterans and lepidopteran pest insects (Liu et al. 2016). Similarly, spider toxin Hv1a fused with a luteovirus coat protein and GNA lectin displayed higher toxicity against aphids and lepidopteran, respectively (Bonning et al. 2014; Fitches et al. 2012). Transgenic rice lines expressing Cry1Ac-ASAL fusion protein impart higher resistance against leaf folder, yellow stem borer, brown plant hopper, and other major lepidopteran pests (Boddupally et al. 2018; Tajne et al. 2014). Furthermore, transgenic expression of ASAL-Hvt (ω-atracotoxin from *Hadronyche versuta*) exhibited 83% mortality against *Phenacoccus solenopsis* (mealybug), whereas each toxin alone can cause only 45% mortality (Javaid et al. 2018). Theoretically, plant R genes can be used for transgenic expression to increase plant defense. R genes can interact with insect Avr proteins for initiating hypersensitivity reactions in attacked plant tissues. However, the extreme specificity of R-Avr interaction confines the range of the target pests. Alternatively, genetic manipulation of harmful plant secondary metabolite pathways for increased production will be a more promising option. Recently, attempts are made to generate GM plants with increased expression of toxic small lipophilic molecules (secondary metabolites) and tannins (Barbehenn and Constabel 2011; Birkett and Pickett 2014).

With the advancement of RNAi technology, it is now feasible to generate GM plants expressing dsRNA targeting key functional genes in the target pests (Douglas 2018b; Price and Gatehouse 2008). Precisely, double-stranded RNA (dsRNA) was delivered to insect cells via oral feeding or feeding of leaves with expressed dsRNA or feeding of leaves with just sprayed dsRNA, where endogenous dicer enzymes cleave them to form small-interfering RNA (siRNA). Subsequently, siRNA aids degradation of target mRNA by guiding the Argonaute protein of RNA-induced silencing complex (RISC) (Scott et al. 2013). RNAi-mediated plant protection provides promising opportunity, which is much more flexible than using protein toxins. RNAi can be designed to be very specific to the target organisms eliminating risk for off-target effects.

The efficiency of RNAi can be increased by *in planta* expression of long hairpin RNA (hpRNA) in the chloroplast via tissue-specific expression promoter to eliminate the risk of possessing by plants' own RNAi machinery (Zhang et al. 2015). However, there is also a constraint of this technology. It does not work well against lepidopteran and hemipteran pests (Scott et al. 2013; Terenius et al. 2011). Recent studies pointed out that accumulation of dsRNA in the endosomes might be one of the reasons for RNAi inefficiency against lepidopterans. Precisely, the dsRNA is trapped inside the acidic bodies (early and late endosomes) that prevent them from further processing by dicer enzymes to siRNA (Yoon et al. 2017). Packaging the dsRNA inside nanoparticles will be an option to escape such problem like trapping in endosomes (Baker et al. 2017).

CRISPR (clustered regularly interspersed short palindromic repeats) gene-editing technology also expands the possibility of the introduction of resistance trait in crops (Georges and Ray 2017; Khatodia et al. 2017). Using CRISPR-CAS9 technology, now it is possible to generate resistance plants against insect pests that already attain resistance. It is possible to transform susceptible allele to resistance allele by introducing desired changes through the CRISPR-CAS9 system. Methods are already available for introducing any modification to all copies of target sequences, which makes CRISPR technology applicable to polyploidy crops. Recently, plants with resistance against insect-vectored geminiviruses are generated using CRISPR technology (Ali et al. 2016). However, there are some concerns related to virus escape due to viral genome mutation leading resistance to CRISPR-CAS9 cleavage (Mehta et al. 2018). Hence, care should be taken to minimize the risk of virus escape when using CRISPR-CAS9-based plant genome modification methods for gaining resistance.

5 Conclusion and Future Perspective

Growing demand for increased crop production demands an ever-greater need for drop protection against pest insects. With the development of plant biotechnology and in-depth knowledge of plant-insect interaction dynamics, it is now possible to adopt superior strategies to increase resistance against target pest without affecting the nontarget organisms. However, such strategies need to take into account various aspects of agricultural practices including crop production schedule, native and invasive pests, and natural enemies. Interestingly, monoculture practices also aid in crop destruction by pests. Global warming parameters are also crucial to the sustainability of the control strategy. Nevertheless, improved genetic technologies such as transgenic, tissue-specific, selectable marker-free expression of entomotoxic proteins such as lectins and others from dietary sources targeting specific pests show massive potential for future crop improvements (Bala et al. 2013a, b; Puchta 2003). Pyramiding diverse entomotoxins with the different mode of action can help in developing GM plants with broad-spectrum resistance against lepidopteran and hemipteran pests (Carrière et al. 2015).

Furthermore, RNAi and state-of-the-art CRISPR-CAS9 technology brings superior ways for eco-friendly crop protection and will undoubtedly aid in the elimination of environmentally aggressive chemicals or pesticides from agroecosystem. RNAi encapsulation within nanoparticles and their controlled release at the target site can make Lepidopteran and Hemipteran insects susceptible to RNAi treatment. Genetically modified plants with engineered secondary metabolite profile are also an emerging alternative. Recent advancement of metagenomics and metatranscriptomics highlight the contribution of the microbiome in shaping up the plant-insect interaction (Douglas 2018a; García-Fraile 2018). They can serve as a potential candidate for future pest management. Besides, members of insect gut microbial communities can also be engineered to deliver target RNAi. However, we need to be careful about the field applications of these new possibilities. Each of them has few limitations that need to overwhelm in advance. The potential nontarget effects, its environmental impact, and allergenicity potential to mammals need thorough evaluation. Finally, yet importantly, we need to accept the reality that insects are under continuous selection pressure to overcome or break the applied control measures. Hence, knowledge-based strategic use of available and upcoming technologies will be the key to winning the battle against insects in the future.

Acknowledgments AR would like to acknowledge the EU project "EXTEMIT – K," No. CZ.02.1.01/0.0/0.0/15_003/0000433 financed by OP RDE, outstanding output project and IGA from Faculty of Forestry and Wood Sciences, Czech University of Life Sciences Prague, for the financial support during the preparation of the book chapter. AC is supported by "EVA4.0," No. CZ.02.1.01 /0.0 /0.0 /16_ 019/0000803 financed by OP RDE. Prof. Sampa Das lab, Bose Institute, Kolkata, India is acknowledged for their contribution in the field of plant lectins and their mode of action, which inspires writing this book chapter.

References

Acharjee S, Sarmah BK (2013) Biotechnologically generating 'super chickpea' for food and nutritional security. Plant Sci 207:108–116

Al Atalah B, De Vleesschauwer D, Xu J, Fouquaert E, Höfte M, Van Damme EJ (2014) Transcriptional behavior of EUL-related rice lectins toward important abiotic and biotic stresses. J Plant Physiol 171(12):986–992

Alfonso-Rubí J, Ortego F, Castañera P, Carbonero P, Díaz I (2003) Transgenic expression of trypsin inhibitor CMe from barley in indica and japonica rice, confers resistance to the rice weevil Sitophilus oryzae. Transgenic Res 12(1):23–31

Ali Z, Ali S, Tashkandi M, Zaidi SS-E-A, Mahfouz MM (2016) CRISPR/Cas9-mediated immunity to geminiviruses: differential interference and evasion. Sci Rep 6:26912

Altpeter F, Diaz I, McAuslane H, Gaddour K, Carbonero P, Vasil IK (1999) Increased insect resistance in transgenic wheat stably expressing trypsin inhibitor CMe. Mol Breed 5(1):53–63

Andrews RE, Faust RM, Wabiko H, Raymond KC, Bulla LA (1987) The biotechnology of Bacillus thuringiensis. Crit Rev Biotechnol 6(2):163–232

Baker S, Volova T, Prudnikova SV, Satish S, Prasad N (2017) Nanoagroparticles emerging trends and future prospect in modern agriculture system. Environ Toxicol Pharmacol 53:10–17

Bala A, Roy A, Behura N, Hess D, Das S (2013a) Insight to the mode of action of Allium sativum leaf agglutinin (ASAL) expressing in T3 rice lines on brown planthopper. Am J Plant Sci 4 (02):400

Bala A, Roy A, Das A, Chakraborti D, Das S (2013b) Development of selectable marker free, insect resistant, transgenic mustard (Brassica juncea) plants using Cre/lox mediated recombination. BMC Biotechnol 13(1):88

Balzarini J, Schols D, Neyts J, Van Damme E, Peumans W, De Clercq E (1991) Alpha-(1-3)-and alpha-(1-6)-D-mannose-specific plant lectins are markedly inhibitory to human immunodeficiency virus and cytomegalovirus infections in vitro. Antimicrob Agents Chemother 35 (3):410–416

Bandyopadhyay S, Roy A, Das S (2001) Binding of garlic (Allium sativum) leaf lectin to the gut receptors of homopteran pests is correlated to its insecticidal activity. Plant Sci 161 (5):1025–1033

Banerjee N, Sengupta S, Roy A, Ghosh P, Das K, Das S (2011) Functional alteration of a dimeric insecticidal lectin to a monomeric antifungal protein correlated to its oligomeric status. PLoS One 6(4):e18593

Barbehenn RV, Constabel CP (2011) Tannins in plant–herbivore interactions. Phytochemistry 72 (13):1551–1565

Barbeta BL, Marshall AT, Gillon AD, Craik DJ, Anderson MA (2008) Plant cyclotides disrupt epithelial cells in the midgut of lepidopteran larvae. Proc Natl Acad Sci 105(4):1221–1225

Barbieri L, Polito L, Bolognesi A, Ciani M, Pelosi E, Farini V et al (2006) Ribosome-inactivating proteins in edible plants and purification and characterization of a new ribosome-inactivating protein from Cucurbita moschata. Biochim Biophys Acta Gen Subj 1760(5):783–792

Barbosa AE, Albuquerque ÉV, Silva MC, Souza DS, Oliveira-Neto OB, Valencia A et al (2010) α-Amylase inhibitor-1 gene from Phaseolus vulgaris expressed in Coffea arabica plants inhibits α-amylases from the coffee berry borer pest. BMC Biotechnol 10(1):44
Barros PR, Stassen H, Freitas MS, Carlini CR, Nascimento MA, Follmer C (2009) Membrane-disruptive properties of the bioinsecticide Jaburetox-2Ec: implications to the mechanism of the action of insecticidal peptides derived from ureases. Biochim Biophys Acta Protein Proteom 1794(12):1848–1854
Becker-Ritt AB, Carlini CR (2012) Fungitoxic and insecticidal plant polypeptides. Pept Sci 98 (4):367–384
Becker-Ritt AB, Portugal CS, Carlini CR (2017) Jaburetox: update on a urease-derived peptide. J Venomous Anim Toxins incl Trop Dis 23(1):32
Bell H, Fitches E, Marris G, Bell J, Edwards J, Gatehouse J, Gatehouse A (2001) Transgenic GNA expressing potato plants augment the beneficial biocontrol of Lacanobia oleracea (Lepidoptera: Noctuidae) by the parasitoid Eulophus pennicornis (Hymenoptera; Eulophidae). Transgenic Res 10(1):35–42
Bell H, Kirkbride-Smith A, Marris G, Edwards J, Gatehouse A (2004) Oral toxicity and impact on fecundity of three insecticidal proteins on the gregarious ectoparasitoid Eulophus pennicornis (Hymenoptera: Eulophidae). Agric For Entomol 6(3):215–222
Birch ANE, Geoghegan IE, Majerus ME, McNicol JW, Hackett CA, Gatehouse AM, Gatehouse JA (1999) Tri-trophic interactions involving pest aphids, predatory 2-spot ladybirds and transgenic potatoes expressing snowdrop lectin for aphid resistance. Mol Breed 5(1):75–83
Birkett MA, Pickett JA (2014) Prospects of genetic engineering for robust insect resistance. Curr Opin Plant Biol 19:59–67
Bloch C, Richardson M (1991) A new family of small (5 kDa) protein inhibitors of insect α-amylases from seeds or sorghum (Sorghum bicolor (L) Moench) have sequence homologies with wheat γ-purothionins. FEBS Lett 279(1):101–104
Boddupally D, Tamirisa S, Gundra SR, Vudem DR, Khareedu VR (2018) Expression of hybrid fusion protein (Cry1Ac:: ASAL) in transgenic rice plants imparts resistance against multiple insect pests. Sci Rep 8(1):8458
Bonning BC, Pal N, Liu S, Wang Z, Sivakumar S, Dixon PM et al (2014) Toxin delivery by the coat protein of an aphid-vectored plant virus provides plant resistance to aphids. Nat Biotechnol 32 (1):102
Boulter D, Edwards GA, Gatehouse AM, Gatehouse JA, Hilder VA (1990) Additive protective effects of different plant-derived insect resistance genes in transgenic tobacco plants. Crop Prot 9(5):351–354
Boyd WC, Shapleigh E (1954) Antigenic relations of blood group antigens as suggested by tests with lectins. J Immunol 73(4):226–231
Brown JC, Hunt RC (1978) Lectins. In: International review of cytology, vol 52. Elsevier, pp 277–349
Caccia S, Van Damme EJ, De Vos WH, Smagghe G (2012) Mechanism of entomotoxicity of the plant lectin from Hippeastrum hybrid (Amaryllis) in Spodoptera littoralis larvae. J Insect Physiol 58(9):1177–1183
Camaroti JRSL, de Almeida WA, do Rego Belmonte B, de Oliveira APS, de Albuquerque Lima T, Ferreira MRA et al (2018) Sitophilus zeamais adults have survival and nutrition affected by Schinus terebinthifolius leaf extract and its lectin (SteLL). Ind Crop Prod 116:81–89
Carlini CR, Grossi-de-Sá MF (2002) Plant toxic proteins with insecticidal properties. A review on their potentialities as bioinsecticides. Toxicon 40(11):1515–1539
Carlini CR, Ligabue-Braun R (2016) Ureases as multifunctional toxic proteins: a review. Toxicon 110:90–109
Carlini CR, Oliveira AE, Azambuja P, Xavier-Filho J, Wells MA (1997) Biological effects of canatoxin in different insect models: evidence for a proteolytic activation of the toxin by insect cathepsin-like enzymes. J Econ Entomol 90(2):340–348
Carrière Y, Crickmore N, Tabashnik BE (2015) Optimizing pyramided transgenic Bt crops for sustainable pest management. Nat Biotechnol 33(2):161

Chagolla-Lopez A, Blanco-Labra A, Patthy A, Sánchez R, Pongor S (1994) A novel alpha-amylase inhibitor from amaranth (Amaranthus hypochondriacus) seeds. J Biol Chem 269 (38):23675–23680

Chakraborti D, Sarkar A, Mondal HA, Das S (2009) Tissue specific expression of potent insecticidal, Allium sativum leaf agglutinin (ASAL) in important pulse crop, chickpea (Cicer arietinum L.) to resist the phloem feeding Aphis craccivora. Transgenic Res 18(4):529–544

Chandra NR, Ramachandraiah G, Bachhawat K, Dam TK, Surolia A, Vijayan M (1999) Crystal structure of a dimeric mannose-specific agglutinin from garlic: quaternary association and carbohydrate specificity. J Mol Biol 285(3):1157–1168

Chandrasekhar K, Vijayalakshmi M, Vani K, Kaul T, Reddy MK (2014) Phloem-specific expression of the lectin gene from Allium sativum confers resistance to the sap-sucker Nilaparvata lugens. Biotechnol Lett 36(5):1059–1067

Chang T, Chen L, Chen S, Cai H, Liu X, Xiao G, Zhu Z (2003) Transformation of tobacco with genes encoding Helianthus tuberosus agglutinin (HTA) confers resistance to peach-potato aphid (Myzus persicae). Transgenic Res 12(5):607–614

Chen MS (2008) Inducible direct plant defense against insect herbivores: a review. Insect Sci 15 (2):101–114

Chen Y, Peumans WJ, Hause B, Bras J, Kumar M, Proost P et al (2002) Jasmonic acid methyl ester induces the synthesis of a cytoplasmic/nuclear chito-oligosaccharide binding lectin in tobacco leaves. FASEB J 16(8):905–907

Chen H, Wilkerson CG, Kuchar JA, Phinney BS, Howe GA (2005) Jasmonate-inducible plant enzymes degrade essential amino acids in the herbivore midgut. Proc Natl Acad Sci 102 (52):19237–19242

Chen H, Gonzales-Vigil E, Wilkerson CG, Howe GA (2007) Stability of plant defense proteins in the gut of insect herbivores. Plant Physiol 143(4):1954–1967

Choi M-S, Kim Y-H, Park H-M, Seo B-Y, Jung J-K, Kim S-T et al (2009) Expression of BrD1, a plant defensin from Brassica rapa, confers resistance against brown planthopper (Nilaparvata lugens) in transgenic Rices. Mol Cells 28(2):131–137

Chougule NP, Bonning BC (2012) Toxins for transgenic resistance to hemipteran pests. Toxins 4 (6):405–429

Cohen E (1993) Chitin synthesis and degradation as targets for pesticide action. Arch Insect Biochem Physiol 22(1–2):245–261

Collard BC, Mackill DJ (2008) Marker-assisted selection: an approach for precision plant breeding in the twenty-first century. Philos Trans R Soc Lond B: Biol Sci 363(1491):557–572

Collinge DB, Kragh KM, Mikkelsen JD, Nielsen KK, Rasmussen U, Vad K (1993) Plant chitinases. Plant J 3(1):31–40

Consiglio A, Grillo G, Licciulli F, Ceci LR, Liuni S, Losito N et al (2011) PlantPIs-an interactive web resource on plant protease inhibitors. Curr Protein Pept Sci 12(5):448–454

Couty A, Poppy GM (2001) Does host-feeding on GNA-intoxicated aphids by Aphelinus abdominalis affect their longevity and/or fecundity? Entomol Exp Appl 100(3):331–337

Cristofoletti PT, Mendonça de Sousa FA, Rahbe Y, Terra WR (2006) Characterization of a membrane-bound aminopeptidase purified from Acyrthosiphon pisum midgut cells. FEBS J 273(24):5574–5588

Da Silva P, Rahioui I, Laugier C, Jouvensal L, Meudal H, Chouabe C et al (2010) Molecular requirements for the insecticidal activity of the plant peptide pea albumin 1 subunit b (PA1b). J Biol Chem 285(43):32689–32694

Dang L, Van Damme EJ (2015) Toxic proteins in plants. Phytochemistry 117:51–64

Das A, Ghosh P, Das S (2018) Expression of Colocasia esculenta tuber agglutinin in Indian mustard provides resistance against Lipaphis erysimi and the expressed protein is non-allergenic. Plant Cell Rep 37(6):849–863

Dayler CS, Mendes PA, Prates MV, Bloch C, Franco OL, Grossi-de-Sá MF (2005) Identification of a novel bean α-amylase inhibitor with chitinolytic activity. FEBS Lett 579(25):5616–5620

De Oliveira CFR, Luz LA, Paiva PMG, Coelho LCBB, Marangoni S, Macedo MLR (2011) Evaluation of seed coagulant Moringa oleifera lectin (cMoL) as a bioinsecticidal tool with potential for the control of insects. Process Biochem 46(2):498–504

del Carmen F-AM, Diaz D, Berbis M, Marcelo F, Cañada J, Jiménez-Barbero J (2012) Protein-carbohydrate interactions studied by NMR: from molecular recognition to drug design. Curr Protein Pept Sci 13(8):816

Dias SC, Franco OL, Magalhaes CP, de Oliveira-Neto OB, Laumann RA, Figueira EL et al (2005) Molecular cloning and expression of an α-amylase inhibitor from rye with potential for controlling insect pests. Protein J 24(2):113–123

dos Santos IS, Carvalho ADO, de Souza-Filho GA, do Nascimento VV, Machado OL, Gomes VM (2010) Purification of a defensin isolated from Vigna unguiculata seeds, its functional expression in Escherichia coli, and assessment of its insect α-amylase inhibitory activity. Protein Expr Purif 71(1):8–15

Douglas AE (2018a) Omics and the metabolic function of insect-microbial symbioses. Curr Opin Insect Sci 29:1–16

Douglas AE (2018b) Strategies for enhanced crop resistance to insect pests. Annu Rev Plant Biol 69:637–660

Dowd PF, Zuo W-N, Gillikin JW, Johnson ET, Boston RS (2003) Enhanced resistance to Helicoverpa zea in tobacco expressing an activated form of maize ribosome-inactivating protein. J Agric Food Chem 51(12):3568–3574

Down RE, Ford L, Woodhouse SD, Raemaekers RJ, Leitch B, Gatehouse JA, Gatehouse AM (2000) Snowdrop lectin (GNA) has no acute toxic effects on a beneficial insect predator, the 2-spot ladybird (Adalia bipunctata L.). J Insect Physiol 46(4):379–391

Down RE, Ford L, Woodhouse SD, Davison GM, Majerus ME, Gatehouse JA, Gatehouse AM (2003) Tritrophic interactions between transgenic potato expressing snowdrop lectin (GNA), an aphid pest (peach–potato aphid; Myzus persicae (Sulz.) and a beneficial predator (2-spot ladybird, Adalia bipunctata L.)). Transgenic Res 12(2):229–241

Du J, Foissac X, Carss A, Gatehouse AM, Gatehouse JA (2000) Ferritin acts as the most abundant binding protein for snowdrop lectin in the midgut of rice brown planthoppers (Nilaparvata lugens). Insect Biochem Mol Biol 30(4):297–305

Duan X, Li X, Xue Q, Abo-EI-Saad M, Xu D, Wu R (1996) Transgenic rice plants harboring an introduced potato proteinase inhibitor II gene are insect resistant. Nat Biotechnol 14(4):494

Dunse K, Stevens J, Lay F, Gaspar Y, Heath R, Anderson M (2010) Coexpression of potato type I and II proteinase inhibitors gives cotton plants protection against insect damage in the field. Proc Natl Acad Sci 107(34):15011–15015

Dutta I, Saha P, Majumder P, Sarkar A, Chakraborti D, Banerjee S, Das S (2005) The efficacy of a novel insecticidal protein, Allium sativum leaf lectin (ASAL), against homopteran insects monitored in transgenic tobacco. Plant Biotechnol J 3(6):601–611

Ewen SW, Pusztai A (1999) Effect of diets containing genetically modified potatoes expressing Galanthus nivalis lectin on rat small intestine. Lancet 354(9187):1353–1354

FAO (2009) How to feed the world 2050: high-level expert forum. Available at: fao.org/fileadmin/templates/wsfs/docs/Issues_papers/HLEF2050_Global_Agriculture.pdf. Accessed on 10 Nov 2014

FAO/WHO (2001) Human vitamin and mineral requirements. Report of a Joint FAO/WHO Expert Consultation, Bangkok, Thailand. Food and Nutrition Division, FAO, Rome, pp 235–247

Felton GW (2005) Indigestion is a plant's best defense. Proc Natl Acad Sci U S A 102 (52):18771–18772

Feng GH, Richardson M, Chen MS, Kramer KJ, Morgan TD, Reeck GR (1996) α-Amylase inhibitors from wheat: amino acid sequences and patterns of inhibition of insect and human α-amylases. Insect Biochem Mol Biol 26(5):419–426

Ferreira-DaSilva CT, Gombarovits MEC, Masuda H, Oliveira CM, Carlini CR (2000) Proteolytic activation of canatoxin, a plant toxic protein, by insect cathepsin-like enzymes. Arch Insect Biochem Physiol 44(4):162–171

Fitches E, Gatehouse AM, Gatehouse JA (1997) Effects of snowdrop lectin (GNA) delivered via artificial diet and transgenic plants on the development of tomato moth (Lacanobia oleracea) larvae in laboratory and glasshouse trials. J Insect Physiol 43(8):727–739

Fitches E, Ilett C, Gatehouse A, Gatehouse L, Greene R, Edwards J, Gatehouse J (2001a) The effects of Phaseolus vulgaris erythro-and leucoagglutinating isolectins (PHA-E and PHA-L) delivered via artificial diet and transgenic plants on the growth and development of tomato moth (Lacanobia oleracea) larvae; lectin binding to gut glycoproteins in vitro and in vivo. J Insect Physiol 47(12):1389–1398

Fitches E, Woodhouse SD, Edwards JP, Gatehouse JA (2001b) In vitro and in vivo binding of snowdrop (Galanthus nivalis agglutinin; GNA) and jackbean (Canavalia ensiformis; Con A) lectins within tomato moth (Lacanobia oleracea) larvae; mechanisms of insecticidal action. J Insect Physiol 47(7):777–787

Fitches E, Wiles D, Douglas AE, Hinchliffe G, Audsley N, Gatehouse JA (2008) The insecticidal activity of recombinant garlic lectins towards aphids. Insect Biochem Mol Biol 38(10):905–915

Fitches EC, Pyati P, King GF, Gatehouse JA (2012) Fusion to snowdrop lectin magnifies the oral activity of insecticidal ω-hexatoxin-Hv1a peptide by enabling its delivery to the central nervous system. PLoS One 7(6):e39389

Foissac X, Loc NT, Christou P, Gatehouse AM, Gatehouse JA (2000) Resistance to green leafhopper (Nephotettix virescens) and brown planthopper (Nilaparvata lugens) in transgenic rice expressing snowdrop lectin (Galanthus nivalis agglutinin; GNA). J Insect Physiol 46 (4):573–583

Follmer C, Real-Guerra R, Wasserman GE, Olivera-Severo D, Carlini CR (2004) Jackbean, soybean and Bacillus pasteurii ureases: biological effects unrelated to ureolytic activity. Eur J Biochem 271(7):1357–1363

Franco OL, Rigden DJ, Melo FR, Grossi-de-Sá MF (2002) Plant α-amylase inhibitors and their interaction with insect α-amylases: structure, function and potential for crop protection. Eur J Biochem 269(2):397–412

Fruttero LL, Moyetta NR, Uberti AF, Grahl MVC, Lopes FC, Broll V et al (2016) Humoral and cellular immune responses induced by the urease-derived peptide Jaburetox in the model organism Rhodnius prolixus. Parasit Vectors 9(1):412

Gaidamashvili M, Ohizumi Y, Iijima S, Takayama T, Ogawa T, Muramoto K (2004) Characterization of the yam tuber storage proteins from Dioscorea batatas exhibiting unique lectin activities. J Biol Chem 279(25):26028–26035

Galvani GL, Fruttero LL, Coronel MF, Nowicki S, Demartini DR, Defferrari MS et al (2015) Effect of the urease-derived peptide Jaburetox on the central nervous system of Triatoma infestans (Insecta: Heteroptera). Biochim Biophys Acta (BBA)-Gen Sub 1850(2):255–262

García-Fraile P (2018) Roles of bacteria in the bark beetle holobiont–how do they shape this forest pest? Ann Appl Biol 172(2):111–125

Garcia-Pino A, Buts L, Wyns L, Imberty A, Loris R (2007) How a plant lectin recognizes high mannose oligosaccharides. Plant Physiol 144(4):1733–1741

Gatehouse AM, Davison GM, Newell CA, Merryweather A, Hamilton WD, Burgess EP et al (1997) Transgenic potato plants with enhanced resistance to the tomato moth, Lacanobia oleracea: growth room trials. Mol Breed 3(1):49–63

Gatehouse AM, Davison GM, Stewart JN, Gatehouse LN, Kumar A, Geoghegan IE et al (1999) Concanavalin A inhibits development of tomato moth (Lacanobia oleracea) and peach-potato aphid (Myzus persicae) when expressed in transgenic potato plants. Mol Breed 5(2):153–165

George BS, Silambarasan S, Senthil K, Jacob JP, Dasgupta MG (2018) Characterization of an insecticidal protein from Withania somnifera against lepidopteran and hemipteran Pest. Mol Biotechnol 60(4):290–301

Georges F, Ray H (2017) Genome editing of crops: a renewed opportunity for food security. GM Crops Food 8(1):1–12

Ghazarian H, Idoni B, Oppenheimer SB (2011) A glycobiology review: carbohydrates, lectins and implications in cancer therapeutics. Acta Histochem 113(3):236–247

Ghosh P, Roy A, Chakraborty J, Das S (2013) Biological safety assessment of mutant variant of Allium sativum leaf agglutinin (mASAL), a novel antifungal protein for future transgenic application. J Agric Food Chem 61(48):11858–11864
Gowda A, Rydel TJ, Wollacott AM, Brown RS, Akbar W, Clark TL et al (2016) A transgenic approach for controlling Lygus in cotton. Nat Commun 7:12213
Grimaldi D, Engel MS, Engel MS (2005) Evolution of the insects. Cambridge University Press, Cambridge
Grossi-de-Sá MF, Pelegrini PB, Vasconcelos IM, Carlini CR, Silva MS (2017) Entomotoxic plant proteins: potential molecules to develop genetically modified plants resistant to insect-pests. Plant Toxins:415–447
Guo P, Wang Y, Zhou X, Xie Y, Wu H, Gao X (2013) Expression of soybean lectin in transgenic tobacco results in enhanced resistance to pathogens and pests. Plant Sci 211:17–22
Gupta S, Das S (2012) Exploring the defensive roles and regulations of GNA domain containing monocot mannose specific lectins. Sci Cult 78:233–241
Harper M, Hopkins T, Czapla T (1998) Effect of wheat germ agglutinin on formation and structure of the peritrophic membrane in European corn borer (Ostrinia nubilalis) larvae. Tissue Cell 30 (2):166–176
Harrison RL, Bonning BC (2010) Proteases as insecticidal agents. Toxins 2(5):935–953
Hegedus D, Erlandson M, Gillott C, Toprak U (2009) New insights into peritrophic matrix synthesis, architecture, and function. Annu Rev Entomol 54:285–302
Hester G, Kaku H, Goldstein IJ, Wright CS (1995) Structure of mannose-specific snowdrop (Galanthus nivalis) lectin is representative of a new plant lectin family. Nat Struct Biol 2 (6):472–479
Hilder V, Powell K, Gatehouse A, Gatehouse J, Gatehouse L, Shi Y et al (1995) Expression of snowdrop lectin in transgenic tobacco plants results in added protection against aphids. Transgenic Res 4(1):18–25
Hogervorst PA, Ferry N, Gatehouse AM, Wäckers FL, Romeis J (2006) Direct effects of snowdrop lectin (GNA) on larvae of three aphid predators and fate of GNA after ingestion. J Insect Physiol 52(6):614–624
Huang Y-H, Colgrave ML, Daly NL, Keleshian A, Martinac B, Craik DJ (2009) The biological activity of the prototypic cyclotide kalata b1 is modulated by the formation of multimeric pores. J Biol Chem. jbc. M109. 003384
Isaacs NW (1995) Cystine knots. Curr Opin Struct Biol 5(3):391–395
Javaid S, Naz S, Amin I, Jander G, Ul-Haq Z, Mansoor S (2018) Computational and biological characterization of fusion proteins of two insecticidal proteins for control of insect pests. Sci Rep 8(1):4837
Jennings C, West J, Waine C, Craik D, Anderson M (2001) Biosynthesis and insecticidal properties of plant cyclotides: the cyclic knotted proteins from Oldenlandia affinis. Proc Natl Acad Sci 98 (19):10614–10619
Jennings CV, Rosengren KJ, Daly NL, Plan M, Stevens J, Scanlon MJ et al (2005) Isolation, solution structure, and insecticidal activity of kalata B2, a circular protein with a twist: do Möbius strips exist in nature? Biochemistry 44(3):851–860
Jiang B, Siregar U, Willeford KO, Luthe DS, Williams WP (1995) Association of a 33-kilodalton cysteine proteinase found in corn callus with the inhibition of fall armyworm larval growth. Plant Physiol 108(4):1631–1640
Kang J-H, Wang L, Giri A, Baldwin IT (2006) Silencing threonine deaminase and JAR4 in Nicotiana attenuata impairs jasmonic acid–isoleucine–mediated defenses against Manduca sexta. Plant Cell 18(11):3303–3320
Kanrar S, Venkateswari J, Kirti P, Chopra V (2002) Transgenic Indian mustard (Brassica juncea) with resistance to the mustard aphid (Lipaphis erysimi Kalt.). Plant Cell Rep 20(10):976–981. https://doi.org/10.1007/s00299-001-0422-z
Kappaun K (2011) Estudos com o Jaburetox: efeito tóxico de E. coli liofilizadas carregadas com o peptídeo e análise da influencia do epitopo V5 na formação de agregados

Kaur R, Kaur N, Gupta AK (2014) Structural features, substrate specificity, kinetic properties of insect α-amylase and specificity of plant α-amylase inhibitors. Pestic Biochem Physiol 116:83–93

Kelemu S, Cardona C, Segura G (2004) Antimicrobial and insecticidal protein isolated from seeds of Clitoria ternatea, a tropical forage legume. Plant Physiol Biochem 42(11):867–873

Khatodia S, Bhatotia K, Tuteja N (2017) Development of CRISPR/Cas9 mediated virus resistance in agriculturally important crops. Bioengineered 8(3):274–279

King A (2017) The future of agriculture. Nature 544(7651):S21–S23

Kitajima S, Kamei K, Taketani S, Yamaguchi M, Kawai F, Komatsu A, Inukai Y (2010) Two chitinase-like proteins abundantly accumulated in latex of mulberry show insecticidal activity. BMC Biochem 11(1):6

Konno K, Hirayama C, Nakamura M, Tateishi K, Tamura Y, Hattori M, Kohno K (2004) Papain protects papaya trees from herbivorous insects: role of cysteine proteases in latex. Plant J 37 (3):370–378

Kumar S, Verma AK, Sharma A, Kumar D, Tripathi A, Chaudhari B et al (2013) Phytohemagglutinins augment red kidney bean (Phaseolus vulgaris L.) induced allergic manifestations. J Proteome 93:50–64

Lacerda A, Vasconcelos ÉAR, Pelegrini PB, Grossi-de-Sa MF (2014) Antifungal defensins and their role in plant defense. Front Microbiol 5:116

Lagarda-Diaz I, Guzman-Partida AM, Urbano-Hernandez G, Ortega-Nieblas MM, Robles-Burgueño MR, Winzerling J, Vazquez-Moreno L (2008) Insecticidal action of PF2 lectin from Olneya tesota (Palo Fierro) against Zabrotes subfasciatus larvae and midgut glycoconjugate binding. J Agric Food Chem 57(2):689–694

Landsteiner K (1990) The specificity of serological reactions. Courier Corporation, New York

Lawrence SD, Novak NG (2006) Expression of poplar chitinase in tomato leads to inhibition of development in Colorado potato beetle. Biotechnol Lett 28(8):593–599

Lay FT, Brugliera F, Anderson MA (2003) Isolation and properties of floral defensins from ornamental tobacco and petunia. Plant Physiol 131(3):1283–1293

Lee SI, Lee S-H, Koo JC, Chun HJ, Lim CO, Mun JH et al (1999) Soybean Kunitz trypsin inhibitor (SKTI) confers resistance to the brown planthopper (Nilaparvata lugens Stål) in transgenic rice. Mol Breed 5(1):1–9

Lehrman A (2007) Does pea lectin expressed transgenically in oilseed rape (Brassica napus) influence honey bee (Apis mellifera) larvae? Environ Biosaf Res 6(4):271–278

Li Y, Chen Z, Wu X, Li S, Jiao G, Wu J et al (1998) Obtaining transgenic cotton plants with cowpea trypsin inhibitor gene. Acta Gossypii Sinica 10(5):237–243

Li HM, Sun L, Mittapalli O, Muir W, Xie J, Wu J et al (2009) Transcriptional signatures in response to wheat germ agglutinin and starvation in Drosophila melanogaster larval midgut. Insect Mol Biol 18(1):21–31

Liu C-L, Tsai C-C, Lin S-C, Wang L-I, Hsu C-I, Hwang M-J, Lin J-Y (2000) Primary structure and function analysis of the Abrus precatorius agglutinin a chain by site-directed mutagenesis Pro199 of amphiphilic α-HELIX H impairs protein synthesis inhibitory activity. J Biol Chem 275(3):1897–1901

Liu SM, Li J, Zhu JQ, Wang XW, Wang CS, Liu SS et al (2016) Transgenic plants expressing the AaIT/GNA fusion protein show increased resistance and toxicity to both chewing and sucking pests. Insect Sci 23(2):265–276

Lopez L, Camas A, Shivaji R, Ankala A, Williams P, Luthe D (2007) Mir1-CP, a novel defense cysteine protease accumulates in maize vascular tissues in response to herbivory. Planta 226 (2):517–527

Lord JM, Spooner RA (2011) Ricin trafficking in plant and mammalian cells. Toxins 3(7):787–801

Lüthi C, Álvarez-Alfageme F, Romeis J (2018) The bean α-amylase inhibitor αAI-1 in genetically modified chickpea seeds does not harm parasitoid wasps. Pest Manag Sci 74(11):2444–2449

Maag D, Dalvit C, Thevenet D, Köhler A, Wouters FC, Vassão DG et al (2014) 3-β-D-Glucopyranosyl-6-methoxy-2-benzoxazolinone (MBOA-N-Glc) is an insect detoxification product of maize 1, 4-benzoxazin-3-ones. Phytochemistry 102:97–105

Macedo MLR, Oliveira CF, Oliveira CT (2015) Insecticidal activity of plant lectins and potential application in crop protection. Molecules 20(2):2014–2033
Maddock S (1991) Expression in maize plants of wheatgerm agglutinin, a novel source of insect resistance. Paper presented at the 3rd International Congress of Plant Molecular Biology, Tucson, Arizona, USA, 1991
Maqbool SB, Riazuddin S, Loc NT, Gatehouse AM, Gatehouse JA, Christou P (2001) Expression of multiple insecticidal genes confers broad resistance against a range of different rice pests. Mol Breed 7(1):85–93
Marchetti S, Delledonne M, Fogher C, Chiaba C, Chiesa F, Savazzini F, Giordano A (2000) Soybean Kunitz, C-II and PI-IV inhibitor genes confer different levels of insect resistance to tobacco and potato transgenic plants. Theor Appl Genet 101(4):519–526
Martinelli AH, Kappaun K, Ligabue-Braun R, Defferrari MS, Piovesan AR, Stanisçuaski F et al (2014) Structure–function studies on jaburetox, a recombinant insecticidal peptide derived from jack bean (Canavalia ensiformis) urease. Biochim Biophys Acta (BBA)-Gen Subj 1840 (3):935–944
Mehta D, Stürchler A, Hirsch-Hoffmann M, Gruissem W, Vanderschuren H (2018) CRISPR-Cas9 interference in cassava linked to the evolution of editing-resistant geminiviruses. BioRxiv:314542
Melander M, Åhman I, Kamnert I, Strömdahl A-C (2003) Pea lectin expressed transgenically in oilseed rape reduces growth rate of pollen beetle larvae. Transgenic Res 12(5):555–567
Michiels K, Van Damme EJ, Smagghe G (2010) Plant-insect interactions: what can we learn from plant lectins? Arch Insect Biochem Physiol 73(4):193–212
Mochizuki A, Nishizawa Y, Onodera H, Tabei Y, Toki S, Habu Y et al (1999) Transgenic rice plants expressing a trypsin inhibitor are resistant against rice stem borers, Chilo suppressalis. Entomol Exp Appl 93(2):173–178
Mohan S, Ma PW, Williams WP, Luthe DS (2008) A naturally occurring plant cysteine protease possesses remarkable toxicity against insect pests and synergizes Bacillus thuringiensis toxin. PLoS One 3(3):e1786
Mondal HA, Chakraborti D, Majumder P, Roy P, Roy A, Bhattacharya SG, Das S (2011) Allergenicity assessment of Allium sativum leaf agglutinin, a potential candidate protein for developing sap sucking insect resistant food crops. PLoS One 6(11):e27716
Mondal H, Roy A, Gupta S, Das S (2012) Exploring the insecticidal potentiality of Amorphophallus paeonifolius tuber agglutinin in hemipteran pest management. Am J Plant Sci 3(06):780
Moreira RDA, Ainouz IL, Oliveira JTAD, Cavada BS (1991) Plant lectins, chemical and biological aspects. Mem Inst Oswaldo Cruz 86:211–218
Morton RL, Schroeder HE, Bateman KS, Chrispeels MJ, Armstrong E, Higgins TJ (2000) Bean α-amylase inhibitor 1 in transgenic peas (Pisum sativum) provides complete protection from pea weevil (Bruchus pisorum) under field conditions. Proc Natl Acad Sci 97(8):3820–3825
Mulinari F (2008) Ureases de Canavalia ensiformis e peptídeo inseticida derivado
Mulinari F, Staniscuaski F, Bertholdo-Vargas L, Postal M, Oliveira-Neto O, Rigden D et al (2007) Jaburetox-2Ec: an insecticidal peptide derived from an isoform of urease from the plant Canavalia ensiformis. Peptides 28(10):2042–2050
Murdock LL, Shade RE (2002) Lectins and protease inhibitors as plant defenses against insects. J Agric Food Chem 50(22):6605–6611
Nagadhara D, Ramesh S, Pasalu I, Rao YK, Sarma N, Reddy V, Rao K (2004) Transgenic rice plants expressing the snowdrop lectin gene (gna) exhibit high-level resistance to the whitebacked planthopper (Sogatella furcifera). Theor Appl Genet 109(7):1399–1405
Nagpure A, Choudhary B, Gupta RK (2014) Chitinases: in agriculture and human healthcare. Crit Rev Biotechnol 34(3):215–232
Oerke E-C (2006) Crop losses to pests. J Agric Sci 144(1):31–43
Ohizumi Y, Gaidamashvili M, Ohwada S, Matsuda K, Kominami J, Nakamura-Tsuruta S et al (2009) Mannose-binding lectin from yam (Dioscorea batatas) tubers with insecticidal properties against Helicoverpa armigera (Lepidoptera: Noctuidae). J Agric Food Chem 57(7):2896–2902

Oliveira AS, Lossio CF, Rangel AJ, Martins MG, Nascimento FE, Andrade MLD et al (2017) Detection, purification and characterization of a lectin from freshwater green algae Spirogyra spp. An Acad Bras Cienc 89(3):2113–2117
Oliveira-Neto OB, Batista JA, Rigden DJ, Franco OL, Falcão R, Fragoso RR et al (2003) Molecular cloning of α-amylases from cotton boll weevil, Anthonomus grandis and structural relations to plant inhibitors: an approach to insect resistance. J Protein Chem 22(1):77–87
Olsen KM, Wendel JF (2013) A bountiful harvest: genomic insights into crop domestication phenotypes. Annu Rev Plant Biol 64:47–70
Pechan T, Ye L, Chang Y-M, Mitra A, Lin L, Davis FM et al (2000) A unique 33-kD cysteine proteinase accumulates in response to larval feeding in maize genotypes resistant to fall armyworm and other Lepidoptera. Plant Cell 12(7):1031–1040
Pechan T, Cohen A, Williams WP, Luthe DS (2002) Insect feeding mobilizes a unique plant defense protease that disrupts the peritrophic matrix of caterpillars. Proc Natl Acad Sci 99 (20):13319–13323
Pelegrini PB, Quirino BF, Franco OL (2007) Plant cyclotides: an unusual class of defense compounds. Peptides 28(7):1475–1481
Pelegrini PB, Lay FT, Murad AM, Anderson MA, Franco OL (2008) Novel insights on the mechanism of action of α-amylase inhibitors from the plant defensin family. Protein Struct Func Bioinf 73(3):719–729
Peumans WJ, Van Damme E (1995) Lectins as plant defense proteins. Plant Physiol 109(2):347
Peumans WJ, Hao Q, van Damme EJ (2001) Ribosome-inactivating proteins from plants: more than RNA N-glycosidases? FASEB J 15(9):1493–1506
Pinto MF, Fensterseifer IC, Migliolo L, Sousa DA, de Capdville G, Arboleda-Valencia JW et al (2012) Identification and structural characterization of novel cyclotide with activity against an insect pest of sugar cane. J Biol Chem 287(1):134–147
Piovesan AR, Martinelli AH, Ligabue-Braun R, Schwartz J-L, Carlini CR (2014) Canavalia ensiformis urease, Jaburetox and derived peptides form ion channels in planar lipid bilayers. Arch Biochem Biophys 547:6–17
Popp J (2011) Cost-benefit analysis of crop protection measures. J Verbr Lebensm 6(1):105–112
Popp J, Pető K, Nagy J (2013) Pesticide productivity and food security. A review. Agron Sustain Dev 33(1):243–255
Postal M, Martinelli AH, Becker-Ritt AB, Ligabue-Braun R, Demartini DR, Ribeiro SF et al (2012) Antifungal properties of Canavalia ensiformis urease and derived peptides. Peptides 38 (1):22–32
Powell K, Gatehouse A, Hilder V, Gatehouse J (1993) Antimetabolic effects of plant lectins and plant and fungal enzymes on the nymphal stages of two important rice pests, Nilaparvata lugens and Nephotettix cinciteps. Entomol Exp Appl 66(2):119–126
Powell KS, Spence J, Bharathi M, Gatehouse JA, Gatehouse AM (1998) Immunohistochemical and developmental studies to elucidate the mechanism of action of the snowdrop lectin on the rice brown planthopper, Nilaparvata lugens (Stal). J Insect Physiol 44(7–8):529–539
Price DR, Gatehouse JA (2008) RNAi-mediated crop protection against insects. Trends Biotechnol 26(7):393–400
Puchta H (2003) Marker-free transgenic plants. Plant Cell Tissue Organ Cult 74(2):123–134
Pusztai A, Ewen SW, Grant G, Peumans WJ, van Damme EJ, Coates ME, Bardocz S (1995) Lectins and also bacteria modify the glycosylation of gut surface receptors in the rat. Glycoconj J 12 (1):22–35
Pusztai A, Koninkx J, Hendriks H, Kok W, Hulscher S, Van Damme EJ et al (1996) Effect of the insecticidal Galanthus nivalis agglutinin on metabolism and the activities of brush border enzymes in the rat small intestine. J Nutr Biochem 7(12):677–682
Quilis J, López-García B, Meynard D, Guiderdoni E, San Segundo B (2014) Inducible expression of a fusion gene encoding two proteinase inhibitors leads to insect and pathogen resistance in transgenic rice. Plant Biotechnol J 12(3):367–377

Rahbé Y, Sauvion N, Febvay G, Peumans WJ, Gatehouse AM (1995) Toxicity of lectins and processing of ingested proteins in the pea aphid Acyrthosiphon pisum. Entomol Exp Appl 76 (2):143–155

Ramachandraiah G, Chandra NR (2000) Sequence and structural determinants of mannose recognition. Proteins Struct Func Bioinf 39(4):358–364

Rani A, Chand S, Thakur N, Nath AK (2018) Alpha-amylase inhibitor from local common bean selection: effect on growth and development of Corcyra cephalonica. J Stored Prod Res 75:35–37

Rao K, Rathore KS, Hodges TK, Fu X, Stoger E, Sudhakar D et al (1998) Expression of snowdrop lectin (GNA) in transgenic rice plants confers resistance to rice brown planthopper. Plant J 15 (4):469–477

Ray DK, Mueller ND, West PC, Foley JA (2013) Yield trends are insufficient to double global crop production by 2050. PLoS One 8(6):e66428

Romeis J, Babendreier D, Wäckers FL (2003) Consumption of snowdrop lectin (Galanthus nivalis agglutinin) causes direct effects on adult parasitic wasps. Oecologia 134(4):528–536

Roy A, Das S (2015) Molecular mechanism underlying the entomotoxic effect of Colocasia esculenta tuber agglutinin against Dysdercus cingulatus. Insects 6(4):827–846

Roy A, Gupta S, Hess D, Das KP, Das S (2014) Binding of insecticidal lectin Colocasia esculenta tuber agglutinin (CEA) to midgut receptors of Bemisia tabaci and Lipaphis erysimi provides clues to its insecticidal potential. Proteomics 14(13–14):1646–1659

Sadeghi A, Broeders S, De Greve H, Hernalsteens JP, Peumans WJ, Van Damme EJ, Smagghe G (2007) Expression of garlic leaf lectin under the control of the phloem-specific promoter Asus1 from Arabidopsis thaliana protects tobacco plants against the tobacco aphid (Myzus nicotianae). Pest Manag Sci 63(12):1215–1223

Sadeghi A, Smagghe G, Broeders S, Hernalsteens J-P, De Greve H, Peumans WJ, Van Damme EJ (2008) Ectopically expressed leaf and bulb lectins from garlic (Allium sativum L.) protect transgenic tobacco plants against cotton leafworm (Spodoptera littoralis). Transgenic Res 17 (1):9

Saha P, Majumder P, Dutta I, Ray T, Roy S, Das S (2006) Transgenic rice expressing Allium sativum leaf lectin with enhanced resistance against sap-sucking insect pests. Planta 223 (6):1329

Schwefel D, Maierhofer C, Beck JG, Seeberger S, Diederichs K, Möller HM et al (2010) Structural basis of multivalent binding to wheat germ agglutinin. J Am Chem Soc 132(25):8704–8719

Scott JG, Michel K, Bartholomay LC, Siegfried BD, Hunter WB, Smagghe G et al (2013) Towards the elements of successful insect RNAi. J Insect Physiol 59(12):1212–1221

Sengupta S, Chakraborti D, Mondal HA, Das S (2010) Selectable antibiotic resistance marker gene-free transgenic rice harbouring the garlic leaf lectin gene exhibits resistance to sap-sucking planthoppers. Plant Cell Rep 29(3):261–271

Sétamou M, Bernal J, Legaspi J, Mirkov T, Legaspi B Jr (2002) Evaluation of lectin-expressing transgenic sugarcane against stalkborers (Lepidoptera: Pyralidae): effects on life history parameters. J Econ Entomol 95(2):469–477

Shang C, Peumans WJ, Van Damme EJ (2014) Occurrence and taxonomical distribution of ribosome-inactivating proteins belonging to the Ricin/Shiga toxin superfamily. In: Ribosome-inactivating proteins: ricin and related proteins, pp 11–27

Sharon N, Lis H (1990) Legume lectins–a large family of homologous proteins. FASEB J 4 (14):3198–3208

Silva CP, Terra WR, Xavier-Filho J, de Sá MFG, Isejima EM, DaMatta RA et al (2001) Digestion of legume starch granules by larvae of Zabrotes subfasciatus (Coleoptera: Bruchidae) and the induction of α–amylases in response to different diets. Insect Biochem Mol Biol 31(1):41–50

Sinha S, Gupta G, Vijayan M, Surolia A (2007) Subunit assembly of plant lectins. Curr Opin Struct Biol 17(5):498–505

Solleti SK, Bakshi S, Purkayastha J, Panda SK, Sahoo L (2008) Transgenic cowpea (Vigna unguiculata) seeds expressing a bean α-amylase inhibitor 1 confer resistance to storage pests, bruchid beetles. Plant Cell Rep 27(12):1841

Stahl E, Hilfiker O, Reymond P (2018) Plant–arthropod interactions: who is the winner? Plant J 93 (4):703–728

Stanisçuaski F, Carlini CR (2012) Plant ureases and related peptides: understanding their entomotoxic properties. Toxins 4(2):55–67

Stanisçuaski F, Ferreira-Dasilva C, Mulinari F, Pires-Alves M, Carlini C (2005) Insecticidal effects of canatoxin on the cotton stainer bug Dysdercus peruvianus (Hemiptera: Pyrrhocoridae). Toxicon 45(6):753–760

Stillmark H (1992) Uber ricin, eines giftiges ferment aus den samen von Ricinus communis L. und Anderson Euphorbiacen, inaugural Disseration, University of Dorpat, Estonia. 1888.(German) IV. A fate of orally administered ricin in rats. J Pharm 15:147–156

Stirpe F (2013) Ribosome-inactivating proteins: from toxins to useful proteins. Toxicon 67:12–16

Sumner JB (1919) The globulins of the jack bean, canavalia ensiformis preliminary paper. J Biol Chem 37(1):137–142

Swamy S, Prasad N, Rao N (2009) Transgenic Bt crops: a major component of integrated pest management–an overview. Indian J Crop Sci 4(1and2):1–10

Tajne S, Boddupally D, Sadumpati V, Vudem DR, Khareedu VR (2014) Synthetic fusion-protein containing domains of Bt Cry1Ac and Allium sativum lectin (ASAL) conferred enhanced insecticidal activity against major lepidopteran pests. J Biotechnol 171:71–75

Taylor ME, Drickamer K (2009) Structural insights into what glycan arrays tell us about how glycan-binding proteins interact with their ligands. Glycobiology 19(11):1155–1162

Taylor SL, Hefle SL (2001) Will genetically modified foods be allergenic? J Allergy Clin Immunol 107(5):765–771

Terenius O, Papanicolaou A, Garbutt JS, Eleftherianos I, Huvenne H, Kanginakudru S et al (2011) RNA interference in Lepidoptera: an overview of successful and unsuccessful studies and implications for experimental design. J Insect Physiol 57(2):231–245

Thevissen K, Terras FR, Broekaert WF (1999) Permeabilization of fungal membranes by plant defensins inhibits fungal growth. Appl Environ Microbiol 65(12):5451–5458

Titarenko E, Chrispeels MJ (2000) cDNA cloning, biochemical characterization and inhibition by plant inhibitors of the α-amylases of the Western corn rootworm, Diabrotica virgifera virgifera. Insect Biochem Mol Biol 30(10):979–990

Van Damme EJ (2014) History of plant lectin research. In: Lectins. Springer, pp 3–13

Van Damme EJ, Peumans WJ, Pusztai A, Bardocz S (1998) Handbook of plant lectins: properties and biomedical applications. Wiley, Chichester

Van Damme EJ, Lannoo N, Fouquaert E, Peumans WJ (2003) The identification of inducible cytoplasmic/nuclear carbohydrate-binding proteins urges to develop novel concepts about the role of plant lectins. Glycoconj J 20(7–8):449–460

Van Damme EJ, Lannoo N, and Peumans WJ (2008) Plant lectins. In Advances in botanical research, vol 48, pp 107–209. Elsevier

Vandenborre G, Groten K, Smagghe G, Lannoo N, Baldwin IT, Van Damme EJ (2009a) Nicotiana tabacum agglutinin is active against lepidopteran pest insects. J Exp Bot 61(4):1003–1014

Vandenborre G, Miersch O, Hause B, Smagghe G, Wasternack C, Van Damme EJ (2009b) Spodoptera littoralis-induced lectin expression in tobacco. Plant Cell Physiol 50(6):1142–1155

Vandenborre G, Van Damme EJ, Ghesquiere B, Menschaert G, Hamshou M, Rao RN et al (2010) Glycosylation signatures in Drosophila: fishing with lectins. J Proteome Res 9(6):3235–3242

Vandenborre G, Smagghe G, Van Damme EJ (2011) Plant lectins as defense proteins against phytophagous insects. Phytochemistry 72(13):1538–1550

Vazquez-Padron RI, Moreno-Fierros L, Neri-Bazán L, Gustavo A, Lopez-Revilla R (1999) Intragastric and intraperitoneal administration of Cry1Ac protoxin from Bacillus thuringiensis induces systemic and mucosal antibody responses in mice. Life Sci 64(21):1897–1912

Vázquez-Padrón RI, Gonzáles-Cabrera J, García-Tovar C, Neri-Bazan L, Lopéz-Revilla R, Hernández M et al (2000) Cry1Ac protoxin from Bacillus thuringiensis sp. kurstaki HD73 binds to surface proteins in the mouse small intestine. Biochem Biophys Res Commun 271 (1):54–58

Vijayan S, Imani J, Tanneeru K, Guruprasad L, Kogel K, Kirti P (2012) Enhanced antifungal and insect α-amylase inhibitory activities of Alpha-TvD1, a peptide variant of Tephrosia villosa defensin (TvD1) generated through in vitro mutagenesis. Peptides 33(2):220–229

Vijayan S, Singh N, Shukla P, Kirti P (2013) Defensin (TvD1) from Tephrosia villosa exhibited strong anti-insect and anti-fungal activities in transgenic tobacco plants. J Pest Sci 86 (2):337–344

Virgilio MD, Lombardi A, Caliandro R, Fabbrini MS (2010) Ribosome-inactivating proteins: from plant defense to tumor attack. Toxins 2(11):2699–2737

Waage J (1997) What does biotechnology bring to integrated pest management? Biotechnol Dev Monit 32:19–21

Walski T, Van Damme EJ, Smagghe G (2014) Penetration through the peritrophic matrix is a key to lectin toxicity against Tribolium castaneum. J Insect Physiol 70:94–101

Wang Z, Guo S (1999) Expression of two insect-resistant genescryIA (b&c)/GNA in transgenic tobacco plants results in added protection against both cotton bollworm and aphids. Chin Sci Bull 44(22):2051–2058

Wang Z, Zhang K, Sun X, Tang K, Zhang J (2005) Enhancement of resistance to aphids by introducing the snowdrop lectin genegna into maize plants. J Biosci 30(5):627–638

Wu J, Luo X, Guo H, Xiao J, Tian Y (2006) Transgenic cotton, expressing Amaranthus caudatus agglutinin, confers enhanced resistance to aphids. Plant Breed 125(4):390–394

Xu D, Xue Q, McElroy D, Mawal Y, Hilder VA, Wu R (1996) Constitutive expression of a cowpea trypsin inhibitor gene, CpTi, in transgenic rice plants confers resistance to two major rice insect pests. Mol Breed 2(2):167–173

Ye S-H, Chen S, Zhang F, Wang W, Tian Q, Liu J-Z et al (2009) Transgenic tobacco expressing Zephyranthes grandiflora agglutinin confers enhanced resistance to aphids. Appl Biochem Biotechnol 158(3):615–630

Yoon J-S, Gurusamy D, Palli SR (2017) Accumulation of dsRNA in endosomes contributes to inefficient RNA interference in the fall armyworm, Spodoptera frugiperda. Insect Biochem Mol Biol 90:53–60

Young NM, Oomen RP (1992) Analysis of sequence variation among legume lectins: a ring of hypervariable residues forms the perimeter of the carbohydrate-binding site. J Mol Biol 228 (3):924–934

Zadoks JC, Schein RD (1979) Epidemiology and plant disease management. Oxford University Press, New York/Oxford

Zavala JA, Baldwin IT (2004) Fitness benefits of trypsin proteinase inhibitor expression in Nicotiana attenuata are greater than their costs when plants are attacked. BMC Ecol 4(1):11

Zhang W, Peumans WJ, Barre A, Astoul CH, Rovira P, Rougé P et al (2000) Isolation and characterization of a jacalin-related mannose-binding lectin from salt-stressed rice (Oryza sativa) plants. Planta 210(6):970–978

Zhang J, Liu F, Yao L, Luo C, Yin Y, Wang G, Huang Y (2012) Development and bioassay of transgenic Chinese cabbage expressing potato proteinase inhibitor II gene. Breed Sci 62 (2):105–112

Zhang J, Khan SA, Hasse C, Ruf S, Heckel DG, Bock R (2015) Full crop protection from an insect pest by expression of long double-stranded RNAs in plastids. Science 347(6225):991–994

Zhu-Salzman K, Zeng R (2015) Insect response to plant defensive protease inhibitors. Annu Rev Entomol 60:233–252

Zhu-Salzman K, Luthe DS, Felton GW (2008) Arthropod-inducible proteins: broad spectrum defenses against multiple herbivores. Plant Physiol 146(3):852–858

Deciphering the Role of Phytoanticipins, Phytoalexins, and Polyphenols in Plant-Insect Defense

Sumanti Gupta and Amit Roy

Abstract

Knowledge of plant-insect interaction is continuously evolving with the coevolution of both interacting partners. In addition, the dynamic environmental factors are playing crucially into the interface chemistry of both the host plant and the attacking herbivorous insect. The present study has made efforts to shed light on the current knowledge of insect behavior during herbivory. Behavioral pattern of insects feeding on various plant hosts revolves around modifying host surveillance and overwhelming their defense mechanisms. On the contrary, hosts pay full attention on strengthening its defense arsenal with adequate and appropriate armors that can mount a timely resistance against the attacking insect. In this context, the molecular role of most important class of phyto-biomolecules referred to as phytoanticipins, phytoalexins, and phenolics that were widely investigated as potential insecticides since early days are elaborated under the light of recent day understanding. Thus, the chapter schematizes the general understanding of plant-insect interaction with an aim at adding to the knowledge of the development of better insect management strategies in the near future.

Keywords

Herbivory · Insect pests · Phenolics · Phytoalexins · Phytoanticipins · Plant defense

S. Gupta (✉)
Department of Botany, Rabindra Mahavidyalaya, Champadanga, WB, India

A. Roy
Faculty of Forestry and Wood Sciences, EXTEMIT-K, Czech University of Life Sciences, Prague, Suchdol, Czech Republic

I. K. Singh, A. Singh (eds.), *Plant-Pest Interactions: From Molecular Mechanisms to Chemical Ecology*, https://doi.org/10.1007/978-981-15-2467-7_13

1 Introduction: Plant-Insect Encounter – The Dynamic Power Game

Interactions between two biotic entities are subjected to continuous variations. These variations are the consequences of constant exposure to opposing selection pressure under a particular ecological alcove. The opposing pressure is directly imposed by the third most crucial factor named "environment." Hence, any interaction involves minimum two living partners and environment in all possible permutations. However, unfortunately, the pivots of all niches, "the plants," have no choice other than being exposed to multiple interacting agents simultaneously. Their sedentary nature compels them to behave as the "spider of the web" that either controls or is controlled by diverse communities of both microbiome and macrobiome (Dicke and Baldwin 2010; Mendes et al. 2011). Microbiome consists of symbiotic microbes like mycorrhiza, endophytes, PGPRs (plant growth-promoting rhizobacteria), and nitrogen-fixing rhizobia, pathogens, and their antagonists, while herbivores; natural enemies of herbivores, like predators and parasitoids; and pollinators of herbivores make up for the macrobiome. Each plant species provides shelter to different organisms (micro or macro) at different trophic levels, all of which together compete within themselves for light and nutrients (Bukovinszky et al. 2008). However, in all cases,the key role in regulating the aftermath of the dynamic interaction both symbiotic and antagonistic is imparted by the environment. The present chapter shall focus on the knowledge acquired on "plant-insect" interaction.

Plants represent the most abundant biomass, while insects share a population of estimated six million (Stam et al. 2014). The term "plant-insect" interaction in present days appears to be more complex than previously envisaged. Earlier understanding of "plant-insect" interaction dealt with interactions taking place in one-to-one fashion involving only two entities. However, in reality, it occurs between multiple organisms at different trophic levels over a particular span of time (Fig. 1). Insects may be classified as generalists and specialists that thrive on a wide range of plant families but feed on specific host plant family (Howe and Jander 2008). Initial mechanism opted by any generalist or polyphagous insect includes inflicting mechanical injury of the host precisely at the site of attempted penetration. However, the magnitude of damage differs both qualitatively and quantitatively according to the feeding tactics adopted by the attacking insect. Coleopterans and lepidopterans cause damage by their mouthparts, while thrips and spider mites insert tube-like projections to extract liquid diet from macerated host tissue. Hemipterans directly pierce their stylet into the phloem tissue regions of the host plant and acquire readily absorbable diet (Scboonhoven et al. 1998).

Plants when under insect attack install different weapons, which traverse through several layers of their defense response (Schuman and Baldwin 2016). All these responses are partitioned into direct and indirect responses. Chemical compounds (secretion of diverse classes of secondary metabolites) and defensive proteins that act as direct repellents, antinutritives, and/or toxic on herbivore insects are termed as direct defensive compounds. Besides, physical barriers such as the presence of

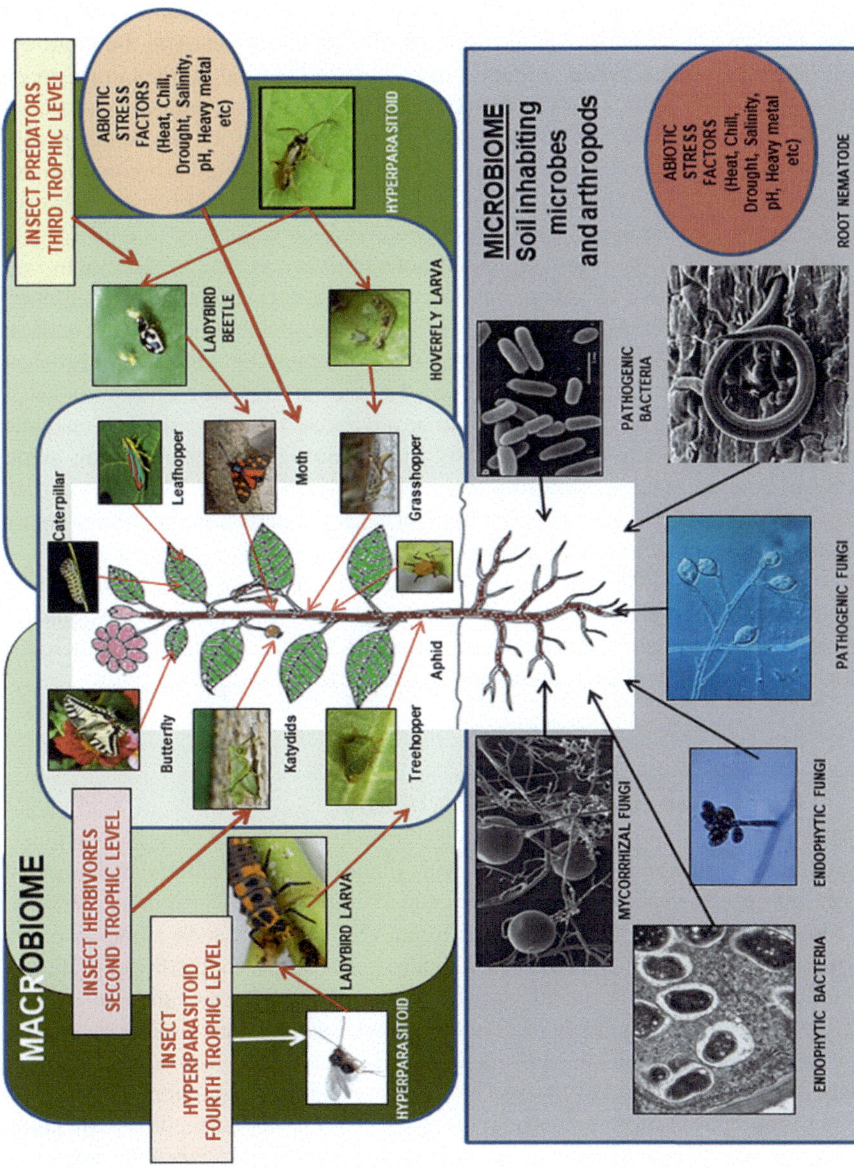

Fig. 1 Illustration of the possible interactions that can take place at a given time at different trophic levels and under the influence of different abiotic factors

trichomes and tough leathery outer layers also behave as direct deterrants of insect herbivory (Kessler and Baldwin 2002).

The second layer of indirect defense is provided by secretion of insect-induced plant volatiles and nectar that lure predators and parasitoids of herbivorous insects (Kessler and Baldwin 2001). Both direct and indirect defenses combine and aim at making the host unpalatable for the residing insect herbivores. On the contrary, insects deploy several survival strategies to overcome physical hurdles imposed by the host, escape host surveillance, detoxify and sequester host-induced poisonous chemicals, and undergo transcriptional reprogramming, all of which are directed to make the infested host body "a haven of peace" for the attacking herbivorous insect (Mello and Silva-Filho 2002). In addition, plant features that confer resistance and/or tolerance against insect pathogens can also be classified according to their expression pattern. Some of them are expressed constitutively as part of developmental phenomenon irrespective of whether an insect threat is present or not. Such constitutive expression of defensive proteins is found in reproductive tissues. In contrast, some specific defenses are mounted at the site of attempted penetration only when insect attack is taking place. These types of defenses are referred to as an induced defense. Induced defenses are considered to be advanced as they need lower resource allotment and can minimize fitness costs by physiologically channelizing sugar reserves to belowground areas during insect attack (Karban et al. 1997).

Thus, although the intricacies of plant-insect interaction involves many compounds both biotic and abiotic, in the following sections, we limit our study to discussing the overall behavioural pattern of herbivorous insects, the signaling events that take place within the host in response to herbivory, the counter-signaling phenomenon operational within the attacking insect, and the role of phytoanticipins, phytoalexins, and phenolics in regulation of host defense during herbivory.

2 Insect Behavior During Herbivory

Approximately, all major insect orders with plant-feeding capability had evolved almost 300 million years ago (Labandeira and Phillips 1996). Partitioning of available food resources and generation of different feeding types such as sap sucking, gall markers, and spore feeders had already been observed by then. With the availability of vascular tissue during the Carboniferous period, phloem-feeding insects evolved. Similarly, chewing insects evolved after the evolution of laminate leaf tissue. Thus, evolution and appearance of specific tissue somehow presided over the advancement of insect diversity. The feeding apparatus (i.e., piercing, sucking, chewing mouthparts) of insects also evolved accordingly based on host plant shifts and aided in niche separation. It can be anticipated from the phylogenetic studies of insect lineages by Mitter et al. (1991) that herbivorous insects are much more diversified than non-herbivorous insects. Thus, the fundamental role of plants is quite evident in endorsing adaptive diversification over million years in herbivore insects through a stepwise coevolutionary arms race (Ehrlich and Raven 1967).

In 1998, Elizabeth A. Bernays stated: "Success seen as different ways to eat without being eaten" (Bernays 1998). Indeed, it is a challenge for insects to eat without being eaten. On the bases of diversity of the host plant chemistry and natural enemies, insect-feeding habit has been diversified over million years of evolutionary history. Being extraordinarily diverse and copious, insect herbivores accumulate almost a quarter of all diversities in eukaryotes. Insect species feeding on many different plants belonging to unrelated plant families are called "generalists," whereas insects with diet breadth restricted to one to few related plant families are considered as "specialists." More precise feeding habits are also observed among insects, i.e., some of them feed on particular plant part such as leaves, whereas others are completely relying on the root, reproductive parts, phloem, or developing seeds for their nutrition. Hence, diversity in herbivore insects primarily results from the variety of host plant diet chemistry that regulates their adaptive feeding behavior. Interestingly, the interaction between evolution and adaptive behavior is bidirectional. Natural selection directly influences insect behavior and adaptive behavioral alterations accumulated by an individual. These, in turn, modify the suite of selective forces, i.e., insects learn to avoid ~~a~~ toxic food or habitat with a high risk of predation and may lead to the selection of favoring new behaviors. Recently, scientists from Max Planck Institute for Chemical Ecology, Jena, Germany, have documented the avoidance behavior of *D. melanogaster* to "geosmin," a toxic substance released by bacteria and mold fungi in putrid food (Stensmyr et al. 2012). Surprisingly, the response to geosmin is mediated by only a single neuron, namely, "ab4B," which carries the neural transmission signal to a specific receptor called "Or56a." Oviposition behavior of the female is also influenced by geosmin. Geosmin acts as a stop signal for flies that prevents flies from eating or depositing eggs on such rotten food materials. This olfactory circuit for geosmin avoidance is evolutionarily conserved among other flies from genus *Drosophila* except for *D. elegans* (Stensmyr et al. 2012). This may be due to the low susceptibility of fresh flowers, the breeding substrate of the species *D. elegans*, to mold growth (Yoshida et al. 2000).

The challenges faced by herbivore insects to fetch nutrition (i.e., nitrogen, phosphorous, sterol) from their host shape the biology of feeding behavior. Plant secondary metabolites, volatiles, and their interplay with other ecological factors significantly influence the selection of specialized feeding behaviors of natural enemies. The exercise of acquiring nutrition from the hosts involved a wide variety of trophic structures such as sucking (i.e., hemipterans) or chewing (i.e., lepidopterans) mouthparts, which are highly specialized in different insect orders, thus restricting diet breadth and feeding behavior in multiple ways. Chewing is closely related to the mandibles. Often, insects feeding on a similar type of tissue or plant part develop morphologically similar mandibles indicating the adaptive significance of these structures. Such convergence of mandible structures is best observed in grasshoppers feeding exclusively on grasses. Mandible structures optimized for cutting and grinding tough grass blades have evolved at least eight times in grasshoppers (Bernays 1991). Interestingly, a wide diversity in mandibular structure also exists in lepidopterans and assists in controlling diet breadth, i.e., caterpillars from two moth families have different mandible structures, one fine-tuned for handling robust and mature leaves and the other suited for tearing and chopping the soft, young leaves (Bernays 1998). Among hemipterans (plant sap-sucking herbivores), species fed on phloem or xylem has developed specialized structures called "stylets" aiding their sap-sucking activity. Intriguingly, further specialization in stylet structure and route to reach phloem sap has been observed among different phloem feeders, i.e., between aphids and planthoppers. Typical aphid stylet passes through the plant tissue toward vascular bundle through a route between plant cells traversing through the layers of polysaccharides of the outer host cell walls. Conversely, planthoppers insert their stylets directly through the parenchyma to the phloem (Bernays 1998).

Unexpectedly, the high degree of phenological complexity in trophic structures is often not correlated well with the ability to the higher digestibility of the host, i.e., the highly specialized mandibles of specialist grass-feeding grasshoppers do not aid better digestion of grass compared to generalist grasshoppers without such highly specialized mandibles (Bernays and Barbehenn 1987). However, considering from another ecological point of view, small organisms have higher predation risk by natural enemies, and during active feeding, the risk of being preyed increased up to 100-fold as described by Bernays (1997). Therefore, the fundamental advantage of extremely specialized mouthparts may be useful for superior handling and rapid ingestion rate of plant tissue or phloem sap aiding better escape rate from the predators. Moreover, herbivore insects, in general, encounter greater challenge in obtaining nitrogen that is primarily required for protein synthesis in the body, due to low levels of nitrogen in most of their host plant diets. Often, insect herbivores opt for compensatory feeding to maintain high protein-to-carbohydrate ratio in their body. Considering the risk of predation, rapid ingestion of food using specialized mouthparts has a considerable adaptive advantage.

Intriguingly, the magnitude of plant defense response against the attacking insects also depends on their feeding behavior. For example, lepidopterans cause rapid damage to plant tissue causing the high magnitude of plant defense response within a short time of the initial attack. Inversely, aphid stylet-mediated feeding generates slow and late plant defense responses (Will et al. 2007). Thus, mode of host feeding is directly correlated with the insect's ability to tolerate low or high magnitude of plant defense response (Fig. 2).

3 Plant Defense Signaling During Insect Attack: All a Matter of the Timely Action

Host defense signaling in plant-insect interaction is as intricate and dynamic as any other host-microbe interaction. In case of plant-pathogen encounter, the microbe-associated molecular patterns (MAMPs) or pathogen-associated molecular patterns (PAMPs) or danger-associated molecular patterns (DAMPs) are recognized by host pattern recognition receptors (PRRs) that bring about pattern-triggered immunity (PTI). PTI effectively nullifies the ill effects of pathogenic devastations in the majority of cases, while some opportunistic and tactful pathogens evade PTI and

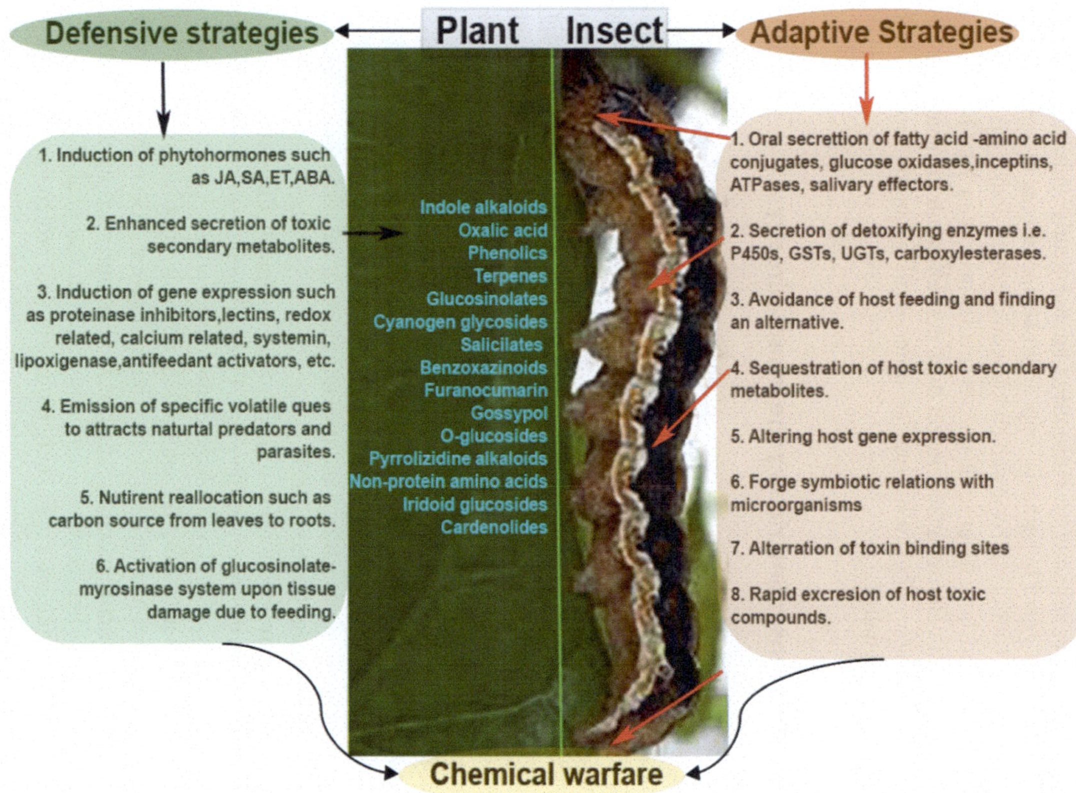

Fig. 2 Illustration of chemical warfare between the plant and its feeder, the herbivore insects

release effectors that struggle to overpower the host immunity and promote effector-triggered susceptibility (ETS). The incompatible host recruits resistant proteins (R proteins) that mount an amplified defense response known as effector-triggered immunity (ETI) which successfully controls the endeavor of the invading pathogen (Jones and Dangl 2006). However, in case of plant-insect interaction, the demarcations of PTI and ETI, as well as the role of PAMP/MAMP/DAMP like insect-specific patterns in inducing host responses, are not clearly understood (Howe and Jander 2008). The "guard hypothesis" appears to be operational in a plant-insect interaction where insect-induced tissue and endogenous protein damage is sensed by the host R proteins and host defense responses are ignited (Van Der Biezen and Jones 1998). Thus, host defense against insect does not follow all the same features as of other encounters. The following sections shall elaborate the overall mechanisms of host signaling.

3.1 Direct Defense: Restricting the Insect Directly

Direct defense means restricting the entry of the herbivorous insect directly by altering the physical and biochemical attributes. Physical features such as the presence of spines and thorns, trichomes, sclerophyllous leaves, granular minerals in plant tissue, divaricating branching pattern, etc. play an essential part in allowing the insect to come at close proximity of the host and choosing it for shelter (Hanley et al. 2007). Sclerophylly reduces the palatability of the host (Hanley et al. 2007). Spinescence prevents the insects from arriving at the host surface (Hanley et al. 2007). Conversely, trichome abundance affects the insect oviposition, larval nutrition, and spatial movement (Handley et al. 2005). Besides, the induction of glandular hair in response to herbivory is also reported to secrete toxic secondary metabolites that straightway deter the growth and reproduction of the herbivores (Sharma et al. 2009). On the other hand, trichome exudates act as extrafloral nectar in attracting scelonid egg parasitoid of squash bug and providing additional resistance to squash plant indirectly (Olson and Nechols 1995).

Host secondary metabolites form anti-insecticidal chemicals that deter, repel, or produce toxic effects on the insect metabolism. Plant phenolics such as phenols are quantitatively as well as qualitatively altered during insect attack. They are known to prime host resistance by activating the antioxidant defensive machinery (War et al. 2012). Besides, phenolics and heteropolymer lignin prevent insect entry by rapid deposition of its biosynthetic materials which is well documented by the enhanced expression of *CAD/CAD*-like genes during herbivory (Barakat et al. 2010). Oxidation of phenols forming quinones covalently binds with leaf proteins and reduces its digestibility, while alkylation of amino acids lessens the food value of plant protein for the feeding insects (Bhonwong et al. 2009). Flavonoids form cytotoxic chemicals and combine with other enzymes and influence the growth and development of insect herbivores (Simmonds 2003). Different flavonoids such as anthocyanins, flavones, flavonols, flavanones, dihydroflavonols, chalcones, aurones, flavan, and proanthocyanidins act as antifeedant of the herbivore insects (Treutter 2006).

Tannins are known to bind to host proteins and reduce their absorption efficiency, directly leading to nutrition depletion of the insect pests and forming of midgut lesions in them (Barbehenn and Constabel 2011). Moreover, tannins are reported to be induced in neighboring leaves of attacked plants and provide systemic host resistance (Peters and Constabel 2002).

Plant's defensive proteins also form a broad class of host resistance proteins that act against insect metabolism. Lectins are ubiquitous carbohydrate binding protein that acts against homopteran, coleopteran, and lepidopteran insects. They are stable toxic proteins that attach to insect epithelium membrane of the digestive tract and disrupt tissue integrity, thus leading to leaching of essential nutrients and causing the death of insects (Vandenborre et al. 2011). Transgenic plants such as brassica, rice, chickpea, etc. expressing mannose-binding lectins isolated from *Allium sativum* leaf have shown promising effects against homopteran insects (Chakraborti et al. 2009; Dutta et al. 2005; Saha et al. 2006). Apart from lectins, proteinase inhibitors bind with insect digestive enzymes and reduce their digestion capability (Lawrence and Koundal 2002). Peroxidases are essential oxidation regulatory enzymes that in association with phenols cause damage to the guts of insects during herbivory (Gulsen et al. 2010). Polyphenol oxidases (PPO) are metalloenzymes that catalyze the oxidation of mono- and ortho-diphenols to quinones that polymerize and react with nucleophilic side chains of amino acids, thus crosslinking essential proteins and making them unavailable for the herbivores (Zhang et al. 2008). Lipoxygenases are another important group of antioxidant enzymes that catalyze peroxidation of polyunsaturated fatty acids (PUFA) into the formation of fatty acid hydroperoxides that are further degraded to transiently expressed reactive oxygen species. These unstable reactive oxygen species induce protein-protein cross-linking and amino acid degradation that prevent protein assimilation. Additionally, lipid peroxidation end products cause antixenosis (Bruinsma et al. 2009) (Fig. 3).

3.2 Indirect Defense: Attracting the Pests of Pests

Similar to direct defenses, indirect defenses are also constitutive or induced or a combination of both, where mechanical injury or elicitors secreted from attacking herbivores play crucial roles in alluring natural enemies of herbivores like parasitoids and predators and indirectly imparting host plant resistance (Maffei 2010). Herbivore-induced plant volatiles (HIPVs) are important lipophilic compounds released from different plant parts both aboveground and belowground and not only attract natural enemies of insects but also act as their feeding deterrents (Dudareva et al. 2006). HIPVs are specific for specific plant-insect interaction. HIPVs include terpenes, ethylene, methyl salicylates, and green leaf volatiles (GLVs) which are the branch products of the oxylipin pathway. Reports suggest that *Spodoptera frugiperda* infestation in rice stimulates the emission of about 30 different HIPVs such as methyl salicylate and methyl benzoates (Yuan et al. 2008). Besides, external applications with a blend of HIPV have also found to prime the host plants against further pest infestation (Kessler et al. 2006). Apart from these,

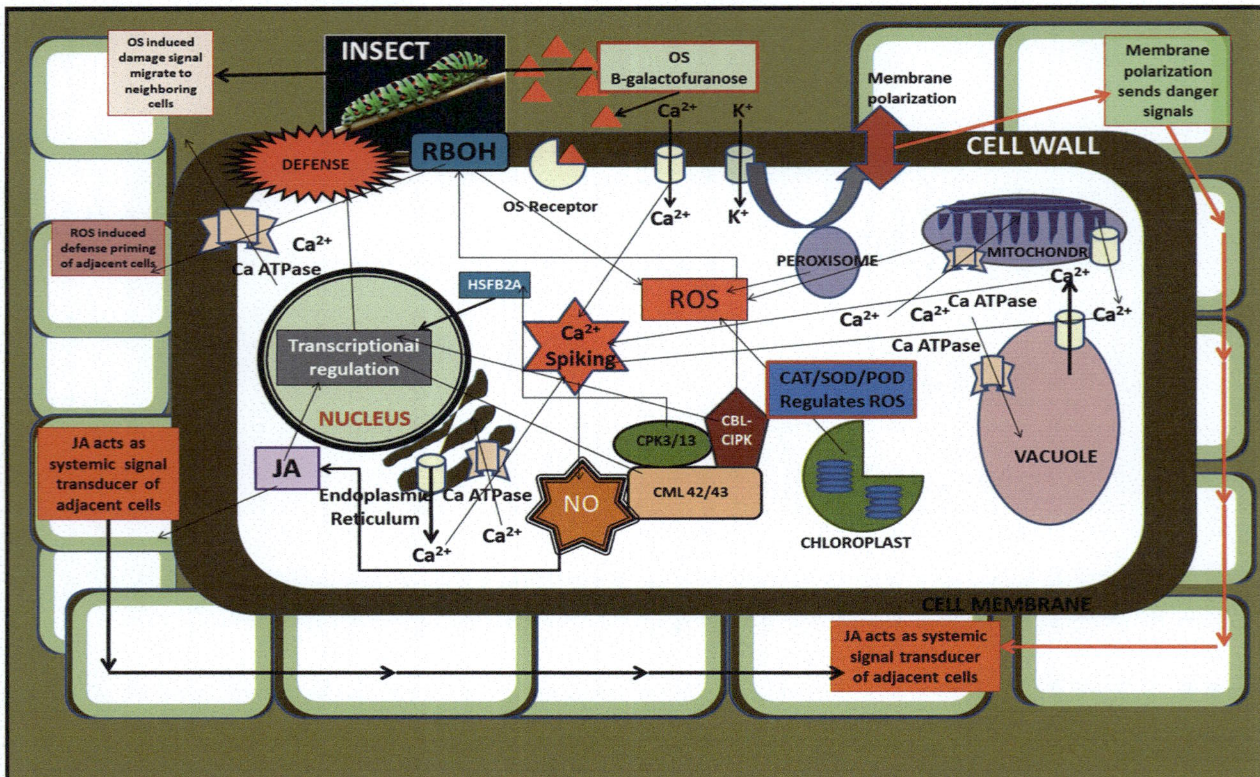

Fig. 3 *Schematic representation illustrates the intracellular localized and systemic signaling in plant cells during insect attack.* Insect oral secretion is accepted by host membrane receptor. OS receptor complex induces the opening of membrane calcium/potassium channels and induction of RBOH. Intracellular calcium influx and RBOH activation add to intracellular calcium spiking and ROS accumulation, respectively. Ca^{2+} spiking leads to increase of NO that regulates

insect oral secretions (fatty acid amino conjugates (FACs)) are also known to elicit the generation of plant volatiles that helps plants resist various insect attacks. Volicitin detected from beet armyworm showed to elicit emission of volatiles in *Zea mays* (Alborn et al. 1997). Other oral secretions like inceptins and caeliferins also hold similar functions (Alborn et al. 2007; Schmelz et al. 2006). Additionally, oral secretion from insects also indirectly influences the host defense signaling pathway by regulating MAPK signaling, jasmonic acid signaling, and accumulating various oxylipins (Schäfer et al. 2011) (Fig. 3).

3.3 Early Defense: Sense Early to Nip in the Bud

Sensing the arrival of an enemy at the doorstep just in time is unarguably the most crucial step of self-defense for any host. Plant-herbivore interaction involves myriad of defensive steps involving several molecular components starting from receptor recognition to signal initiation that occurs when the herbivore successfully overcomes the host physical barrier, sits on it and inserts its feeding organ into the host interior (Zebelo and Maffei 2014).The mechanical injury caused in host due to insertion of the foreign feeding part or due to insect's oral secretion (OS) brings about a drastic change in the plasma transmembrane potential (V_m) of the pierced host cell. The electric signal generated from the damaged cell or tissue travels systemically at different rates conveying different danger messages (Volkov et al. 2013). V_m depolarization is coupled downstream with an increase in cytoplasmic calcium spikes, change in ion channel activity, MAPK (mitogen-activated protein kinase) activation and protein phosphorylation, NADPH oxidase expression, and ROS (reactive oxygen species) and RNS (reactive nitrogen species) bursts (Maffei et al. 2007a). Transmission of the electrical signal is known to be transmitted via action potential (APs), variation potential (VPs), and/or system potentials (SPs). APs are a transitory change in electrical potentials that sense stimuli from environmental stressors leading to both intercellular and intracellular communication by activating voltage-gated ion channels. APs generated by herbivory travel from the site of origin at a rate of about 40 cm s^{-1}. APs are known to subsequently trigger VPs at the receiver cells. However, the direct generation of SPs following herbivory still lacks documentation (Zebelo et al. 2012). Besides, how the electrical signals are propagated to distant parts also remains an open question. The role of plasmodesmata containing PDLPs (plasmodesmata-located proteins), which are class 1 membrane proteins with receptor-like activities, appears to serve centrally in propagating distant systemic signals (Amari et al. 2010). Interestingly, recent

Fig. 3 (continued) transcriptional reprogramming and ultimately host defense against attacking insect. Intracellular Ca^{2+} homeostasis is maintained by organellar (mitochondrial, chloroplastic, ER, vacuolar) and membrane calcium channel proteins and Ca ATPase. Systemic danger signals upon insect penetration are transmitted to neighboring cells by insect OS, intracellular ROS elevations, NO-induced JA expression, and potassium-induced membrane depolarization

reports have emphasized on the capability of the plant to differentiate between mechano-vibrations caused during chewing or piercing and/or during ripples caused by abiotic factors such as wind or water. Such feature has opened up the possibility of the existence of a vibration regulated signaling pathway that could also explain the long-distance transfer of electric signals during insect attack (Appel and Cocroft 2014) by transmitters, signal generators to modulators, enzymes, etc. However, for any interaction, the herbivore-induced Vm variations are caused by several candidate ions such as calcium (Ca2+), proton (H+), potassium (K+), and chlorine (Cl-). Insect feeding leads to remarkable calcium influx in the near vicinity of damaged regions which are believed to trigger further downstream defense responses (Howe and Jander 2008). Calcium sensors such as calmodulin (CaM) and CaM-like protein (CML) play a significant role in calcium signaling following herbivory (Batistič and Kudla 2012; Du et al. 2011).

Also, calcineurin B-like (CBL) protein and calcium-dependent protein kinase (CDPK) also behave as calcium sensors and activate downstream defense signaling pathways following insect attack. CBLs after sensing calcium signatures form complexes with calcium-interacting protein kinases (CIPK) and regulated other signal transduction cross talks (Yu et al. 2014). On the other hand, CDPKs control the production of plant defensin genes as well as provide negative feedback roles during herbivory (Kanchiswamy et al. 2010). Calcium homeostasis is well maintained by Ca^{2+}ATPases that manage calcium efflux subsequently after the influx that takes place within milliseconds after insect feeding induced membrane depolarization (Maffei et al. 2007b). K^+ acts as another modulator of V_m that enters the cells as rectifier following the rapid influx of Ca^{2+} post-herbivory. CBL-CIPK complex is known to regulate the cellular inflow of K^+ (Cuéllar et al. 2013).

CBL-CIPK complex is also known to trigger plasma membrane NADPH oxidase (RBOH), which produces ROS like superoxide radical anions ($0_2^{-\cdot}$). These ROS molecules are dismutated to hydrogen peroxide (H_2O_2), which operates as a signaling molecule and triggers immune responses in adjoining cells. Dismutation and efficient scavenging of H_2O_2 by the composite action of a cellular antioxidant system comprising of superoxide dismutase (SOD), peroxidase, catalase, and ascorbate peroxidase (APX) takes place in a well-coordinated fashion. The increase of cytosolic calcium is also coupled with production of RNS mainly nitric oxide (NO). NO adduct S-nitrosoglutathione (GSNO) levels are reduced by S-nitrosoglutathione reductase (GSNOR) forming oxidized glutathione disulfide (GSSG) and ammonia (NH_3). GSNOR plays a vital role in resistance against herbivory and triggering JA-mediated immune signaling in infested host plants (Wünsche et al. 2011) (Fig. 3).

3.4 Role of Phytohormones: Regulating Many Spheres of Defense and Growth

Jasmonates (JA) primarily regulate the defense signaling during herbivory and different trophic levels, prime the plants for direct and indirect defense, and also help in transmission of systemic signals (Howe and Jander 2008). Wounding accumulates JA at the site in less than 30 mins. JA is synthesized from linolenic acid through the octadecanoid pathway. Chewing of insects causes dioxygenation of linoleic and linolenic acid, leading to the formation of 12-oxophytodieonic acid (OPDA). This OPDA is transported to peroxisome where it is converted to JA by OPDA reductase 3. JA conjugation with isoleucine (JA-Ile) mediates direct defense against caterpillars (Kang et al. 2006). Besides, methyl jasmonate (MeJA) also regulates resistance against insects. E3 ligase-mediated ubiquitination of regulatory proteins controls JA signaling by modulating the transcription of jasmonate-responsive genes (Devoto and Turner 2005). JA regulates cytosolic calcium levels and ROS levels during herbivory (Walter et al. 2007).Besides, JA is also known to alter CDPK transcript level and induce defense enzymes like peroxidase and polyphenol oxidase (Ulloa et al. 2002; War et al. 2012).

Salicylic acid (SA) is reported to induce greater defense against sucking and piercing insect compared to chewing ones. They are regulated by the regulatory protein non-expressor of pathogenesis-related gene1 (*NPR1*). Activation of *NPR1* by binding with *TGA* transcription factors takes place following the accumulation of stress-induced SA through the redox-dependent pathway (Pieterse and Van Loon 2004). SA regulates not only localized but also systemic responses (Peng et al. 2004). Both SA and JA control the emission of plant volatiles that triggers the colonization of natural enemies (De Boer et al. 2004). However, SA and JA are reported to act antagonistically (Maffei et al. 2007a). Ethylene acts both synergistically and antagonistically with JA in imparting defense against herbivores directly or indirectly (Horiuchi et al. 2003).

3.5 Transgenerational Immunity: Carrying Ahead the Good Things

Transgenerational immunity is an induced immunity that the offsprings inherit from their mother plants that had been previously exposed to various stress factors both abiotic and biotic. Such filial plants when exposed to insect attack show immunity that is part of their transgenerational memory (Agrawal 2001). Although knowledge of inherited immunity is still at infancy, some important examples like offsprings of *Arabidopsis* plants exposed to abiotic stress showed enhanced resistance to herbivory. Besides, *Raphanus raphanistrum* damaged by *Phytophthora rapae* showed increased resistance to insect attack in their next filial generation (Agrawal 2002). There are increasing reports where stress factors lead to DNA methylation that hold prominent chances of being transferred to the offsprings as part of epigenetic changes. Mobile siRNAs are reported to have an active role in causing epigenetic alterations and modulation of different defense signaling pathways (Espinas et al. 2016; Schmelz et al. 2006). However, still, much investigation needs to be performed regarding what type of chromosomal alterations take place during insect attack and to what extent they are confined only to the parental generations and which portions of changes are likely to pass on to their offsprings.

3.6 How Do Plants React to Insect Community?

A single plant is capable of sheltering hundreds of organisms belonging to diverse members of microbiome and/or macrobiome. Thus, under natural ecological habitats and at a given time, plants respond simultaneously not only to the individual class of insects or insect herbivores but also to a diverse community of insects (Stam et al. 2014). Plants are known to genotypically influence their own morphological, physiological, and chemical traits that bring about a gross change in their phenotypic appearance. These changes in appearance directly not only manipulate the density and composition of herbivores but also control the gathering of parasitoids and hyperparasitoid communities on the host plants (Whitham et al. 2006). Transcriptomic alterations are reported to take place in host plant when attacked by insects individually. However, little is known about the conjugate effect that results when a single host is attacked by multiple attackers. However, recent reports highlight that the effect is a lot more than being merely additive (Rodriguez-Saona et al. 2010). Besides, the transcriptomic changes occurring during a particular plant-insect interaction is also known to add to the priming of the said host against subsequent attacks by other similar or dissimilar insect communities of the different trophic levels (Stam et al. 2014). This entire transcriptomic signaling within the host plant that ranges from controlling individual insect population to community of insects, predators, parasitoids, and hyperparasitoids is intricately under the fine-tune balance of phytohormonal cross talk involving primarily SA (salicylic acid), JA (jasmonic acid), ET (ethylene), etc. (Pieterse et al. 2012). Thus, the plant's phenotype is believed to decide the nature, density, and time span of its insect communities that have a long-term influence on building the ecological environment surrounding the host plant.

4 Insect Response to Plant Defense: Coevolutionary Arms Race

Plants develop sophisticated, multilayered defense mechanisms composed of physical and chemical defenses against herbivores to compensate their sessile nature (Gatehouse 2002; Howe and Jander 2008). The physical barrier attempts to confine the entry of the herbivore to the host or minimize the damage during herbivory. Alternatively, the chemical defenses (i.e., primarily secondary toxic metabolites) reduce the digestibility of the food source and target vital physiological processes inside insects to limit their growth, survival, and fecundity (Howe and Jander 2008). Conversely, the success of an insect's life solely depends on their ability to obtain nutrients from the toxin-laden host plant tissues. Hence, the coevolutionary tug-of-war between insect and their host plants is fundamentally shaped by the biochemical reactions of the plant tissues with the alimentary canal secretions of the herbivore. Herbivores evolved various strategies to surpass the entomotoxic impact of their food using adaptive mechanisms such as avoidance, excretion, detoxification, or sequestration of plant secondary metabolites (Howe and Herde 2015). In the following section, some of these strategies will be discussed briefly to give a comprehensive outline of the chemical warfare underneath the ongoing evolutionary arms race between herbivore and its host plants.

4.1 Resistance Through Avoidance

Host plant choice and insect feeding behavior play a primary role in insect's choice-making process for the selection of a plant to feed on or not. Indeed, an insect can skip feeding on certain plants as soon as they distinguish them as entomotoxic by contact, olfactory, or visual cues (Chapman 2003). Often, such avoidance mechanisms are genetically predetermined due to previous experience. For instance, the reason underlying the avoidance of egg-laying on nonhost or unsuitable plants by the female seed beetles is genetically determined (Fox et al. 2004).In black vine weevil (*Otiorhynchus sulcatus*), previous host feeding experience influences the oviposition choice of the female adult but not the feeding preference (Coyle et al. 2011). Hence, oviposition preference often does not correlate well with larval host feeding performance. There are multiple instances against the classical ecological hypothesis that "mothers know the best," a principle suggesting that the host plant choice for oviposition in the females is influenced by their offspring survival rate on the given plant (Birke and Aluja 2017; Heidel-Fischer and Vogel 2015). The insect can also feed on plant parts with less toxin concentration (i.e., niche shift) or explore the plant in particular developmental stage where they have manageable secondary metabolite concentration (i.e., phenological shift) (Nealis and Nault 2005).

Taste such as bitterness can be a primary parameter for food avoidance in polyphagous herbivores. Ecologically, such responses modify insect feeding behavior and thus predation rate. For example, solitarious-phase locusts avoid plants containing hyoscyamine, a bitter tasting alkaloid, whereas gregarious-phase locusts prefer those (Despland and Simpson 2005). Gregarious-phase locusts feeding on hyoscyamine develop dark coloration and become tasteless, thus getting survival benefit due to avoidance by predators. Bitterness-signaling pathways are highly diverse in an insect. Ironically, often insects cannot differentiate between natural bitter compounds and entomotoxic bitter compounds due to activation of the same signaling pathways inside them. For example, tobacco hornworm (*Manduca sexta*) cannot distinguish between salicin (nontoxic phenolic compound) and toxic alkaloid such as caffeine (Glendinning et al. 2002). Hence, avoiding plants based on bitterness may not be a good strategy always as it increases the evolutionary cost of avoidance.

4.2 Alteration of Plant Defense

Insects can also modify plant defenses before or during feeding. Deactivation of plant defense is another useful strategy for some insects that fed on hosts with the secretory canals. For example, late-instar larvae of the monarch butterfly (*Danaus plexippus*) cut the leaf vein before feeding to avoid highly toxic cardenolides from milkweed latex (Helmus and Dussourd 2005). Vein cutting is quite a common strategy among other lepidopterans (Dussourd 2003). However, suppression of plant defenses can also be achieved by other means at the molecular level. For example, often insect effector molecules present in the saliva suppress plant defenses, i.e., oral secretions of tobacco earworm (*Helicoverpa zea*) reduce nicotine-mediated defense of tobacco (*Nicotiana tabacum*) using glucose oxidase, the salivary enzyme (Musser et al. 2002). The same enzyme in another lepidopteran insect (*Spodoptera exigua*) saliva reduces the expression of regulatory genes at a transcriptomic level in *Medicago truncatula* that is involved in early plant defense signaling (Bede et al. 2006). In an extreme scenario, insect herbivores do not only modify the plant defense elicitor but also produce an antagonistic form, i.e., caterpillar *Anticarsia gemmatalis* (legume specialist) alters the plant defense elicitor inceptin and produces an antagonistic form indicating the role of oral secretion in the adaptation to host plants within lepidopterans.

Alternatively, hemipteran insect (i.e., aphids, whitefly) effectors can play a significant role in host colonization. Effector proteins from aphid saliva can prevent sieve element sealing during feeding to get uninterrupted access to the phloem. They can minimize the usual occlusion response of the host by interacting with the calcium molecules (Will et al. 2007). Precisely, *Aphis* saliva contains calcium-binding proteins to disperse forisomes, a calcium-dependent protein from plant origin engaged in sieve tube occlusion. Plants generate ROS (reactive oxygen species) due to damage caused by aphid feeding. Some of the aphid salivary effector proteins such as Mp55 from green peach aphid (*Myzus persicae*) reduce ROS production, thus assisting better performance of aphids on the hosts (Elzinga et al. 2014). Moreover, transgenic expression of aphid effector genes (i.e., Me10, Me23) in the host enhances aphid fecundity by unknown mechanisms (Atamian et al. 2013). Aphids can also use their effectors as a decoy to trigger less effective defense response and thus suppress ones that are more effective. For example, potato aphid (*M. euphorbiae*) secretes a glutathione S-transferase (Me47) effector molecule that induces PAMP-triggered immunity (PTI) in tomato and ultimately supports colonization (Kettles and Kaloshian 2016).

4.3 Metabolic Resistance: Detoxification of Plant Allelochemicals

Metabolic resistance is one of the primary resistance mechanisms observed in herbivores. It is a common defense mechanism relying on detoxifying enzymatic systems such as cytochromes and glutathione S-transferases (GST) that protect the insect by detoxifying plant secondary metabolites (Heidel-Fischer and Vogel 2015; Rane et al. 2016; War et al. 2012). Some of these enzymes give broad-spectrum resistance to insects against various toxic plant compounds (Isman 2006). Insect usually detoxifies plant compounds in different distinctive phases: phase I and phase II. In phase I, plant compounds are hydrolyzed or oxidized mainly by cytochrome P450s or carboxylesterases (functionalization step), and then the lipophilic xenobiotics are modified to hydrophilic compounds via glycosylation process (phase II) by two other groups of enzymes, namely, GSTs and UDP-glucosyl transferase (UGTS) (Berenbaum and Johnson 2015; Bues et al. 2005; Despres et al. 2007; Sonoda and Tsumuki 2005). However, metabolic resistance mechanisms in herbivore insects and the evolutionary forces underneath such sophisticated mechanisms are not yet clearly understood. With recent advancements in next-generation sequencing technology, we have started understanding the molecular mechanisms underlying such resistance in herbivores. A noteworthy observation from several transcriptional profiling experiments with model and non-model insect herbivores is the physiological plasticity within the gut in response to diverse plant allelochemicals. Induction of a specific set of detoxification enzymes is one of the key measures documented in various studies (Celorio-Mancera et al. 2012; Giraudo et al. 2015).

Detoxifying enzymes is primarily composed of three main subfamilies: cytochrome P450s, GSTs, and carboxylesterases (COEs). Among these, the role of cytochrome P450s (heme-thiolate enzymes) in insect adaptation to plant allelochemicals is studied extensively. For instance, cytochrome P450 (primarily CYP6B gene family)-mediated detoxification of furanocoumarin helps *Depressaria pastinacella* (parsnip webworm) and *Papilio polyxenes* (black swallowtail) to overcome the furanocoumarin-mediated defense of their host (i.e., wild parsnip) (Li et al. 2001; Li et al. 2003). Similarly, CYP6AE14 in *H. armigera* is documented to be linked with the tolerance of the toxic compound gossypol present in the cotton plant (Mao et al. 2007). Tobacco-adapted strain of aphid *Myzus persicae* overexpresses CYP6CY3 to detoxify nicotine (Bass et al. 2013). A tissue-specific expression study with *Spodoptera frugiperda* (fall armyworm) in response to various plant allelochemicals indicated the involvement of cytochrome P450s in the adaptation of generalist lepidopterans against a myriad of plant defense compounds (Giraudo et al. 2015).

The GST superfamily enzymes are involved in plant xenobiotic detoxification or sequestration process through accelerating the reaction involving the addition of glutathione moiety to harmful plant compounds that enhance their water solubility and assist their excretion from insect body (Enayati et al. 2005). Overproduction of GSTs is assumed to be involved in herbivore tolerance to toxic glucosinolates and isothiocyanates of Brassicaceae hosts (Francis et al. 2005). The greater diversity of GSTs and P450s observed in generalist herbivores compared to specialists can be correlated with their phenotypic plasticity and the evolution of extended diet breadth (Francis et al. 2001, 2005).

UGTs detoxify plant allelochemicals by conjugating sugar moiety (glycosylation) and make them water-soluble. Thus, UGTs assists detoxification of plant allelochemicals inside the insect body. Like other vital enzymes (i.e., P450s, GSTs), UGTs are ubiquitous with more than 310 putative UGTs already documented in insects (Ahn et al. 2012). However, the specific role of UGTs in xenobiotic detoxification is recognized very recently. Lepidopteran species such as *H. armigera*, *H.zea*, and *H.assulta* detoxify capsaicin (compound acts as feeding and oviposition deterrent) produced from chili peppers through glycosylation by UGTs

(Ahn et al. 2011a, 2011b). Similarly, benzoxazinoids (BXDs) from maize plant are also detoxified through glycosylation in maize specialist *Spodoptera frugiperda* (Maag et al. 2014; Roy et al. 2016). BXDs and its corresponding aglucones cause growth retardation and induce mortality in herbivore insects. The reglycosylation of benzoxazinoid DIMBOA (stereochemical configuration alteration: S to R form) inside *S. frugiperda* gut makes them resistant against the plant and insect ß-glucosidase, and thus formation of toxic aglucones is escaped inside the insect gut (Wouters et al. 2014). Thus, glycosylation by UGTs comes out to be a common detoxification strategy among many lepidopterans. However, it is worth mentioning here that feeding on plant tissues with toxic secondary metabolites targeting various physiological processes in insects may result in metabolic trade-offs between detoxification and nutrition acquisition in the gut (Fig. 2).

4.4 Target Site Resistance, Excretion, and Sequestration of Plant Toxins

Insects can achieve resistance by altering the target site and thus avoid the toxicity of plant allelochemicals. For instance, a specific amino acid change of Na/K-ATPase in monarch butterfly helps them to survive on toxic cardenolide-containing host plants (Holzinger et al. 1992). Four other insect orders show different mutations in Na/K-ATPase resulting in insensitivity of target site against cardenolides indicating toward a convergent molecular mechanism (Dobler et al. 2012). Excretion of toxic plant compound out of the body often helps insect to survive on toxin-laden plant food. In *Drosophila melanogaster*, OATP58Db (a specific anion-transporting polypeptide) assists in rapid excretion of ouabain, a plant-derived toxic cardiac glycoside, outside their body making them resistant to the compound (Torrie et al. 2004). Recently, the function of ABC transporters in the elimination of toxic plant xenobiotics and pesticides is documented (Merzendorfer 2014). Specific transporters such as ABC transporters regulate the spatiotemporal dynamics of plant allelochemical accumulation inside insect body (Strauss et al. 2013). Numerous insect herbivores can also sequester (selective uptake and accumulate) plant secondary metabolites and use them against their predators (Beran et al. 2014; Bridges et al. 2002; Ratzka et al. 2002). For instances, flea beetle constructs their glucosinolate-myrosinase system using their host plant defense metabolites (Beran et al. 2014). The glucosinolate-myrosinase system is a host defense system where the substrate glucosinolate and the enzyme myrosinase come into contact only upon tissue damage caused by insect herbivores, leading to the production of toxic hydrolysis products.

4.5 Horizontal Gene Transfer: Contribution of Microbes to Insect Adaptation

The gaining of new enzymatic function through horizontal gene transfer (HGT) is a promising alternative for insects to detoxify toxic plant compounds. Some butterflies and the spider mite (*Tetranychus urticae*) observed to possess a ß-cyanoalanine synthase originated from bacteria through HGT. ß-cyanoalanine synthase detoxifies the toxic cyanogenic glycosides by changing them to nontoxic ß-cyanoalanine (Wybouw et al. 2014). Herbivore beetles degrade plant cell wall that is made up of a complex mixture of plant carbohydrates using plant cell wall-degrading enzymes (PCWDE). Herbivores acquire PCWDE through HGT events from bacteria or fungi (Kirsch et al. 2014; Pauchet and Heckel 2013). Insect ß-fructofuranosidases that help in breaking down of plant sucrose are also obtained from independent HGT events from bacteria (Daimon et al. 2008; Zhao et al. 2014). In summary, insects also generate a number of counter-defense responses against multilayered plant defenses which make plant-insect evolutionary arms race very dynamic (Fig. 2). Recent advancements in transcriptomics, proteomics, metagenomics, and gene-editing techniques such as RNAi, CRISPR-Cas9, gene drive, and their use in the field of entomology will help to map the fate of plant defense compounds precisely inside insect body and disentangle the insect responses against them even better than before.

5 Role of Constitutive Chemical Compounds: The Phytoanticipins

Secondary metabolites are considered to be biomolecules of prime importance that directly influence host immunity. These compounds do not alter normal growth and development but lessen the palatability of the tissue from where they are produced (Howe and Jander 2008). Alteration in tissue quality has a direct impact on the herbivores decision to colonize the said host or not. Some secondary metabolites are constitutively expressed throughout the lifespan and at specific locations of the host. These metabolites are referred to as the “phytoanticipins” that anticipate the entry of foreign enemy and trigger defense signaling (War et al. 2012). Phytoanticipins are primarily activated by β-glucosidase during insect attack that triggers the liberation of biocidal aglycone metabolites (Morant et al. 2008).

Glucosinolates are believed to represent the largest class of phytoanticipins that affect plant-insect interaction (Hopkins et al. 2009). Mostly occurring in Brassicaceae members, their chemical structure constitutes of the β-thioglucose moiety, sulfonated oxime moiety, and a changeable side chain. The nomenclature and grouping of glucosinolates depend on the amino acid constituent present in the variable side chain (Fahey et al. 2001). Indole glucosinolates constitute 10%, aliphatic amino acid containing glucosinolates 50%, aromatic amino acid containing glucosinolates 10% while rest glucosinolates contain various amino acids. Conformational change of glucosinolates occurs due to chain elongation, oxidation, or hydroxylation of side chains. Such structural changes are genetically controlled and are influenced by plant-insect interaction (Grubb and Abel 2006). Even subtle structural changes like the addition of an extra OH at C2 position yield products that have biologically opposing activities (Matusheski et al. 2006). Such findings were supported by studies conducted on generalists *Mamestra brassicae* that performed well on *Barbarea vulgaris* yielding gluconasturtiin, while their larvae instantly died on plants expressing glucobarbarin, an OH added substituent of the glucosinolate gluconasturtiin (van Leur et al. 2008). Although the bioactivity of glucosinolate is often found when they are in intact form, in the majority of cases, the enzyme myrosinase hydrolyzes them to convert into an active form. A thioglucosidase, myrosinase, is stored in special myrosinase cells occurring in all plant organs (Rask et al. 2000). Damage of host tissue during insect feeding releases host glucosinolates from cell vacuoles which come in contact with myrosinase and yields several toxic and pungent products such as isothiocyanates, nitriles, and oxazolidinethiones (Bones and Rossiter 2006). pH, concentration of Fe^{2+}, presence of epithiospecific protein (ESP), or epithiospecific modifier protein (ESMI) are important contributors to glucosinolate-myrosinase reaction (Burow et al. 2006). Glucosinolate composition varies greatly both qualitatively and quantitatively across and within species with *Arabidopsis thaliana* reported to have 34 different types of glucosinolates among 39 ecotypes with minimum overlap among profiles (Kliebenstein et al. 2001). Furthermore, a single plant also varies in glucosinolate composition both spatially and temporally. Such variation is believed to be a result of variable insect infestation across plant tissue of a single plant (Van Dam et al. 2009). Host glucosinolates and their hydrolytic products impose negative selection pressure on the generalist herbivores by creating growth depression, host tissue unpalatability, reduced food efficiency, hypertrophy and goiter formation of the insect gland, liver lesion, and necrosis of insect tissue (Anilakumar et al. 2006). All such negative growth parameters are linked to antibiosis and/or antixenosis, a resistance mechanism utilized by the plants to prevent pest colonization (Hopkins et al. 2009). However, glucosinolates are also believed to have kairomonal roles, which are exploited by the specialist herbivores that utilize glucosinolates as their oviposition and feeding stimuli (Wittstock et al. 2003). Thus, variable roles, composition, and quantity of glucosinolates in and among host species still leave this area of research open in a better understanding of plant-herbivore interaction.

Another phytoanticipin that significantly contributes to the study of plant-insect interaction is benzoxazinoids (BXDs), having the skeleton of benzoxazinones and their derivatives. Degradation of benzoxazinones yielding benzaxizolinones is also reported to be lethal to insects (Wouters et al. 2016). Although the knowledge of BXDs as insect deterrents are comparatively new, several studies have already started documenting about BXDs as an antifeedant, insecticidal, and antimicrobial and as having allelopathic activities (Niemeyer 2009). This group of nitrogen-containing secondary metabolites is present in many kinds of grass and members of Poaceae including agriculturally essential crops like maize, wheat, and rye. Besides some dicots like Acanthaceae, Ranunculaceae, Plantaginaceae, and Lamiaceae, members also contain BXD (Frey et al. 2009). The release of stored BXDs is similar to that of glucosinolates that occur during mechanical injury caused by insect feeding (Pentzold et al. 2014). Besides, interspecies and intraspecies quantitative and qualitative difference of BXDs also makes them analogous to glucosinolates (Dafoe et al. 2011).The mode of BXDs in acting antagonistically against insect pests and imparting host defense is not clearly understood. However, few features such as electrophilic nature of open form and nitrenium ions, the capacity of hydroxamic acid to complex with metal ions, and reactivity of degraded products of BXDs attribute to the toxic effects of BXDs (Wouters et al. 2016). Harmful effects of BXDs were studied against lepidopteran *Ostrinia nubilalis* and *Spodoptera frugiperda* (Roy et al. 2016; Wouters et al. 2014). Besides, BXDs also proved to be effective against aphid *Sitobion avenae and Rhopalosiphum padi* (Bravo et al. 2004). Even then, the variability of BXDs makes the study challenging which requires an integrated approach to arrive at understanding the specific role of host BXDs at a given time point in any plant-insect interaction.

6 Role of Inducible Chemical Compounds: The Phytoalexins and Phenolics

Phytoalexins are critical components of plant's induced defense, which trigger their self-expression only when needed. Thus, timely expression of these particular chemical compounds does not impose any additional energy charges on the host (War et al. 2012). The role of phytoalexins against herbivore resistance is relatively less discussed than other biotic stress factors. However, the role of secondary metabolites such as isoflavonoids, terpenoids, and alkaloids in herbivore resistance is widely studied. Interestingly, these class of secondary metabolites behave as phytoalexins during insect attack (War et al. 2012). The present section shall brief the role of these compounds in imparting host resistance against insect attack.

Isoflavonoids are considered as phytoalexins. However, flavonoids are referred to as phenolics, with isoflavonoid being a 3-phenylchroman member of flavonoids (Dixon and Pasinetti 2010). However, both flavonoids and isoflavonoids guard the host plant by deterring the growth and development of the herbivore (Simmonds 2003). Isoflavonoid licoisoflavone A and B, luteone, and wighteone have been reported to be an insect deterrent (Lane et al. 1987). Judaicin and maackain isolated from wild members of chickpea have been reported to work as antifeedant against *Helicoverpa armigera*, *Spodoptera littoralis*, and *Spodoptera frugiperda*, respectively (Simmonds and Stevenson 2001). Terpenes also form a large class of secondary metabolites that act as insecticidal toxins. Pyrethroids from *Chrysanthemum* show neurotoxic effects against herbivores and parasitoids. Alpha- and beta-pinene, limonene, and myrcene found in resin ducts of conifers are reported to have toxic effects against beetles and other wide range of insects (Turlings et al. 1995). Sesquiterpene costunolides of family compositae act as strong feeding repellent for insects (Picman 1986). The diterpene abetic acid of pines and leguminous trees physically blocks insect feeding by forming chemical complexes with resin (Bradley et al. 1992). Phytoecdysone, a triterpene, disrupts insect molting that proves to be detrimental for insect development (Slama 1980). Besides, other triterpenes like azadirachtin from *Azadirachta indica* exert toxic effects on herbivores (Mordue and Blackwell 1993). Additionally, the polyterpene rubber occurring in laticiferous tissue acts as an insect repellent (Eisner et al. 1996). Alkaloids are nitrogenous secondary metabolites that are frequently found in herbaceous dicots and few monocots and gymnosperms (Hegnauer 1988). Colchicine from *Colchicum autumnale* is toxic to *Apis mellifera* as it stalls the microtubule polymerization by combining with it and blocking mitosis. Sanguinarine from *Sanguinaria canadensis* affects neurotransmission and DNA synthesis of insects (Mithöfer and Boland 2012). Nicotine also behaves as a neurotoxin causing continuous neuronal excitation that results in paralysis of feeding insects (Dewey and Xie 2013). Besides, the pyrrolizidine alkaloids (PA) are known to exert toxic effects on many insect herbivores and predators (Matsuura and Fett-Neto 2017). PAs occur in two forms: free tertiary base form and *N*-oxide form. The *N*-oxide form is broken down into tertiary form in insect gut which is absorbed and further reduced to highly toxic pyrrols with the help of insect P450 monooxygenase enzymes (Macel 2011). PA seneciphyline from *Senecio* sp. deterred the performance of *Acyrthosiphon pisum.* However, structural stereo-isomeric PAs were reported to exert variable deterrent effects on the insect species with N-oxide being less effective than tertiary base-type PA (Macel et al. 2005). Furthermore, specialist insects were reported to detoxify and sequester PAs, while generalists showed growth and performance deterioration when exposed to host PAs (Macel 2011).

Plant phenolics are essential secondary metabolites that consist of different structural forms, all arising primarily from the shikimate-phenylpropanoid-flavonoid pathways (Lattanzio et al. 2006). Several phenolics act as feeding obstacles for insect herbivores such as salicylates in *Salix* leaves slowing down the growth of *Operophtera brumata* (Simmonds 2003). Hypercalin A found in pollen and developing seeds of *Hypericum calycinum* proved to be noxious to *Utetheisa ornatrix* (Gronquist et al. 2001).The chlorogenic acid of willow plants acted as antifeedant against beetle *Lochmaea caprea* (Simmonds 2003). The role of isoflavonoids as antifeedants of herbivorous insects has already being discussed in the section where the role of phytoalexins has been briefed. Besides, increased concentration of flavonoid aglycones such as quercetin, kaempferol, and isorhamnetin of *Vigna unguiculata* showed to reduce reproductive rates of *Aphis craccivora* (Lattanzio et al. 2000). Lignans (leptostachyol acetate) isolated from roots of herbaceous perennial plant *Phryma leptostachya* showed insecticidal properties against many lepidopterans (Park et al. 2005). Besides, magnolol isolated from *Magnolia virginiana* showed poisonous effects against larvae of moth *Callosamia promethea* (Harborne 2001). Among other phenolics, tannins form an important class that serves as insecticides. They reduce the palatability of host tissue by adding an astringent effect, form complexes with insect's digestive proteins, and hamper their digestion and act as enzyme inactivators of insects (Shirley 1998). Cowpea proanthocyanidins (condensed tannins) showed resistance to colonization by cowpea weevil, *Callosobruchus maculatus* (Lattanzio et al. 2005).

Thus, the insecticidal properties of phytoalexins and plant phenolics are prominent, but spatial and temporal expressional sketch and exact mode of action of these induced secondary metabolites still need extensive experimentations. Moreover, rapid variations in both host chemical profiles and insect physiology with changing ecological conditions demand detailed and case-specific studies on delineating the role of these secondary metabolites in imparting host defense against insect herbivores.

7 Conclusion

It is evident from past reports that plant-insect interaction is a highly complex phenomenon that engages different direct and indirect biotic and abiotic factors simultaneously. It is almost impossible to analyze the role of all the entities at a given spatiotemporal scale, especially when discussing a specific plant-insect interaction in one-to-one fashion. Besides this, environment is a key regulatory factor for any interaction; the slightest alteration of its components is believed to bring about substantial changes in the behavior and metabolism of the one-to-one interacting partners. However, to unveil still hidden intricacies of plant-insect interaction, more and more case studies and community studies are needed which shall help build ecological interaction models and help to approach toward generalized conclusions. Hence, plant-insect interaction research is still in its infancy paving its way for interactional science ahead.

Acknowledgments AR would like to acknowledge the EU project "EXTEMIT – K," No. CZ.02.1.01/0.0/0.0/15_003/0000433, financed by OP RDE and IGA, Faculty of Forestry and Wood Sciences, Czech University of Life Sciences, Prague, for their financial support during the preparation of the book chapter.

References

Agrawal AA (2001) Transgenerational consequences of plant responses to herbivory: an adaptive maternal effect? Am Nat 157(5):555–569

Agrawal AA (2002) Herbivory and maternal effects: mechanisms and consequences of transgenerational induced plant resistance. Ecology 83(12):3408–3415

Ahn S-J, Badenes-Pérez FR, Heckel DG (2011a) A host-plant specialist, Helicoverpa assulta, is more tolerant to capsaicin from Capsicum annuum than other noctuid species. J Insect Physiol 57(9):1212–1219

Ahn SJ, Badenes-Pérez FR, Reichelt M, Svatoš A, Schneider B, Gershenzon J, Heckel DG (2011b) Metabolic detoxification of capsaicin by UDP-glycosyltransferase in three Helicoverpa species. Arch Insect Biochem Physiol 78(2):104–118

Ahn S-J, Vogel H, Heckel DG (2012) Comparative analysis of the UDP-glycosyltransferase multigene family in insects. Insect Biochem Mol Biol 42(2):133–147

Alborn H, Turlings T, Jones TH, Stenhagen G, Loughrin J, Tumlinson J (1997) An elicitor of plant volatiles from beet armyworm oral secretion. Science 276(5314):945–949

Alborn HT, Hansen TV, Jones TH, Bennett DC, Tumlinson JH, Schmelz EA, Teal PE (2007) Disulfooxy fatty acids from the American bird grasshopper Schistocerca americana, elicitors of plant volatiles. Proc Natl Acad Sci 104(32):12976–12981

Amari K, Boutant E, Hofmann C, Schmitt-Keichinger C, Fernandez-Calvino L, Didier P et al (2010) A family of plasmodesmal proteins with receptor-like properties for plant viral movement proteins. PLoS Pathog 6(9):e1001119

Anilakumar K, Khanum F, Bawa A (2006) Dietary role of glucosinolate derivatives: a review. J Food Sci Technol Mysore 43(1):8–17

Appel HM, Cocroft R (2014) Plants respond to leaf vibrations caused by insect herbivore chewing. Oecologia 175(4):1257–1266

Atamian HS, Chaudhary R, Cin VD, Bao E, Girke T, Kaloshian I (2013) In planta expression or delivery of potato aphid Macrosiphum euphorbiae effectors Me10 and Me23 enhances aphid fecundity. Mol Plant-Microbe Interact 26(1):67–74

Barakat A, Bagniewska-Zadworna A, Frost CJ, Carlson JE (2010) Phylogeny and expression profiling of CAD and CAD-like genes in hybrid Populus (P. deltoides× P. nigra): evidence from herbivore damage for subfunctionalization and functional divergence. BMC Plant Biol *10* (1):100

Barbehenn RV, Constabel CP (2011) Tannins in plant–herbivore interactions. Phytochemistry 72 (13):1551–1565

Bass C, Zimmer CT, Riveron JM, Wilding CS, Wondji CS, Kaussmann M et al (2013) Gene amplification and microsatellite polymorphism underlie a recent insect host shift. Proc Natl Acad Sci 110(48):19460–19465

Batistič O, Kudla J (2012) Analysis of calcium signaling pathways in plants. Biochim Biophys Acta Gen Subj 1820(8):1283–1293

Bede JC, Musser RO, Felton GW, Korth KL (2006) Caterpillar herbivory and salivary enzymes decrease transcript levels of Medicago truncatula genes encoding early enzymes in terpenoid biosynthesis. Plant Mol Biol 60(4):519–531

Beran F, Pauchet Y, Kunert G, Reichelt M, Wielsch N, Vogel H et al (2014) Phyllotreta striolata flea beetles use host plant defense compounds to create their own glucosinolate-myrosinase system. Proc Natl Acad Sci 111(20):7349–7354

Berenbaum MR, Johnson RM (2015) Xenobiotic detoxification pathways in honey bees. Curr Opin Insect Sci 10:51–58

Bernays EA (1991) Evolution of insect morphology in relation to plants. Philos Trans R Soc Lond B 333(1267):257–264

Bernays E (1997) Feeding by lepidopteran larvae is dangerous. Ecol Entomol 22(1):121–123

Bernays EA (1998) Evolution of feeding behavior in insect herbivores. Bioscience 48(1):35–44

Bernays E, Barbehenn R (1987) Nutritional ecology of grass foliage-chewing insects. In: Nutritional ecology of insects, mites, spiders and related invertebrates. Wiley, New York, pp 147–175

Bhonwong A, Stout MJ, Attajarusit J, Tantasawat P (2009) Defensive role of tomato polyphenol oxidases against cotton bollworm (Helicoverpa armigera) and beet armyworm (Spodoptera exigua). J Chem Ecol 35(1):28–38

Birke A, Aluja M (2017) Do mothers really know best? Complexities in testing the preference-performance hypothesis in polyphagous frugivorous fruit flies. *Bull Entomol Res* 108:1–11

Bones AM, Rossiter JT (2006) The enzymic and chemically induced decomposition of glucosinolates. Phytochemistry 67(11):1053–1067

Bradley DJ, Kjellbom P, Lamb CJ (1992) Elicitor-and wound-induced oxidative cross-linking of a proline-rich plant cell wall protein: a novel, rapid defense response. Cell 70(1):21–30

Bravo HR, Copaja SV, Argandoña VH (2004) Chemical basis for the antifeedant activity of natural hydroxamic acids and related compounds. J Agric Food Chem 52(9):2598–2601

Bridges M, Jones AM, Bones AM, Hodgson C, Cole R, Bartlet E et al (2002) Spatial organization of the glucosinolate–myrosinase system in brassica specialist aphids is similar to that of the host plant. Proc R Soc Lond B Biol Sci 269(1487):187–191

Bruinsma M, Posthumus MA, Mumm R, Mueller MJ, van Loon JJ, Dicke M (2009) Jasmonic acid-induced volatiles of Brassica oleracea attract parasitoids: effects of time and dose, and comparison with induction by herbivores. J Exp Bot 60(9):2575–2587

Bues R, Bouvier J, Boudinhon L (2005) Insecticide resistance and mechanisms of resistance to selected strains of Helicoverpa armigera (Lepidoptera: Noctuidae) in the south of France. Crop Prot 24(9):814–820

Bukovinszky T, van Veen FF, Jongema Y, Dicke M (2008) Direct and indirect effects of resource quality on food web structure. Science 319(5864):804–807

Burow M, Markert J, Gershenzon J, Wittstock U (2006) Comparative biochemical characterization of nitrile-forming proteins from plants and insects that alter myrosinase-catalysed hydrolysis of glucosinolates. FEBS J 273(11):2432–2446

Celorio-Mancera MDP, Heckel DG, Vogel H (2012) Transcriptional analysis of physiological pathways in a generalist herbivore: responses to different host plants and plant structures by the cotton bollworm, Helicoverpa armigera. Entomol Exp Appl 144(1):123–133

Chakraborti D, Sarkar A, Mondal HA, Das S (2009) Tissue specific expression of potent insecticidal, Allium sativum leaf agglutinin (ASAL) in important pulse crop, chickpea (Cicer arietinum L.) to resist the phloem feeding Aphis craccivora. Transgenic Res 18(4):529–544

Chapman R (2003) Contact chemoreception in feeding by phytophagous insects. Annu Rev Entomol 48(1):455–484

Coyle DR, Clark KE, Raffa KF, Johnson SN (2011) Prior host feeding experience influences ovipositional but not feeding preference in a polyphagous insect herbivore. Entomol Exp Appl 138(2):137–145

Cuéllar T, Azeem F, Andrianteranagna M, Pascaud F, Verdeil JL, Sentenac H et al (2013) Potassium transport in developing fleshy fruits: the grapevine inward K+ channel VvK1. 2 is activated by CIPK–CBL complexes and induced in ripening berry flesh cells. Plant J 73 (6):1006–1018

Dafoe NJ, Huffaker A, Vaughan MM, Duehl AJ, Teal PE, Schmelz EA (2011) Rapidly induced chemical defenses in maize stems and their effects on short-term growth of Ostrinia nubilalis. J Chem Ecol 37(9):984

Daimon T, Taguchi T, Meng Y, Katsuma S, Mita K, Shimada T (2008) β-Fructofuranosidase genes of the silkworm, Bombyx mori INSIGHTS INTO ENZYMATIC ADAPTATION OF B. MORI TO TOXIC ALKALOIDS IN MULBERRY LATEX. J Biol Chem 283(22):15271–15279

De Boer JG, Posthumus MA, Dicke M (2004) Identification of volatiles that are used in discrimination between plants infested with prey or nonprey herbivores by a predatory mite. J Chem Ecol 30(11):2215–2230

Despland E, Simpson SJ (2005) Food choices of solitarious and gregarious locusts reflect cryptic and aposematic antipredator strategies. Anim Behav 69(2):471–479

Despres L, David J-P, Gallet C (2007) The evolutionary ecology of insect resistance to plant chemicals. Trends Ecol Evol 22(6):298–307

Devoto A, Turner JG (2005) Jasmonate-regulated Arabidopsis stress signalling network. Physiol Plant 123(2):161–172

Dewey RE, Xie J (2013) Molecular genetics of alkaloid biosynthesis in Nicotiana tabacum. Phytochemistry 94:10–27

Dicke M, Baldwin IT (2010) The evolutionary context for herbivore-induced plant volatiles: beyond the 'cry for help'. Trends Plant Sci 15(3):167–175

Dixon RA, Pasinetti GM (2010) Flavonoids and isoflavonoids: from plant biology to agriculture and neuroscience. Plant Physiol 154(2):453–457

Dobler S, Dalla S, Wagschal V, Agrawal AA (2012) Community-wide convergent evolution in insect adaptation to toxic cardenolides by substitutions in the Na, K-ATPase. Proc Natl Acad Sci 109(32):13040–13045

Du L, Yang T, Puthanveettil S, Poovaiah B, Luan S (2011) Calmodulin-binding proteins in plants: decoding of calcium signal through calmodulin. Coding and Decoding of Calcium Signals in Plants. Springer-Verlag, Berlin, pp 177–233

Dudareva N, Negre F, Nagegowda DA, Orlova I (2006) Plant volatiles: recent advances and future perspectives. Crit Rev Plant Sci 25(5):417–440

Dussourd DE (2003) Chemical stimulants of leaf-trenching by cabbage loopers: natural products, neurotransmitters, insecticides, and drugs. J Chem Ecol 29(9):2023–2047

Dutta I, Saha P, Majumder P, Sarkar A, Chakraborti D, Banerjee S, Das S (2005) The efficacy of a novel insecticidal protein, Allium sativum leaf lectin (ASAL), against homopteran insects monitored in transgenic tobacco. Plant Biotechnol J 3(6):601–611

Ehrlich PR, Raven PH (1967) Butterflies and plants. Sci Am 216(6):104–114

Eisner T, Meinwald J, Wyatt TD (1996) Chemical ecology: the chemistry of biotic interaction. Nature 381(6580):289–289

Elzinga DA, De Vos M, Jander G (2014) Suppression of plant defenses by a Myzus persicae (green peach aphid) salivary effector protein. Mol Plant-Microbe Interact 27(7):747–756
Enayati AA, Ranson H, Hemingway J (2005) Insect glutathione transferases and insecticide resistance. Insect Mol Biol 14(1):3–8
Espinas NA, Saze H, Saijo Y (2016) Epigenetic control of defense signaling and priming in plants. Front Plant Sci 7:1201
Fahey JW, Zalcmann AT, Talalay P (2001) The chemical diversity and distribution of glucosinolates and isothiocyanates among plants. Phytochemistry 56(1):5–51
Fox C, Stillwell R, Amarillo-S A, Czesak M, Messina FJ (2004) Genetic architecture of population differences in oviposition behaviour of the seed beetle Callosobruchus maculatus. J Evol Biol 17 (5):1141–1151
Francis F, Lognay G, Wathelet J-P, Haubruge E (2001) Effects of allelochemicals from first (Brassicaceae) and second (Myzus persicae and Brevicoryne brassicae) trophic levels on Adalia bipunctata. J Chem Ecol 27(2):243–256
Francis F, Vanhaelen N, Haubruge E (2005) Glutathione S-transferases in the adaptation to plant secondary metabolites in the Myzus persicae aphid. Arch Insect Biochem Physiol 58 (3):166–174
Frey M, Schullehner K, Dick R, Fiesselmann A, Gierl A (2009) Benzoxazinoid biosynthesis, a model for evolution of secondary metabolic pathways in plants. Phytochemistry 70 (15–16):1645–1651
Gatehouse JA (2002) Plant resistance towards insect herbivores: a dynamic interaction. New Phytol 156(2):145–169
Giraudo M, Hilliou F, Fricaux T, Audant P, Feyereisen R, Le Goff G (2015) Cytochrome P450s from the fall armyworm (Spodoptera frugiperda): responses to plant allelochemicals and pesticides. Insect Mol Biol 24(1):115–128
Glendinning JI, Davis A, Ramaswamy S (2002) Contribution of different taste cells and signaling pathways to the discrimination of "bitter" taste stimuli by an insect. J Neurosci 22 (16):7281–7287
Gronquist M, Bezzerides A, Attygalle A, Meinwald J, Eisner M, Eisner T (2001) Attractive and defensive functions of the ultraviolet pigments of a flower (Hypericum calycinum). Proc Natl Acad Sci 98(24):13745–13750
Grubb CD, Abel S (2006) Glucosinolate metabolism and its control. Trends Plant Sci 11(2):89–100
Gulsen O, Eickhoff T, Heng-Moss T, Shearman R, Baxendale F, Sarath G, Lee D (2010) Characterization of peroxidase changes in resistant and susceptible warm-season turfgrasses challenged by Blissus occiduus. Arthropod Plant Interact 4(1):45–55
Handley R, Ekbom B, Ågren J (2005) Variation in trichome density and resistance against a specialist insect herbivore in natural populations of Arabidopsis thaliana. Ecol Entomol 30 (3):284–292
Hanley ME, Lamont BB, Fairbanks MM, Rafferty CM (2007) Plant structural traits and their role in anti-herbivore defence. Perspect Plant Ecol Evol Syst 8(4):157–178
Harborne JB (2001) Twenty-five years of chemical ecology. Nat Prod Rep 18(4):361–379
Hegnauer R (1988) Biochemistry, distribution and taxonomic relevance of higher plant alkaloids. Phytochemistry 27(8):2423–2427
Heidel-Fischer HM, Vogel H (2015) Molecular mechanisms of insect adaptation to plant secondary compounds. Curr Opin Insect Sci 8:8–14
Helmus MR, Dussourd DE (2005) Glues or poisons: which triggers vein cutting by monarch caterpillars? Chemoecology 15(1):45–49
Holzinger F, Frick C, Wink M (1992) Molecular basis for the insensitivity of the monarch (Danaus plexippus) to cardiac glycosides. FEBS Lett 314(3):477–480
Hopkins RJ, van Dam NM, van Loon JJ (2009) Role of glucosinolates in insect-plant relationships and multitrophic interactions. Annu Rev Entomol 54:57–83

Horiuchi J-I, Arimura G-I, Ozawa R, Shimoda T, Dicke M, Takabayashi J, Nishioka T (2003) Lima bean leaves exposed to herbivore-induced conspecific plant volatiles attract herbivores in addition to carnivores. Appl Entomol Zool 38(3):365–368

Howe GA, Herde M (2015) Interaction of plant defense compounds with the insect gut: new insights from genomic and molecular analyses. Curr Opin Insect Sci 9:62–68

Howe GA, Jander G (2008) Plant immunity to insect herbivores. Annu Rev Plant Biol 59:41–66

Isman MB (2006) Botanical insecticides, deterrents, and repellents in modern agriculture and an increasingly regulated world. Annu Rev Entomol 51:45–66

Jones JD, Dangl JL (2006) The plant immune system. Nature 444(7117):323

Kanchiswamy CN, Takahashi H, Quadro S, Maffei ME, Bossi S, Bertea C et al (2010) Regulation of Arabidopsis defense responses against Spodoptera littoralis by CPK-mediated calcium signaling. BMC Plant Biol 10(1):97

Kang J-H, Wang L, Giri A, Baldwin IT (2006) Silencing threonine deaminase and JAR4 in Nicotiana attenuata impairs jasmonic acid–isoleucine–mediated defenses against Manduca sexta. Plant Cell 18(11):3303–3320

Karban R, Agrawal AA, Mangel M (1997) The benefits of induced defenses against herbivores. Ecology 78(5):1351–1355

Kessler A, Baldwin IT (2001) Defensive function of herbivore-induced plant volatile emissions in nature. Science 291(5511):2141–2144

Kessler A, Baldwin IT (2002) Plant responses to insect herbivory: the emerging molecular analysis. Annu Rev Plant Biol 53(1):299–328

Kessler A, Halitschke R, Diezel C, Baldwin IT (2006) Priming of plant defense responses in nature by airborne signaling between Artemisia tridentata and Nicotiana attenuata. Oecologia 148 (2):280–292

Kettles GJ, Kaloshian I (2016) The potato aphid salivary effector Me47 is a glutathione-S-transferase involved in modifying plant responses to aphid infestation. Front Plant Sci 7:1142

Kirsch R, Gramzow L, Theißen G, Siegfried BD, Heckel DG, Pauchet Y (2014) Horizontal gene transfer and functional diversification of plant cell wall degrading polygalacturonases: key events in the evolution of herbivory in beetles. Insect Biochem Mol Biol 52:33–50

Kliebenstein DJ, Kroymann J, Brown P, Figuth A, Pedersen D, Gershenzon J, Mitchell-Olds T (2001) Genetic control of natural variation in Arabidopsis glucosinolate accumulation. Plant Physiol 126(2):811–825

Labandeira CC, Phillips T (1996) A carboniferous insect gall: insight into early ecologic history of the Holometabola. Proc Natl Acad Sci 93(16):8470–8474

Lane G, Sutherland O, Skipp R (1987) Isoflavonoids as insect feeding deterrents and antifungal components from root ofLupinus angustifolius. J Chem Ecol 13(4):771–783

Lattanzio V, Arpaia S, Cardinali A, Di Venere D, Linsalata V (2000) Role of endogenous flavonoids in resistance mechanism of Vigna to aphids. J Agric Food Chem 48(11):5316–5320

Lattanzio V, Terzano R, Cicco N, Cardinali A, Venere DD, Linsalata V (2005) Seed coat tannins and bruchid resistance in stored cowpea seeds. J Sci Food Agric 85(5):839–846

Lattanzio V, Lattanzio VM, Cardinali A (2006) Role of phenolics in the resistance mechanisms of plants against fungal pathogens and insects. Phytochem Adv Res 661(2):23–67

Lawrence PK, Koundal KR (2002) Plant protease inhibitors in control of phytophagous insects. Electron J Biotechnol 5(1):5–6

Li W, Berenbaum M, Schuler M (2001) Molecular analysis of multiple CYP6B genes from polyphagous Papilio species. Insect Biochem Mol Biol 31(10):999–1011

Li W, Schuler MA, Berenbaum MR (2003) Diversification of furanocoumarin-metabolizing cytochrome P450 monooxygenases in two papilionids: specificity and substrate encounter rate. Proc Natl Acad Sci 100(suppl 2):14593–14598

Maag D, Dalvit C, Thevenet D, Köhler A, Wouters FC, Vassão DG et al (2014) 3-β-D-Glucopyranosyl-6-methoxy-2-benzoxazolinone (MBOA-N-Glc) is an insect detoxification product of maize 1, 4-benzoxazin-3-ones. Phytochemistry 102:97–105

Macel M (2011) Attract and deter: a dual role for pyrrolizidine alkaloids in plant–insect interactions. Phytochem Rev 10(1):75–82

Macel M, Bruinsma M, Dijkstra SM, Ooijendijk T, Niemeyer HM, Klinkhamer PG (2005) Differences in effects of pyrrolizidine alkaloids on five generalist insect herbivore species. J Chem Ecol 31(7):1493–1508

Maffei M (2010) Sites of synthesis, biochemistry and functional role of plant volatiles. S Afr J Bot 76(4):612–631

Maffei ME, Mithöfer A, Boland W (2007a) Before gene expression: early events in plant–insect interaction. Trends Plant Sci 12(7):310–316

Maffei ME, Mithöfer A, Boland W (2007b) Insects feeding on plants: rapid signals and responses preceding the induction of phytochemical release. Phytochemistry 68(22–24):2946–2959

Mao Y-B, Cai W-J, Wang J-W, Hong G-J, Tao X-Y, Wang L-J et al (2007) Silencing a cotton bollworm P450 monooxygenase gene by plant-mediated RNAi impairs larval tolerance of gossypol. Nat Biotechnol 25(11):1307

Matsuura HN, Fett-Neto AG (2017) Plant alkaloids: main features, toxicity, and mechanisms of action. In: *Plant toxins*. Springer, Dordrecht, pp 243–261

Matusheski NV, Swarup R, Juvik JA, Mithen R, Bennett M, Jeffery EH (2006) Epithiospecifier protein from broccoli (Brassica oleracea L. ssp. italica) inhibits formation of the anticancer agent sulforaphane. J Agric Food Chem 54(6):2069–2076

Mello MO, Silva-Filho MC (2002) Plant-insect interactions: an evolutionary arms race between two distinct defense mechanisms. Braz J Plant Physiol 14(2):71–81

Mendes R, Kruijt M, De Bruijn I, Dekkers E, van der Voort M, Schneider JH et al (2011) Deciphering the rhizosphere microbiome for disease-suppressive bacteria. Science 1203980

Merzendorfer H (2014) ABC transporters and their role in protecting insects from pesticides and their metabolites. In: *Advances in insect physiology*, vol 46. Elsevier, Oxford, pp 1–72

Mithöfer A, Boland W (2012) Plant defense against herbivores: chemical aspects. Annu Rev Plant Biol 63:431–450

Mitter C, Farrell B, Futuyma DJ (1991) Phylogenetic studies of insect-plant interactions: insights into the genesis of diversity. Trends Ecol Evol 6(9):290–293

Morant AV, Jørgensen K, Jørgensen C, Paquette SM, Sánchez-Pérez R, Møller BL, Bak S (2008) β-Glucosidases as detonators of plant chemical defense. Phytochemistry 69(9):1795–1813

Mordue A, Blackwell A (1993) Azadirachtin: an update. J Insect Physiol 39(11):903–924

Musser RO, Hum-Musser SM, Eichenseer H, Peiffer M, Ervin G, Murphy JB, Felton GW (2002) Herbivory: caterpillar saliva beats plant defences. Nature 416(6881):599

Nealis VG, Nault JR (2005) Seasonal changes in foliar terpenes indicate suitability of Douglas-fir buds for western spruce budworm. J Chem Ecol 31(4):683–696

Niemeyer HM (2009) Hydroxamic acids derived from 2-hydroxy-2 H-1, 4-benzoxazin-3 (4 H)-one: key defense chemicals of cereals. J Agric Food Chem 57(5):1677–1696

Olson D, Nechols J (1995) Effects of squash leaf trichome exudates and honey on adult feeding, survival, and fecundity of the squash bug (Heteroptera: Coreidae) egg parasitoid Gryon pennsylvanicum (Hymenoptera: Scelionidae). Environ Entomol 24(2):454–458

Park I-K, Shin S-C, Kim C-S, Lee H-J, Choi W-S, Ahn Y-J (2005) Larvicidal activity of lignans identified in Phryma leptostachya var. asiatica roots against three mosquito species. J Agric Food Chem 53(4):969–972

Pauchet Y, Heckel DG (2013) The genome of the mustard leaf beetle encodes two active xylanases originally acquired from bacteria through horizontal gene transfer. Proc R Soc Lond B Biol Sci 280(1763):20131021

Peng J, Deng X, Jia S, Huang J, Miao X, Huang Y (2004) Role of salicylic acid in tomato defense against cotton bollworm, Helicoverpa armigera Hubner. Z Naturforsch C 59(11–12):856–862

Pentzold S, Zagrobelny M, Rook F, Bak S (2014) How insects overcome two-component plant chemical defence: plant β-glucosidases as the main target for herbivore adaptation. Biol Rev 89 (3):531–551

Peters DJ, Constabel CP (2002) Molecular analysis of herbivore-induced condensed tannin synthesis: cloning and expression of dihydroflavonol reductase from trembling aspen (Populus tremuloides). Plant J 32(5):701–712

Picman AK (1986) Biological activities of sesquiterpene lactones. Biochem Syst Ecol 14 (3):255–281

Pieterse CM, Van Loon L (2004) NPR1: the spider in the web of induced resistance signaling pathways. Curr Opin Plant Biol 7(4):456–464

Pieterse CM, Van der Does D, Zamioudis C, Leon-Reyes A, Van Wees SC (2012) Hormonal modulation of plant immunity. Annu Rev Cell Dev Biol 28:489–521

Rane RV, Walsh TK, Pearce SL, Jermiin LS, Gordon KH, Richards S, Oakeshott JG (2016) Are feeding preferences and insecticide resistance associated with the size of detoxifying enzyme families in insect herbivores? Curr Opin Insect Sci 13:70–76

Rask L, Andréasson E, Ekbom B, Eriksson S, Pontoppidan B, Meijer J (2000) Myrosinase: gene family evolution and herbivore defense in Brassicaceae. Plant Mol Biol 42(1):93–114

Ratzka A, Vogel H, Kliebenstein DJ, Mitchell-Olds T, Kroymann J (2002) Disarming the mustard oil bomb. Proc Natl Acad Sci 99(17):11223–11228

Rodriguez-Saona CR, Musser RO, Vogel H, Hum-Musser SM, Thaler JS (2010) Molecular, biochemical, and organismal analyses of tomato plants simultaneously attacked by herbivores from two feeding guilds. J Chem Ecol 36(10):1043–1057

Roy A, Walker W III, Vogel H, Chattington S, Larsson M, Anderson P et al (2016) Diet dependent metabolic responses in three generalist insect herbivores Spodoptera spp. Insect Biochem Mol Biol 71:91–105

Saha P, Majumder P, Dutta I, Ray T, Roy S, Das S (2006) Transgenic rice expressing Allium sativum leaf lectin with enhanced resistance against sap-sucking insect pests. Planta 223 (6):1329

Scboonhoven L, Jermy T, Van Loon J (1998) Insect-plant biology: from physiology to evolution. Stanley Thornes, Cheltenham, UK

Schäfer M, Fischer C, Meldau S, Seebald E, Oelmüller R, Baldwin IT (2011) Lipase activity in insect oral secretions mediates defense responses in Arabidopsis thaliana. Plant Physiol 156:1520–1534

Schmelz EA, Carroll MJ, LeClere S, Phipps SM, Meredith J, Chourey PS et al (2006) Fragments of ATP synthase mediate plant perception of insect attack. Proc Natl Acad Sci 103(23):8894–8899

Schuman MC, Baldwin IT (2016) The layers of plant responses to insect herbivores. Annu Rev Entomol 61:373–394

Sharma H, Sujana G, Rao DM (2009) Morphological and chemical components of resistance to pod borer, Helicoverpa armigera in wild relatives of pigeonpea. Arthropod Plant Interact 3 (3):151–161

Shirley BW (1998) Flavonoids in seeds and grains: physiological function, agronomic importance and the genetics of biosynthesis. Seed Sci Res 8(4):415–422

Simmonds MS (2003) Flavonoid–insect interactions: recent advances in our knowledge. Phytochemistry 64(1):21–30

Simmonds MS, Stevenson PC (2001) Effects of isoflavonoids from Cicer on larvae of Heliocoverpa armigera. J Chem Ecol 27(5):965–977

Slama K (1980) Animal hormones and antihormones in plants. Biochem Physiol Pflanz 175 (3):177–193

Sonoda S, Tsumuki H (2005) Studies on glutathione S-transferase gene involved in chlorfluazuron resistance of the diamondback moth, Plutella xylostella L.(Lepidoptera: Yponomeutidae). Pestic Biochem Physiol 82(1):94–101

Stam JM, Kroes A, Li Y, Gols R, van Loon JJ, Poelman EH, Dicke M (2014) Plant interactions with multiple insect herbivores: from community to genes. Annu Rev Plant Biol 65:689–713

Stensmyr MC, Dweck HK, Farhan A, Ibba I, Strutz A, Mukunda L et al (2012) A conserved dedicated olfactory circuit for detecting harmful microbes in Drosophila. Cell 151 (6):1345–1357

Strauss AS, Peters S, Boland W, Burse A (2013) ABC transporter functions as a pacemaker for sequestration of plant glucosides in leaf beetles. elife 2

Torrie LS, Radford JC, Southall TD, Kean L, Dinsmore AJ, Davies SA, Dow JA (2004) Resolution of the insect ouabain paradox. Proc Natl Acad Sci U S A 101(37):13689–13693

Treutter D (2006) Significance of flavonoids in plant resistance: a review. Environ Chem Lett 4 (3):147

Turlings T, Loughrin JH, Mccall PJ, Röse U, Lewis WJ, Tumlinson JH (1995) How caterpillar-damaged plants protect themselves by attracting parasitic wasps. Proc Natl Acad Sci 92 (10):4169–4174

Ulloa RM, Raíces M, MacIntosh GC, Maldonado S, Téllez-Iñón MT (2002) Jasmonic acid affects plant morphology and calcium-dependent protein kinase expression and activity in Solanum tuberosum. Physiol Plant 115(3):417–427

Van Dam NM, Tytgat TO, Kirkegaard JA (2009) Root and shoot glucosinolates: a comparison of their diversity, function and interactions in natural and managed ecosystems. Phytochem Rev 8 (1):171–186

Van Der Biezen EA, Jones JD (1998) Plant disease-resistance proteins and the gene-for-gene concept. Trends Biochem Sci 23(12):454–456

van Leur H, Vet LE, Van der Putten WH, van Dam NM (2008) Barbarea vulgaris glucosinolate phenotypes differentially affect performance and preference of two different species of lepidopteran herbivores. J Chem Ecol 34(2):121–131

Vandenborre G, Smagghe G, Van Damme EJ (2011) Plant lectins as defense proteins against phytophagous insects. Phytochemistry 72(13):1538–1550

Volkov AG, Vilfranc CL, Murphy VA, Mitchell CM, Volkova MI, O'Neal L, Markin VS (2013) Electrotonic and action potentials in the Venus flytrap. J Plant Physiol 170(9):838–846

Walter A, Mazars C, Maitrejean M, Hopke J, Ranjeva R, Boland W, Mithöfer A (2007) Structural requirements of jasmonates and synthetic analogues as inducers of Ca2+ signals in the nucleus and the cytosol of plant cells. Angew Chem Int Ed 46(25):4783–4785

War AR, Paulraj MG, Ahmad T, Buhroo AA, Hussain B, Ignacimuthu S, Sharma HC (2012) Mechanisms of plant defense against insect herbivores. Plant Signal Behav 7(10):1306–1320

Whitham TG, Bailey JK, Schweitzer JA, Shuster SM, Bangert RK, LeRoy CJ et al (2006) A framework for community and ecosystem genetics: from genes to ecosystems. Nat Rev Genet 7 (7):510

Will T, Tjallingii WF, Thönnessen A, van Bel AJ (2007) Molecular sabotage of plant defense by aphid saliva. Proc Natl Acad Sci 104(25):10536–10541

Wittstock U, Kliebenstein DJ, Lambrix V, Reichelt M, Gershenzon J (2003) Chapter five glucosinolate hydrolysis and its impact on generalist and specialist insect herbivores. In: *Recent advances in phytochemistry*, vol 37. Elsevier, New York, pp 101–125

Wouters FC, Reichelt M, Glauser G, Bauer E, Erb M, Gershenzon J, Vassão DG (2014) Reglucosylation of the benzoxazinoid DIMBOA with inversion of stereochemical configuration is a detoxification strategy in lepidopteran herbivores. Angew Chem 126(42):11502–11506

Wouters FC, Gershenzon J, Vassão DG (2016) Benzoxazinoids: reactivity and modes of action of a versatile class of plant chemical defenses. J Braz Chem Soc 27(8):1379–1397

Wünsche H, Baldwin IT, Wu J (2011) S-Nitrosoglutathione reductase (GSNOR) mediates the biosynthesis of jasmonic acid and ethylene induced by feeding of the insect herbivore Manduca sexta and is important for jasmonate-elicited responses in Nicotiana attenuata. J Exp Bot 62 (13):4605–4616

Wybouw N, Dermauw W, Tirry L, Stevens C, Grbić M, Feyereisen R, Van Leeuwen T (2014) A gene horizontally transferred from bacteria protects arthropods from host plant cyanide poisoning. *Elife*:3, e02365

Yoshida T, Chen H, Toda M, Kimura M, Davis A (2000) New host plants and host plant use for Drosophila elegans Bock and Wheeler 1972. Drosoph Inf Serv 83:18–21

Yu Q, An L, Li W (2014) The CBL–CIPK network mediates different signaling pathways in plants. Plant Cell Rep 33(2):203–214

Yuan JS, Köllner TG, Wiggins G, Grant J, Degenhardt J, Chen F (2008) Molecular and genomic basis of volatile-mediated indirect defense against insects in rice. Plant J 55(3):491–503

Zebelo SA, Maffei ME (2014) Role of early signalling events in plant–insect interactions. J Exp Bot 66(2):435–448

Zebelo SA, Matsui K, Ozawa R, Maffei ME (2012) Plasma membrane potential depolarization and cytosolic calcium flux are early events involved in tomato (Solanum lycopersicon) plant-to-plant communication. Plant Sci 196:93–100

Zhang S-Z, Hua B-Z, Zhang F (2008) Induction of the activities of antioxidative enzymes and the levels of malondialdehyde in cucumber seedlings as a consequence of Bemisia tabaci (Hemiptera: Aleyrodidae) infestation. Arthropod Plant Interact 2(4):209–213

Zhao C, Doucet D, Mittapalli O (2014) Characterization of horizontally transferred β-fructofuranosidase (ScrB) genes in Agrilus planipennis. Insect Mol Biol 23(6):821–832

Microbial Influence on Plant–Insect Interaction

Amrita Chakraborty and Amit Roy

Abstract

Plants and insects live in a microbial world, and the co-existence have shaped their ecology and evolution. These microbial allies play an essential role in the health, well-being, and vigor of their hosts and are often considered as "hidden players" in plant–insect interaction. The present chapter attempts to cover the contribution of microbes as drivers of plant–insect interaction where the microbial companions directly or indirectly influence the plant–insect interaction. The chapter also emphasizes the diversity of microbial communities linked with both plants and insects and their contribution toward plant–insect interaction from an ecological standpoint. It further deals with the recent updates on the use of microorganisms in pest management and the implications of microbes as a toolbox in future IPM strategies.

Keywords

Microbes · Metagenomics · Hidden allies · Toolbox · Host–microbe interaction · Microbe-associated molecular patterns (MAMP) · Symbiosis · Mycorrhiza-induced resistance (MIR) · Herbivore-induced plant volatiles (HIPVs)

A. Chakraborty (✉)
Faculty of Forestry and Wood Sciences, EVA 4.0, Czech University of Life Sciences, Prague, Czech Republic

Department of Plant Protection Biology, Swedish University of Agricultural Sciences, Alnarp, Sweden
e-mail: chakraborty@fld.czu.cz

A. Roy
Faculty of Forestry and Wood Sciences, Extemit-K, Czech University of Life Sciences, Prague, Czech Republic

I. K. Singh, A. Singh (eds.), *Plant-Pest Interactions: From Molecular Mechanisms to Chemical Ecology*, https://doi.org/10.1007/978-981-15-2467-7_14

1 Introduction

The co-existence of plants and insects evolved over 400 million years and shaped the ecosystem. Both plants and insects are engaged in an arms race where plant defenses against insect herbivory, while insects evolve strategies to overwhelm them. Phytophagous insects (generalists and/or specialists) attack diverse plant species for herbivory. In response, plants produce an array of defensive compounds known as plant secondary metabolites to cope with their enemies. Glucosinolates, alkaloids, terpenoids, and phenolics are classic secondary metabolites serving as defensive compounds (Papadopoulou and van Dam 2017), which are either constitutively expressed in plants or induced in response to herbivory (Wu and Baldwin 2010). To counteract these plant defenses, insects secrete elicitors with their salivary secretions that decoy the defense responses (Fürstenberg-Hägg et al. 2013) or detoxify them (Ceja-Navarro et al. 2015).

In the last two decades, biologists have been keen on exploring the role of microbes in shaping the ecology of plants and animals. Studies have revealed that microbial communities associated with plants and insects play an essential role in health, well-being, and vigor of their hosts and are often considered as "hidden players" in plant–insect interaction (Douglas 2018; Sugio et al. 2014); (Biere and Bennett 2013). Various microbial communities (e.g., mycorrhizal fungi, rhizobacteria, root endophytic fungi) promote plant growth and protect them against a wide range of diseases by inducing resistance in systemic tissues (Induced Systemic Resistance – ISR) (Pineda et al. 2010; Van Wees et al. 2008). Furthermore, microbes have profound effects on insect feeding efficiency by helping to digest food or detoxifying entomotoxic compounds, modulating host growth, development, behavior, etc. Microbial allies of insects affect the plant defense mechanisms by either suppressing or counteracting the plant defense response (Sugio et al. 2014; Zhu et al. 2014). Microbes may engage in altering plant metabolisms and/or defense systems that significantly impact the plant–insect interaction either benefiting the plants or insects. Recent advances in high-throughput omics technology have opened up fascinating research area with the possibilities to conduct global analysis on the composition and functional capabilities of microbial symbionts that may contribute to the health and fitness of their host. The fast-moving scientific developments offer excellent potential for in-depth investigation of the fundamental processes and manipulation of the microbiota for effective microbial therapies.

The present chapter attempts to cover the role of microbes as drivers of plant–insect interaction where microbial associates directly or indirectly influence the plant–insect interaction. This chapter centers on the diversity of microbial communities and their contribution to plant–insect interaction from an ecological perspective. It further deals with the recent updates on the use of microorganisms in pest management and the implications of microbes as a toolbox in future IPM strategies.

2 Microbial Contribution in Shaping the Tri-Trophic Interactions in an Ecosystem

A diverse spectrum of microbes are often allied to plants and insects, and the nature of their association may vary from pathogenic to mutualistic interaction depending on underlying ecological factors. Symbiotic microorganisms live in the close interface with the host either permanently or for a considerable part of host's life cycle and play a key role in their diversification and evolutionary stability (Salem et al. 2015). Most of the intracellular symbionts show maternal inheritance where the symbionts are vertically transmitted from mother to the offspring. The horizontally transmitted symbionts are however transmitted directly from the environment or other conspecific or heterospecific host individuals (Kikuchi et al. 2007). The pathogenic interactions may also shift to the beneficial relationship over the course of time such as *Wolbachia* infection in *Drosophila simulans*, leading to an increase in the fecundity over uninfected females (Weeks et al. 2007).

Apart from the two-way interactions between microbes and their hosts (plants or insects), microbes are also engaged in a multi-trophic interaction where microbes interact with plants and insects simultaneously (Biere and Bennett 2013; Biere and Tack 2013). For example, the aphid–barley interaction depends on the interacting aphid species and bacteria present in rhizosphere (Tétard-Jones et al. 2007, 2012). The development of next-generation sequencing (NGS) technologies has enabled assessing the microbial diversity in different ecosystems. Meta-genomic and meta-transcriptomic sequencing combined with bioinformatic tools have assisted in exploring the taxonomic and functional diversity of hitherto hidden microbial association in a given environment (Douglas 2018).

2.1 Microbial Diversity Allied with Plants

Plants harbor diverse microbial communities in different compartments such as the rhizosphere (near the roots), phyllosphere (plant surface like leaves), and endosphere (within the plant and root tissues) (Andreote and e Silva 2017). Microbial communities associated with plants, both belowground and aboveground, benefit their host by aiding in the better uptake of nutrients from the soil for plant growth, increased tolerance to environmental stress (saline stress, drought, and occurrence of heavy metals) (Pineda et al. 2010), and protection against the pathogen (Bulgarelli et al. 2013). Some microbes are also capable of synthesizing plant growth–promoting hormones (Contreras-Cornejo et al. 2009; Van Loon 2007).

2.1.1 Microbes Enhancing Plant Growth and Nutrient Uptake

Plant-associated microbes such as the nitrogen-fixing bacteria (*Rhizobium, Bradyrhizobium, Mesorhizobium, Ensifer* (*Sinorhizobium*), and *Azorhizobium*) are widely studied for its interaction with the host and biogeochemical function (Batterman et al. 2013). The symbiotic association with plants helps in the better nutrient uptake and enables fixing atmospheric nitrogen required for plant growth.

These endosymbionts form nitrogen-fixing nodules in the roots of the leguminous plants on expressing the rhizobial nodule–forming (*nod*) and nitrogen-fixing (*nif*) genes which are generally located on "symbiosis islands" (Ling et al. 2016). By definition, "symbiosis islands" are the mobile, integrative, conjugative elements that carry genes that enable them to expunge from the chromosome to form closed circular molecule that eventually conjugate and recombine into recipient chromosomes through horizontal gene transfer (HGT) from bacteria present in soil to host leguminous plants (Haskett et al. 2016; Ling et al. 2016). These mobile elements bear various novel traits such as antibiotic resistance, virulence, biofilm formation, degradation of aromatic compounds, and symbiosis (Ling et al. 2016; Okubo et al. 2016). For example, the transfer of the "symbiosis island" of 500 kb from *Mesorhizobium loti* to *Lotus corniculatus* and its integration into a phenylalanine-tRNA gene of the host plant chromosome resulted in root nodule formation and nitrogen-fixation in lotus plant (Ramsay and Ronson 2015). A similar example of such symbiotic interaction was documented in *S. rostrate-Azorhizobium caulinodans* system where the "symbiosis island" of *A. caulinodans* on integration to glycine-tRNA gene of *S. rostrata*-induced host nodulation (Ling et al. 2016).

Another interesting symbiotic association, observed between microbes present in plants and soil, is the arbuscular mychorrhizal symbiosis (Hammer et al. 2014; Richardson et al. 2009) where arbuscular mychorrhizal fungi (AMF, obligate biotrophs) belonging to the phylum Glomeromycota colonize on the cortical cells of the plant root. The AMF profit from the host carbon compounds to obtain metabolic energy (Gianinazzi et a. 2010), and in exchange caters better uptake of water and mineral nutrients (such as nitrogen and phosphorous) (Baum et al. 2015; Gutjahr and Parniske 2013) leading to increased host plant biomass, higher tolerance to abiotic stress (salinity, drought, heavy metals) (Singh et al. 2011) and protection against plant diseases (Bernardo et al. 2017). Enhanced plant growth determines increased food supply and improved nutrient quality for the herbivores and in turn influences the plant–insect interaction. Conversely, beneficial microbes accelerate the plant regrowth after herbivory by facilitating the nutrient and water uptake (Herman et al. 2008; Kempel et al. 2009; Kula et al. 2005).

2.1.2 Microbes-Induced Resistance in Plants

Plant defenses against pathogen attack can be either constitutively expressed in plants (passive resistance) or induced after the infection or herbivore attack (induced resistance). Microbes such as plant growth–promoting rhizobacteria (PGPR) and fungi (PGPF) (Segarra et al. 2009; Van Wees et al. 2008) as well as mycorrhizal and endophytic fungi (Stein et al. 2008) often initiate induced systemic resistance (ISR) to mitigate biotic and abiotic stresses in plants (Pozo and Azcón-Aguilar 2007; Shikano et al. 2017; Trillas and Segarra 2009). Induced systemic resistance involves activation of jasmonic acid and ethylene signaling pathways (Van der Ent et al. 2009) either by priming of plant defense genes in response to pathogen or insect attack (Conrath et al. 2001) or on interaction between non-pathogenic microbes with the plant roots. For instance, the establishment of arbuscular mychorrhizal symbiosis activates and boosts plant basal defense mechanisms on pathogen attack through

Mycorrhiza-induced Resistance (MIR) (Pozo and Azcón-Aguilar 2007; Song et al. 2015). Based on transcriptomic and proteomic profiling, Fiorilli et al. (2018) demonstrated that the wheat–AMF association was not only benefiting the wheat plant in mineral nutrition but also protecting them against the pathogen (*Xanthomonas translucens*). Mycorrhizal colonization on plant root induces systemic defense responses via microbe-associated molecular patterns (MAMPs) recognition (Zamioudis and Pieterse 2012). It is interesting to note that the host plant initially perceives this mycorrhizal interaction as a putative pathogen by plant MAMP-recognition receptors and activates MAMP-triggered immunity (MIT) response as the first line of defense to prevent further invasion (Jones and Dangl 2006; Millet et al. 2010). This MIT response induced by mycorrhizal invasion results in transcriptional and hormonal changes that leads to the accumulation of hydrolytic enzymes (chitinase, glucanase), reactive oxygen species in roots, and activation of phenylpropanoid metabolism (García-Garrido and Ocampo 2002; Pozo and Azcón-Aguilar 2007) in the host plant, leading to the establishment of the symbiosis (Schouteden et al. 2015).

Interestingly, plant pathogens also invade plant tissues through stomata and thus stomatal closure is a part of innate immunity (Melotto et al. 2006). *Pseudomonas syringae* overwhelms this innate defense by the release of phytotoxin Coronatine (COR) (Zheng et al. 2012) that activates the Jasmonic acid signaling pathway, enabling the reopening of the stomatal pores. Plant-associated microbes have been reported to influence plant metabolic processes to block pathogen invasion (Kumar et al. 2012). A recent study documented that the plant growth–promoting fungi *Penicillium simplicissium* induces systemic resistance to protect *Arabidopsis thaliana* against the pathogen *Pseudomonas syringae* by altering the plant metabolic processes (Desclos-Theveniau et al. 2012; Du et al. 2014). The MYB44 gene product of endophytic fungi acts as stomata-specific enhancer of the plant Abscisic acid (ABA) signaling pathway that promotes stomatal closure thereby by blocking the entry of pathogen through stomata (Hieno et al. 2016; Montillet et al. 2013).

2.1.3 Microbial Toxin Production Against Insects

Microbes colonizing on plants can produce toxic compounds that can be harmful for insects during herbivory (Bizzarri and Bishop 2008; Monnerat et al. 2009). The crystal-like proteins (delta- endotoxins) produced by gram-positive bacteria *Bacillus thuringiensis* (Bt) on sporulation is known to have insecticidal activity (Palma et al. 2014). These endotoxins constitute Cry (crystal) and Cyt (cytosolic) group of proteins that interact synergistically to have a potential insecticidal effect (Butko 2003). The inactive Cry protoxins are ingested and proteolytically cleaved to yield shorter active toxins of 55-60 kDa in the insect midgut (Bravo et al. 2007). There are several models proposed for the mechanism behind the toxicity of Cry protein (Jurat-Fuentes and Crickmore 2017). It is generally believed that the active toxic molecules bind to specific receptors such as cadherin-like proteins, glycosylphophatidylinositol (GPI)-anchored Alkaline phosphatase (ALP), and GPI -anchored aminopeptidase (APN) at the surface of midgut, forming pores on the membrane, increasing its permeability, and disrupting the transmembrane ionic gradient, resulting in cell lysis and insect death (Pigott and Ellar 2007). Alternatively, the interaction of the Cry toxin with cadherin receptor triggers a signaling cascade involving protein G, adenylate cyclase and protein kinase A, as well as induces the activation of mitogen-activated kinases such as MAPK p38-triggering cell apoptosis (Zhang et al. 2006). More recently, Portugal et al. (Portugal et al. 2017) demonstrated that the binding of Cry toxin to specific receptors activates the phosphorylation of MAPK p38, disrupting the calcium ion influx through pore formation, which leads to cell death. Among different groups of Cry toxins (de Maagd et al. 2003), Cry1A toxin binds to cadherin protein receptors of most lepidopteran species (such as *Manduca sexta*, *Bombyx mori*, *Heliothis virescens*, *Helicoverpa armigera*, *Pectinophora gossypiella*, and *Ostrinia nubilalis*) (Pigott and Ellar 2007). The use of Cry toxins into transgenic crops for targeted and effective pest control has significantly reduced the use of chemical insecticides (Bravo et al. 2011). Another well-characterized rhizospheric bacteria, *Pseudomonas protegens,* secretes an antimicrobial compound that on ingestion promotes apoptosis in insects (Haas and Keel 2003; Loper and Gross 2007). In addition to the endotoxins, insecticidal proteins are also secreted during bacterial vegetative growth phase known as Vip (vegetative insecticidal proteins) (de Maagd et al. 2003; Estruch et al. 1996) that can be categorized into four groups Vip1, Vip2, Vip3, and Vip4 depending on their amino acid sequences (Chakroun et al. 2016; Zack et al. 2017). Vip1 and Vip2 are binary insecticidal proteins that are toxic to Coleopteran and Hemipteran species (Bi et al. 2015; Chakroun et al. 2016) whereas Vip3 targets against Lepidopteran insects (Song et al. 2016). However, Vip4 protein is not yet reported to have insecticidal activity (Chakroun et al. 2016).

2.2 Microbial Diversity Allied with Insects

Insect-associated microbes colonize mostly in the external cuticle and the gut. However, they can breach the exoskeleton and the gut to gain access to the hemocoel and within the specialized insect cells. Microbial communities present in insect influence several aspects of insect ecology, behavior, and physiology such as responses to the utilization of plant nutrients, immunity, reproduction, detoxification of defensive plant compounds, and protection against natural enemies (Oliver et al. 2010).

2.2.1 Microbes Providing Essential Nutrients

Insects along with other animals are unable to synthesize the essential amino acids and co-factors obligatory for many metabolic enzymes to function. In addition to these essential amino acids, insects cannot synthesize sterols that contribute to membrane architecture (Behmer and Nes 2003). Most insects derive these essential nutrients from their diet while feeding on plant sap. These insects constitute large populations of specific microorganisms localized in specialized cells, called bacteriocytes, within their body. The microbial symbionts present in most of the plant sap–feeding insects (hemipterans) are transmitted vertically from mother to their offspring via ovaries to the cytoplasm of each egg at oviposition that provides essential amino acids and/or vitamin cofactors that are limiting in host diet. Furthermore, the gut symbionts that are deposited externally over the eggs are acquired by the feeding offsprings (Buchner 1965). For instance, the bacterial endosymbiont, *Buchnera aphidicola,* in aphids is a model for such association where the bacterial symbiont is localized in metabolically active *Bucherna* cells and uses the insect body as its habitat and in return provides essential nutrients to its host (Wang et al. 2018). Upon the elimination of the symbiotic bacteria by the antibiotic treatment, the capability to synthesize essential amino acids is lost in aphids (Douglas et al. 2001).

Similarly, the symbiotic association of gall midges with fungi is essential for invading the plant stem to access the vascular tissue for nutrients and development of gall (Rohfritsch 2008). The close association of termites with microbes is also one of the best-studied symbiotic relationships in insects. The ability of the termites to harness and feed on nitrogen-deficient wood-based diet is due to the presence of unique consortium of microbes living in the termite gut. The microbial cellulolytic enzymes play a critical role in the digestion by enhancing its digestive efficiency (Peterson and Scharf 2016). The presence of mutualistic gut symbiont *Erwinia dacicola* in the olive fruit fly *Bactrocera oleae* benefits the host by providing essential amino acids and protease enzymes to digest the food (Capuzzo et al. 2005). Furthermore, in ants, cockroaches, and termites microbial allies recycle the nitrogenous wastes to essential amino acids (Douglas 2015). The resident microbes often produce glucosyl hydrolase that degrades plant cellulose and hemicellulose to short-chain fatty acids to provide a readily available nutrient source to insect host (Berenbaum 1980; Calderón-Cortés et al. 2012).

2.2.2 Microbes Influencing Insect Immunity

Microbes often contribute to insect innate immunity wherein gut microbes regulate the expression of immune genes (Johnson 2015a). The bacterial symbionts such as *Wolbachia pipientis*, *Spiroplasma* species, and *Hamiltonella defensa* either vertically or horizontally transmitted to the host have shown to influence host immunity (Engel and Moran 2013). For example, the facultative symbiont *H. defensa* protects aphids against the parasitoid through bacteriophage-encoded gene expression (Oliver et al. 2010). Similarly, *Spiroplasma* present in *Drosophila neotestacea* imparts resistance against the parasitic nematode (Jaenike et al. 2010). Furthermore, in mosquitoes, the gut microbiota activates the innate epithelial immunity against *Plasmodium* infection whereas the elimination of the microbiota renders the mosquitoes susceptible to infection (Dong et al. 2009). Over the years, *Wolbachia* infection is considered parasitic to insects as they contribute to cytoplasmic incompatibility, leading to reproductive disruption. Nevertheless, studies suggested that *Wolbachia* infection in *Drosophila* also deliberates antiviral protection leading to the higher survival of the flies (Johnson 2015b). The presence of *Wolbachia* in *Drosophila* induce antiviral resistance against a wide range of RNA viruses (Dicistroviridae, Nodaviridae, Flaviviridae, Togaviridae, and Reoviridae) but not DNA viruses (Teixeira et al. 2008). The reduction of the viral load and the anti-viral protection is reported to be due to the competition between the symbiotic bacteria and the viruses for the cellular resources (Caragata et al. 2013). Alternatively, the symbiotic association of *Wolbachia* affects the reactive oxygen species (ROS) levels, a key player in the insect immune system. The increase in ROS levels stimulates the Toll pathway, imparting anti-viral protection (Pan et al. 2012). Other studies reported the proliferation of *Wolbachia* inside the insect body suppresses the viral infection by inducing host immune responses by stimulating the miRNA expression (Hussain et al. 2011), thereby leading to cell death in insect (Brackney 2017; Terradas and McGraw 2017). However, the exact mechanism behind the antiviral protection by *Wolbachia* is still poorly understood (Yixin et al. 2017) (Fig. 1).

2.2.3 Microbes Influencing Detoxification of Plant-Defensive Compounds

Microbes are considered as the drivers promoting plant specialization in herbivorous insects (Janson et al. 2008). The acquisition of symbiotic microbes enabled the different sap-feeding insects to colonize on several plants. Plant-defensive secondary metabolites (terpenoids, phenolics, alkaloids, glucosinolates, and alliinins) are an essential determinant in plant–insect interaction. The ability of the insects to detoxify these toxic plant allelochemicals is often attributed to microorganisms associated with insects (Boone et al. 2013; Douglas 2013; García-Fraile 2018; Howe et al. 2018). Some herbivores often neutralize toxic phenolic compounds by increasing their gastrointestinal mucus production, by recruiting the gut microorganisms for degradation, and/or by secreting phenol-binding proteins in the saliva (Dearing et al. 2005). Insect symbionts can inhibit or counteract the host plant defenses through the direct or indirect production of enzymes targeting plant-defensive compounds (Broderick et al. 2004; Dowd and Shen 2011). The symbiotic fungus, *Leucocoprinus gongylophorus*, is present in the nest of the leaf-cutting ant, *Acromyrmex echinatior*, and aids to overwhelm plant-defensive phenolic compounds. Precisely, leaf-cutting ants preferentially feed on the fungal hyphae called gongylidia that expresses the laccase coding genes. On ingestion, the laccase molecules pass through the gut of these ants, released on defecation onto the ingested plant materials, and degrade plant defense compounds, for example, flavonoids and tannins (De Fine Licht et al. 2013).

Similarly, Ceja-Navarro et al. (2015) demonstrated the role of gut microbiome of the coffee berry borer (*Hypothenemus hampei*) in the detoxification of toxic plant alkaloid (Caffeine). *H. hampei* is a devastating insect pest of coffee that resulted in 80% crop loss on infestation. Caffeine, produced by the coffee plants, acts as a defense mechanism in response to herbivory. Interestingly, the coffee borer *H. hampei* possesses a core gut microbiota that is responsible for detoxification of caffeine in the insect gut and supports in the survival of the insect in a hostile environment. The gut bacteria such as *Pseudomonas* possess caffeine demethylase genes that aid in caffeine detoxification. Upon treatment with antibiotic that confiscates the insect gut microflora, eliminates the caffeine degradation ability of the beetle. However, the re-inoculation of *Pseudomonas* strain re-establishes the caffeine detoxification ability, thus certifying the pivotal role of microbial associates

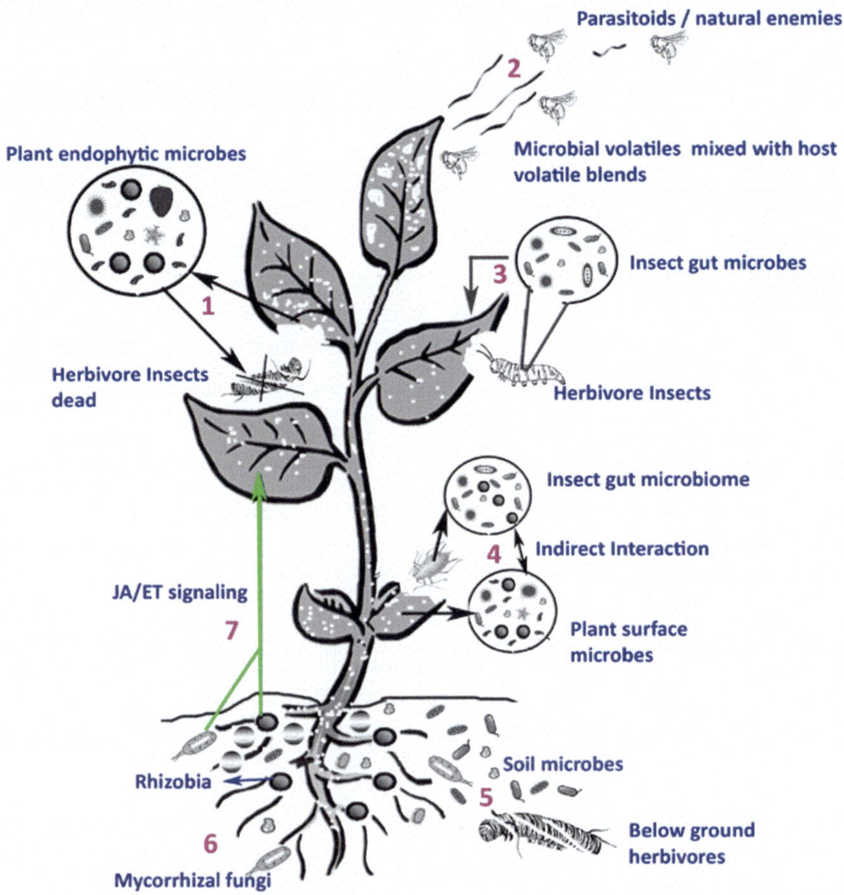

Fig. 1 The tri-trophic interactions between plant-microbe-insect. (1) Plant endophytic microbes producing toxin against insect herbivore. (2) Microbial volatiles mixed with host plant volatiles attract parasitoids. (3) Insect gut microbiome enables detoxifying defensive plant compounds. (4) Indirect interaction between the plant endophytes and insect gut microbiome wherein the insect ultimately confiscates the plant defenses. (5) Soil microbes produce different metabolites that are detrimental to belowground herbivores. (6) Plant growth–promoting rhizobia as well as mycorrhizal fungi interact with plant roots, influencing the JA/ ET signaling pathways inducing resistance against aboveground herbivores

in caffeine degradation. Similarly, gut microbiota of the velvet bean caterpillar *Anticarsia gemmatalis* is involved in the production of serine and cysteine proteases and contributes to the insect's tolerance to dietary protease inhibitors in soy plant (Pilon et al. 2013). Another interesting study showed that in the mountain pine beetle, *Dendroctonus ponderosae,* females initiate mass colonization through the production of aggregation pheromone trans-verbenol (Vité and Pitman 1968) by confiscating the host plant defense mechanisms. The trans-verbenol is a product of

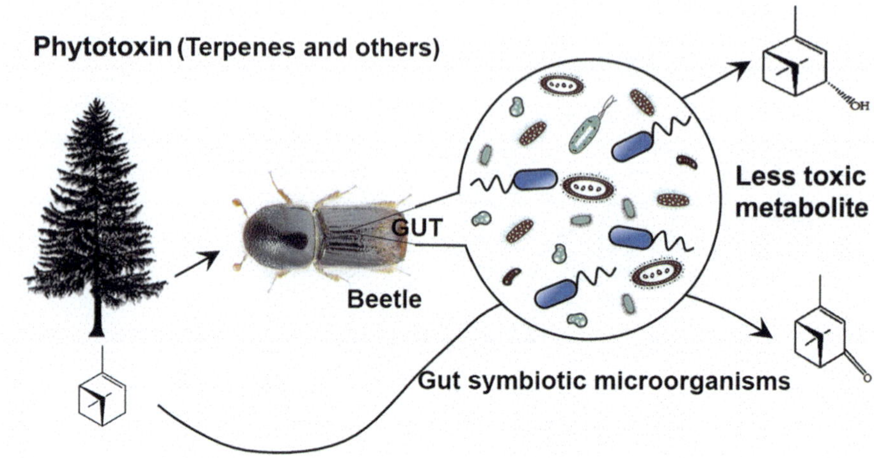

Fig. 2 Figure illustrating the contribution of the bark beetle gut microbiome in plant toxin detoxification

oxidative degradation of plant secondary defense compound monoterpene a-pinene (Renwick et al. 1976).

Recent studies have reported symbiont-mediated terpene degradation and verbenone production in beetles (Fig. 2) (Berasategui et al. 2017; Cao et al. 2018). Shotgun DNA sequencing on the gut microbiome of the mountain pine beetle revealed the presence of terpene-degrading bacteria belonging to genera *Pseudomonas, Rahnella, Serratia,* and *Burkholderia* (Adams et al. 2013).

2.2.4 Impact on Insect Pheromone Production and Reproduction

Pheromones are chemical compounds that serve as cues/signals for communication between individuals of the same species. These chemical compounds are involved in courtship, mating, defense, trail marking, aggregation, kin recognition, etc. (Howard and Blomquist 2005; Regnier and Law 1968). Though pheromones are generally encoded by insect gene, studies have shown that host-associated microbes also play a significant role in modulating their host chemical profiles, mating preference, and social behavior (Engl and Kaltenpoth 2018). Such modulation of the chemical signals occur either directly by influencing the biosynthetic pathway of pheromone production (Marshall et al. 2016) or by manipulating the host metabolic pool and allocating resources into pheromone production (Engl et al. 2018). The microbial symbiont in saw-toothed grain beetle *Oryzaephilus surinamensis* modulates the cuticle synthesis, resulting in the thinner cuticle and thereby rendering the beetles more susceptible to desiccation (Engl et al. 2018). Several insects exhibit reduced attractiveness and fecundity on the disruption of resident microbes with antibiotic treatment. This suggests a microbial role in mate choice and sexual communication (Ben-Yosef et al. 2008). For instance, the disruption of gut microbiota by the administration of antibiotics in Tephritid fruit fly, *Ceratitis capitata*, showed increased oviposition rates of females under nutritional stress as well as prolonged mating latency in males on a standard diet (Ben-Yosef et al. 2008). The reduced attractiveness of the female oriental fruit flies, *Bactrocera dorsalis,* to the males on antibiotic treatment could be reversed by the re-establishment of the gut microbiota into the female flies through feeding (Engl and Kaltenpoth 2018). Another fascinating example showed that Drosophila flies reared on different diets exhibit a strong mating preference where flies fed on the same diet preferred to mate with each other but not with files reared on a different diet (Sharon et al. 2010). Such preference in mate choice could be mediated by diet itself or by diet-associated gut microbial shifts. Sharon et al. (2010) showed evidence of resident gut bacteria influencing the mating preference *in D. melanogaster,* which could be abolished by antibiotic treatment. Interestingly, the lost preference for mate choice could be re-established by infecting the axenic *D. melanogaster* flies with the microbiota of the healthy flies on a diet. However, this study was controversial as several researchers tried to replicate the experiment that had conflicting results where the assortative mating pattern was only observed in inbreed fly lines before the transfer to different diets (Najarro et al. 2015). Others showed no stability of the results within the replicates (Arbuthnott et al. 2016). Thus, extensive research is needed to elucidate the factors influencing the mating preference and success in *D. melanogaster*.

Microbes-associated with insects often manipulate host reproduction by feminizing genetic males, inducing parthenogenesis or male killing, and by inducing cytoplasmic incompatibility (i.e., reproductive sterility when infected males mate with uninfected or infected females with a different symbiont strain) (Hughes et al. 2012; Miller and Schneider 2012; Werren et al. 2008). *Wolbachia, Arsenophonus, Cardinium, Rickettsia,* and *Spiroplasma* are among the universal reproductive manipulators that influence host reproduction (Engelstädter and Hurst 2009). *Spiroplasma* in pea aphid induces male killing to prevent competition with the infected females and avoidance of inbreeding depression (Simon et al. 2011). Increased female bias in infected female whiteflies was observed due to the invasion of *Rickettsia* (Himler et al. 2011) that swayed the population dynamics of whiteflies. The virus LbFv decreases the competitive ability of the parasitoid *Leptopilina boulardi* to infect *Leptopilina heterotoma* by manipulating the reproductive behavior of the parasitoid (Patot et al. 2012). Reproductive manipulators may serve as novel targets to be exploited in the development of alternative control strategies. These reproductive manipulators indirectly impact the plant–insect interactions by regulating the population dynamics and in so doing minimize the genetic diversity and/or recombination rates in infected species (Engelstädter and Hurst 2009) which in turn influence their co-evolutionary dynamics and functioning of ecological networks (Ferrari and Vavre 2011).

2.3 Crosstalk in Signaling Pathways – Decoy of Plant Defenses

Plants are armed with a plethora of defense mechanisms to combat against insect and pathogen attack. These defensive mechanisms are either constitutively present or activated upon insect or pathogen invasion (Pieterse and Dicke 2007). On perceiving the pathogen or insect attack, plants initially retort through its primary immune response and also activate effective systemic broad-spectrum resistance known as induced resistance against attackers (Walters et al. 2007). The phytohormones – salicylic acid (SA), jasmonic acid (JA), and ethylene (ET) – are documented as key players in the regulation of plant defense signaling pathways (Koornneef and Pieterse 2008). In response to pathogen or insect attack, plants emit alarm signals with the production of SA, JA, and ET that contributes to plant defense response. SA-mediated defense responses are generally induced by microbial pathogens whereas insect invasion is usually dissuaded by JA/ET-mediated defenses (Kessler and Baldwin 2002; Thomma et al. 2001). However, in nature plants often encounter raid by different aggressors (pathogens and or herbivores) either simultaneously or by consequent invasion (Stout et al. 2006). Therefore, the crosstalk between the defense signaling pathways delivers a powerful defensive mechanism. These signaling pathways can be either mutually antagonistic or synergistic that allows the plant to combat against its invaders (Bostock 2005). Intriguingly, insect herbivores and pathogens have evolved to decoy the plant defenses for their own benefit by overwhelming the defense mechanisms modulating the plant's signaling network (Pieterse and Dicke 2007). Herbivores often exploit its symbionts to overwhelm the anti-herbivore defenses by dodging the plant perception (Giron and Glevarec 2014; Sugio et al. 2014). For instance, the bacteria present in oral secretion of the Colorado potato beetle, *Leptinotarsa decemlineata*, activate the plant defense response through the stimulation of the SA signaling pathway as a response to microbial pathogen attack which in turn downregulates the JA anti-herbivore response, ensuring improved larval growth (Chung et al. 2013). A similar example, herbivorous silverleaf whitefly nymphs (*Bemisia tabaci*), activates SA signaling pathway as a decoy strategy to overcome JA-mediated defense to enhance larval performance (Zarate et al. 2007). Microbial pathogens often have the ability to produce phytohormones or their functional mimics and thereby manipulate plant signaling network (Robert-Seilaniantz et al. 2007). For instance, *P. syringae* bacteria produce a potent mimic of JA-Ile called coronatine that activates JA-Ile responses and suppresses SA-dependent defenses, resulting in enhanced pathogen growth (Nomura et al. 2005). The induction of SA signaling pathways results in the activation of pathogenesis-related protein encoding genes having antimicrobial activity (van Loon et al. 2006). Some of the prominent molecular players in the crosstalk between SA/JA signaling pathways are the regulatory protein NONEXPRESSOR OF PR GENES1 (NPR1), WRKY transcription factors, glutaredoxin GRX480, and Mitogen-activated protein (MAP) kinases. These regulatory components are essential for the activation of the SA signaling pathway which in turn suppresses the JA-induced response, resulting in overcoming plant defense against herbivore attack (Koornneef and Pieterse 2008).

2.4 Soil Microbial Diversity Influencing the Plant–Insect Interaction

Apart from the plant- and insect-associated microbes, the soil microbial community also plays a crucial role not only in enhancing plant growth and increased tolerance to abiotic stress but also in influencing aboveground insect herbivores through biochemical changes in plant-mediated mechanisms (Pineda et al. 2017). For example, the foliar-feeding *Aphis jacobaea* population depends on the soil microbial communities of its host plant ragwort (*Senecio jacobaea*). A different consortium of free-living soil-borne microbes influences the concentration of amino acids in the plant phloem sap, thereby affecting the aphid population (Kos et al. 2015). Additionally, inoculation of the distinct microbiome in soil manipulates the leaf metabolome of *Arabidopsis,* making it resistant against caterpillar *Trichoplusia ni* (Badri et al. 2013).

Intriguingly, belowground microbes have been shown to influence plant–insect interaction by modulating herbivore-induced plant volatile (HIPV) emission (Pineda et al. 2015). Plants in response to herbivore attack emit varieties of volatile organic compounds (HIPVs) in order to attract the potential predator. For example, the volatiles emitted by *Nerium oleander* plants in response to *Aphis nerii* attack signal the predator *Chrysoperla carnea* which could be altered by the presence or absence of soil microbial communities (Benítez et al. 2017). It was interesting to note *C. carnea* females preferred the HIPV blend emitted from plants grown on soil inoculated with microbes to those emitted from plants grown on control sterile soil (Benítez et al. 2017).

Furthermore, certain beneficial soil microbes can synthesize the phytohormones that enhance the plant growth and can mitigate abiotic stress (salinity, drought, heavy metals) (Egamberdieva et al. 2011; Egamberdieva et al. 2017; Liu et al. 2013). For example, root-colonizing soil bacterium *B. licheniformis* can synthesize indole-acetic-acid (IAA), which promotes wheat plant under saline stress (Singh and Jha 2016). Recent studies have demonstrated a linkage between the leaf microbiome and soil microbial communities (Pineda et al. 2017), wherein belowground microbial entities impact the aboveground insect herbivory as well as the composition of symbiotic "phytobiome" (i.e., plant microbiome). For instance, entomopathogenic fungi (*Beauveria bassiana* and *Metarhizium anisopliae*) that are typically present in soil colonize in different parts of broad bean (*Vicia faba*) plant and enhance plant growth as well as exhibit resistance against insects by translocating nitrogen to the plant from the insect cadavers via their fungal mycelia (Behie et al. 2012; Jaber and Enkerli 2016). Another example of such interaction is observed by a fungus *Trichoderma*, thought to be restricted to the soil, have now been known to colonize on the leaves and can suppress insect pests such as thrips (Muvea et al. 2014).

In the quest for crop protection, there is extensive use of insecticides that pose a threat of insecticide resistance (Whalon et al. 2008). One of the common organophosphorus insecticides used worldwide is fenitrothion that targets acetylcholine esterases and exhibits insect-specific toxicities (Stenersen 2004). Extensive application of such insecticides have led to an increased population of fenitrothion-degrading microbes in the soil that convert the toxic fenitrothion to non-toxic 3-methyl-4-nitrophenol and utilize it for their growth (Itoh et al. 2018). *Riptortus pedestris* (bean bug), a severe pest of leguminous crops, harbors *Burkholderia* in its midgut in sac-like tissues called "crypts" during its larval second instar stage that enables the bean bug to circumvent the toxic compounds, conferring insecticide resistance (Kikuchi et al. 2005). Notably, such symbiotic association ensures not only host survival but also an increase in body size, growth, and higher fecundity of the host (Kikuchi et al. 2005) .

2.5 Role of Microbial Volatiles in Plant–Insect Interaction

Similar to plants and animals, microbes also emit a plethora of volatile organic compounds (VOCs) in the course of their metabolic processes (Bitas et al. 2013). These compounds are usually lipophilic in nature that belong to the class of alcohols, aldehydes, esters, terpenoids, thiols, and fatty acid derivatives and have low molecular weight (<300 g mol^{-1}), low boiling point, and high vapor pressure (0.01 kPa at 20 °C) (Kanchiswamy et al. 2015a, b). The volatile compounds are perceived from a distance as chemical signals to communicate with each other and contribute significantly in multitrophic interaction (Schulz-Bohm et al. 2017). Over the years, the role of microbial volatile compounds (mVOCs) in plant physiology has gained attention. mVOCs affect hormonal balance, metabolism, sugar concentration, and the acquisition of essential nutrients in plants, thereby inducing growth and regulating stress response. For instance, volatiles released from *Bacillus subtilis* have been shown to stimulate growth and salt tolerance in *Arabidopsis thaliana* (Ryu et al. 2003; Zhang et al. 2008a). The underpinning mechanism behind the contribution of VOC in *A thaliana* was demonstrated using proteome analysis in combination with other biochemical experiments (Kwon et al. 2010). The VOCs released by *B. subtilis* upregulates the iron-regulated transporter 1 (IRT1) gene expression, facilitating the iron uptake from soil. Iron is an essential micronutrient in photosynthesis. Its increased uptake enhances the photosynthesis efficiency and the chlorophyll content thus, inducing plant growth (Fincheira and Quiroz 2018). Salt tolerance in *A. thaliana* in response to mVOCs resulted in the regulation of HKT1 gene that encodes high-affinity Na^{+} transporter (Zhang et al. 2008b). Similarly, *Pseudomonas chlororaphis* releases 2, 3- butanediol that induces shoot growth and confers resistance in the tobacco leaves against the soft-rot pathogen *Erwinia carotovora* (Han et al. 2006). The VOC-mediated resistance requires JA/SA/ET signaling pathways (Farag et al. 2013). Apart from the synergistic effect, the microbes also influence antagonistically to plants. Some bacterial species belonging to genera *Burkholderia, Pseudomonas, Serratia, Chromobacterium* release a wide array of volatiles that exhibit phytotoxicity and inhibit plant growth (Bailly and Weisskopf 2012; Kai et al. 2009).

Microbial volatiles are equally crucial to insects and their natural enemies. Insects rely on olfactory cues to locate their host as food resource or as oviposition site and exhibit defense against pathogens (Davis et al. 2013). For instance, the gut bacteria in locust produce antimicrobial phenolic compounds to protect against other microbial pathogens as well as aggregation pheromone "guaiacol" that promotes mating in locust (Dillon et al. 2000, 2002). In *Rhagoletis pomonella* (apple maggot fly) the oviposition behavior is influenced by the release of volatiles emitted by *Enterobacter agglomerans* present on the fruit (Lauzon et al. 1998). Insects are often attracted to fermented fruit that is inhabited by microbes. The mVOCs emitted enable insects to locate their food source (DeVries 1987). Yeast volatiles have also been reported to modulate sexual behavior and mating in *Drosophila melanogaster* (Gorter et al. 2016). Not only this, microbial volatiles contribute significantly to tri-trophic interaction (Hulcr et al. 2005). The volatiles released by plants or microbes associated with plants provide cues for the natural enemies to locate attacked plants (Hulcr et al. 2005). Interestingly, the yeast volatiles deployed by *Ogataea pini* inhibit the growth of entomopathogenic fungus *(B. bassiana)* on bark beetle. Understanding mVOCs and its role in plant–insect interaction provides a great platform to develop novel, eco-friendly, cost-effective, sustainable pest management strategies (Bitas et al. 2013).

3 Microbes as a Toolbox: Integration of Microbes in Pest Management

The world population is predicted to upsurge from a present population of 6 billion to 9 billion in 30 years, and the need for increased food production to meet the demands of the ever-increasing population is a major challenge (Lacey et al. 2015). Approximately 42% of the total crop loss is caused by pest infestation and is anticipated to rise to 83% without any crop protection (Oerke and Dehne 2004). In the quest of increasing crop yield, farmers have embraced a wide range of conventional pesticides such as organochlorines, organophosphates, carbamates, and pyrethroids. The use of chemical pesticides to control devastating pest has been undoubtedly a great success but suffers from many limitations. Extensive pesticide usage and the constant evolutionary dynamics of insects have led to the selection for pesticide resistance in target species as well as killed a number of non-target beneficial insect species, including pollinators and natural enemies. These chemicals often pollute the surface water and are harmful to birds, humans, and domestic and aquatic animals (Usta 2013). It is high time to reduce the use of chemical pesticides, so as not to gamble with the ecosystem, and to choose an eco-friendly alternative to pest control. As discussed earlier, microbes play a crucial role in host physiology and traits and contribute significantly to plant–insect interaction. Harnessing the potential of the microbes as a toolbox in controlling pests is indeed a smarter alternative approach toward sustainable IPM strategy for crop protection. The development of biocontrol against insect pests by exploiting the microbial potential has progressed tremendously over the last 20 years. However, the European legislation is making continuous efforts to promote the use of biopesticides through policies to restrict the broad-spectrum chemical pesticide practice and ban certain pesticides, but still, it holds no more than 3% of the total global pesticide market (Lacey et al. 2015). The use of microbial entomopathogens in agriculture is an excellent substitute for chemical fertilizer. Several entomopathogenic microbes are available in the global market as microbial control agents (MCAs) (Lacey et al. 2015). These entomopathogenic microbes easily invade the insect body while feeding where they multiply and confiscate the host, ultimately causing insect death.

Most of the commercially available biopesticides target one specific pest, and although it is advantageous for the safety of the environment and the non-target species, such low range of effectivity has restricted biopesticides to a niche market (Lacey et al. 2015). For example, entomopathogenic viral formulations are commercially available to control insect pests such as codling moth, *Cydia pomonella*, that are highly selective for the target pest and often sensitive to environmental conditions such as solar radiation (Lacey et al. 2008). To achieve commercially successful biopesticides, improvements are needed on the insecticidal activity spectra, persistence to environmental variations, and delivery to target-specific sites of pest occurrence and should be cost-effective (Glare et al. 2012).

Interestingly, by transferring microbial symbionts from one insect species to another species that do not harbor such microbes naturally can have a drastic effect on the insect physiology and behavior. For instance, *Wolbachia* isolated from Drosophila and introduced to mosquitoes by injecting into the *A. aegypti* embryos has remarkably reduced the virus load and viral transmission by mosquitoes (Fraser et al. 2017). Moreover, in addition to suppression of viral transmission, *Wolbachia* infection also causes cytoplasmic incompatibility, leading to reproductive disruption and population reduction (Ferguson et al. 2015; Joshi et al. 2017). This target-specific control strategy could be an effective alternative to control disease outbreaks. Another strategy could either be the mass release of sterile male insect or release of *Wolbachia* infected incompatible females into the environment, resulting in reproductive disruption, thereby controlling the pest population (Nikolouli et al. 2018). However, these strategies suffer certain drawbacks. The mass production of sterile insects is often challenging and not cost-effective. Moreover, environmental factors such as temperature change might have an effect on anti-viral protection and cytoplasmic incompatibility imparted by *Wolbachia* (Ross et al. 2017). Additionally, anti-viral protection depends on the bacterial load that considerably affects the physiology and fitness of the insect host (Martinez et al. 2015).

Megacopta punctatissima, a soybean crop pest, utilizes its gut bacterial symbiont *Ishikawaella* to thrive on soybean. However, a closely related species *M. cribraria* shows high mortality on soybean. Administrating *Ishikawaella* from *M. punctatissima* into the newly hatched nymphs of *M. cribraria* enabled successfully thriving on soybean whereas *M. punctatissima* lost the ability to survive on soybean (Hosokawa et al. 2007). This suggests that the potential of microbial symbionts could be used as an approach for the manipulation of insect host range.

The use of genetically modified (GM) crop variety is expressing microbial endotoxins or inducing RNA interference (RNAi) to target-specific insect species also holds excellent potential against pest control (Zhang et al. 2017). However, it is not feasible to engineer all vulnerable crop varieties as polyphagous pests have a broad host range. Insects do not only attack for feeding but also vectors plant pathogens. An alternative approach to this could be genetically modifying the microbes to deliver RNA interference to the insect by knocking down the genes essential for insect metabolic processes (Whitten et al. 2016). The delivery of dsRNA for RNA interference can be easily achieved through genetically modifying the microbes that are invariably ingested by the insects where it can proliferate in the gut and spread through feces. For example, administration of the genetically modified bacterial strain, expressing dsRNA against insect α-tubulin gene, to the western flower thrips (*F. occidentalis*) significantly increases the insect mortality (Whitten et al. 2016). However, RNAi technique holds high potential to control insect pest population, though a fundamental problem exists, i.e., dissemination of genetically modified microbes to non-target hosts through horizontal transfer. However, the development of highly specific dsDNA for RNAi to target genes of a particular pest species can mitigate the limitation (Arora and Douglas 2017).

Microbial symbionts are an integral part of the insect life cycle that often influence different aspects: host physiology, behavior, immunity, and reproduction. The elimination of these obligate microbial partners could be a promising strategy to control insect pests. The use of antimicrobial peptides such as melittin, cecropin, or toxin proteins to target obligate symbionts would compromise the insect pest. However, the delivery of such antimicrobial agents to a specific site to target gut symbionts is a challenge. Nevertheless, Husseneder et al. (2016) used genetically engineered *Kluyveromyces lactics* (as a microbial delivery vehicle) that expresses melittin against the termite *Coptotermes formosanus,* which resulted in the elimination of termite gut symbiont thereby losing its cellulose degrading capability.

Furthermore, manipulating the genetic pool of the microorganisms for specific expression in different habitat could provide a much safer strategy to target pest insects. In particular, various entomopathogenic microbes (*Metarhzium and Photorhabdus*) have been identified to possess promoters that express toxin gene only in insect habitat (Fang et al. 2011; Münch et al. 2008). Several bacterial suicidal genes are available that degrade in a non-permissive habitat (Li and Wu 2009). The encapsulation of the microbes enables microbial release on insect feeding and in insect gut under particular environmental conditions (such as a change in pH, hydrostatic pressure, or high protease activity) (Arora et al. 2015).

4 Conclusion and Future Perspective

To satisfy the ever-augmenting demands of the growing population, the need for increased food production and crop protection is a major challenge. An army of researchers has been engaged over the years in the development of robust IPM strategies, but most cropping systems to date are hugely dependent on chemical pesticides (Stenberg 2017). There is a clear need for a holistic approach for sustainable pest management as well as to minimize the associated risks. The recent development in technologies has opened up new dimensions in crop protection. The advent of genomics and next-generation sequencing has made it practicable to explore the full spectrum of microbial diversity as there are no longer "hidden players" in plant–insect interaction. The recent advancement in omics technologies is anticipated to have a considerable impact on the development of biocontrol strategies by harvesting the knowledge in the interaction between insects and their microbial allies. The characterization of microbial diversity together with metabolic fingerprinting plays a crucial role in an in-depth understanding of host–microbe interaction (Douglas 2018). Exploiting microbial partners can serve as a potential candidate for future pest management. Furthermore, recent advancements in RNAi and CRISPR-Cas9 technology have led to breakthroughs in agriculture by manipulating host-associated microorganisms as control strategies against pest insects (Arora and Douglas 2017; Gao 2018). The recent genetic engineering of gut microbiota in honeybee through state-of-the-art CRISPR-CAS9 technology has proven to be an excellent toolkit to characterize and manipulate the gut microbiome in insect host physiology (Leonard et al. 2018). The reproductive alteration mediated by bacterium *Wolbachia* by inducing cytoplasmic incompatibility in the host insect also serves as a potent strategy for pest control (Arora and Douglas 2017). However, it is essential to consider the risk associated with the release of genetically modified microbes to the environment. The application of antagonistic microbes is an alternative eco-friendly approach toward crop protection where the antagonistic microbe competes and/or inhibits the growth of plant pathogens (Feichtmayer et al. 2017). Not only microbes but also microbial-volatile compounds (mVOCs) are potential candidates in biocontrol (Bailly and Weisskopf 2017). It is important to consider that each of these strategies has its limitations that need to be considered in advance. Insects and microbes have a relatively short generation time and are in a constant evolutionary race to overwhelm our control endeavors. The continuous improvement of existing strategies and development of new avenues are pivotal to get rid of crop losses due to insect infestation in future.

Acknowledgments AC would like to acknowledge "EVA4.0", No. CZ.02.1.01 /0.0 /0.0 /16_019/0000803 financed by OP RDE" and IGA, Faculty of Forestry and Wood Sciences, Czech University of Life Sciences, Prague for the financial support during the preparation of the book chapter. AR is supported by the EU project "EXTEMIT - K," No. CZ.02.1.01/0.0/0.0/15_003/0000433 financed by OP RDE and IGA, Faculty of Forestry and Wood Sciences, Czech University of Life Sciences, Prague.

References

Adams AS, Aylward FO, Adams SM, Erbilgin N, Aukema BH, Currie CR et al (2013) Mountain pine beetles colonizing historical and naive host trees are associated with a bacterial community highly enriched in genes contributing to terpene metabolism. Appl Environ Microbiol, AEM:00068–00013

Andreote FD, e Silva M d CP (2017) Microbial communities associated with plants: learning from nature to apply it in agriculture. Curr Opin Microbiol 37:29–34

Arbuthnott D, Levin TC, Promislow DE (2016) The impacts of Wolbachia and the microbiome on mate choice in Drosophila melanogaster. J Evol Biol 29(2):461–468

Arora, A. K., & Douglas, A. E. (2017). Hype or opportunity? Using microbial symbionts in novel strategies for insect pest control. J Insect Physiol, 103, 10-17. doi:https://doi.org/10.1016/j.jinsphys.2017.09.011

Arora AK, Forshaw A, Miller TA, Durvasula R (2015) A delivery system for field application of paratransgenic control. BMC Biotechnol 15(1):59

Badri DV, Zolla G, Bakker MG, Manter DK, Vivanco JM (2013) Potential impact of soil microbiomes on the leaf metabolome and on herbivore feeding behavior. New Phytol 198 (1):264–273

Bailly A, Weisskopf L (2012) The modulating effect of bacterial volatiles on plant growth: current knowledge and future challenges. Plant Signal Behav 7(1):79–85

Bailly A, Weisskopf L (2017) Mining the Volatilomes of plant-associated microbiota for new biocontrol solutions. Front Microbiol 8:1638. https://doi.org/10.3389/fmicb.2017.01638

Batterman SA, Hedin LO, Van Breugel M, Ransijn J, Craven DJ, Hall JS (2013) Key role of symbiotic dinitrogen fixation in tropical forest secondary succession. Nature 502(7470):224

Baum C, El-Tohamy W, Gruda N (2015) Increasing the productivity and product quality of vegetable crops using arbuscular mycorrhizal fungi: a review. Sci Hortic 187:131–141

Behie S, Zelisko P, Bidochka M (2012) Endophytic insect-parasitic fungi translocate nitrogen directly from insects to plants. Science 336(6088):1576–1577

Behmer ST, Nes WD (2003) Insect sterol nutrition and physiology: a global overview. Adv Insect Physiol 31(1)

Benítez E, Paredes D, Rodríguez E, Aldana D, González M, Nogales R et al (2017) Bottom-up effects on herbivore-induced plant defences: a case study based on compositional patterns of rhizosphere microbial communities. Sci Rep 7(1):6251

Ben-Yosef M, Jurkevitch E, Yuval B (2008) Effect of bacteria on nutritional status and reproductive success of the Mediterranean fruit fly Ceratitis capitata. Physiol Entomol 33(2):145–154

Berasategui A, Salem H, Paetz C, Santoro M, Gershenzon J, Kaltenpoth M, Schmidt A (2017) Gut microbiota of the pine weevil degrades conifer diterpenes and increases insect fitness. Mol Ecol 26(15):4099–4110

Berenbaum M (1980) Adaptive significance of midgut pH in larval Lepidoptera. Am Nat 115 (1):138–146

Bernardo L, Morcia C, Carletti P, Ghizzoni R, Badeck FW, Rizza F, Terzi V (2017) Proteomic insight into the mitigation of wheat root drought stress by arbuscular mycorrhizae. J Proteome 169:21–32

Bi Y, Zhang Y, Shu C, Crickmore N, Wang Q, Du L et al (2015) Genomic sequencing identifies novel Bacillus thuringiensis Vip1/Vip2 binary and Cry8 toxins that have high toxicity to Scarabaeoidea larvae. Appl Microbiol Biotechnol 99(2):753–760

Biere A, Bennett AE (2013) Three-way interactions between plants, microbes and insects. Funct Ecol 27(3):567–573

Biere A, Tack AJ (2013) Evolutionary adaptation in three-way interactions between plants, microbes and arthropods. Funct Ecol 27(3):646–660

Bitas V, Kim H-S, Bennett JW, Kang S (2013) Sniffing on microbes: diverse roles of microbial volatile organic compounds in plant health. Mol Plant-Microbe Interact 26(8):835–843

Bizzarri M, Bishop A (2008) The ecology of Bacillus thuringiensis on the phylloplane: colonization from soil, plasmid transfer, and interaction with larvae of Pieris brassicae. Microb Ecol 56 (1):133–139

Boone CK, Keefover-Ring K, Mapes AC, Adams AS, Bohlmann J, Raffa KF (2013) Bacteria associated with a tree-killing insect reduce concentrations of plant defense compounds. J Chem Ecol 39(7):1003–1006

Bostock RM (2005) Signal crosstalk and induced resistance: straddling the line between cost and benefit. Annu Rev Phytopathol 43:545–580

Brackney DE (2017) Implications of autophagy on arbovirus infection of mosquitoes. Curr Opin Insect Sci 22:1–6

Bravo A, Gill SS, Soberón M (2007) Mode of action of Bacillus thuringiensis Cry and Cyt toxins and their potential for insect control. Toxicon 49(4):423–435. https://doi.org/10.1016/j.toxicon.2006.11.022

Bravo A, Likitvivatanavong S, Gill SS, Soberón M (2011) Bacillus thuringiensis: a story of a successful bioinsecticide. Insect Biochem Mol Biol 41(7):423–431. https://doi.org/10.1016/j.ibmb.2011.02.006

Broderick NA, Raffa KF, Goodman RM, Handelsman J (2004) Census of the bacterial community of the gypsy moth larval midgut by using culturing and culture-independent methods. Appl Environ Microbiol 70(1):293–300

Buchner P (1965) Endosymbiosis of animals with plant microorganims

Bulgarelli D, Schlaeppi K, Spaepen S, van Themaat EVL, Schulze-Lefert P (2013) Structure and functions of the bacterial microbiota of plants. Annu Rev Plant Biol 64:807–838

Butko P (2003) Cytolytic toxin Cyt1A and its mechanism of membrane damage: data and hypotheses. Appl Environ Microbiol 69(5):2415–2422. https://doi.org/10.1128/AEM.69.5.2415-2422

Calderón-Cortés N, Quesada M, Watanabe H, Cano-Camacho H, Oyama K (2012) Endogenous plant cell wall digestion: a key mechanism in insect evolution. Annu Rev Ecol Evol Syst 43:45–71

Cao Q, Wickham JD, Chen L, Ahmad F, Lu M, Sun J (2018) Effect of oxygen on Verbenone conversion from cis-Verbenol by gut facultative anaerobes of Dendroctonus valens. Front Microbiol 9:464

Capuzzo C, Firrao G, Mazzon L, Squartini A, Girolami V (2005) 'Candidatus Erwinia dacicola', a coevolved symbiotic bacterium of the olive fly Bactrocera oleae (Gmelin). Int J Syst Evol Microbiol 55(4):1641–1647

Caragata EP, Rancès E, Hedges LM, Gofton AW, Johnson KN, O'Neill SL, McGraw EA (2013) Dietary cholesterol modulates pathogen blocking by Wolbachia. PLoS Pathog 9(6):e1003459

Ceja-Navarro JA, Vega FE, Karaoz U, Hao Z, Jenkins S, Lim HC et al (2015) Gut microbiota mediate caffeine detoxification in the primary insect pest of coffee. Nat Commun 6:7618

Chakroun M, Banyuls N, Bel Y, Escriche B, Ferré J (2016) Bacterial vegetative insecticidal proteins (Vip) from entomopathogenic bacteria. Microbiol Mol Biol Rev 80(2):329–350

Chung SH, Rosa C, Scully ED, Peiffer M, Tooker JF, Hoover K et al (2013) Herbivore exploits orally secreted bacteria to suppress plant defenses. Proc Nat Acad Sci:201308867

Conrath U, Thulke O, Katz V, Schwindling S, Kohler A (2001) Priming as a mechanism in induced systemic resistance of plants. Eur J Plant Pathol 107(1):113–119

Contreras-Cornejo HA, Macías-Rodríguez L, Cortés-Penagos C, López-Bucio J (2009) Trichoderma virens, a plant beneficial fungus, enhances biomass production and promotes lateral root growth through an auxin-dependent mechanism in Arabidopsis. Plant Physiol 149 (3):1579–1592

Davis TS, Crippen TL, Hofstetter RW, Tomberlin JK (2013) Microbial volatile emissions as insect semiochemicals. J Chem Ecol 39(7):840–859

De Fine Licht HH, Schiøtt M, Rogowska-Wrzesinska A, Nygaard S, Roepstorff P, Boomsma JJ (2013) Laccase detoxification mediates the nutritional alliance between leaf-cutting ants and fungus-garden symbionts. Proc Natl Acad Sci U S A 110(2):583–587. https://doi.org/10.1073/pnas.1212709110

de Maagd RA, Bravo A, Berry C, Crickmore N, Schnepf HE (2003) Structure, diversity, and evolution of protein toxins from spore-forming entomopathogenic bacteria. Annu Rev Genet 37 (1):409–433

Dearing MD, Foley WJ, McLean S (2005) The influence of plant secondary metabolites on the nutritional ecology of herbivorous terrestrial vertebrates. Annu Rev Ecol Evol Syst 36:169–189

Desclos-Theveniau M, Arnaud D, Huang T-Y, Lin GJ-C, Chen W-Y, Lin Y-C, Zimmerli L (2012) The Arabidopsis lectin receptor kinase LecRK-V. 5 represses stomatal immunity induced by Pseudomonas syringae pv. tomato DC3000. PLoS Pathogens 8(2):e1002513

DeVries PJ (1987) The butterflies of Costa Rica and their natural history

Dillon RJ, Vennard CT, Charnley AK (2000) Pheromones: exploitation of gut bacteria in the locust. Nature 403(6772):851

Dillon R, Vennard C, Charnley A (2002) A note: gut bacteria produce components of a locust cohesion pheromone. J Appl Microbiol 92(4):759–763

Dong Y, Manfredini F, Dimopoulos G (2009) Implication of the mosquito midgut microbiota in the defense against malaria parasites. PLoS Pathog 5(5):e1000423. https://doi.org/10.1371/journal.ppat.1000423

Douglas AE (2013) Microbial brokers of insect-plant interactions revisited. J Chem Ecol 39 (7):952–961

Douglas AE (2015) Multiorganismal insects: diversity and function of resident microorganisms. Annu Rev Entomol 60:17–34

Douglas AE (2018) Omics and the metabolic function of insect-microbial symbioses. Curr Opin Insect Sci

Douglas AE, Minto LB, Wilkinson TL (2001) Quantifying nutrient production by the microbial symbionts in an aphid. J Exp Biol 204(2):349

Dowd PF, Shen SK (2011) *The contribution of symbiotic yeast to toxin resistance of the cigarette beetle (Lasioderma serricorne).* Entomol Exp Appl 56(3):241–248

Du M, Zhai Q, Deng L, Li S, Li H, Yan L et al (2014) Closely related NAC transcription factors of tomato differentially regulate stomatal closure and reopening during pathogen attack. The Plant Cell 114:128272

Egamberdieva D, Kucharova Z, Davranov K, Berg G, Makarova N, Azarova T et al (2011) Bacteria able to control foot and root rot and to promote growth of cucumber in salinated soils. Biol Fertil Soils 47(2):197–205

Egamberdieva D, Wirth S, Jabborova D, Räsänen LA, Liao H (2017) Coordination between Bradyrhizobium and Pseudomonas alleviates salt stress in soybean through altering root system architecture. J Plant Interact 12(1):100–107

Engel P, Moran NA (2013) The gut microbiota of insects–diversity in structure and function. FEMS Microbiol Rev 37(5):699–735

Engelstädter J, Hurst GD (2009) The ecology and evolution of microbes that manipulate host reproduction. Annu Rev Ecol Evol Syst 40:127–149

Engl T, Kaltenpoth M (2018) Influence of microbial symbionts on insect pheromones. Nat Prod Rep 35(5):386–397

Engl T, Eberl N, Gorse C, Krüger T, Schmidt TH, Plarre R et al (2018) Ancient symbiosis confers desiccation resistance to stored grain pest beetles. Mol Ecol 27(8):2095–2108

Estruch JJ, Warren GW, Mullins MA, Nye GJ, Craig JA, Koziel MG (1996) Vip3A, a novel Bacillus thuringiensis vegetative insecticidal protein with a wide spectrum of activities against lepidopteran insects. Proc Natl Acad Sci 93(11):5389–5394

Fang W, Vega-Rodríguez J, Ghosh AK, Jacobs-Lorena M, Kang A, Leger RJS (2011) Development of transgenic fungi that kill human malaria parasites in mosquitoes. Science 331 (6020):1074–1077

Farag MA, Zhang H, Ryu C-M (2013) Dynamic chemical communication between plants and Bacteria through airborne signals: induced resistance by bacterial volatiles. J Chem Ecol 39 (7):1007–1018. https://doi.org/10.1007/s10886-013-0317-9

Feichtmayer J, Deng L, Griebler C (2017) Antagonistic microbial interactions: contributions and potential applications for controlling pathogens in the aquatic systems. Front Microbiol 8:2192. https://doi.org/10.3389/fmicb.2017.02192

Ferguson NM, Kien DTH, Clapham H, Aguas R, Trung VT, Chau TNB, McGraw EA (2015) Modeling the impact on virus transmission of Wolbachia-mediated blocking of dengue virus infection of Aedes aegypti. Sci Trans Med 7(279):279ra237–279ra237

Ferrari J, Vavre F (2011) Bacterial symbionts in insects or the story of communities affecting communities. Philos Trans R Soc London B: Biol Sci 366(1569):1389–1400

Fincheira P, Quiroz A (2018) Microbial volatiles as plant growth inducers. Microbiol Res

Fiorilli V, Vannini C, Ortolani F, Garcia-Seco D, Chiapello M, Novero M, Bagnaresi P (2018) Omics approaches revealed how arbuscular mycorrhizal symbiosis enhances yield and resistance to leaf pathogen in wheat. Sci Rep 8(1):9625

Fraser JE, De Bruyne JT, Iturbe-Ormaetxe I, Stepnell J, Burns RL, Flores HA, O'Neill SL (2017) Novel Wolbachia-transinfected Aedes aegypti mosquitoes possess diverse fitness and vector competence phenotypes. PLoS Pathog 13(12):e1006751

Fürstenberg-Hägg J, Zagrobelny M, Bak S (2013) Plant defense against insect herbivores. Int J Mol Sci 14(5):10242–10297

Gao C (2018) The future of CRISPR technologies in agriculture. Nat Rev Mol Cell Biol 19:275. https://doi.org/10.1038/nrm.2018.2
García-Fraile P (2018) Roles of bacteria in the bark beetle holobiont–how do they shape this forest pest? Ann Appl Biol 172(2):111–125
García-Garrido JM, Ocampo JA (2002) Regulation of the plant defence response in arbuscular mycorrhizal symbiosis. J Exp Bot 53(373):1377–1386
Gianinazzi S, Gollotte A, Binet M-N, van Tuinen D, Redecker D, Wipf D (2010) Agroecology: the key role of arbuscular mycorrhizas in ecosystem services. Mycorrhiza 20(8):519–530
Giron D, Glevarec G (2014) Cytokinin-induced phenotypes in plant-insect interactions: learning from the bacterial world. J Chem Ecol 40(7):826–835
Glare T, Caradus J, Gelernter W, Jackson T, Keyhani N, Köhl J et al (2012) Have biopesticides come of age? Trends Biotechnol 30(5):250–258. https://doi.org/10.1016/j.tibtech.2012.01.003
Gorter JA, Jagadeesh S, Gahr C, Boonekamp JJ, Levine JD, Billeter J-C (2016) The nutritional and hedonic value of food modulate sexual receptivity in Drosophila melanogaster females. Sci Rep 6:19441. https://doi.org/10.1038/srep19441
Gutjahr C, Parniske M (2013) Cell and developmental biology of arbuscular mycorrhiza symbiosis. Annu Rev Cell Dev Biol 29:593–617
Haas D, Keel C (2003) Regulation of antibiotic production in root-colonizing Pseudomonas spp. and relevance for biological control of plant disease. Annu Rev Phytopathol 41(1):117–153
Hammer EC, Balogh-Brunstad Z, Jakobsen I, Olsson PA, Stipp SL, Rillig MC (2014) A mycorrhizal fungus grows on biochar and captures phosphorus from its surfaces. Soil Biol Biochem 77:252–260
Han SH, Lee SJ, Moon JH, Park KH, Yang KY, Cho BH et al (2006) GacS-dependent production of 2R, 3R-butanediol by Pseudomonas chlororaphis O6 is a major determinant for eliciting systemic resistance against Erwinia carotovora but not against Pseudomonas syringae pv. tabaci in tobacco. Mol Plant-Microbe Interact 19(8):924–930
Haskett TL, Terpolilli JJ, Bekuma A, O'Hara GW, Sullivan JT, Wang P et al (2016) Assembly and transfer of tripartite integrative and conjugative genetic elements. Proc Natl Acad Sci 113 (43):12268–12273
Herman M, Nault B, Smart C (2008) Effects of plant growth-promoting rhizobacteria on bell pepper production and green peach aphid infestations in New York. Crop Prot 27(6):996–1002
Hieno A, Naznin HA, Hyakumachi M, Higuchi-Takeuchi M, Matsui M, Yamamoto YY (2016) Possible involvement of MYB44-mediated stomatal regulation in systemic resistance induced by Penicillium simplicissimum GP17-2 in Arabidopsis. Microbes Environ 31(2):154–159
Himler AG, Adachi-Hagimori T, Bergen JE, Kozuch A, Kelly SE, Tabashnik BE et al (2011) Rapid spread of a bacterial symbiont in an invasive whitefly is driven by fitness benefits and female bias. Science 332(6026):254–256
Hosokawa T, Kikuchi Y, Shimada M, Fukatsu T (2007) Obligate symbiont involved in pest status of host insect. Proc R Soc Lond B Biol Sci 274(1621):1979–1984
Howard RW, Blomquist GJ (2005) Ecological, behavioral, and biochemical aspects of insect hydrocarbons. Annu Rev Entomol 50(1):371–393. https://doi.org/10.1146/annurev.ento.50.071803.130359
Howe M, Keefover-Ring K, Raffa KF (2018) Pine engravers carry bacterial communities whose members reduce concentrations of host monoterpenes with variable degrees of redundancy, specificity, and capability. Environ Entomol 47(3):638–645
Hughes DP, Brodeur J, Thomas F (2012) Host manipulation by parasites. Oxford University Press, Oxford
Hulcr J, Pollet M, Ubik K, Vrkoc J (2005) Exploitation of kairomones and synomones by Medetera spp.(Diptera: Dolichopodidae), predators of spruce bark beetles. Eur J Entomol 102(4):655
Hussain M, Frentiu FD, Moreira LA, O'Neill SL, Asgari S (2011) Wolbachia uses host microRNAs to manipulate host gene expression and facilitate colonization of the dengue vector Aedes aegypti. Proc Natl Acad Sci 108(22):9250–9255

Husseneder C, Donaldson JR, Foil LD (2016) Genetically engineered yeast expressing a lytic peptide from Bee Venom (Melittin) kills symbiotic protozoa in the gut of Formosan subterranean termites. PLoS One 11(3):e0151675

Itoh H, Tago K, Hayatsu M, Kikuchi Y (2018) Detoxifying symbiosis: microbe-mediated detoxification of phytotoxins and pesticides in insects. Nat Prod Rep 35(5):434–454

Jaber LR, Enkerli J (2016) Effect of seed treatment duration on growth and colonization of Vicia faba by endophytic Beauveria bassiana and Metarhizium brunneum. Biol Control 103:187–195

Jaenike J, Unckless R, Cockburn SN, Boelio LM, Perlman SJ (2010) Adaptation via symbiosis: recent spread of a Drosophila defensive symbiont. Science 329(5988):212–215

Janson EM, Stireman JO III, Singer MS, Abbot P (2008) Phytophagous insect–microbe mutualisms and adaptive evolutionary diversification. Evol Int J Org Evol 62(5):997–1012

Johnson K (2015a) The impact of Wolbachia on virus infection in mosquitoes. Viruses 7 (11):5705–5717

Johnson KN (2015b) Bacteria and antiviral immunity in insects. Curr Opin Insect Sci 8:97–103

Jones JD, Dangl JL (2006) The plant immune system. Nature 444(7117):323

Joshi D, Pan X, McFadden MJ, Bevins D, Liang X, Lu P et al (2017) The maternally inheritable Wolbachia wAlbB induces refractoriness to Plasmodium berghei in Anopheles stephensi. Front Microbiol 8:366

Jurat-Fuentes JL, Crickmore N (2017) Specificity determinants for cry insecticidal proteins: insights from their mode of action. J Invertebr Pathol 142:5–10

Kai M, Haustein M, Molina F, Petri A, Scholz B, Piechulla B (2009) Bacterial volatiles and their action potential. Appl Microbiol Biotechnol 81(6):1001–1012

Kanchiswamy CN, Malnoy M, Maffei ME (2015a) Bioprospecting bacterial and fungal volatiles for sustainable agriculture. Trends Plant Sci 20(4):206–211

Kanchiswamy CN, Malnoy M, Maffei ME (2015b) Chemical diversity of microbial volatiles and their potential for plant growth and productivity. Front Plant Sci 6:151

Kempel A, Brandl R, Schädler M (2009) Symbiotic soil microorganisms as players in aboveground plant–herbivore interactions–the role of rhizobia. Oikos 118(4):634–640

Kessler A, Baldwin IT (2002) Plant responses to insect herbivory: the emerging molecular analysis. Annu Rev Plant Biol 53(1):299–328

Kikuchi Y, Meng X-Y, Fukatsu T (2005) Gut symbiotic bacteria of the genus Burkholderia in the broad-headed bugs Riptortus clavatus and Leptocorisa chinensis (Heteroptera: Alydidae). Appl Environ Microbiol 71(7):4035–4043

Kikuchi Y, Hosokawa T, Fukatsu T (2007) Insect-microbe mutualism without vertical transmission: a stinkbug acquires a beneficial gut symbiont from the environment every generation. Appl Environ Microbiol 73(13):4308–4316

Koornneef A, Pieterse CM (2008) Cross talk in defense signaling. Plant Physiol 146(3):839–844

Kos M, Tuijl MA, de Roo J, Mulder PP, Bezemer TM (2015) Species-specific plant–soil feedback effects on above-ground plant–insect interactions. J Ecol 103(4):904–914

Kula AA, Hartnett DC, Wilson GW (2005) Effects of mycorrhizal symbiosis on tallgrass prairie plant–herbivore interactions. Ecol Lett 8(1):61–69

Kumar AS, Lakshmanan V, Caplan JL, Powell D, Czymmek KJ, Levia DF, Bais HP (2012) Rhizobacteria Bacillus subtilis restricts foliar pathogen entry through stomata. Plant J 72 (4):694–706

Kwon YS, Ryu C-M, Lee S, Park HB, Han KS, Lee JH et al (2010) Proteome analysis of Arabidopsis seedlings exposed to bacterial volatiles. Planta 232(6):1355–1370

Lacey LA, Thomson D, Vincent C, Arthurs SP (2008) Codling moth granulovirus: a comprehensive review. Biocontrol Sci Tech 18(7):639–663. https://doi.org/10.1080/09583150802267046

Lacey L, Grzywacz D, Shapiro-Ilan D, Frutos R, Brownbridge M, Goettel M (2015) Insect pathogens as biological control agents: back to the future. J Invertebr Pathol 132:1–41

Lauzon C, Sjogren R, Wright S, Prokopy R (1998) Attraction of Rhagoletis pomonella (Diptera: Tephritidae) flies to odor of bacteria: apparent confinement to specialized members of Enterobacteriaceae. Environ Entomol 27(4):853–857

Leonard SP, Perutka J, Powell JE, Geng P, Richhart DD, Byrom M, Moran NA (2018) Genetic engineering of bee gut microbiome bacteria with a toolkit for modular assembly of broad-host-range plasmids. ACS Synth Biol 7(5):1279–1290

Li Q, Wu Y-J (2009) A fluorescent, genetically engineered microorganism that degrades organophosphates and commits suicide when required. Appl Microbiol Biotechnol 82 (4):749–756

Ling J, Wang H, Wu P, Li T, Tang Y, Naseer N et al (2016) Plant nodulation inducers enhance horizontal gene transfer of Azorhizobium caulinodans symbiosis island. Proc Natl Acad Sci 113 (48):13875–13880

Liu F, Xing S, Ma H, Du Z, Ma B (2013) Cytokinin-producing, plant growth-promoting rhizobacteria that confer resistance to drought stress in Platycladus orientalis container seedlings. Appl Microbiol Biotechnol 97(20):9155–9164

Loper JE, Gross H (2007) Genomic analysis of antifungal metabolite production by Pseudomonas fluorescens Pf-5. In: New perspectives and approaches in plant growth-promoting rhizobacteria research. Springer, Dordrecht, pp 265–278

Marshall D, Jackson T, Unelius CR, Wee S, Young S, Townsend R, Suckling D (2016) Morganella morganii bacteria produces phenol as the sex pheromone of the New Zealand grass grub from tyrosine in the colleterial gland. Sci Nat 103(7-8):59

Martinez J, Ok S, Smith S, Snoeck K, Day JP, Jiggins FM (2015) Should symbionts be nice or selfish? Antiviral effects of Wolbachia are costly but reproductive parasitism is not. PLoS Pathog 11(7):e1005021

Melotto M, Underwood W, Koczan J, Nomura K, He SY (2006) Plant stomata function in innate immunity against bacterial invasion. Cell 126(5):969–980

Miller WJ, Schneider D (2012) Endosymbiotic microbes as adaptive manipulators of arthropod behavior and natural driving sources of host speciation. Host Manipulation by Parasites, 119–137

Millet YA, Danna CH, Clay NK, Songnuan W, Simon MD, Werck-Reichhart D, Ausubel FM (2010) Innate immune responses activated in Arabidopsis roots by microbe-associated molecular patterns. Plant Cell 22(3):973–990

Monnerat RG, Soares CM, Capdeville G, Jones G, Martins ÉS, Praça L et al (2009) Translocation and insecticidal activity of Bacillus thuringiensis living inside of plants. Microb Biotechnol 2 (4):512–520

Montillet J-L, Leonhardt N, Mondy S, Tranchimand S, Rumeau D, Boudsocq M et al (2013) An abscisic acid-independent oxylipin pathway controls stomatal closure and immune defense in Arabidopsis. PLoS Biol 11(3):e1001513

Münch A, Stingl L, Jung K, Heermann R (2008) Photorhabdus luminescens genes induced upon insect infection. BMC Genomics 9(1):229

Muvea AM, Meyhöfer R, Subramanian S, Poehling H-M, Ekesi S, Maniania NK (2014) Colonization of onions by endophytic fungi and their impacts on the biology of Thrips tabaci. PLoS One 9(9):e108242

Najarro MA, Sumethasorn M, Lamoureux A, Turner TL (2015) Choosing mates based on the diet of your ancestors: replication of non-genetic assortative mating in Drosophila melanogaster. Peer J 3:e1173

Nikolouli K, Colinet H, Renault D, Enriquez T, Mouton L, Gibert P et al (2018) Sterile insect technique and Wolbachia symbiosis as potential tools for the control of the invasive species Drosophila suzukii. J Pest Sci:1–15

Nomura K, Melotto M, He S-Y (2005) Suppression of host defense in compatible plant–Pseudomonas syringae interactions. Curr Opin Plant Biol 8(4):361–368

Oerke E-C, Dehne H-W (2004) Safeguarding production—losses in major crops and the role of crop protection. Crop Prot 23(4):275–285

Okubo T, Piromyou P, Tittabutr P, Teaumroong N, Minamisawa K (2016) Origin and evolution of nitrogen fixation genes on symbiosis islands and plasmid in Bradyrhizobium. Microbes Environ 31(3):260–267

Oliver KM, Degnan PH, Burke GR, Moran NA (2010) Facultative symbionts in aphids and the horizontal transfer of ecologically important traits. Annu Rev Entomol 55:247–266
Palma L, Muñoz D, Berry C, Murillo J, Caballero P (2014) Bacillus thuringiensis toxins: an overview of their biocidal activity. Toxins 6(12):3296–3325
Pan X, Zhou G, Wu J, Bian G, Lu P, Raikhel AS, Xi Z (2012) Wolbachia induces reactive oxygen species (ROS)-dependent activation of the Toll pathway to control dengue virus in the mosquito Aedes aegypti. Proc Natl Acad Sci 109(1):E23–E31
Papadopoulou GV, van Dam NM (2017) Mechanisms and ecological implications of plant-mediated interactions between belowground and aboveground insect herbivores. Ecol Res 32 (1):13–26
Patot S, Allemand R, Fleury F, Varaldi J (2012) An inherited virus influences the coexistence of parasitoid species through behaviour manipulation. Ecol Lett 15(6):603–610
Peterson BF, Scharf ME (2016) Lower termite associations with microbes: synergy, protection, and interplay. Front Microbiol 7:422
Pieterse CM, Dicke M (2007) Plant interactions with microbes and insects: from molecular mechanisms to ecology. Trends Plant Sci 12(12):564–569
Pigott CR, Ellar DJ (2007) Role of receptors in Bacillus thuringiensis crystal toxin activity. Microbiol Mol Biol Rev 71(2):255–281
Pilon F, Visôtto L, Guedes R, Oliveira M (2013) Proteolytic activity of gut bacteria isolated from the velvet bean caterpillar Anticarsia gemmatalis. J Comp Physiol B 183(6):735–747
Pineda A, Zheng S-J, Van Loon JJ, Pieterse CM, Dicke M (2010) Helping plants to deal with insects: the role of beneficial soil-borne microbes. Trends Plant Sci 15(9):507–514
Pineda A, Soler R, Pozo MJ, Rasmann S, Turlings TC (2015) Above-belowground interactions involving plants, microbes and insects. Front Plant Sci 6:318
Pineda A, Kaplan I, Bezemer TM (2017) Steering soil microbiomes to suppress aboveground insect pests. Trends Plant Sci 22(9):770–778
Portugal L, Muñóz-Garay C, de Castro DLM, Soberón M, Bravo A (2017) Toxicity of Cry1A toxins from Bacillus thuringiensis to CF1 cells does not involve activation of adenylate cyclase/ PKA signaling pathway. Insect Biochem Mol Biol 80:21–31
Pozo MJ, Azcón-Aguilar C (2007) Unraveling mycorrhiza-induced resistance. Curr Opin Plant Biol 10(4):393–398
Ramsay J, Ronson C (2015) Genetic regulation of symbiosis island transfer in Mesorhizobium loti. In Biological Nitrogen Fixation
Regnier FE, Law JH (1968) Insect pheromones. J Lipid Res 9(5):541–551
Renwick J, Hughes P, Krull I (1976) Selective production of cis-and trans-verbenol from (−)-and (+)-alpha by a bark beetle. Science 191(4223):199–201
Richardson AE, Barea J-M, McNeill AM, Prigent-Combaret C (2009) Acquisition of phosphorus and nitrogen in the rhizosphere and plant growth promotion by microorganisms. Plant Soil 321 (1-2):305–339
Robert-Seilaniantz A, Navarro L, Bari R, Jones JD (2007) Pathological hormone imbalances. Curr Opin Plant Biol 10(4):372–379
Rohfritsch O (2008) Plants, gall midges, and fungi: a three-component system. Entomol Exp Appl 128(1):208–216
Ross PA, Wiwatanaratanabutr I, Axford JK, White VL, Endersby-Harshman NM, Hoffmann AA (2017) Wolbachia infections in Aedes aegypti differ markedly in their response to cyclical heat stress. PLoS Pathog 13(1):e1006006
Ryu C-M, Farag MA, Hu C-H, Reddy MS, Wei H-X, Paré PW, Kloepper JW (2003) Bacterial volatiles promote growth in Arabidopsis. Proc Natl Acad Sci 100(8):4927–4932
Salem H, Florez L, Gerardo N, Kaltenpoth M (2015) An out-of-body experience: the extracellular dimension for the transmission of mutualistic bacteria in insects. Proc R Soc B 282 (1804):20142957

Schouteden N, De Waele D, Panis B, Vos CM (2015) Arbuscular mycorrhizal fungi for the biocontrol of plant-parasitic nematodes: a review of the mechanisms involved. Front Microbiol 6:1280

Schulz-Bohm K, Martín-Sánchez L, Garbeva P (2017) Microbial volatiles: small molecules with an important role in intra- and inter-kingdom interactions. Front Microbiol 8(2484). https://doi.org/10.3389/fmicb.2017.02484

Segarra G, Van der Ent S, Trillas I, Pieterse C (2009) MYB72, a node of convergence in induced systemic resistance triggered by a fungal and a bacterial beneficial microbe. Plant Biol 11 (1):90–96

Sharon G, Segal D, Ringo JM, Hefetz A, Zilber-Rosenberg I, Rosenberg E (2010) Commensal bacteria play a role in mating preference of Drosophila melanogaster. Proc Natl Acad Sci U S A 107(46):20051–20056. https://doi.org/10.1073/pnas.1009906107

Shikano I, Rosa C, Tan C-W, Felton GW (2017) Tritrophic interactions: microbe-mediated plant effects on insect herbivores. Annu Rev Phytopathol 55:313–331

Simon J-C, Boutin S, Tsuchida T, Koga R, Le Gallic J-F, Frantz A et al (2011) Facultative symbiont infections affect aphid reproduction. PLoS One 6(7):e21831

Singh RP, Jha PN (2016) A halotolerant bacterium Bacillus licheniformis HSW-16 augments induced systemic tolerance to salt stress in wheat plant (Triticum aestivum). Front Plant Sci 7:1890

Singh LP, Gill SS, Tuteja N (2011) Unraveling the role of fungal symbionts in plant abiotic stress tolerance. Plant Signal Behav 6(2):175–191

Song Y, Chen D, Lu K, Sun Z, Zeng R (2015) Enhanced tomato disease resistance primed by arbuscular mycorrhizal fungus. Front Plant Sci 6:786

Song F, Lin Y, Chen C, Shao E, Guan X, Huang Z (2016) Insecticidal activity and histopathological effects of Vip3Aa protein from Bacillus thuringiensis on Spodoptera litura. J Microbiol Biotechnol

Stein E, Molitor A, Kogel K-H, Waller F (2008) Systemic resistance in Arabidopsis conferred by the mycorrhizal fungus Piriformospora indica requires jasmonic acid signaling and the cytoplasmic function of NPR1. Plant Cell Physiol 49(11):1747–1751

Stenberg JA (2017) A conceptual framework for integrated pest management. Trends Plant Sci 22 (9):759–769

Stenersen J (2004) Chemical pesticides mode of action and toxicology. CRC Press, Boca Raton

Stout MJ, Thaler JS, Thomma BP (2006) Plant-mediated interactions between pathogenic microorganisms and herbivorous arthropods. Annu Rev Entomol 51:663–689

Sugio A, Dubreuil G, Giron D, Simon J-C (2014) Plant–insect interactions under bacterial influence: ecological implications and underlying mechanisms. J Exp Bot 66(2):467–478

Teixeira L, Ferreira Á, Ashburner M (2008) The bacterial symbiont Wolbachia induces resistance to RNA viral infections in Drosophila melanogaster. PLoS Biol 6(12):e1000002

Terradas G, McGraw EA (2017) Wolbachia-mediated virus blocking in the mosquito vector Aedes aegypti. Curr Opin Insect Sci 22:37–44

Tétard-Jones C, Kertesz MA, Gallois P, Preziosi RF (2007) Genotype-by-genotype interactions modified by a third species in a plant-insect system. Am Nat 170(3):492–499

Tétard-Jones C, Kertesz MA, Preziosi RF (2012) Identification of plant quantitative trait loci modulating a rhizobacteria-aphid indirect effect. PLoS One 7(7):e41524

Thomma BP, Penninckx IA, Cammue BP, Broekaert WF (2001) The complexity of disease signaling in Arabidopsis. Curr Opin Immunol 13(1):63–68

Trillas M, Segarra G (2009) Interactions between nonpathogenic fungi and plants. Adv Bot Res 51:321–359

Usta C (2013) Microorganisms in biological pest control—a review (bacterial toxin application and effect of environmental factors). In Current progress in biological research: Intech

Van der Ent S, Van Wees SC, Pieterse CM (2009) Jasmonate signaling in plant interactions with resistance-inducing beneficial microbes. Phytochemistry 70(13–14):1581–1588

Van Loon L (2007) Plant responses to plant growth-promoting rhizobacteria. In: New perspectives and approaches in plant growth-promoting rhizobacteria research. Springer, Dordrecht, pp 243–254

van Loon LC, Rep M, Pieterse CM (2006) Significance of inducible defense-related proteins in infected plants. Annu Rev Phytopathol 44:135–162

Van Wees SC, Van der Ent S, Pieterse CM (2008) Plant immune responses triggered by beneficial microbes. Curr Opin Plant Biol 11(4):443–448

Vité J, Pitman G (1968) Bark beetle aggregation: effects of feeding on the release of pheromones in Dendroctonus and Ips. Nature 218(5137):169

Walters D, Newton A, Lyon G (2007) Induced resistance for plant defence. Wiley Online Library

Wang D, Huang Z, He H, Wei C (2018) Comparative analysis of microbial communities associated with bacteriomes, reproductive organs and eggs of the cicada Subpsaltria yangi. Arch Microbiol 200(2):227–235

Weeks AR, Turelli M, Harcombe WR, Reynolds KT, Hoffmann AA (2007) From parasite to mutualist: rapid evolution of Wolbachia in natural populations of Drosophila. PLoS Biol 5(5): e114

Werren JH, Baldo L, Clark ME (2008) Wolbachia: master manipulators of invertebrate biology. Nat Rev Microbiol 6(10):741

Whalon ME, Mota-Sanchez D, Hollingworth RM (2008) Global pesticide resistance in arthropods. Cabi

Whitten MM, Facey PD, Del Sol R, Fernández-Martínez LT, Evans MC, Mitchell JJ et al (2016) Symbiont-mediated RNA interference in insects. Proc R Soc B 283(1825):20160042

Wu J, Baldwin IT (2010) New insights into plant responses to the attack from insect herbivores. Annu Rev Genet 44:1–24

Yixin HY, Seleznev A, Flores HA, Woolfit M, McGraw EA (2017) Gut microbiota in Drosophila melanogaster interacts with Wolbachia but does not contribute to Wolbachia-mediated antiviral protection. J Invertebr Pathol 143:18–25

Zack MD, Sopko MS, Frey ML, Wang X, Tan SY, Arruda JM et al (2017) Functional characterization of Vip3Ab1 and Vip3Bc1: two novel insecticidal proteins with differential activity against lepidopteran pests. Sci Rep 7(1):11112

Zamioudis C, Pieterse CM (2012) Modulation of host immunity by beneficial microbes. Mol Plant-Microbe Interact 25(2):139–150

Zarate SI, Kempema LA, Walling LL (2007) Silverleaf whitefly induces salicylic acid defenses and suppresses effectual jasmonic acid defenses. Plant Physiol 143(2):866–875

Zhang X, Candas M, Griko NB, Taussig R, Bulla LA (2006) A mechanism of cell death involving an adenylyl cyclase/PKA signaling pathway is induced by the Cry1Ab toxin of Bacillus thuringiensis. Proc Natl Acad Sci 103(26):9897–9902

Zhang H, Xie X, Kim MS, Kornyeyev DA, Holaday S, Paré PW (2008a) Soil bacteria augment Arabidopsis photosynthesis by decreasing glucose sensing and abscisic acid levels in planta. Plant J 56(2):264–273

Zhang H, Kim M-S, Sun Y, Dowd SE, Shi H, Paré PW (2008b) Soil bacteria confer plant salt tolerance by tissue-specific regulation of the sodium transporter HKT1. Mol Plant-Microbe Interact 21(6):737–744

Zhang J, Khan SA, Heckel DG, Bock R (2017) Next-generation insect-resistant plants: RNAi-mediated crop protection. Trends Biotechnol 35(9):871–882. https://doi.org/10.1016/j.tibtech.2017.04.009

Zheng X-Y, Spivey NW, Zeng W, Liu P-P, Fu ZQ, Klessig DF et al (2012) Coronatine promotes Pseudomonas syringae virulence in plants by activating a signaling cascade that inhibits salicylic acid accumulation. Cell Host Microbe 11(6):587–596

Zhu F, Poelman EH, Dicke M (2014) Insect herbivore-associated organisms affect plant responses to herbivory. New Phytol 204(2):315–321

Microbe-Plant-Insect Interactions: A Comparative Dissection of Interactome

Rahul Arora and Garima Malik

Abstract

Plants being the producers of food encounter various pests and pathogens, including microbes (virus, bacteria, fungi, protists, nematodes), insect and vertebrates with diverse modes of attack. Although some organisms have a mutualistic/symbiotic association with plants, others are harmful, may greatly impair plant productivity and threaten overall food security. In nature, plants and its enemies have coevolved for an interdependent co-existent. The plant defence mechanisms may vary when it encounters different attackers; however, they have evolved their defence mechanisms with a high degree of overlap to keep its resources allocated in an orderly fashion. The tripartite interactions between the microbe-plant-insect (MPI) form the basis for the plants to host the room for evolution and structure the communities of the interacting organisms along with the development of the host-vector relationship. Since, all the three interacting organisms (MPI) form a single system, the defence generated by plants is usually modified and may affect the microbes and insects up to a different degree, ranging from beneficial to detrimental. The early defensive strategies developed by plants against microbes and insects are almost similar and may involve the same mechanism; however, the difference lies at the molecular level. In this chapter, the biochemical and molecular aspect of defence mechanism regulating these interactions has been presented and discussed to gain an insight of practical applications for improving plant productivity and the plant immunity.

R. Arora
Division of Biosciences, University College London, London, UK

G. Malik (✉)
R.G. (P.G.) College, C.C.S University, Meerut, Uttar Pradesh, India

I. K. Singh, A. Singh (eds.), *Plant-Pest Interactions: From Molecular Mechanisms to Chemical Ecology*, https://doi.org/10.1007/978-981-15-2467-7_15

Keywords

Coevolution · DAMPS · Defence mechanisms · HAMPS · MAMPS Microbe-plant-insect interactions · Phytohormones

1 Introduction

The sunlight fixed by plants via photosynthesis drives almost all life processes on our planet. Plants are the fundamental players in a complex food web in which almost all organisms, either directly or indirectly, rely upon the plant's resources. Being primary producers, plants get frequently exposed to a large group of beneficial and detrimental organisms and are under constant risk of infection by phytopathogens armed with various elicitors and effector molecules that aid them to colonise their host. The survival of plant species in an ever-changing environment depends on their ability to regulate their physiological processes to maximise their fitness (Schfer et al. 2015). Phytopathogens are organisms that can inhabit the internal environment of plants or live on the plant surface and can compromise the well-being of the plant. There is a wide range of phytopathogens such as bacteria, fungi, virus, and insects, covering diverse levels of host specificity; some are host species specific, while others with an extensive host range. Phytopathogens are usually classified into three categories: biotrophs, hemibiotrophs and necrotrophs. Biotrophic pathogens feed, grow and reproduce on living plant tissue and keep their host alive. On the other hand, pathogens that kill the host tissue at the beginning of the infection by producing toxins and tissue-degrading enzymes and feed on the dead tissue are known as necrotrophic pathogens. Some pathogens are hemibiotrophs, such that the host on they attack, remains viable in early stages of infection but die as the infection progresses toward the later stages.

The dynamic and continuing battles of coevlolution between plants and pathogens have lead to the advancement of extremely specific and highly sophisticated assault strategies by the attackers and in turn likewise a wide array of finely tuned defence mechanisms in the host to guard themselves against the stress caused by the enormous number of potential pathogens. However, plants are able to perceive invading pathogens, mount efficient defences, and remain healthy for most of their life cycle (Jones and Dangl 2006). Plant-pathogen interactions have shaped the plant immune system over the course of an endless process of coevolution. Plants are defiant to most pathogens due to an intricate defence signalling network which is flexible to perceive and act in response to an episode of attack by an invader. Along with various morphological barriers (glandular trichomes, thick cuticle, etc.), reactive oxygen species (ROS), defensive proteins, secondary metabolites and volatile organic compounds (VOCs), they further enhance their defence strategies (Pieterse et al. 2013). The role of phytohormones such as salicylic acid (SA), Jasmonic acid (JA) and ethylene (ET) is well- known in plant defense response during both biotic and abiotic stress. When attacked by a pathogen, plants retort by producing specific mix of phytohormones that varies significantly in amount, composition and timing. The varied response generated by the highly specific nature of phytohormones contributes to the specificity of the plant's defence response that are effective against different types of pathogens (Reymond and Farmer 1998; De Vos et al. 2005). The MPI interaction can impact the plant defence response in total and, accordingly, manipulate insect behaviour or microbe infection and thereby plant's fitness (Karban et al. 1987; Simon and Hilker 2003). Most often both herbivores and pathogens attack simultaneously and thus require a synchronisation of the defence responses from plants. The MPI interactions are complex and thus make it hard to unravel defence-related phytochemical and gene functions by utilising different genetic approaches and biochemical pathway analysis (Franco et al. 2017). Therefore, studies involving microbe-plant-insect interactions would inevitably improve our understanding of how the parallel defence mechanisms employed by plants protect them against a diverse group of multiple enemies. The major class of phytohormones that participate in MPI tripartite interactions are JA, SA and ET. These plant hormones interact with each other either in a mutualistic or an antagonistic manner and thus act as a great line of evidence supporting the interactive network of chemicals, transcription factors (TF) and other molecular chaperones linking the microbial defence signalling with herbivore defence signalling. The role of other phytohormones, namely, abscisic acid (ABA), auxin (IAA), brassinosteroids (BRs), cytokinin (CK) and gibberellic acid (GA), is comparatively less known and is a promising field of investigation (Wielkopolan and Obrępalska-Stęplowska 2016).

Numerous studies dealing with the plant-microbe interaction and plant-insect interaction are available, but very less amount of work has been done in the light of multitrophic interactions considering how important they are in the maintenance of the food web. Recent research although, has been focusing on studying the effect of microbes, plants and insects as one system under different physical, chemical and biological conditions such as soil composition, nutrition availability, role of detritivores, etc. (Johnson and Rasmann 2015). This chapter is an attempt to cover the concept of tripartite interactions taking place between MPI from an evolutionary and chemical ecology view with an emphasis on comparative analysis of how plant defence varies with attacks from microbes and insects. Understanding how plants defend themselves is crucial for the development of engineered strategies for pest resistances in economically important crop plant species. A new insight into the MPI interaction and their defence mechanisms would help scientists to work towards fulfilling the dream of sustainable crop protection through plant self-defence in the near future. This chapter deals with the defence mechanisms in terrestrial plants against microbial pathogen and insect herbivory, highlighting appreciable differences, parallels and cross talks between respective defence mechanisms.

2 Plant Interactions with Microbes and Insects

Plants are a complex system in themselves as they connect two different ecosystems, below the ground and above the ground. A large number of studies have been conducted on plant interactions with different pathogens, restricting their interactions either with a single microbe or a specific insect species, either in greenhouse or in laboratory settings. Although, the situations are not idealistic. Plants have to face multiple stresses, both biotic and abiotic stresses concurrently, such as the attack from bacterial and soil dwelling herbivores along with leaf-chewing insects and various fungal pathogens above and below the ground in harsh environmental conditions, such as extreme heat or salt or heavy metal stresses, etc. In this section, we discuss the plant-insect interactions, influenced by various microbes. The abiotic factors substantially influence these interactions, but the research in this field is still in its infancy.

2.1 Bacterial Influence on Plant-Insect Interactions

The rhizosphere microbiome consists of both beneficial and pathogenic bacteria. Due to herbivore and other pathogen attack, the plant shapes its rhizosphere microbiome in order to protect itself not just from the below ground attackers but also from the above-ground enemies. The roots, thus, change its chemical composition depending on the mutualistic or antagonistic interactions further making the insect-host interactions to be more complex (Berendsen et al. 2012). The nitrogen (N)-fixing bacteria have developed a mutualistic relationship with the leguminous plants, supplying them with N in exchange for carbon (C), but this association has also increased the cases of nodule herbivory. Insects belonging to the classes Coleoptera and Diptera feed on root nodules, thus completing their nutritional requirements (Johnson and Rasmann 2015). For example, *S.lineatus* and *S. hispidulus* are usually considered as an obligate nodule feeders as they feed on nodules during early developmental stages but later shift onto feeding on other parts of the roots (Quinn and Hower 1986). Whether this kind of association is useful for plants or not is not known. But it can be hypothesised that there might either be some beneficial or neutral relationship between the two, as plants of *Trifolium repens* are known to attract nodule herbivores by increasing the concentration of formononetin, an isoflavonoid in its roots which acts as a chemical cue for the insects and helps them to locate the roots (Johnson et al. 2005). However, more investigations are required to understand the behaviour of this MPI interaction.

Although the role of these interactions is not fully understood, it has also been found that the symbiotic associations of bacteria below the ground can make the plant more susceptible towards above-ground herbivory. For example, the Colorado potato beetles use the symbiotic bacteria of the plant to inhibit the JA-associated defence and can make the plant more susceptible for chewing herbivores (Bruce 2015). Similarly, the flagellin protein released by the *Pseudomonas* bacteria triggers the JA-SA cross talk and inhibit the plant to show defence against beetles and thus help them to show increased performance (Chung et al. 2013). In a similar fashion, the above-ground herbivore, *Bemisia tabaci*, the phloem-feeding whitefly shows resistance against the root bacterial pathogen *Ralstonia solanacearum* while feeding on sweet pepper (Yang et al. 2011b). But at the same time, it activates a beneficial change in the constitution of the rhizosphere microbiome by increasing the population of gram-positive bacteria. Likewise, *Myzus persicae* (aphid) when feed on the pepper plants, also modifies its below ground or the root microbiome by increasing the population of *Bacillus subtilis,* the beneficial bacteria and reducing the population of the pathogenic bacteria, *R. solanacearum* (Lee et al. 2012).

The best example to understand the effect of plant microbes on insect herbivory is the case of *Bacillus thuringiensis* (Bt). These bacteria are usually found in the phyllosphere and near the surface of the soil. When insects such as *Diabrotica virgifera virgifera* feed on plants, along with the plant parts, they also ingest these bacteria. Bacteria release the Cry proteins, which upon activation, disrupt the insect's gut membrane and kill the insects, thus acting as an insecticide (Bizzarri and Bishop 2008; Wielkopolan and Obręplaska-Stęplowska 2016). Due to limited number of studies focusing on the plant-bacteria-herbivore defence, it is inconclusive to say what role does each organism play to fine-tune the defence and ensure their survival. Therefore, more studies are required focusing on this area to better understand the role of different bacteria in plant-insect interaction systems.

2.2 Fungal Influence on Plant-Insect Interactions

Fungi itself show a wide spectrum of relationship with plants, ranging from being in a mutualistic relationship to being pathogenic. Plants use different defence strategies to protect themselves from the pathogenic fungus and contrary to this form a beneficial or symbiotic relationship with the non-pathogenic fungus. But in crop fields and forests, these plants are attacked by both, the fungi and the insects either at the same time or one after the other. These complex interactions together shape the final defensive traits of plants to each of its enemies. For example, the plants *Arachis hypogaea* and *Theobroma cacao* produce AhPR-10 and TcPR-10, respectively, that kills the pathogenic fungus *Fusarium oxysporum* and *Moniliophthora perniciosa* by actively transporting these proteins into the fungal hyphae without disrupting their membranes and prevent their growth and infection (Chadha and Das 2006; Pungartnik et al. 2009). Additionally, plants also synthesise a number of secondary metabolites that reduce the nutritional value of the plant and prevent herbivory to cause damage, thus becoming resistant to herbivores in an indirect manner; this has been discussed later in the chapter. The interactions between fungi-plant-insect are more complex, and hence fungi interfere with the plant and insect system, respectively, to fine-tune their interactions.

Plant-associated fungi is mainly classified into five different types: biotrophic, necrotrophic, epiphyte, endophyte and mycorrhizal. The most widely studied of these fungi is the mycorrhizae, which is known to alter insects' performance and preference both positively and negatively (Fernandez-Conradi et al. 2018). Fungal infections usually alter the visual as well as chemical cues of the plant and thus sometimes attract or repel the herbivore insects. The arbuscular mycorrhizal fungi (AMF) live in a symbiotic or mutualistic association beneath the ground with the roots, thus, altering the plant defence pattern either by supplying chemical constituents for the synthesis of a large variety of metabolites or by participating in defence-associated signalling via molecular machinery (Fernandez-Conradi et al. 2018). The negative influence of AMF as seen on plant insect interaction system of the leaf miner insects, *Chromatomyia syngenesiae* on the plant *Leucanthemum vulgare* for the first time was reported in 2003 as an example of parasitism (Gange et al. 2003). Later on, it was proposed that the performance of mono and oligophagous chewers is positively correlated with AMF, whereas there is a negative correlation for polyphagous chewers. On the other hand, for piercing-sucking insects, phloem feeders have usually been reported to be positively affected, whereas the mesophyll feeders are negatively affected by these symbionts (Fernandez-Conradi et al. 2018). Few such examples highlighting positive interactions showing increased insect performance are between the necrotrophic fungus *Phoma destructive* and the leaf beetle *Cassida rubiginosa* on *Cirsium arvense* plant (Kruess 2002), the aphid *Euceraphis betulae* on silver birch tree infected with necrotroph *Marssonina betulae* and others (Johnson et al. 2003).

Apart from the above-mentioned positive interactions, pathogenic fungi affect the plant-insect interaction negatively as well. The negative impact on insects includes emigration of insects from infected parts to noninfected parts, decreased rate of larval development and increase in mortality rate of adult insects. Apart from emigration, the insects also show a deviation in their oviposition site and rate (Franco et al. 2017). All these changes are usually induced because of the toxic chemical species produced by the pathogenic fungi, such as *Fusarium culmorum, F. Avenaceum* and *Alternaria brassicae* producing deoxynivalenol, enniatins and destruxins, respectively (Guo et al. 2014; Franco et al. 2017). Examples highlighting such negative interactions are usually seen in the aphids, *Aphis fabae*, while they feed on the *Vicia faba* plant, infected with the pathogen *Botrytis cinerea*. The aphids generally show a decreased rate of development, survival and fecundity (Fernandez-Conradi et al. 2018). Similarly, the development of the beetle *Cassida rubiginosa* has been observed to be negatively affected when the larvae is observed to develop on *Puccinia punctiformis*-infected *Cirsium arvense* weed (Kluth et al. 2002).

In all the examples so far, it is seen how fungi affects the plant-inset interaction. But in a real setting in a crop field, these interactions can go the other way around. Plants have developed another defence strategy by introducing a defence protein, which is synthesised on insect attack but is antifungal in nature. This is seen on the attack by the insect sugarcane borer, *Diatraea saccharalis* on the sugarcane plant. The damage caused by this insect creates the wounding sites on the plant body, which acts as the potential gates for the entry of two different fungi, namely, *Fusarium verticillioides* and *Colletotrichum falcatum*. The plant on the other hand to ensure its survival and no further damage synthesises a defence protein known as SUGAR-WIN. This protein changes the morphology of the attacking fungi and ultimately kills them to prevent further damage and acting as the first line of defence (Medeiros et al. 2012).

2.3 Viral Influence on Plant-Insect Interaction

Similarly, like bacteria and fungi, a number of virus particles are known to affect the plant-insect interactions. The virus particles upon infection usually interfere with the genome of the host plant, thus altering its metabolism promoting physiological changes giving rise to new and attractive phenotypes. A similar case is seen where the pea plant infected with *pea enation mosaic virus* (PEMV) attracts the insect *Acyrthosiphon pisum* due to attractive phenotypes created as a consequence of altered genetic system upon infection (Hodge and Powell 2010). Similarly, the *Beet Yellow Virus* (BYV) and *Beet Western Yellow Virus* (BWYV)-infected leaves of sugar beet attract the *Myzus persicae*, the peach aphid and ensure the effective transmission of virus particles (Tack and Dicke 2013). Other examples of aphid insects preferring to either settle or feed on virus-infected plants include aphid settling on cardamom plants infected with *Cardamom bushy dwarf virus* (CBDV) (Ghosh et al. 2016) and the insect *Frankliniella occidentalis* usually settling on tomatoes infected with *tomato spotted wilt virus* (TSWV) (Maris et al. 2004). The viruses usually alter the phenotype in such a way that they should attract more insects in order to spread faster. Therefore, these viruses usually live in a symbiotic relationship with the insects in their bodies following different modes of transmission (Dietzgen et al. 2016).

Although, these viruses use plants as hosts and insects as vectors to further spread their infection and damage neighbouring plants eradicating the entire crop fields, plants have also developed their own defence responses specific to viruses. For example, the protein CaPR-10 from *Capsicum annuum* is a pathogenesis-related protein and is known to have antiviral action against the *tobacco mosaic virus* (TMV) (Park et al. 2013). A lot of studies have been conducted on plant-virus and the virus-insect systems, but we still lack the understanding between these three organisms as one system.

The interactions as discussed between the microbes (bacteria, fungi and virus), plants and insects as one system under the influence of complex environmental factors have taken a long time to evolve. Plants being the connecting link have shown their relevance to be a ball room for the coevolution of traits specific to the plant-microbe and the plant-insect systems. Although being enemies of the plant, both microbes and insects have been driven by the evolutionary forces and have been found to maintain the tripartite interactions between the organisms both below and above the ground. The question, how coevolution has shaped these interactions has been addressed in the next section.

3 Plants: A Connecting Link Between Microbe-Insect Interactions

Plants host a large and a diverse group of organisms, such as different species of bacteria, fungi, virus, nematodes, herbivorous insects and detritivores. These species of microbes and insects continuously attack their host plant and modify their defence response, but on the other hand, they also interact with each other, thus modifying their behaviour towards the host. Therefore, plants act as a connecting link between two different ecosystems, the one below the ground constituting the soil microbiome and the roots as one system and the other above the ground which is in contact with abiotic factors as another system (Fig. 1).

The below-ground rhizosphere consists of plant roots which are in contact with both beneficial and pathogenic microbes which apart from interacting with the host plant also interact with the soil-inhabiting herbivores. The phyllosphere or the above-ground plant system is a home to a number of microbes including viruses, bacteria and fungi and insects, such as phloem feeders and leaf chewers (Sugio et al.

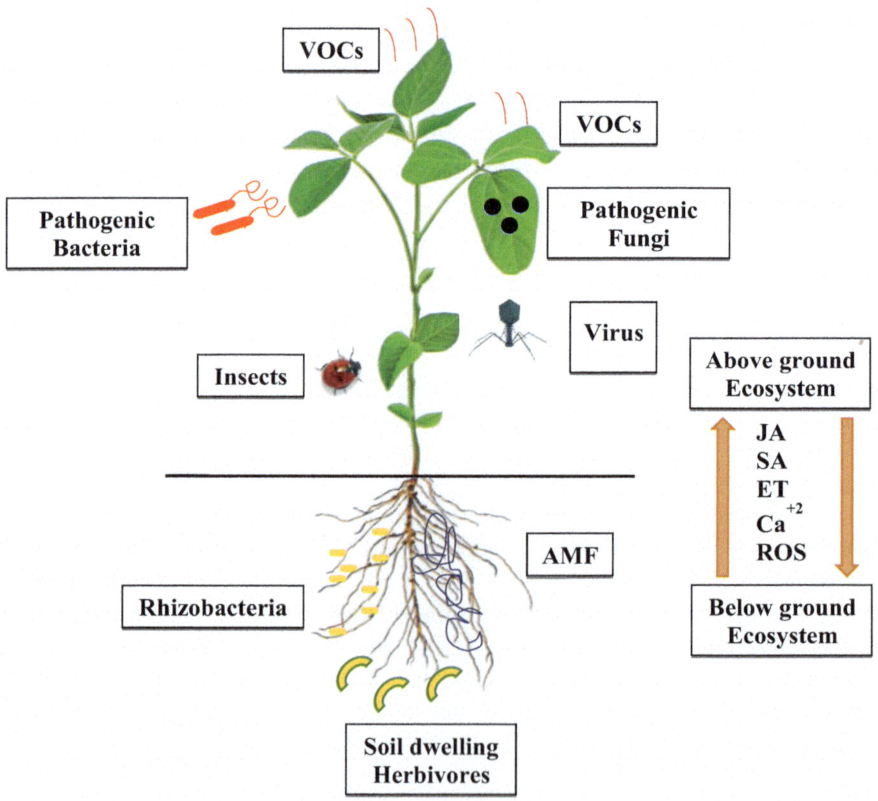

Fig. 1 The above-ground below-ground interactions between pathogenic and beneficial microbes and insects shaping the host plant defence

2015). These insects interact with the host modifying their above-ground defence response but also interact with the microbes and help them to further spread their infection, thus acting as vectors. Over a course of 400 million years of interrelated relationships, the host plants and their associated microbes and insects have evolved genetically, socially and behaviourally (Sugio et al. 2015). Apart from their coevolution, they have also developed some epigenetic memory to ensure their survival and fitness (Bruce 2015).

The process of coevolution in context of these tripartite and/or multitrophic levels has given rise to the concept of systems community, where the complex interactions taking place between the microbes and insects both below and above ground have shaped the community ecology to be investigated from a whole new perspective of systems biology and its associated interactome, thus highlighting mutualism, antagonism and neutral interactions between different communities of hosts and their enemies.

3.1 Evolution of Plant Interactions with Microbes

Due to a complex and dynamic structure of the ecosystem, plants are bound to interact with the microbes in an ever-changing fashion (Occhipinti 2013). The plant-microbe interactions below the ground are different from ones taking place above the ground. The beneficial or symbiotic associations are known to occur between plant and its microbe, including rhizobacteria that forms the root nodules and the symbiotic association of AMF. Both of these associations are mutualistic as they help the host and the microbe to thrive by a mutual exchange of N and phosphorus (P) from the microbe against C from the plants (Occhipinti 2013). Since, the mycorrhizal and plant associations are quite ancient and they are believed to have coevolved with the land plants beneath the soil surface and hence form an excellent living fossil evidence to understand this mutualistic behaviour (Brundrett 2002). The association of the rhizobacteria is also ancient, but it has been suggested that the root nodule associations have evolved from the mycorrhizal plant associations via genetic links between the bacterial and the fungal microbes (Kistner and Parniske 2002). These symbiotic associations are known to help plant to defend itself against the below-ground pathogens using chemical defence strategies, helping the host plant to excrete root exudates to prevent damage by pathogenic microbes.

Pathogenic microbes, on the other hand, have developed an intricate relationship with their respective host plants. The pathogenicity of the microbes has evolved due to a number of contributing factors, such as the genotype of the host and the microbe, epistatic interactions between their genomes, ploidy levels, etc. (Woolhouse et al. 2002). The gene-for-gene model suggests that the selection pressure on both the host and the microbe is operating in opposite directions. This has shaped the counter-defensive strategies in the microbes with progress in defence mechanism development in the host plants. This evolutionary force has given rise to both the virulent and nonvirulent pathogens and the trait of resistance against them in the plants (Occhipinti 2013). Hence, selection pressure and the struggle for fitness have given rise to complex gene structures in the host plant suggesting resistance of the host to be polygenic (Burdon and Thrall 2009). However, a substantial amount of promising research is required to understand and interpret the interactions leading to virulence and resistance trait development in different host plants against different species of microbes both below and above the ground.

3.2 Evolution of Plant Interactions with Insects

Insects that feed on plant parts are known as herbivorous insects. The interplay of interactions taking place between the plants and the insects can be dated back to as ancient as the interactions between the host plants and the associated microbes. Evolutionary forces of genotype selection, allele fixation and mutations have shaped the antagonistic relationship between the insects and the host, thus, shaping their defence strategies throughout the course of evolution preventing their extinction (Occhipinti 2013).

Plants being intelligent have distributed their defensive strategies into different categories, namely, tolerance, resistance, overcompensation and phenological escape from herbivores (Agrawal 2000; Occhipinti 2013). The traits of tolerance and resistance against herbivores act alternatively. Resistance, as suggested above in the plant-microbe interaction, is a polygenic trait and has evolved due to the pressure of evolutionary forces, but the trait of tolerance in plants has evolved due to a complex interplay between the evolutionary forces operating both on the insect and the plant with a significant contribution of environmental factors that affect the growth rate of the plant, its photosynthetic capacity and its novel and sophisticated defence strategies under the mixed conditions of biotic and abiotic stresses (Agrawal 2000). Since, a plant witnesses multiple enemies at the same time, hence its defensive response usually lies towards the middle of the spectrum of tolerance and resistance traits (Occhipinti 2013). In other words, resistance and tolerance complement each other to fine-tune the final defence. A similar phenomenon is seen in the plant *D. stramonium,* which is attacked by a generalist insect, *Epitrix parvula* against which the plant shows resistance by synthesising secondary metabolites and a specialist insect, *Lema daturaphila* towards which the plant shows tolerance, thus, indicating a mixed tolerance-resistant defensive strategy (Carmona and Fornoni 2013; Turley et al. 2013).

On one hand, the evolution of defence traits in plants has given rise to sophisticated molecular mechanisms of sensing the attack and triggering a response; on the other hand, it has also developed a counter defensive mechanism in insects. Herbivores have developed a strategy to sequester the defensive chemicals produced by plants and to use these metabolites further to protect themselves from predators (Cogni et al. 2012). This coevolution between plants and insects has also developed in the below-ground ecosystems. Nematodes and soil-dwelling herbivorous insects attack the host plant by locating their roots using carbon dioxide, glutamic acid, carbohydrates, volatiles and other signals, whereas plants have also evolved a defensive mechanism by masking the carbon dioxide emission during respiration from roots and other phagostimulating signals (Johnson and Gregory 2006; Hiltpold et al. 2011; Occhipinti 2013). These observations suggest that there has been a constant fight between the plants and the insects that has not only shaped the sophisticated defence mechanism in each one of them but has also given rise to a spatial and temporal pattern of the communities of both the plant and the attacker under a dynamic environment on an evolutionary scale.

3.3 Evolutionary Interactions Shape the Spatial and Temporal Pattern of the Interacting Communities

Evolutionary forces of selection and mutation have lead the microbial, plant and the insect communities to attain a structure in concordance with the dynamic environment and to develop sophisticated defence strategies in all the interacting organisms. The spatial pattern shows an entire continuum of coevolution of the multi-trophic interactions because of the geographical isolation of interacting species under the influence of different abiotic conditions. This suggests that sympatric species are adapted to a range of similar species of microbes and insects due to long-term exposure to common enemies, whereas the allopatric species are not (Occhipinti 2013; Garrido et al. 2012). In comparison with the spatial pattern, the temporal pattern coevolution is discrete. The congruent phylogenies in the evolution of the interacting communities evidently suggest that the interacting lineages must have diversified at specific points in time, as implied by the coevolution of trait specific alleles participating in interactions of the host and the enemies (Cavender-Bares et al. 2009; Dinnage et al. 2012).

The spatial dimension in the tripartite interactions has evolved an ability of discrimination of plant parts affected and not affected by the pathogens in herbivorous insects which has altered their emigration rates and thus the structuring of their communities on the host plant. For example, the stripped cucumber beetles are usually found in more concentration on virus-infected leaves, whereas the leaf beetles, *O. elongate*, are usually found on the healthy and rust-free *Adenostyles* plants (Tack and Dicke 2013). In the former case, the cucumber beetles act as vectors and help the virus to spread in the adjacent plants. The time scale of the interacting MPI plays a crucial role in developing different defensive strategies against each other. The evolution of systemic-acquired resistance (SAR) mechanism over the time has developed its pattern of developing and receding in a certain time period (Tack and Dicke 2013). The pathogen infection on the host plant affects the interacting insect at different time scales, such as from the time of onset of infection in the host plant to many years. One such example is seen in the case of *Pristiphora erichsonii*, the larch sawfly. When their larvae feed on the seedlings of *Larix decidua*, the European larch infected with the needle cast fungus a year before, the larvae quickly leave the seeds and feed on the uninfected ones (Krause and Raffa 1992).

Thus, it is evident from the above examples that coevolution of the tripartite interactions not only evolves the traits of defence in all the interacting systems but also shapes the structure of their respective communities. These communities not only have the positive effect on the plant as seen in the case of larch sawfly larvae but also have a negative effect as an insect feeding on pathogen-infected plant part becomes vector for the pathogens and thus spread the infection, damaging the entire crop fields.

4 Insects as Vectors for Microbial Infection

Insects act as vectors for microbial infection to spread to a larger population of plants in the crop fields. These herbivorous insects transfer either the insect- specific pathogen living inside the insect body to the plant, thus altering the defence or by interacting with the pathogen that has infected the plant previously. In both the cases, the interaction between the MPI systems modifies the behaviour of all the interacting organisms.

A number of bacterial species are transferred by their insect vectors, namely, phytoplasmas that are vectored by the insects belonging to the taxa Psylliidae and Fulgoroidea and Enterobacteriales, such as *Erwinia* spp. which are vectored by the insects from the taxa Chrysomelidae and others (Eigenbrode et al. 2018). These bacterial pathogens infect the host plant, thus changing their physiology and morphology, attracting insects to feed on them and carry the microbes with them and transfer to a healthy plant while feeding. For example, the phytoplasma strain found in the aster yellow witches' broom disease is transferred by its vector *Macrosteles quadrilineatus*, a leafhopper insect species. Apart from these changes, these bacterial strains also interfere with indirect defence strategies of the host plant to derive benefit for themselves (Bai et al. 2009). Another phytoplasma strain is known to increase the emission of a sesquiterpene, β-caryophyllene, in apples that attracts the vector *Cacopsylla picta* and helps in the transmission of *Candidatus phytoplasma mali* infection (Sugio et al. 2015).

As seen above, the bacteria use insects as their vectors, so does the pathogenic fungal species. They can use insects' outer body for their transfer, such as in case of the weed *Cirsium arvense* infected with the rust fungus, *Puccinia punctiformis* by the insect vectors such as *Aphis fabae* and *Cassida rubiginosa*. These insects carry the spores of this fungus on their legs and wings and transfer them on healthy plants while feeding on them (Kluth et al. 2002). In other cases, the fungus also lives inside the body of the insect and interfere with their system thus modifying their behaviour. One such example is seen in the interaction of parasitic fungus *O. unilateralis* with the carpenter ants, *Camponotus* spp. that feed on leaf margins in rainforests and sometimes also on twigs found in temperate woods. The fungus interferes with the ants' physiological system and makes them wander like zombies. These ants usually during the solar cue climb up to 25 cms height on the plant where the fungus kills them by erupting from their head (Shang et al. 2015). This sophisticated mechanism adapted by the fungus for its transmission using insects as vectors depicts how tripartite interactions have been shaped by the evolutionary forces.

Similar to bacteria and fungi, viruses also use insect vectors for their efficient transfer from an infected to a healthy plant. For example, the aphid *Myzus persicae* is a commonly known vector for the cucumber mosaic virus (CMV). The efficient transmission of this virus not only uses the viral RNA but also employs the coat proteins and its associated domains via aphids (Dietzgen et al. 2016). Likewise, the viruses belonging to the class of Luteoviridae use phloem-feeding insects belonging to the taxa Aphididae for their transfer to a healthy plant. The virus belonging to this taxon passes through the gut and hemocoel to the salivary glands in insect body and gets transferred to the plant via insect saliva that enters the plant system while plant sap ingestion (Dietzgen et al. 2016).

5 Host Plant Defences Against Microbes and Herbivores

Plants have an innate immune system which involves multiple constitutive and inducible defence responses that are active at many different stages of colonisation by the invader species (Ton et al. 2009). Prerequisite to the initiation of defence response is identification of the presence of pathogenic danger, which is detected by plant cells through the recognition of pathogen signals called elicitors. The perception of general elicitors such as pathogen-associated molecular patterns (PAMPs) or microbe-associated molecular patterns (MAMPs) via plasma membrane-localised-specific pattern recognition receptors (PRRs) triggers a wide array of reactions that culminates into activation of basal defence response, also called as PAMP-triggered immunity (PTI) (Chisholm et al. 2006; Jones and Dangl 2006; Nürnberger and Kemmerling 2006; Nicaise et al. 2009). Basal defence is marked by ion fluxes, activation of mitogen-activated protein kinase (MAPK) cascades, phosphorylation/dephosphorylation of proteins, synthesis of signalling molecules such as SA, JA and ET, accumulation of ROS, deposition of localised callose, etc. (Hou et al. 2009). PTI is able to stop majority of hostile microbes; however, virulent pathogens repress the defence response by means of effectors (also called avirulence/Avr protein) that enable further colonisation of pathogens. To counteract effectors, plants have evolved different mechanisms to sense effector-induced perturbations through intracellular resistance proteins (RPs) and generate an appropriate response against them (Boyes et al. 1998; McDowell and Woffenden 2003). This effector-triggered immunity (ETI) often induces a hypersensitive response (HR) or a small area of localised cell death at the infection site that blocks the virulent pathogen growth. The initiation of local immune responses leads to the activation of systemic immune responses that renders the host less susceptible to successive pathogen attacks. Although there are many similarities in the defence mechanisms of plants against a plethora of pathogens, they may activate different types of defences, depending upon the organism they are interacting with.

5.1 Structural Barriers Are Different in Plants for Microbes and Herbivores

Structural barriers are the first line of defence developed by the plants to prevent their invasion by pathogens and attack by herbivores. Thick cuticles and waxes prevent entry of a pathogen in a plant, but apart from this, other structures such as glandular trichomes, hooked trichomes and thorns or spines prevent the incidences of herbivory. These modified structures usually provide the plant with direct defence strategies. The bacterial pathogens usually cannot degrade the thick cuticle of the leaves, and hence they use natural openings in plants to invade them, such as stomata, lenticels and hydathodes (Agrios 2005). For example, the potato scab bacteria, *Streptomyces scabies*, enter the tubers through lenticels, although, suberisation of lenticels usually reduces its size to prevent their entry (Khatri et al. 2010). *Xanthomonas campestris* enter the plant via hydathodes (McElhaney et al. 1998). *Pseudomonas tabaci* that causes wild fire of tobacco gains entry to the host plant via stomata (Lee et al. 2013). Plants on the other hand make sure to prevent this and hence keep its aperture usually small or sometimes, they adapt to open less frequently. Bacteria cannot breach the thick cuticle and cell wall, but fungal pathogens are usually capable of doing this. Fungi releases a mix of cuticle and cell wall-degrading enzymes, such as chitinases, cutinases, cellulases, pectinases, etc. that aid the fungal pathogens to infect the plant (Knogge 1996). Viruses, on the other hand, usually enter the plant with the help of phloem feeder insects that ingest on sap usually transfer the virus particles into the plant. Once inside, they get transported to the different parts of the plant (Dietzgen et al. 2016). For below-ground entry, bacteria and fungi usually gain entry by root hairs (A'Bear et al. 2014).

Insects on the other hand do not require such intricate mechanisms to invade the plant. Epicuticular waxes are usually present as a film on the surface of cuticle, and thick cuticular folds of most vascular plants enhance the slipperiness which strongly reduces the ability of non-specialised insects to stick and settle on the leaf surfaces (Müller 2007; Prüm et al. 2013). Cell wall fortification and sclerophylly (i.e. the toughened or hardened leaves) not only interfere with the penetration of plant tissues by specialised mouthparts of the piercing and sucking insects but also increase the mandibular wear in biting and chewing herbivores. Thus, this diminishes the palatability and digestibility of the tissues, in turn reducing herbivore feeding (Raupp 1985; Clissold et al. 2004). The trichomes also function as physical defensive structures against several agricultural pests, such as aphids and leafhoppers (Pillemer and Tingey 1976). The glandular trichomes mostly exude sticky or viscous secretions that not only discourage herbivore feeding but also help in entrapping insects. This in turn sometimes attract predatory enemies of the herbivores thus, enhancing the floras' indirect defences (Wagner 1991; Krimmel and Pearse 2013). Many non-glandular trichomes are hooked at various angles and function as entrapment devices that are capable of spearing many insects, entangle them and obstruct their feeding behaviour (Riddick and Simmons 2014). The hook-like trichomes present in the members of the Passifloraceae family, *Passiflora adenopoda*, effectively prevent the damage caused by heliconiine butterfly larvae, one of its major classes of herbivore by trapping and killing its larvae (Gilbert 1971).

Since, microbes usually cannot migrate on their own, therefore, for their efficient transfer, they depend on herbivorous insects feeding on different plant parts. Thus, insects act as vectors which inoculate the healthy plants with microbes, and they seek entry into the plant body through the wounds created by insects while feeding on leaves, shoot or root. Therefore, it is certain that plants have developed a wide array of different morphological structures to protect them from their enemies, but in a likewise manner, insects and microbes have also developed various strategies to breach them. Once the physical barrier is removed and the plant cells' intracellular environment is exposed to the outer environment, it triggers early signalling events such as electrical signalling to initiate the process of indirect defence by recruiting molecular machinery that interacts with each other, such as kinases and other molecules, thus beginning the process of gene regulation and secondary metabolite synthesis to prevent further damage and to compensate for the loss.

5.2 Early and Late Signalling Events in Plant Defence

When physical barrier is breached, the plant senses the site of attack and triggers some early signalling events, such as generation of electrical signals and its propagation throughout the plant body to prevent further damage. Other early responses involve the Ca^{2+} ion flux and changes in the levels of calcium-associated proteins, ROS, etc. (Gilroy et al. 2016). These events are responsible for the activation of the downstream kinase signalling cascade that merges with the activation and/or suppression of further downstream signalling pathways involving phytohormones. The phytohormone and other molecular entities present inside the cell activate the defensive genes including the ones coding for toxic compounds, antinutritive chemicals, defence-associated proteins (such as chitinases, proteinases, etc.) and VOCs (Pangesti et al. 2013).

5.3 Electrical Signalling, Calcium Influx and Reactive Oxygen Species

Whether a plant is attacked by a microbe or an insect, plants trigger electrical signal as its first line of defence in order to intimate the entire plant far from the actual wounding site to divert the resources towards defence rather than towards growth. The electrical signals are broadly classified into two types, namely, the action potentials (AP) and the variation potentials (VP) (Dziubińska et al. 2001). The action potentials spread rapidly, covering a distance from few mms to several cms/sec, thus propagating in a localised area. On the other hand, the variation potentials cover a longer distance as they travel via vascular bundles; thus, their regulation depends on the hydraulic pressure and the associated apoplastic ionic pool (Zimmermann et al. 2009). The events of alternate polarisation and depolarisation of the membrane lead to its propagation throughout the membrane and are further helped by the plasmodesmata and its associated proteins to travel from one cell to the next (Gilroy et al. 2016). The mechanism of electrical signal generation and its propagation during an episode of attack by a microbe and by a herbivore are essentially the same, except for the fact that the stimulus and its recognition by the plant differ. For example, in case of different microbes such as bacteria, a receptor is generally present in the plant cell membrane that recognises the protein flagellin as a MAMP signal, similar is the case when a viral capsid protein is recognised or the fungal proteins, such as chitin (Nürnberger and Scheel 2001). But on the other hand, when an herbivore attacks, stimuli such as an insect crawl, oviposition secretions or the oral secretion, are recognised as external stimulus, and hence, the herbivore-associated molecular pattern (HAMP) is recognised, and the plant, to defend itself, triggers herbivore defence signalling mechanisms (Arimura et al. 2011).

Similar to electrical signalling, the role of calcium and ROS remains the same in case of an attack by either enemy. There is a high overlap in early defence strategies in both the cases. The involvement of Ca^{2+} channels leading to the influx, involvement of proton pumps, potassium channels and other channels is identical that sets up the membrane potential (V_m). The generation of ROS as well as reactive nitrogen species (RNS) has also been observed during an attack by either of the plant enemies (Nürnberger and Scheel 2001). In both the cases, the participation of interacting proteins, such as calmodulin and calmodulin like proteins, has been observed, thus indicating that plants have intelligently evolved and have developed to play smart and outpower its enemies by defending itself against multiple enemies at the same time without getting metabolically exhausted or without compromising its growth to its extremities (Zebelo and Maffei 2015).

5.4 MAMPs, HAMPs, DAMPs and the Kinases in Early Signalling

The term MAMPs is used interchangeably with PAMPs which usually consist of specific bacterial, fungal and viral proteins which are recognised by the PRRs. The MAMPs usually include proteins such as flagellin, lipopolysaccharides, peptidoglycans, chitins, β-glucans, viral coat proteins, etc. Though nematodes are also classified as pathogens, their recognition is associated with nematode-associated molecular patterns (NAMPs) (Choi and Klessig 2016). On the other hand, the HAMPs are usually present in insect saliva, such as glucose oxidase (GOX) and ATPases or in regurgitant such as, fatty acid amino acid conjugates (FACs), inceptins, bruchins, etc. Apart from MAMPs and HAMPs produced by microbes and herbivores, plants also synthesise compounds to trigger plant immunity (Acevedo et al. 2015). These molecular compounds are known as damage-associated molecular patterns or DAMPs. The class of DAMPs usually contains compounds such as systemin, oligogalacturonides, plant elicitor peptides, etc. A recently recognised DAMP molecule is the *At*HMGB3 from *Arabidopsis thaliana* (Choi and Klessig 2016). They work in conjugation with other signalling molecules to generate a defence response, for example, systemin-based signalling which triggers JA synthesis in tomatoes during a herbivore attack.

The integration of these molecular patterns can provide new insight into the tripartite interactions of MPI as one system. One such example has been found where the bacteria living in symbiotic association with the Colorado potato beetle help the beetle larvae to develop by getting benefit from the plant (Chung et al. 2013).

Kinases form another important class of signal transduction proteins, acting as a connecting link between the receptor recognition events and the TF interactions for defence-associated gene expression. Various classes of kinases are present in the plant cells, such as calcium-dependent protein kinases (CDPKs), MAPKs and its variants, such as MAPKKK, MAPKK, MAPK, etc., wound-induced protein kinases (WIPKs), SA-induced protein kinases (SIPKs), etc. (Nürnberger and Scheel 2001). These kinases are conserved in the MAMPs and HAMPs triggered signals. For example, flagellin (MAMP) in *Arabidopsis* and FACs (HAMPs) in *Nicotiana* are known to activate CDPKs, MAPKs, WIPKs and SIPKs (Bonaventure et al. 2011). Although, the specific kinases interacting in MAMP/HAMP with the PRRs are different, for example, in case of MAMPs, the kinases known to get activated are CDPK 4,5,6 and 11 and MKK 3,4,5 and 6, whereas the specific kinases interacting in case of HAMPs are CBL-interacting protein kinases (CBL-CIPK) and other calcium-dependent protein kinases, such as CPK 3 and 13 apart from WIPKs and SIPKs, which is a class of kinases common to both types of defence- associated signal transduction (Bonaventure et al. 2011). Therefore, this suggests that there is a considerable overlap in the mechanism of signal transduction in cases of defence response generation against both the enemies, but the difference lies at the molecular level of interaction.

5.5 Phytohormone Signalling: Pathogens vs. Herbivores, Is It Different?

When plants are attacked by its enemies, they usually show a response in either of the two ways, i.e., by SAR or induced systemic resistance (ISR). The ISR response is triggered usually by JA signalling and is effective against necrotrophic pathogens and tissue chewing insects, such as caterpillars and is also effective against thrips. On the other hand, SAR, whose activation involves SA biosynthesis, provides immunity to the plant against biotrophic pathogens and piercing-sucking herbivores, such as whiteflies and aphids via SA signalling pathway (Pangesti et al. 2013). Hence, it is clear that depending upon the attacker, plants manage to divert their resources effectively by triggering appropriate phytohormone signalling pathway. ET and ABA are known as the interacting hormones to the JA and SA pathways and have been discussed under appropriate headings.

5.6 Phytohormone Signalling: Defence Against Pathogens

JA induces a great number of defence-associated responses, such as synthesis of glucosinolates, proteinase inhibitors (PIs) and alkaloids and also plays a role in trichome formation (Howe and Jander 2008). The JA-signalling is known to set ISR and is stimulated by root-associated microbes, thus defending the plant below ground, but it also defends the plant against above-ground herbivory (Pangesti et al. 2013). The JA has been reported to be effective against the leaf fungus, *Alternaria brassicicola,* which creates necrotic lesions on the leaves. The attack by this necrotrophic fungus triggered the expression of the plant defensin 1.2 *(PDF1.2)* gene which is a JA and ET-responsive gene and Hevein-like preproprotein (*HEL*), an ET- responsive gene (De Vos et al. 2005). It is proven that the JA-pathway functions with coronatine-insensitive 1 (COI1), and hence it has been reported that *Arabidopsis* plants, mutant for the *coi1*, show an increased susceptibility towards the fungi *Botrytis cinerea* and *Plectosphaerella cucumerina* (Antico et al. 2012).

SA-signalling has also been found to be effective against the bacterial pathogen, *Pseudomonas syringae*. The SA-signalling is triggered via SA-biosynthesis by the enzyme isochorismate synthase (ICS). The downstream signalling involving oxidative burst is mediated by the enzyme NPR1 (non-expression of pathogenesis-related protein 1) (Groen et al. 2013). But if the same plant is attacked by a generalist insect, such as *Trichoplusia ni*, the cabbage looper, the plant defence strategy changes. In this case, the pathogen, *P. syringae*, synthesises a phytotoxin coronatine, COR. This molecule mimics the compound JA-Ile (jasmonic acid-isoleucine conjugate). Since, JA and SA work antagonistically, therefore COR hinders the plant to develop SAR and instead drives the immunity towards developing systemic resistance against insects (Cui et al. 2005). Although, it has also been found in a study that *P. syringae* also induced the expression of pathogenesis-related 1 *(PR-1),* vegetative storage protein 2 *(VSP2)* and *HEL*, which are SA, JA and ET-responsive genes, respectively, in *Arabidopsis* (De Vos et al. 2005). It has been observed that SA is responsive against other pathogenic bacteria as well, such as *Pectobacterium carotovorum* and prevents the biofilm formation.

The role of JA and SA in plant defence against a number of viruses has been observed. For example, the infection on a host plant by *potato virus X* & *Y* (PVX, PVY) is repressed by the JA-responsive genes (Alazem and Lin 2015). Similarly, the attack by the *tobacco mosaic virus* (TMV) also increases the SA concentration and induces the expression of *PR*-genes (Vlot et al. 2009). Contrary to this, ET instead of defending the host plant helps the symptoms to develop, for example, in the case of *Cauliflower Mosaic Virus* (CaMV) infection on *Arabidopsis* plant. The viruses have also evolved a strategy to not just alter the ET pathway but JA pathway as well (Alazem and Lin 2015). The *Begomovirus*, *Tomato Yellow Curl Leaf China Virus*, suppresses the JA-signalling by using its two virulent factors, namely, C2 and βC1. This suppression of the JA-signalling makes the tomato more susceptible towards the silverleaf whitefly insect, *Bemisia tabaci* (Kazan and Lyons 2014). This behaviour of virus ensures its efficient transmission. It is clear from the above examples how plants defend themselves against a number of pathogens by using their phytohormones. But on the other side, these microbes have also evolved numerous ways to cope up with the wide array of defence mechanisms.

5.7 Phytohormone Signalling: Defence Against Herbivores

The plants are attacked by chewing insects and the piercing-sucking insects. They both trigger a different kind of defence response. The following examples with *Pieris rapae*, as a chewing insect model and *Frankliniella occidentalis* and *Myzus persicae*, as cell and phloem-feeding insect model, respectively, depict how plants defend themselves against these different sets of insects (De Vos et al. 2005). The cabbage white butterfly usually attacks plants of the family Brassicaceae or Cruciferae, and it has been observed that on attack by this leaf chewer, plant defends itself by triggering the JA-signalling pathway (Van Loon et al. 2000). In another study with *Arabidopsis* as the experimental model plant, the insect induced the expression of the gene VSP2. Similarly, *P. rapae* also feed on the leaves of *Solanum lycopersicum*, tomato plant which is deficient of JA-production. On the other hand, when *F. occidentalis* was made to feed on *Arabidopsis*, the cell content feeder also induced the JA-signalling defence mechanism, but in this case, instead of VSP2, PDF1.2 was expressed (De Vos et al. 2005). Since PDF1.2 is modulated by both JA and ET, therefore, it can be hypothesised that response to the cell content feeding insect might involve JA-ET cross talk. On similar grounds, when *M. persicae* was allowed to attack the plant, the SA and JA-responsive genes were expressed. This might highlight the involvement of SA-JA antagonism. But when the same plant is made to infect the tomato plant expressing the JA-responsive gene constitutively, the aphid cannot thrive for very long (Moran and Thompson 2001; Li 2002).

Therefore, in conclusion, it is seen that plants essentially trigger the same signalling pathway when attacked either by a microbes or an insect, but the difference lies in the cross talk. Also, depending upon the host-enemy interaction, these signalling pathways come into action. Apart from the three main pathways, i.e., JA, SA and ET, other hormones also integrate either synergistically or antagonistically with them to fine-tune the defence. Their role in defence makes the next part of this topic along with the integration of pathways.

5.8 Phytohormone Signalling: Role of ET in Defence and the Integration of Different Signalling Pathways

The role of ET in defence signalling is an enigma. On one hand, ET integrates with the JA-pathway and expresses the defence-associated genes against microbes and insects, and on the other hand, induces systemic-induced susceptibility (SIS) in the plants for its enemies (Groen et al. 2013; De Vos et al. 2005). For example, the hemibiotrophs such as *Phytophthora parasitica* induce ET signalling, enhancing pathogen proliferation. This implies that the pathogen-encoded effectors interfere with the ET signalling pathway. But contrary to this, when ETI is activated by the plant, ET signalling works synergistically with the SA-pathway and triggers antimicrobial response (Mur et al. 2009; Wi et al. 2012).

TF and other molecular machinery integrate the JA-pathway with ET and ABA pathways. The MYC2 and ethylene response factor (ERF) are the two branches of JA-pathway which are antagonistic to each other. They activate antiherbivore and antifungal defence responses, respectively (Pangesti et al. 2013; Groen et al. 2013). The MYC2 branch integrates with ABA synergistically, whereas the ERF branch integrates with ET (Abe 2003; Lorenzo 2003). The ERF branch bifurcates to ERF1 and octadecanoid- responsive element-binding factor (ORA59) pathways. The interaction between the ET and JA activates the JA-ET-dependent genes, such as PDF1.2 (De Vos et al. 2005). ET interacts with SA in an antagonistic manner. ET interacts with the transcriptional machinery and inhibits the transcription of ICS1 gene (Chen et al. 2009). Similarly, SA interacts with JA antagonistically. The antagonism between SA and JA occurs at the protein level with ORA59 (Van der Does et al. 2013). This TF, responsive for both JA and ET signalling pathways, belongs to the APETALA2 (AP2)/ERF family of TF (Lorenzo 2003). NPR1 is also involved in the SA-JA cross talk as seen in *Arabidopsis npr1* mutants; the plants lacked JA-mediated gene expression in SAR pathway (Spoel 2003). Apart from these interactions, the ET and ABA interactions also take place in antagonism. If ABA is deficient in a plant during a biotic stress, ET usually increases and inhibits the invasion of roots. Such negative interaction between them is also responsible for the disruption of mycorrhizal associations (Martín-Rodríguez et al. 2011).

5.9 Phytohormone Signalling: Role of Other Phytohormones in Defence and Their Integration

The role of other phytohormones, namely, GA, CK, Auxin (IAA) and BRs in defence against pathogens and herbivores, is in their preliminary stages of investigation. Although, they are also known to connect with the JA pathway, however, the molecular details are not yet fully discovered.

5.9.1 Defence Against Pathogens

Auxin or indole acetic acid (IAA) is known to play a crucial role in plant growth and development. However, its role in plant defence has been recently highlighted during a microbe and herbivore attack. Although, auxin usually makes the plant susceptible to pathogen attack such as by *Xanthomonas oryzae*, *Magnaporthe oryzae*, etc. but it also provides resistance against necrotrophic fungal pathogens, namely, *Alternaria brassicicola* and *Plectosphaerella cucumerina* (Llorente et al. 2008; Kazan and Lyons 2014). Similarly, ABA is a stress hormone, with major role in defence against abiotic stresses, such as salt, cold and drought, but it also makes the plant susceptible to microbes, namely, *B. cinerea* and *F. oxysporum* and resistant to the microbes, such as *Pythium irregulare* and *Cochliobolus miyabeanus* (Kazan and Lyons 2014). The role of CK on the other hand is contradictory. It helps the pathogens to infect the plant and progress the disease symptoms. When the level of

CK in the plant is reduced, it has been observed that this decrease in CK levels helps the mycorrhizal fungus to grow well in association with the roots (Kiba et al. 2011). The role of CK in bacteria-mediated plant growth has also been seen with the bacteria *Bacillus megaterium* in *Arabidopsis* (Ortíz-Castro et al. 2008). Although, the bacterial effector such as HopQ1 from *P. syringae* triggers the activation of CK-pathway that further helps the infection to progress and develop symptoms (Hann et al. 2014). The receptor BAK1 or BRI1-associated kinase 1 is involved in BR-signalling. BRs also play a similar role to that of auxins during plant defence, thus making the plant both resistant and susceptible to a few classes of pathogens. For example, the *P. syringae* effectors, AvrPto and AvrPtoB, inhibit the PTI (Kazan and Lyons 2014). Similarly, the rice plants which show hyperaccumulation of GA due to being mutated in *gid1* show increased susceptibility towards the fungal pathogen *Pyricularia grisea* (Tanaka et al. 2006).

5.10 Defence Against Herbivores

The less studied area of auxins is its role in defence against herbivory. But a few examples highlighting the same are available. For example, the levels of auxin or IAA are increased in plants on attack by the gall-feeding insects, but on the other hand, in the wild *Nicotiana attenuata* plant, the levels of IAA decrease within a course of 3 days of herbivory (Tooker and de Moraes 2011; Mapes and Davies 2001). Similarly, a few examples evident of the participation of ABA in herbivory defence are found in plants such as *Zea mays* in which the ABA levels rise after its infestation by the insect *Diabrotica virgifera virgifera*, the western corn rootworm (Erb et al. 2009); in *Arabidopsis* followed by wounding; and by oral secretions due to desert locust, *Schistocerca gregaria* (Schafer et al. 2011). For CK, in the *N. attenuata* plant, the FAC elicitors from the insects have been found to regulate the levels of CK-related transcript level (Hui 2003). However, interestingly, the CK levels are also regulated by the leaf-mining moth *Phyllonorycter blancardella* in association with symbiotic bacteria. They usually induce metabolically active green areas to overcome the food supply in a yellow senescent leaf. This phenomenon is known as the green island phenotype and is specific to the regulation of CK in the tripartite interaction of bacteria-plant-insect (Sugio et al. 2015). Likewise, for the hormone BRs, in tomato, they are usually seen to antagonise the JA-mediated density of trichomes and the accumulation of defence metabolites, thus interfering with the mechanical defences making the plant more susceptible to the chewing herbivores (Campos et al. 2009). For the GA-associated defence in herbivory, it usually acts to provide the plant with over compensatory growth, such as in cases of caterpillar herbivory (Trumble et al. 1993; Arab and Trigo 2011). But it also acts in integration with other pathways to fine-tune the defence. The integration of these pathways can be understood as under.

5.11 Integration of Different Signalling Pathways with JA as the Master Regulator

The expression of defence-related genes is not directly regulated by the JA-Ile. The gene regulation is controlled by a protein known as JAZ, a jasmonate ZIM-domain protein. In the absence of JA-Ile, the JAZ protein is usually bound to the TF MYC2 thus inhibiting the recruitment of polymerase and the transcription machinery (Boter et al. 2004; Lorenzo 2004). As JAZ proteins interact with MYC2 during JA-mediated response, similarly, the DELLA proteins work to integrate the GA hormonal pathway with JA by interacting with the JAZ proteins. The integration of GA with JA is antagonistic which comes into play when GA is present in the cell. The GA molecules bind to the DELLA protein leading to its ubiquitination and 26S proteasome-mediated degradation. This releases the JAZ protein to which MYC2 easily binds as its competitive binder is no longer available thus inhibiting JA-response (Fürstenberg-Hägg et al. 2013).

The point of integration of ABA with JA is the expression of a helix-loop-helix TF. Both the hormones synergistically induce the expression of MYC2- dependent genes during wound response; this MYC2 protein then encodes for the TF which acts both as an activator and a repressor of JA-mediated expression of defence-associated genes (Anderson 2004). The protein TOPLESS (TPL) is a corepressor that interacts with the JAZ protein via another protein, NOVEL INTERACTOR OF JAZ (NINJA) and helps to sequester the MYC2 TF thus inhibiting the JA-mediated gene response to herbivory (Pauwels et al. 2010). The TPL also interacts with EAR-motif protein, an ET-responsive element-binding factor. This represses the transcription of a number of genes involved in different hormonal pathways (Szemenyei et al. 2008). The TPL proteins also interact with the NINJA- related proteins as a part of a complex with other proteins. These interactions induce the negative regulation of transcription regulators by mediating the ABA-induced degradation of the regulator proteins (Erb et al. 2012).

At the molecular level, the JAZ and MYC2 proteins show co-regulation by JA and auxin, providing evidence towards a possible cross talk between the two hormones (Grunewald et al. 2009). Similar to ABA signalling, the EAR-motif proteins with AUX/IAA domain, the negative regulator of auxin-induced response interacts with TOPLESS protein, suggesting it to be a central point of integration of many signalling pathways (Pauwels et al. 2010). The hormone CK is also involved in the biosynthesis of JA, thus providing a possible connecting link between the hormones. For example, in the hybrid *Poplar* plant, the CK treatment induced the JA-burst in leaves upon wounding which further initiated a cascade of JA-biosynthesis (Dervinis et al. 2010). BRs are also known to integrate with other defence mechanisms at multiple levels. An example of such integration is visible in *N. attenuata* in which the BAK1 is silenced. The silencing of this gene leads to the reduction of wound and herbivory-induced JA and JA-Ile levels and has also reduced the reduction of JA-induced trypsin proteinase inhibitor (TPI) activity (Yang et al. 2011a). The cross talk for all the phytohormones in defence against pathogens and herbivores is represented in Fig. 2.

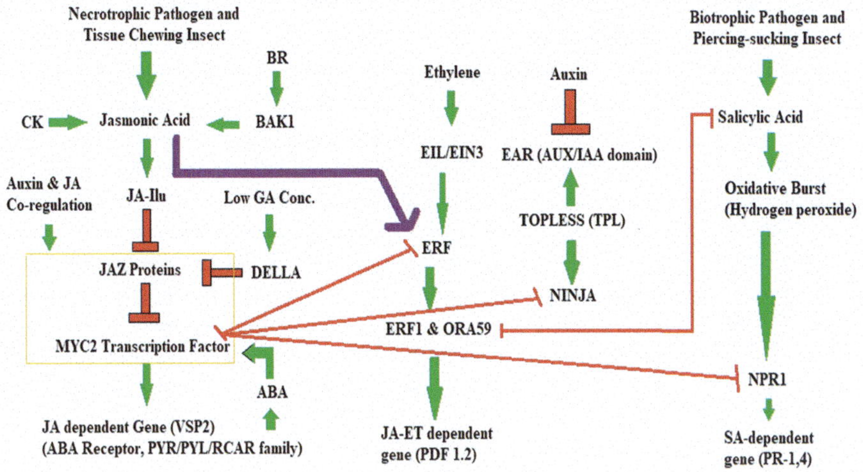

Fig. 2 Integration of phytohormone signalling network in plants against pathogens and herbivores

The above integrative network of phytohormones with JA being the master regulator is visible in all cases of defences against microbes and herbivores. Apart from this, when SA-JA antagonism takes place, sometimes, depending upon the specificity of the host-attacker interaction, SA pathways become the major player in plant immunity. Nevertheless, these interactions depict the importance of these pathways in implementing and strengthening the attack triggered immunity by the synthesis of VOCs and other phytotoxins. How plants use these volatiles and other compounds against its attackers to protect themselves has been addressed in the next section.

5.12 Secondary Metabolites: The Defensive Chemical Species

There is a significant overlap in the hormonal signaling pathways that are activated by particular types of microbes and insects; however the pattern of downstream commencement of chemical defence synthesis and their associated mechanism of action is highly precise for the particular plant-attacker combination (De Vos et al. 2005). Plant VOCs are the chief moderator of indirect defence. When plants sense herbivore-related elicitors, they generate a complex blend of VOCs that attract predators, parasites and other natural enemies which attack herbivorous host insects and thereby reducing herbivory (Aljbory and Chen 2018). Plant microbes are also competent of triggering VOCs, although it has not been studied in detail as that of induction by herbivorous insects.

In response to perceived volatile signals, plants activate various defence genes and produce secondary metabolites or prime their defences against pests (Shulaev et al. 1997; Arimura et al. 2000; Heil and Kost 2006). Major components of VOCs, the green leaf volatiles (GLVs), participate in plant defence against herbivores, bacterial and fungal microbes (Shiojiri et al. 2006). Similarly, a mixture of monoterpenes, sesquiterpenes or GLVs are known to discourage oviposition by females of numerous Lepidopteran insects (Holopainen and Gershenzon 2010). Likewise, monoterpenoids (e.g. geraniol) can also induce apoptosis-like cell death, which is caused as a defence reaction against bacterial infection (Jabs et al. 1996; Izumi et al. 1999). A universal GLV, (Z)-3-hexenol, induced by mechanical damage or in leaf-miner-damaged plants, is suggested as the most common damage attractant. It is known to help parasitoids trace their prey or the host plant (Wei et al. 2007). With exposure to (Z)-3-hexenol, the upregulation of various genes involved in direct and indirect defences has been reported in maize (ul Hassan et al. 2015). In some studies, GLVs were reported to exhibit bactericidal activity against both Gram-positive and Gram-negative bacteria (Croft et al. 1993; Nakamura and Hatanaka 2002). In addition, GLVs also serve as signals used for intra- and interplant communication and play a crucial role in priming, whereby damaged plant prepares the neighbouring plants to defend themselves against a future episodes of attack (Gershenzon 2007; Frost et al. 2008; Goellner and Conrath 2008). The oilseed rape plant also synthesises many volatiles, but of all, linalool and 2-phenylethanol attract the honey bees and thus help in pollination. This depicts the mutualistic behaviour of plants and insects (Bruce 2015).

The co-infestations of insects and microbes can drastically change the pattern of VOC emission. For example, the *Brassica rapa* plant showed differences in plant VOC emission patterns when the plant was assaulted by the specialist insect herbivore *Pieris brassicae* alone as compared to a dual attack by the insect and fungus *Erysiphe cruciferarum* (the powdery mildew) (Desurmont et al. 2016). In dual assault on a plant by a chewing insect and a microbe, a necrotrophic pathogen usually increases VOC emission in comparison with insect assault alone, contrarily, a biotrophic pathogen decrease VOC emission. Necrotrophs normally activate JA signaling that is involved in triggering the production of a large number of volatiles, ensuing augmentation of herbivore-induced volatile emission, whereas biotrophs are usually signaled through a SA-dependent pathway that inhibits the JA pathway, leading to reduction in volatile emission (Ponzio et al. 2013; Biere and Bennett 2013). These observations proved that the large number of signals and aggressors leads to differential defence responses in plant; therefore, they should be studied with a more integrated and holistic approach.

Apart from VOCs, plants are also known to synthesise a number of toxic compounds that are effective against specific enemies of the host plant. For example, the pigeon pea plant synthesises protease inhibitors which are effective against *Helicoverpa armigera* (Parde et al. 2012). The plants of the Solanaceae family secrete O-acyl sugars and 7-epizingiberene in glandular trichomes which prevents the insects from damaging the plants (Schilmiller et al. 2012; Bleeker et al. 2012). When attacked by a herbivore, plants also synthesise another class of proteins, known as herbivore-induced plant volatiles (HIPVs). Angiosperms or flowering plants produce two most widely distribute homoterpenes, namely, 4,8,12-trimethyltrideca-1,3,7,11-tetraene (TMTT) and 4,8-dimethylnona-1,3,7-triene (DNMT) (Bruce 2015). Roots also produce a variety of exudates that protect the plant from further attack and simultaneously also change the root microbiome. For example, the rhizosphere microbes of the plant *Zea mays* are prevented to attack the roots by *Pseudomonas putida* due to the secretion of an antimicrobial benzoxazinoid, namely, DIMBOA or 2,4-dihydroxy-7-methoxy-2H-1,4-benzoxazine-3(4H)-one (Neal et al. 2012). Another example of root secretion is seen in the plant *Arabidopsis thaliana* when attacked by *Pseudomonas syringae*. The plant roots on attack release malic acid which attracts *B. subtilis*, thus modifying the rhizosphere microbiome (Rudrappa et al. 2008). A case of volatiles linking the above-ground and below-ground effects is seen in *Thuja occidentalis*. When this plant is attacked by *Otiorhynchus sulcatus*, roots secrete the volatiles that attract *H. megidis*, an entomopathogenic nematode (Van Tol et al. 2001).

Plants also synthesise antinutritional compounds, thus reducing the quality of the plant and making them unsuitable for the insects to feed upon, thus reducing herbivory. Two sugar-binding proteins or lectins are synthesised by the *Phaseolus vulgaris* plant, namely, agglutinin and arcelin which are effective against the insects *C. maculatus* and *Z. subfasciatus*, respectively (Gatehouse and Gatehouse 1998; Osborni et al. 1988). Similarly, the PIs synthesised by the PR-genes (PR-6) is a class of toxic enzymes that inhibit the activity of proteases in insect gut. For example, cysteine proteases are effective against the proteases present inside the *Chrysomela tremulae* gut (Wielkopolan and Obrępalska-Stęplowska 2016); another PI, the trypsin-papain inhibitor, is synthesised by the *Pithecellobium dumosum* plant seeds and is effective against the insects *C. maculatus* and *Z. subfasciatus* (Oliveira et al. 2007). Likewise, the α-amylase inhibitors from wheat inhibit the activity of the α-amylase enzyme in insects belonging to the classes Tenebrionidae, Curculionidae and Silvanidae (Wielkopolan and Obrępalska-Stęplowska 2016).

6 Conclusion

Ecological interactions between microbes, plants and insects are complex and dynamic in nature. In this chapter, we discussed the complexity of MPI interactions and focussed on plant-microbe influence on plant-insect behaviour and vice versa. We also discussed the ecological implications and evolutionary importance of the tripartite interactions between microbes, plants, and insects. Even though there may be many similarities in the plant defence mechanisms against diverse microbes and insects, there are some elementary differences between their interactions. Though, various facets of plant-microbe and plant-insect interactions have long been studied, a thorough knowledge of the multifarious defences naturally employed by plants during MPI interactions is also crucial to understand the cross talk between microbes, plants and insects for the survival of agricultural and horticultural plants and for the development of engineered disease resistance and/or tolerance traits in plants against multiple pathogens. Connecting this information to environmental conditions would certainly provide a way for developing a favourable and friendly environment for plant growth. To have a better insight in the functioning and the involvement of various biomolecules in plant defence-associated overlap, tritrophic studies with multiple system need to be carried out to further understand the interacting and constantly evolving nature of the MPI system. Such investigative analysis will be crucial in understanding how plants prioritise their defence responses and also cope with multiple stresses simultaneously. Further studies in this field will help scientists to work upon manipulating the MPI interactions to promote crop yields and food security, which has received very little attention from a tri-trophic perspective.

References

A'Bear AD, Johnson SN, Jones TH (2014) Putting the "upstairs-downstairs" into ecosystem service: what can aboveground-belowground ecology tell us? Biol Control. https://doi.org/10.1016/j.biocontrol.2013.10.004

Abe H (2003) Arabidopsis AtMYC2 (bHLH) and AtMYB2 (MYB) function as transcriptional activators in Abscisic acid signaling. Plant Cell Online. https://doi.org/10.1105/tpc.006130

Acevedo FE, Rivera-Vega LJ, Chung SH, Ray S, Felton GW (2015) Cues from chewing insects – the intersection of DAMPs, HAMPs, MAMPs and effectors. Curr Opin Plant Biol. https://doi.org/10.1016/j.pbi.2015.05.029

Agrawal AA (2000) Overcompensation of plants in response to herbivory and the by-product benefits of mutualism. Trends Plant Sci. https://doi.org/10.1016/S1360-1385(00)01679-4

Agrios GN (2005) How plants defend themselves against pathogens. Plant Pathol. https://doi.org/10.1016/B978-0-08-047378-9.50012-9

Alazem M, Lin NS (2015) Roles of plant hormones in the regulation of host-virus interactions. Mol Plant Pathol. https://doi.org/10.1111/mpp.12204

Aljbory Z, Chen MS (2018) Indirect plant defense against insect herbivores: a review. Insect Sci. https://doi.org/10.1111/1744-7917.12436

Anderson JP (2004) Antagonistic interaction between Abscisic acid and Jasmonate-Ethylene signaling pathways modulates defense gene expression and disease resistance in Arabidopsis. Plant Cell Online 16(12):3460–3479. https://doi.org/10.1105/tpc.104.025833

Antico CJ, Colon C, Banks T, Ramonell KM (2012) Insights into the role of jasmonic acid-mediated defenses against necrotrophic and biotrophic fungal pathogens. Front Biol. https://doi.org/10.1007/s11515-011-1171-1

Arab A, Trigo JR (2011) Host plant invests in growth rather than chemical defense when attacked by a specialist herbivore. J Chem Ecol 37(5):492–495. https://doi.org/10.1007/s10886-011-9955-y

Arimura GI, Tashiro K, Kuhara S, Nishioka T, Ozawa R, Takabayashi J (2000) Gene responses in bean leaves induced by herbivory and by herbivore-induced volatiles. Biochem Biophys Res Commun. https://doi.org/10.1006/bbrc.2000.3672

Arimura GI, Ozawa R, Maffei ME (2011) Recent advances in plant early signaling in response to herbivory. Int J Mol Sci. https://doi.org/10.3390/ijms12063723

Bai X, Correa VR, Toruño TY, Ammar E-D, Kamoun S, Hogenhout SA (2009) AY-WB Phytoplasma secretes a protein that targets plant cell nuclei. Mol Plant-Microbe Interact. https://doi.org/10.1094/MPMI-22-1-0018

Berendsen RL, Pieterse CMJ, Bakker PAHM (2012) The rhizosphere microbiome and plant health. Trends Plant Sci. https://doi.org/10.1016/j.tplants.2012.04.001

Biere A, Bennett AE (2013) Three-way interactions between plants, microbes and insects. Funct Ecol. https://doi.org/10.1111/1365-2435.12100

Bizzarri MF, Bishop AH (2008) The ecology of Bacillus thuringiensis on the phylloplane: colonization from soil, plasmid transfer, and interaction with larvae of Pieris brassicae. Microb Ecol. https://doi.org/10.1007/s00248-007-9331-1

Bleeker PM, Mirabella R, Diergaarde PJ, VanDoorn A, Tissier A, Kant MR, Prins M, de Vos M, Haring MA, Schuurink RC (2012) Improved herbivore resistance in cultivated tomato with the sesquiterpene biosynthetic pathway from a wild relative. Proc Natl Acad Sci 109 (49):20124–20129. https://doi.org/10.1073/pnas.1208756109

Bonaventure G, VanDoorn A, Baldwin IT (2011) Herbivore-associated elicitors: FAC signaling and metabolism. Trends Plant Sci. https://doi.org/10.1016/j.tplants.2011.01.006

Boter M, Ruíz-Rivero O, Abdeen A, Prat S (2004) Conserved MYC transcription factors play a key role in jasmonate signaling both in tomato and Arabidopsis. Genes Dev 18(13):1577–1591. https://doi.org/10.1101/gad.297704

Boyes DC, Nam J, Dangl JL (1998) The Arabidopsis thaliana RPM1 disease resistance gene product is a peripheral plasma membrane protein that is degraded coincident with the hypersensitive response. Proc Natl Acad Sci U S A. https://doi.org/10.1073/pnas.95.26.15849

Bruce TJA (2015) Interplay between insects and plants: dynamic and complex interactions that have coevolved over millions of years but act in milliseconds. J Exp Bot. https://doi.org/10.1093/jxb/eru391

Brundrett MC (2002) Coevolution of roots and mycorrhizas of land plants. New Phytol. https://doi.org/10.1046/j.1469-8137.2002.00397.x

Burdon JJ, Thrall PH (2009) Coevolution of plants and their pathogens in natural habitats. Science. https://doi.org/10.1126/science.1171663

Campos ML, De Almeida M, Rossi ML, Martinelli AP, Litholdo Junior CG, Figueira A, Rampelotti-Ferreira FT, Vendramim JD, Benedito VA, Pereira Peres LE (2009) Brassinosteroids interact negatively with jasmonates in the formation of anti-herbivory traits in tomato. J Exp Bot 60(15):4347–4361. https://doi.org/10.1093/jxb/erp270

Carmona D, Fornoni J (2013) Herbivores can select for mixed defensive strategies in plants. New Phytol. https://doi.org/10.1111/nph.12023

Cavender-Bares J, Kozak KH, Fine PVA, Kembel SW (2009) The merging of community ecology and phylogenetic biology. Ecol Lett. https://doi.org/10.1111/j.1461-0248.2009.01314.x

Chadha P, Das RH (2006) A pathogenesis related protein, AhPR10 from peanut: an insight of its mode of antifungal activity. Planta 225(1):213–222. https://doi.org/10.1007/s00425-006-0344-7

Chen H, Xue L, Chintamanani S, Germain H, Lin H, Cui H, Cai R, Zuo J, Tang X, Li X, Guo H, Zhou J-M (2009) ETHYLENE INSENSITIVE3 and ETHYLENE INSENSITIVE3-LIKE1 repress SALICYLIC ACID INDUCTION DEFICIENT2 expression to negatively regulate plant innate immunity in Arabidopsis. Plant Cell Online 21(8):2527–2540. https://doi.org/10.1105/tpc.108.065193

Chisholm ST, Coaker G, Day B, Staskawicz BJ (2006) Host-microbe interactions: shaping the evolution of the plant immune response. Cell 124(4):803–814. https://doi.org/10.1016/j.cell.2006.02.008

Choi HW, Klessig DF (2016) DAMPs, MAMPs, and NAMPs in plant innate immunity. BMC Plant Biol. https://doi.org/10.1186/s12870-016-0921-2

Chung SH, Rosa C, Scully ED, Peiffer M, Tooker JF, Hoover K, Luthe DS, Felton GW (2013) Herbivore exploits orally secreted bacteria to suppress plant defenses. Proc Natl Acad Sci. https://doi.org/10.1073/pnas.1308867110

Clissold FJ, Sanson GD, Read J (2004) Indigestibility of plant cell wall by the Australian plague locust, Chortoicetes terminifera. Entomologia Experimentalis et Applicata. https://doi.org/10.1111/j.0013-8703.2004.00192.x

Cogni R, Trigo JR, Futuyma DJ (2012) A free lunch? No cost for acquiring defensive plant pyrrolizidine alkaloids in a specialist arctiid moth (Utetheisa ornatrix). Mol Ecol. https://doi.org/10.1111/mec.12086

Croft K, Juttner F, Slusarenko AJ (1993) Volatile products of the lipoxygenase pathway evolved from Phaseolus vulgaris (L.) leaves inoculated with Pseudomonas syringae pv phaseolicola. Plant Physiol. https://doi.org/10.1104/pp.101.1.13

Cui J, Bahrami AK, Pringle EG, Hernandez-Guzman G, Bender CL, Pierce NE, Ausubel FM (2005) Pseudomonas syringae manipulates systemic plant defenses against pathogens and herbivores. Proc Natl Acad Sci. https://doi.org/10.1073/pnas.0409450102

De Vos M, Van Oosten VR, Van Poecke RMP, Van Pelt JA, Pozo MJ, Mueller MJ, Buchala AJ, Métraux J-P, Van Loon LC, Dicke M, Pieterse CMJ (2005) Signal signature and transcriptome changes of *Arabidopsis* during pathogen and insect attack. Mol Plant-Microbe Interact 18 (9):923–937. https://doi.org/10.1094/MPMI-18-0923

Dervinis C, Frost CJ, Lawrence SD, Novak NG, Davis JM (2010) Cytokinin primes plant responses to wounding and reduces insect performance. J Plant Growth Regul 29(3):289–296. https://doi.org/10.1007/s00344-009-9135-2

Desurmont GA, Xu H, Turlings TCJ (2016) Powdery mildew suppresses herbivore-induced plant volatiles and interferes with parasitoid attraction in Brassica rapa. Plant Cell Environ 39 (9):1920–1927. https://doi.org/10.1111/pce.12752

Dietzgen RG, Mann KS, Johnson KN (2016) Plant virus-insect vector interactions: current and potential future research directions. Viruses. https://doi.org/10.3390/v8110303

Dinnage R, Cadotte MW, Haddad NM, Crutsinger GM, Tilman D (2012) Diversity of plant evolutionary lineages promotes arthropod diversity. Ecol Lett. https://doi.org/10.1111/j.1461-0248.2012.01854.x

Dziubińska H, Trębacz K, Zawadzki T (2001) Transmission route for action potentials and variation potentials in Helianthus annuus L. J Plant Physiol 158(9):1167–1172. https://doi.org/10.1078/S0176-1617(04)70143-1

Eigenbrode SD, Bosque-Pérez N, Davis TS (2018) Insect-borne plant pathogens and their Vectors: Ecology, evolution, and complex interactions. Annu Rev Entomol. https://doi.org/10.1146/annurev-ento-020117-043119

Erb M, Flors V, Karlen D, De Lange E, Planchamp C, D'Alessandro M, Turlings TCJ, Ton J (2009) Signal signature of aboveground-induced resistance upon belowground herbivory in maize. Plant J 59(2):292–302. https://doi.org/10.1111/j.1365-313X.2009.03868.x

Erb M, Meldau S, Howe GA (2012) Role of phytohormones in insect-specific plant reactions. Trends Plant Sci. https://doi.org/10.1016/j.tplants.2012.01.003

Fernandez-Conradi P, Jactel H, Robin C, Tack AJM, Castagneyrol B (2018) Fungi reduce preference and performance of insect herbivores on challenged plants. Ecology. https://doi.org/10.1002/ecy.2044

Franco FP, Moura DS, Vivanco JM, Silva-Filho MC (2017) Plant–insect–pathogen interactions: a naturally complex ménage à trois. Curr Opin Microbiol. https://doi.org/10.1016/j.mib.2017.04.007

Frost CJ, Mescher MC, Carlson JE, De Moraes CM (2008) Plant defense priming against herbivores: getting ready for a different Battle. Plant Physiol 146(3):818–824. https://doi.org/10.1104/pp.107.113027

Fürstenberg-Hägg J, Zagrobelny M, Bak S (2013) Plant defense against insect herbivores. Int J Mol Sci. https://doi.org/10.3390/ijms140510242

Gange AC, Brown VK, Aplin DM (2003) Multitrophic links between arbuscular mycorrhizal fungi and insect parasitoids. Ecol Lett. https://doi.org/10.1046/j.1461-0248.2003.00540.x

Garrido E, Andraca-Gómez G, Fornoni J (2012) Local adaptation: simultaneously considering herbivores and their host plants. New Phytol. https://doi.org/10.1111/j.1469-8137.2011.03923.x

Gatehouse AMR, Gatehouse JA (1998) Identifying proteins with insecticidal activity: use of encoding genes to produce insect-resistant transgenic crops. Pestic Sci. https://doi.org/10.1002/(SICI)1096-9063(199802)52:2<165::AID-PS679>3.0.CO;2-7

Gershenzon J (2007) Plant volatiles carry both public and private messages. Proc Natl Acad Sci U S A. https://doi.org/10.1073/pnas.0700906104

Ghosh A, Das A, Vijayanandraj S, Mandal B (2016) Cardamom bushy dwarf virus infection in large cardamom alters plant selection preference, life stages, and fecundity of aphid vector, micromyzus kalimpongensis (hemiptera: Aphididae). Environ Entomol. https://doi.org/10.1093/ee/nvv161

Gilbert LE (1971) Butterfly-plant coevolution: has Passiflora adenopoda won the selectional race with heliconiine butterflies? Science. https://doi.org/10.1126/science.172.3983.585

Gilroy S, Białasek M, Suzuki N, Górecka M, Devireddy AR, Karpiński S, Mittler R (2016) ROS, calcium, and electric signals: key mediators of rapid systemic signaling in plants. Plant Physiol 171(3):1606–1615. https://doi.org/10.1104/pp.16.00434

Goellner K, Conrath U (2008) Priming: It's all the world to induced disease resistance. In: *Sustainable disease management in a European context*. Springer, Dordrecht, pp 233–242. https://doi.org/10.1007/978-1-4020-8780-6_3

Groen SC, Whiteman NK, Bahrami AK, Wilczek AM, Cui J, Russell JA, Cibrian-Jaramillo A, Butler IA, Rana JD, Huang GH, Bush J, Ausubel FM, Pierce NE (2013) Pathogen-triggered ethylene signaling mediates systemic-induced susceptibility to herbivory in Arabidopsis. Plant Cell 25(11):4755–4766. https://doi.org/10.1105/tpc.113.113415

Grunewald W, Vanholme B, Pauwels L, Plovie E, Inzé D, Gheysen G, Goossens A (2009) Expression of the Arabidopsis jasmonate signalling repressor JAZ1/TIFY10A is stimulated by auxin. EMBO Rep 10(8):923–928. https://doi.org/10.1038/embor.2009.103

Guo Z, Döll K, Dastjerdi R, Karlovsky P, Dehne HW, Altincicek B (2014) Effect of fungal colonization of wheat grains with Fusarium spp. on food choice, weight gain and mortality of meal beetle larvae (Tenebrio molitor). PLoS One. https://doi.org/10.1371/journal.pone.0100112

Hann DR, Domínguez-Ferreras A, Motyka V, Dobrev PI, Schornack S, Jehle A, Felix G, Chinchilla D, Rathjen JP, Boller T (2014) The Pseudomonas type III effector HopQ1 activates cytokinin signaling and interferes with plant innate immunity. New Phytol. https://doi.org/10.1111/nph.12544

Heil M, Kost C (2006) Priming of indirect defences. Ecol Lett 9(7):813–817. https://doi.org/10.1111/j.1461-0248.2006.00932.x

Hiltpold I, Erb M, Robert CAM, Turlings TCJ (2011) Systemic root signalling in a belowground, volatile-mediated tritrophic interaction. Plant Cell Environ. https://doi.org/10.1111/j.1365-3040.2011.02327.x

Hodge S, Powell G (2010) Conditional facilitation of an aphid vector, Acyrthosiphon pisum, by the plant pathogen, pea enation mosaic virus. J Insect Sci. https://doi.org/10.1673/031.010.14115

Holopainen JK, Gershenzon J (2010) Multiple stress factors and the emission of plant VOCs. Trends Plant Sci. https://doi.org/10.1016/j.tplants.2010.01.006

Hou S, Yang Y, Zhou JM (2009) The multilevel and dynamic interplay between plant and pathogen. Plant Signal Behav. https://doi.org/10.4161/psb.4.4.8155

Howe GA, Jander G (2008) Plant immunity to insect herbivores. Annu Rev Plant Biol. https://doi.org/10.1146/annurev.arplant.59.032607.092825

Hui D (2003) Molecular interactions between the specialist herbivore Manduca sexta (Lepidoptera, Sphingidae) and its natural host Nicotiana attenuata: V. Microarray analysis and further characterization of large-scale changes in Herbivore-Induced mRNAs. Plant Physiol 131 (4):1877–1893. https://doi.org/10.1104/pp.102.018176

Izumi S, Takashima O, Hirata T (1999) Geraniol is a potent inducer of apoptosis-like cell death in the cultured shoot primordia of Matricaria chamomilla. Biochem Biophys Res Commun. https://doi.org/10.1006/bbrc.1999.0813

Jabs T, Dietrich RA, Dangl JL (1996) Initiation of runaway cell death in an Arabidopsis mutant by extracellular superoxide. Science 273(5283):1853–1856. https://doi.org/10.1126/science.273.5283.1853

Johnson SN, Gregory PJ (2006) Chemically-mediated host-plant location and selection by root-feeding insects. Physiol Entomol. https://doi.org/10.1111/j.1365-3032.2005.00487.x

Johnson SN, Rasmann S (2015) Root-feeding insects and their interactions with organisms in the rhizosphere. Annu Rev Entomol. https://doi.org/10.1146/annurev-ento-010814-020608

Johnson SN, Douglas AE, Woodward S, Hartley SE (2003) Microbial impacts on plant-herbivore interactions: the indirect effects of a birch pathogen on a birch aphid. Oecologia. https://doi.org/10.1007/s00442-002-1139-6

Johnson SN, Gregory PJ, Greenham JR, Zhang X, Murray PJ (2005) Attractive properties of an isoflavonoid found in white clover root nodules on the clover root weevil. J Chem Ecol. https://doi.org/10.1007/s10886-005-6355-1

Jones JDG, Dangl JL (2006) The plant immune system. Nature 444(7117):323–329. https://doi.org/10.1038/nature05286

Karban R, Adamchak R, Schnathorst WC (1987) Induced resistance and interspecific competition between spider mites and a vascular wilt fungus. Science 235(4789):678–680. https://doi.org/10.1126/science.235.4789.678

Kazan K, Lyons R (2014) Intervention of Phytohormone pathways by pathogen effectors. Plant Cell. https://doi.org/10.1105/tpc.114.125419

Khatri BB, Tegg RS, Brown PH, Wilson CR (2010) Infection of potato tubers with the common scab pathogen streptomyces scabiei in a soil-less system. J Phytopathol. https://doi.org/10.1111/j.1439-0434.2009.01637.x

Kiba T, Kudo T, Kojima M, Sakakibara H (2011) Hormonal control of nitrogen acquisition: roles of auxin, abscisic acid, and cytokinin. J Exp Bot. https://doi.org/10.1093/jxb/erq410

Kistner C, Parniske M (2002) Evolution of signal transduction in intracellular symbiosis. Trends Plant Sci. https://doi.org/10.1016/S1360-1385(02)02356-7

Kluth S, Kruess A, Tscharntke T (2002) Insects as vectors of plant pathogens: mutualistic and antagonistic interactions. Oecologia. https://doi.org/10.1007/s00442-002-1016-3

Knogge W (1996) Fungal infection of plants. Plant Cell. https://doi.org/10.1016/0160-9327(89)90012-4

Krause SC, Raffa KF (1992) Comparison of insect, fungal, and mechanically induced defoliation of larch: effects on plant productivity and subsequent host susceptibility. Oecologia. https://doi.org/10.1007/BF00317699

Krimmel BA, Pearse IS (2013) Sticky plant traps insects to enhance indirect defence. Ecol Lett 16(2):219–224. https://doi.org/10.1111/ele.12032

Kruess A (2002) Indirect interaction between a fungal plant pathogen and a herbivorous beetle of the weed Cirsium arvense. Oecologia. https://doi.org/10.1007/s00442-001-0829-9

Lee B, Lee S, Ryu CM (2012) Foliar aphid feeding recruits rhizosphere bacteria and primes plant immunity against pathogenic and non-pathogenic bacteria in pepper. Ann Bot. https://doi.org/10.1093/aob/mcs055

Lee S, Yang DS, Uppalapati SR, Sumner LW, Mysore KS (2013) Suppression of plant defense responses by extracellular metabolites from Pseudomonas syringae pv. Tabaci in Nicotiana benthamiana. BMC Plant Biol. https://doi.org/10.1186/1471-2229-13-65

Li C (2002) Resistance of cultivated tomato to cell content-feeding herbivores is regulated by the Octadecanoid-signaling pathway. Plant Physiol. https://doi.org/10.1104/pp.005314

Llorente F, Muskett P, Sánchez-Vallet A, López G, Ramos B, Sánchez-Rodríguez C, Jordá L, Parker J, Molina A (2008) Repression of the auxin response pathway increases Arabidopsis susceptibility to necrotrophic fungi. Mol Plant 1(3):496–509. https://doi.org/10.1093/mp/ssn025

Lorenzo O (2003) ETHYLENE RESPONSE FACTOR1 integrates signals from ethylene and Jasmonate pathways in plant defense. Plant Cell Online 15(1):165–178. https://doi.org/10.1105/tpc.007468

Lorenzo O (2004) JASMONATE-INSENSITIVE1 encodes a MYC transcription factor essential to discriminate between different Jasmonate-regulated defense responses in Arabidopsis. Plant Cell Online 16(7):1938–1950. https://doi.org/10.1105/tpc.022319

Mapes CC, Davies PJ (2001) Indole-3-acetic acid and ball gall development on Solidago altissima. New Phytol 151:195–202. https://doi.org/10.1046/j.1469-8137.2001.00161.x

Maris PC, Joosten NN, Goldbach RW, Peters D (2004) Tomato spotted wilt virus infection improves host suitability for its vector Frankliniella occidentalis. Phytopathology. https://doi.org/10.1094/PHYTO.2004.94.7.706

Martín-Rodríguez JÁ, León-Morcillo R, Vierheilig H, Ocampo JA, Ludwig-Müller J, García-Garrido JM (2011) Ethylene-dependent/ethylene-independent ABA regulation of tomato plants colonized by arbuscular mycorrhiza fungi. New Phytol. https://doi.org/10.1111/j.1469-8137.2010.03610.x

McDowell JM, Woffenden BJ (2003) Plant disease resistance genes: recent insights and potential applications. Trends Biotechnol. https://doi.org/10.1016/S0167-7799(03)00053-2

McElhaney R, Alvarez AM, Kado CI (1998) Nitrogen limits Xanthomonas campestris pv. Campestris invasion of the host xylem. Physiol Mol Plant Pathol. https://doi.org/10.1006/pmpp.1997.0131

Medeiros AH, Franco FP, Matos JL, de Castro PA, Santos-Silva LK, Henrique-Silva F, Goldman GH, Moura DS, Silva-Filho MC (2012) *Sugarwin* : a sugarcane insect-induced gene with Antipathogenic activity. Mol Plant-Microbe Interact. https://doi.org/10.1094/MPMI-09-11-0254

Moran PJ, Thompson G a (2001) Molecular responses to aphid feeding in *Arabidopsis* in relation to plant defense pathways. Plant Physiol. https://doi.org/10.1104/pp.125.2.1074

Müller C (2007) Plant-insect interactions on Cuticular surfaces. Annu Plant Rev. https://doi.org/10.1002/9780470988718.ch13

Mur LAJ, Lloyd AJ, Cristescu SM, Harren FJM, Hall MA, Smith AR (2009) Biphasic ethylene production during the hypersensitive response in arabidopsis: a window into defense priming mechanisms? Plant Signal Behav. https://doi.org/10.4161/psb.4.7.8904

Nakamura S, Hatanaka A (2002) Green-leaf-derived C6-aroma compounds with potent antibacterial action that act on both gram-negative and gram-positive bacteria. J Agric Food Chem 50(26):7639–7644. https://doi.org/10.1021/jf025808c

Neal AL, Ahmad S, Gordon-Weeks R, Ton J (2012) Benzoxazinoids in root exudates of maize attract pseudomonas putida to the rhizosphere. PLoS One. https://doi.org/10.1371/journal.pone.0035498

Nicaise V, Roux M, Zipfel C (2009) Recent advances in PAMP-triggered immunity against Bacteria: Pattern Recognition Receptors Watch over and Raise the Alarm. Plant Physiol. https://doi.org/10.1104/pp.109.139709

Nürnberger T, Kemmerling B (2006) Receptor protein kinases--pattern recognition receptors in plant immunity. Trends Plant Sci. https://doi.org/10.1016/j.tplants.2006.09.005

Nürnberger T, Scheel D (2001) Signal transmission in the plant immune response. Trends Plant Sci. https://doi.org/10.1016/S1360-1385(01)02019-2

Occhipinti A (2013) Plant coevolution: evidences and new challenges. J Plant Interact. https://doi.org/10.1080/17429145.2013.816881

Oliveira AS, Migliolo L, Aquino RO, Ribeiro JKC, Macedo LLP, Andrade LBS, Bemquerer MP, Santos EA, Kiyota S, de Sales MP (2007) Purification and characterization of a trypsin-papain inhibitor from Pithecelobium dumosum seeds and its in vitro effects towards digestive enzymes from insect pests. Plant Physiol Biochem. https://doi.org/10.1016/j.plaphy.2007.08.002

Ortíz-Castro R, Valencia-Cantero E, López-Bucio J (2008) Plant growth promotion by Bacillus megaterium involves cytokinin signaling. Plant Signal Behav. https://doi.org/10.4161/psb.3.4.5204

Osborni TC, Alexander DC, Sun SSM, Cardona C, Bliss FA (1988) Insecticidal activity and lectin homology of Arcelin seed protein. Science. https://doi.org/10.1126/science.240.4849.207

Pangesti N, Pineda A, Pieterse CMJ, Dicke M, van Loon JJA (2013) Two-way plant mediated interactions between root-associated microbes and insects: from ecology to mechanisms. Front Plant Sci:4. https://doi.org/10.3389/fpls.2013.00414

Parde VD, Sharma HC, Kachole MS (2012) Protease inhibitors in wild relatives of pigeonpea against the cotton bollworm/legume pod borer, Helicoverpa armigera. Am J Plant Sci. https://doi.org/10.4236/ajps.2012.35076

Park CJ, Kim KJ, Shin R, Park JM, Shin YC, Paek KH (2013, 2004) Erratum: pathogenesis-related protein 10 isolated from hot pepper functions as a ribonuclease in an antiviral pathway. Plant J:186–198. https://doi.org/10.1111/tpj.12282

Pauwels L, Barbero GF, Geerinck J, Tilleman S, Grunewald W, Pérez AC, Chico JM, Bossche RV, Sewell J, Gil E, García-Casado G, Witters E, Inzé D, Long JA, De Jaeger G, Solano R, Goossens A, Goossens A (2010) NINJA connects the co-repressor TOPLESS to jasmonate signalling. Nature 464(7289):788–791. https://doi.org/10.1038/nature08854

Pieterse CMJ, Poelman EH, Van Wees SCM, Dicke M (2013) Induced plant responses to microbes and insects. Front Plant Sci. https://doi.org/10.3389/fpls.2013.00475

Pillemer EA, Tingey WM (1976) Hooked trichomes: a physical plant barrier to a major agricultural pest. Science 193(4252):482–484. https://doi.org/10.1126/science.193.4252.482

Ponzio C, Gols R, Pieterse CMJ, Dicke M (2013) Ecological and phytohormonal aspects of plant volatile emission in response to single and dual infestations with herbivores and phytopathogens. Funct Ecol. https://doi.org/10.1111/1365-2435.12035

Prüm B, Florian Bohn H, Seidel R, Rubach S, Speck T (2013) Plant surfaces with cuticular folds and their replicas: influence of microstructuring and surface chemistry on the attachment of a leaf beetle. Acta Biomater. https://doi.org/10.1016/j.actbio.2013.01.030

Pungartnik C, da Silva AC, de Melo SA, Gramacho KP, de Mattos Cascardo JC, Brendel M, Micheli F, da Silva Gesteira A (2009) High-affinity copper transport and Snq2 export permease of saccharomyces cerevisiae modulate cytotoxicity of PR-10 from Theobroma cacao. Mol Plant-Microbe Interact 22(1):39–51. https://doi.org/10.1094/MPMI-22-1-0039

Quinn MA, Hower AA (1986) Effects of root nodules and taproots on survival and abundance of Sitona hispidulus (Coleoptera: Curculionidae) on Medicago sativa. Ecol Entomol. https://doi.org/10.1111/j.1365-2311.1986.tb00318.x

Raupp MJ (1985) Effects of leaf toughness on mandibular wear of the leaf beetle, Plagiodera versicolora. Ecol Entomol. https://doi.org/10.1111/j.1365-2311.1985.tb00536.x

Reymond P, Farmer EE (1998) Jasmonate and salicylate as global signals for defense gene expression. Curr Opin Plant Biol. https://doi.org/10.1016/S1369-5266(98)80264-1

Riddick EW, Simmons AM (2014) Do plant trichomes cause more harm than good to predatory insects? Pest Manag Sci. https://doi.org/10.1002/ps.3772

Rudrappa T, Czymmek KJ, Pare PW, Bais HP (2008) Root-secreted malic acid recruits beneficial soil Bacteria. Plant Physiol. https://doi.org/10.1104/pp.108.127613

Schafer M, Fischer C, Meldau S, Seebald E, Oelmuller R, Baldwin IT (2011) Lipase activity in insect Oral secretions mediates defense responses in Arabidopsis. Plant Physiol 156 (3):1520–1534. https://doi.org/10.1104/pp.111.173567

Schfer M, Meza-Canales ID, Navarro-Quezada A, Brtting C, Vankov R, Baldwin IT, Meldau S (2015) Cytokinin levels and signaling respond to wounding and the perception of herbivore elicitors in Nicotiana attenuata. J Integr Plant Biol. https://doi.org/10.1111/jipb.12227

Schilmiller AL, Charbonneau AL, Last RL (2012) Identification of a BAHD acetyltransferase that produces protective acyl sugars in tomato trichomes. Proc Natl Acad Sci. https://doi.org/10.1073/pnas.1207906109

Shang Y, Feng P, Wang C (2015) Fungi that infect insects: altering host behavior and beyond. PLoS Pathog. https://doi.org/10.1371/journal.ppat.1005037

Shiojiri K, Kishimoto K, Ozawa R, Kugimiya S, Urashimo S, Arimura G, Horiuchi J, Nishioka T, Matsui K, Takabayashi J (2006) Changing green leaf volatile biosynthesis in plants: an approach for improving plant resistance against both herbivores and pathogens. Proc Natl Acad Sci 103 (45):16672–11667. https://doi.org/10.1073/pnas.0607780103

Shulaev V, Silverman P, Raskin I (1997) Airborne signalling by methyl salicylate in plant pathogen resistance. Nature. https://doi.org/10.1038/385718a0

Simon M, Hilker M (2003) Herbivores and pathogens on willow: do they affect each other? Agric For Entomol. https://doi.org/10.1046/j.1461-9563.2003.00189.x

Spoel SH (2003) NPR1 modulates cross-talk between salicylate- and Jasmonate-dependent defense pathways through a novel function in the cytosol. Plant Cell Online 15(3):760–770. https://doi.org/10.1105/tpc.009159

Sugio A, Dubreuil G, Giron D, Simon JC (2015) Plant-insect interactions under bacterial influence: ecological implications and underlying mechanisms. J Exp Bot. https://doi.org/10.1093/jxb/eru435

Szemenyei H, Hannon M, Long JA (2008) TOPLESS mediates auxin-dependent transcriptional repression during Arabidopsis embryogenesis. Science 319(5868):1384–1386. https://doi.org/10.1126/science.1151461

Tack AJM, Dicke M (2013) Plant pathogens structure arthropod communities across multiple spatial and temporal scales. Funct Ecol. https://doi.org/10.1111/1365-2435.12087

Tanaka N, Matsuoka M, Kitano H, Asano T, Kaku H, Komatsu S (2006) gid1, a gibberellin-insensitive dwarf mutant, shows altered regulation of probenazole-inducible protein (PBZ1) in response to cold stress and pathogen attack. Plant Cell Environ. https://doi.org/10.1111/j.1365-3040.2005.01441.x

Ton J, Flors V, Mauch-Mani B (2009) The multifaceted role of ABA in disease resistance. Trends Plant Sci. https://doi.org/10.1016/j.tplants.2009.03.006

Tooker JF, de Moraes CM (2011) Feeding by a gall-inducing caterpillar species alters levels of indole-3-acetic and abscisic acid in Solidago altissima (Asteraceae) stems. Arthropod Plant Interact 5(2):115–124. https://doi.org/10.1007/s11829-010-9120-5

Trumble JT, Kolodny-Hirsch DM, Ting IP (1993) Plant compensation for arthropod herbivory. Annu Rev Entomol 38(1):93–119. https://doi.org/10.1146/annurev.en.38.010193.000521

Turley NE, Godfrey RM, Johnson MTJ (2013) Evolution of mixed strategies of plant defense against herbivores. New Phytol. https://doi.org/10.1111/nph.12103

ul Hassan MN, Zainal Z, Ismail I (2015) Green leaf volatiles: biosynthesis, biological functions and their applications in biotechnology. Plant Biotechnol J 13(6):727–739. https://doi.org/10.1111/pbi.12368

Van der Does D, Leon-Reyes A, Koornneef A, Van Verk MC, Rodenburg N, Pauwels L, Goossens A, Körbes AP, Memelink J, Ritsema T, Van Wees SC, Pieterse CMJ (2013) Salicylic acid suppresses Jasmonic acid signaling downstream of SCFCOI1-JAZ by targeting GCC promoter motifs via transcription factor ORA59. Plant Cell 25(2):744–761. https://doi.org/10.1105/tpc.112.108548

Van Loon JJA, De Boer JG, Dicke M (2000) Parasitoid-plant mutualism: parasitoid attack of herbivore increases plant reproduction. Entomologia Experimentalis et Applicata. https://doi.org/10.1023/A:1004032225239

Van Tol RWHM, Van Der Sommen ATC, Boff MIC, Van Bezooijen J, Sabelis MW, Smits PH (2001) Plants protect their roots by alerting the enemies of grubs. Ecol Lett. https://doi.org/10.1046/j.1461-0248.2001.00227.x

Vlot AC, Dempsey DA, Klessig DF (2009) Salicylic acid, a multifaceted hormone to combat disease. Annu Rev Phytopathol. https://doi.org/10.1146/annurev.phyto.050908.135202

Wagner GJ (1991) Secreting glandular Trichomes: more than just hairs. Plant Physiol. https://doi.org/10.1104/pp.96.3.675

Wei J, Wang L, Zhu J, Zhang S, Nandi OI, Kang L (2007) Plants attract parasitic wasps to defend themselves against insect pests by releasing hexenol. PLoS One 2(9). https://doi.org/10.1371/journal.pone.0000852

Wi SJ, Ji NR, Park KY (2012) Synergistic biosynthesis of biphasic ethylene and reactive oxygen species in response to Hemibiotrophic Phytophthora parasitica in tobacco plants. Plant Physiol. https://doi.org/10.1104/pp.112.194654

Wielkopolan B, Obrępalska-Stęplowska A (2016) Three-way interaction among plants, bacteria, and coleopteran insects. Planta. https://doi.org/10.1007/s00425-016-2543-1

Woolhouse MEJ, Webster JP, Domingo E, Charlesworth B, Levin BR (2002) Biological and biomedical implications of the co-evolution of pathogens and their hosts. Nat Genet. https://doi.org/10.1038/ng1202-569

Yang DH, Hettenhausen C, Baldwin IT, Wu J (2011a) BAK1 regulates the accumulation of jasmonic acid and the levels of trypsin proteinase inhibitors in Nicotiana attenuata's responses to herbivory. J Exp Bot 62(2):641–652. https://doi.org/10.1093/jxb/erq298

Yang JW, Yi HS, Kim H, Lee B, Lee S, Ghim SY, Ryu CM (2011b) Whitefly infestation of pepper plants elicits defence responses against bacterial pathogens in leaves and roots and changes the below-ground microflora. J Ecol. https://doi.org/10.1111/j.1365-2745.2010.01756.x

Zebelo SA, Maffei ME (2015) Role of early signalling events in plant-insect interactions. J Exp Bot. https://doi.org/10.1093/jxb/eru480

Zimmermann MR, Maischak H, Mithofer A, Boland W, Felle HH (2009) System potentials, a novel electrical Long-distance Apoplastic signal in plants, induced by wounding. Plant Physiol 149 (3):1593–1600. https://doi.org/10.1104/pp.108.133884

Simplified Perspective of Complex Insect–Plant Interactions

Anupam Varshney Sharma and Vachaspati Mishra

Abstract

Scientific literature pertaining to the investigations on insect–plant interactions spans more than a century. This is a challenging frontier area today as it was for the pioneers, and it would continue to be so for researchers in their pursuit to help elucidate the complex relationship between the insects and plants. Despite the ready availability of exhaustive literature on this subject, the mechanisms of insect–plant interactions are still not completely understood. Insect–plant interaction is an extremely rich subject that transcends several disciplines of science and has far-reaching implications, especially in the management of ecosystem and crop protection. The interaction between pests and plants starts at the interface of plasma membrane and in response to perception of a pest and release of herbivore-associated molecular patterns (HAMPs); plants respond quickly by setting up the electrical signalling followed by depolarization of membrane, leading to increase in Ca^{2+} ion concentration and activation of calcium-sensing proteins. Further, this interaction is primarily governed by various signalling mechanisms, such as mitogen-activated kinase (MAP-kinase), jasmonic acid (JA), salicylic acid (SA) and ethylene (ET)-based pathways that regulate changes in gene and protein expression leading to synthesis of defensive compounds. Plants defend themselves not only by direct means but also by indirect means, wherein plants emit volatiles to attract natural enemies of the herbivores. Herein, we summarize the molecular and ecological aspects of complex insect–plant interactions to enable researchers to direct their course of action towards addressing them for making a meaningful contribution in this field, which will

A. V. Sharma (✉)
Department of Zoology, Hindu College, University of Delhi, Delhi, India
e-mail: avsharma@hindu.du.ac.in

V. Mishra
Acharya Narendra Dev College (University of Delhi), New Delhi, India

I. K. Singh, A. Singh (eds.), *Plant-Pest Interactions: From Molecular Mechanisms to Chemical Ecology*, https://doi.org/10.1007/978-981-15-2467-7_16

have far reaching implications in the success of insect pest management programs.

Keywords

Insect–plant interactions · Co-evolution · Chemical ecology · Insect pest management

1 Introduction

Studies on insect–plant interactions comprise an immensely exciting area of research in the field of Life Sciences. It has been built upon the strong foundational work of the pioneers in this area, and has motivated and challenged the biologists to undertake further studies to gain insights especially pertaining to its wider ecological context. Various types of insect–plant interactions have been reported in scientific literature, and these are believed to be dated approximately 420 million years before present (Devonian period in the geological scale). And, subsequent to the appearance of flowering plants (angiosperms) nearly 125 million years ago, these interactions became more complex (Calatayud et al. 2018). This topic has assumed greater significance, especially in respect of the vital role played by insects in pollination, and their other valuable services to the ecosystem. In fact, the earliest studies of insect–plant relationships from about 200 years before present have elucidated the significant role of insects as the major pollinators.

As per the scientific literature, approximately half of all known species of multicellular organisms on Earth are the plants and insects. It is believed that the present day phytophagous insects and the plants they feed on are, in fact, the outcome of the process of co-evolution that started about 400 million years before present, because of which varying degrees of insect–plant interactions have been recorded, and described later in the chapter. One of the major interactions between insects and plants is that of *mutualism*, as exemplified by the well-known aspect of entomophily, and it constitutes a major evolutionary advancement in insect–plant interactions, leading to diversification of insects and plants (Labandeira et al. 2007; Peñalver et al. 2012). The other kind of insect–plant relationship is *antagonistic*, in which the former feeds either completely or partially on the latter. It is quite detrimental to the host plant but beneficial to the insect and is commonly referred to as herbivory. Finally, the extreme kind of insect–plant interactions is *insectivory* wherein many examples of carnivorous plants that consume insects have been reported (Renner and Specht 2013).

The perusal of scientific literature on insect–plant interactions reveals that this subject transcends boundaries of almost all scientific disciplines. Investigations of such interactions, in particular between crops and herbivorous insects, have led to greater understanding of plant chemistry, insect physiology and behaviour and ecology. The process of co-evolution between insects and plants has played a very important role in the development and selection of characters, facilitating their adaptive radiation and evolution. These characters are being studied today in every branch of Biology, from Entomology, Insect behaviour, Biochemistry, Genetics, Molecular Biology and Ecology to Evolutionary Biology. The understanding of insect–plant interactions has been further strengthened with the unravelling of diverse molecular processes regulating the entire system. Molecular evidences of involvement of several genes and proteins have further structured and fine-tuned the way these interactions were perceived (Erb and Reymond 2019). Recently, the ability to establish molecular phylogenies with the help of latest technological advancements has renewed the interest of researchers to comprehend the co-evolution of plants and insects. Significant advances in the recent past have opened enormous possibilities for plant genetic engineers. Above all, the enormous economic losses in agricultural ecosystems and serious implications on food security due to excessive damage caused by phytophagous insects pests have drawn considerable attention of researchers to this area of research (Bruce 2010). According to Lipke and Fraenkel (1956), the study of insect–plant relationships constitutes '*the very heart of agricultural entomology*'. Studies on insect–plant interactions assume a lot of importance in relation to nature conservation in wild habitats, and it is well known that the co-evolutionary forces involved therein drive speciation and increased biodiversity (Bruce 2015). Further research demands studies aimed at comprehending the molecular mechanisms of insect–plant interactions. This will invariably unfold many unexplored mysteries of novel molecules that may have tremendous applications in agricultural and medicinal sectors.

The major question that drives the research on insect–plant interactions is ***"What factors govern the interactions between the insects and plants?"***

Green plants cover most of the terra firma on planet Earth. Having endowed with the ability to harness solar energy through the process of photosynthesis, green plants have acquired the very special status of 'primary producers' in the living world. It is these plants that sustain all other trophic levels in the intricate food-webs on planet Earth, and they serve as major conduits via which the energy flows from one trophic level to the other. At the same time, phytophagous insects comprise one of the major groups of 'primary consumers', and form an important connecting link between the primary producers and secondary consumers (particularly those who feed on these insects) in the terrestrial ecosystem. In other words, the phytophagous insects perform an extremely invaluable ecosystem service not only to all other living organisms, but to the plants as well. And, the extent of diversification of flowering plants that we see today (approximately 275,000 species) can be attributed to adaptive radiation, believed to be a necessary prerequisite for co-evolving with their insect herbivores (Yuan et al. 2013). Fossil records indicate that pollination originated about 250 million years before present (Labandeira 2013). Similarly, the enormous diversity of phytophagous insects which represent nearly a quarter of all terrestrial macroscopic biodiversity can be attributed to their generally specific association with angiosperms (Kergoat et al. 2017).

1.1 What Do the Insects Need the Plants for and How Do They Reach Their Host Plants Amongst the Other Species of Plants in the Diverse, Dynamic and Complex Ecosystem?

The phytophagous insects are under great selection pressure to find quality hosts and enhance their fitness by selecting the suitable plants from the unsuitable ones in the 'dynamic ecosystem' stated above (Bruce 2015). Pioneering works by Jean-Henri Fabre have provided vital information on insect behaviour and ecophysiology, and chemical communication in insects (Fabre 1879). Similarly, important aspects related to insect olfaction were revealed in the research studies by Viallanes (1887). The sensorial perception in insects inside their environment was first described in these studies (von Frisch 1953). That the plant chemicals play a very crucial role in governing their interactions with insects was revealed in the exhaustive studies by Dethier (1941). Saxena (1969) was the first to describe the phases of an insect's establishment on the plant. The different steps of an insect in search of its host plant were described by Bell (1990).

Insects tend to approach plants for procuring food (plants being a very good food/energy resource) and/or shelter for themselves or for securing a comfortable site for the development of their progeny and subsequent establishment of their population. But in most of the cases, the primary reason seems to be the procurement of nutritious food (which is directly available from the 'nature's kitchen' or 'ready-to-eat food served on the platter', so to say!) for their own growth, development and survival to adulthood (sexual maturity) so as to be able to perpetuate their species.

1.2 Adaptive Radiation of Phytophagous Insects

The phytophagous insects comprising nearly 43% of all extant insect species primarily belong to the five major orders: Lepidoptera, Coleoptera, Hemiptera, Orthoptera, and Hymenoptera, three of these being holometabolous or endopterygotes. Quite diverse and specialized types of feeding habits of these insects have been recognized (in relation to their mouthparts and ecological niches); and these range from foliage feeding, leaf mining, seed-eating, fruit-eating, stem-boring, root feeding, detritus feeding, etc. It is very pertinent to draw the attention of the reader to a very important aspect of feeding behaviour in relation to the morphological and anatomical modifications in the mouthparts of the larval and adult stages of a holometabolous insect species. This enables them to exploit entirely different food resources compatible with their distinct mouthparts. From the evolutionary point of view, it has led to significant reduction in their competition (even with their own progeny) for food and ecological niches. In other words, the holometabolous insects are better able to exploit the mutually exclusive food resources (compatible with their mouthparts) to establish themselves in their respective feeding zones, with absolutely no conflict of interest with the feeding preferences of each other. Moreover, there is a quiescent non-feeding stage (lasting for nearly 4-7 days) in the life cycle of such insects, giving them sufficient time to remodel their feeding apparatus and develop wings and other necessary structures to enable them to confront the new challenges as they embark on their journey outside of the protective puparia. It is from these aspects pertaining to the adaptive significance of the anatomical modifications in the mouthparts of insects, their behaviour and life cycle that we can draw some critical inputs, and incorporate them in pest management programs in a very effective manner.

1.3 Characteristics of Plants Governing the Interaction with Phytophagous Insects

The plants also interact with insects in many ways, and as such many of the physical or chemical traits of the plants seem to have co-evolved in relation to the beneficial or harmful/destructive insects. And, for utilizing the beneficial services of insects which mainly act as pollinating agents, guard the plants against other herbivores, provide ready source of fertilizers (after their death) for plants, etc., the plants need to exhibit certain attractive characteristics (visual and/or olfactory stimuli). In evolutionary terms, it is due to the selection of those traits which favoured the beneficial interaction between the two. It needs to be emphasized here that the reproductive fitness of plants is increased by attraction of pollinators (Haverkamp et al. 2016). According to Darwin, this perfect mutual adaptation of flowers and pollinators was the result of an evolutionary process, more aptly referred to as co-evolution. The intricacies of co-evolution between insects and plants have been described in the classic review by Ehrlich and Raven (1964). In respect of mutualism between plants and pollinating insects, it was emphasized by Crane et al. (1995) that diversification of flowers constitutes a feature that in turn may drive diversification of pollinating insects, allowing the radiation of angiosperms and of pollinating insects. More recent research studies on pollinators and co-evolution are being carried out in Bill Hansson's lab in the Max Planck Institute for Chemical Ecology, Jena, Germany (Haverkamp et al. 2016).

In a manner contrasting to that stated above, adverse physical traits (such as the spines, trichomes, waxy cuticle, etc.) and chemicals (repellents, antifeedants, growth inhibitors, oviposition deterrents, etc.) develop in plants as defence mechanism/s against insect damage. While phytophagous insects have been adapting to exploit their host plants, the plants have simultaneously been evolving defensive mechanisms to counteract herbivore attack (Anderson and Mitchell-Olds 2011; Johnson 2011). Interestingly though, in insect–plant mutualistic pollination relationship, plants at the molecular level have developed tissue-specific diurnal rhythmicity in the secretion of their secondary metabolites to regulate the attraction of pollinators, on one hand, and indirectly deter the larvae of the pollinators from feeding on the plant, on the other hand. In *Nicotiana attenuata*, *TPS 38* gene produces (*E*)-α-bergamotene which when expressed at night time attracts *M. sexta* but their expression in leaves at day time attracts the predators of its larvae (Zhou et al. 2017). According to Pedigo (1999), the ancestors of modern-day crop plants accumulated through natural selection, many physical and chemical traits that comprise a wide range of sophisticated defence systems against phytophagous insects during the process of their co-evolution with them.

Plants respond to herbivory through various morphological, biochemical and molecular mechanisms to counter/offset the effects of herbivore attack. The biochemical mechanisms of defence against the herbivores are wide-ranging and highly dynamic. The defensive compounds may either constitute an integral part of the chemical defence system of the plant or they may be produced in response to the damage caused to its tissues by herbivores. Such compounds act by virtue of their ability to affect feeding, growth and survival of herbivores. In addition, plants also release volatile organic compounds (VOCs) which serve to attract the natural enemies of the herbivores, thus augmenting their protection from herbivores. These strategies may either act one at a time or simultaneously in conjunction with each other. More research needs to be carried out to precisely understand such mechanisms. Many more relevant examples can be obtained from the existing literature on the subject, but they have not been reproduced here with changing space and time.

If we look at the entire case from the insects' perspective, their counterargument seems to be that it is not as easy a task to directly consume the 'ready-to-eat food served on the platter' (as stated above). It is true that the insect herbivores face many problems associated with feeding on plants that are not experienced by other herbivores, scavengers living in soil, or by the predators. Some of the major problems faced by the phytophagous insects are stated below:

1. Insects need to secure attachment on most suitable part of the host plant before they can initiate feeding on plant tissues, and thus they need to overcome many physical hurdles such as spines, trichomes, slippery surface (due to protective waxy coating on foliage), etc.
2. Insects are greatly vulnerable to desiccation as they remain exposed on plant tissue while feeding, and/or oviposition.
3. Vegetarian diet (excluding seeds) is nutritionally inferior in proteins, sterol and vitamin content in comparison to the diet from animal or microbial origin; hence, feeding on plant tissues is quite a costly proposition for phytophagous insects.
4. Above all, it needs to be remembered that the plants are not passive 'acceptors' of herbivory. They have rather developed physical/morphological (spines, spicules, trichomes, waxy cuticle, etc.) and chemical (repellents, feeding deterrents, antifeedants, growth inhibitors, oviposition deterrents, etc.) defence mechanisms to counteract herbivory. Thus, the phytophagous insects need to fight out the challenges posed by plants, besides facing the adverse climatic conditions and their natural enemies in the field, to earn their food (bread and butter!). It can now be well imagined what kind of hurdles on the way to plant feeding have been successfully overcome by the ancestors of the present day insects, before adaptive radiation and speciation.

1.4 Historical Account of Insect–Plant Interactions

One of the latest Oxford Bibliographies on Insect-plant Interactions by Calatayud et al. (2018) is an exhaustive account of publications on all aspects related to insect–plant interactions, and it has served as the base for the author to incorporate the recent findings on the topic in this chapter. Broadly speaking, there are three main types of insect–plant interactions, viz., *mutualism*, *antagonism* or *commensalism*. In *mutualism*, both the partners associated in the interaction benefit from each other and neither gets harmed. For instance, entomophily or pollination by insects, guarding of plants by insects or seed dispersal (e.g., plant/ant systems) exhibits mutualism. In *antagonistic relationships*, one counterpart benefits and the other harmed in the process, as exemplified by herbivory (especially in relation to severe damage caused by insect pests). At the same time some insectivorous or carnivorous plants such as Venus flytrap or *Dionaea* sp., pitcher plant or *Nepenthes* sp., etc. have also been reported in the literature. In *commensalism*, one counterpart benefits but the other is not harmed. The relationship of the larvae of Monarch butterfly with certain species of milkweeds to store cardiac glycosides for defensive purposes is one of the best examples of commensalism (Calatayud et al. 2018).

A brief account of the pioneering studies that laid a strong foundation for the modern studies on insect–plant interactions is stated below. Hopkins (1916) was amongst the great pioneers to explain the mechanism of host plant fidelity in phytophagous insects. It was Snelling (1941) who first defined the defence mechanisms of plants against herbivorous insects. Painter (1951) first categorised the mechanisms of host plant resistance as *Antixenosis*, *Antibiosis* and *Tolerance*. The role and importance of secondary plant metabolites in insect–plant interactions was first described by Fraenkel (1959). Details pertaining to the sensory physiology of the gustatory systems of insects were provided by the studies by Schoonhoven and Dethier (1966). May R. Berenbaum and Art R. Zangerl at the University of Illinois are regarded amongst the first Major scholars of insect–plant interactions in the early twenty-first century in the area of chemical ecology of insect–plant interactions/detoxification of plant defences/co-evolution. And, one of the most renowned scientists having carried out tremendous research on plant–herbivore interactions in the 1980s was Elizabeth (Liz) Bernays at the University of Arizona. Finally, dame Miriam Rothschild is renowned as the grand-old lady in caterpillar–plant interactions. The honour of publishing the first synthesis of insect–plant interactions goes to Fritz and Simms (1992).

The complex relationships between phytophagous insects and plants are the result of a long common evolution. The appearance and evolution of terrestrial plants have strongly influenced evolution of insects; similarly, pressures exerted by insects on plants contributed towards diversification of plants. Recently, Schatz et al. (2017) showed how fossil records, which contain different types of vital indicators, have enabled the reconstitution of the history of relationships between plants and insects. Exhaustive studies on fossil plant–insect associations by Labandeira (2013) suggest that insects have been feeding on plants for the past 400 million years. Studies by Nel (1997) show that the morphological characteristics of buccal pieces of fossilized insects provide direct evidence of feeding habits as shown in the traces of phytophagy on fossilized plants (Labandeira and Sepkoski Jr 1993). Exhaustive studies by Labandeira (2006) and Schatz et al. (2017) provide details about different steps and periods of the evolution of insect–plant interactions, and the main distinct phases of evolution of interactions between arthropods and plants.

The history of insects' diversification is inseparable from that of plants. During the evolution of these organisms, the gradual diversification of plants led insects to develop different adaptations at various complexity levels to utilize the plants in spite of the barriers they built to resist their aggressions. The major adaptive strategies of phytophagous insects to combat plant defence mechanisms have been discussed in the references stated below:

1. Morphological (Bernays et al. 1991).
2. Chemical (presence of repellents, antifeedants, oviposition deterrents, etc.) (Varshney 1998; Varshney and Singh 2000; Varshney et al. 2003, 2007).
3. Behavioural (cf. host/resource selection, Thompson and Pellmyr 1991).
4. Physiological (cf. synchronization of reproductive cycle between insects and plants, (Seger and Brockmann 1987). Initially, insect strategies for plant exploitation certainly focused on ensuring insect reproduction and development (van Veen et al. 2006).
5. Cooperation with other organisms, in particular conspecifics and microorganisms, using the same plant resource for the benefit of both users (Paine et al. 1997).
6. Optimal utilization of nutrients from their host plants to ensure their protection from adverse abiotic and biotic conditions (natural enemies, competitors) by adopting different strategies:
   (a) Manipulation of plants to reorient their metabolisms to insect's needs (e.g., effectors (Hogenhout and Jorunn 2011), and plant growth regulators or phytohormones (Zhang et al. 2016).
   (b) Remodelling of the host plant ranging from its ultrastructure to anatomy, thereby affecting its primary and secondary metabolites (Lieutier et al. 2017).
   (c) Utilization of its toxic chemical constituents by the plant as a formidable defence against predators (Reichstein et al. 1968; Duffey 1980).

2 Important Considerations in Relation to the Hurdles on the Way, and Prospects of Future Studies on Insect–Plant Interactions

Studies on insect–plant interactions have far-reaching implications, especially with respect to the applied aspects. The phytophagous insects being the chief pests of crops and stored products need to be more strategically managed as they seem to have nullified all the available pest control measures so far. It is now extremely important to comprehend and target the nuances of the complex relationships between insects and plants that facilitate the development of pests. Such knowledge assumes great significance for someone interested to create biologically safe control strategies for preventing insect pest outbreaks (Schoonhoven et al. 2005).

Besides the fascinating aspects of studies on insect–plant interactions, another major factor which drives such studies is the serious concern of agricultural scientists for tremendous economic loss due to agricultural pests, with annual worldwide crop losses reported to be hundreds of billions of dollars (Kerchev et al. 2012). Although modern advances in genetic engineering and genomics have opened up unlimited possibilities for inserting specific traits for host plant resistance into crops and for teasing apart complex host plant resistant traits (Gordon and Waterhouse 2007; Prado et al. 2014) yet it needs to be noted that widespread planting of transgenic crops, such as those expressing *Bacillus thuringiensis* Berliner (Bt) endotoxins, creates strong selection pressure on pest populations to overcome the resistance factor (Follett 2017) and would turn all such efforts redundant in near future.

Several studies have demonstrated that research into novel aspects of insect–plant interactions (tritrophic interactions) may provide improved alternatives for controlling insect pest populations. For instance, Tumlinson et al. (1993) have reported that following the damage caused due to feeding by moth larvae, corn plants release a new complex of odorants into the air. Some of these odorous chemicals seem to attract the parasitic wasps, which in turn parasitize these larvae. Such strategies may help augment the pest management programs.

One of the very significant implications arising from the studies on insect–plant interactions could be using insects as biological control agents for weeds. Some insects have been reported to feed voraciously on weeds, and a classical example of cactus moth or prickly pear moth from Argentina, *Cactoblástis cactòrum* (Berg) employed to control the noxious weed or the prickly pear cacti (*Opuntia* sp.) in Australia, has been well documented in the literature. Similar success stories of biological control of weeds by this moth in other parts of the world such as South Africa and some Caribbean islands have also been reported by Kevan and Shipp (2011).

The research findings demonstrating the role of plant chemicals/volatiles in attracting or repelling insects can be put to use to manage crop pests in a more strategic and effective manner. Follett (2017) has reported that volatiles emanated in the air close to the plant surface, may act as repellents for insects and affect host selection by phytophagous insects. In such cases, intercropping of host plants with highly aromatic non-host plants may interfere with host-locating mechanism of the insects, thereby acting as an effective strategy to prevent pest attack. It has been successfully demonstrated by Carvalho et al. (2017) that intercropping of tomato plants with basil and coriander can help protect it from the attack and damage by whitefly, *Bemisia tabaci* (Gennadius) (Hemiptera: Aleyrodidae).

3 Important Considerations on Molecular Mechanisms Underlying Herbivory and Insect Adaptations

With the growing population as ever, it is highly important to focus on crop production improvements through safe and sustainable means by reducing our reliance on pesticides. In this direction, it is important to learn molecular aspects of plant defence mechanisms and insect herbivory. Herbivory is the most important biotic stress factors in plant–insect interaction that determines early establishment of the plants. A proper understanding of the adaptations by insect pests to the plant defensive traits will equip us better to develop and deploy management strategies to outsmart the insect pests (War et al. 2018). During the ongoing process of insect–plant co-evolution, there occurs a tug of war between plants and herbivorous insects that manifest through the release of chemicals from the insects as well as the host plants that act as weapons and messengers both. This interaction is intricate, and one of its major outcomes is the development of myriad of morphological and biochemical defensive traits in plants and the counter-adaptive features in insects as well (War et al. 2018). The biochemical interactions are dynamic in nature and hence find better address in experimental analysis in current literature. The various plant defensive traits evolved during this intricate interaction include toxic furanocoumarins, toxic amino acids, trichomes, lignin and latex (War et al. 2018). Furthermore, known diverse molecular interactions also exist that process plant and insect herbivory; and it is generally perceived that herbivore-associated and damage-associated molecular patterns (HAMP & DAMP) find their expression primarily in this process, effecting their expression via receptors associated with calcium, reactive oxygen species, MAP kinase cascade (Erb and Reymond 2019) and octadecanoid and phenylpropanoid pathways mediated by jasmonic acid (JA) and salicylic acid (SA), respectively (Scott et al. 2010; Zhao et al. 2009). A number of plant defensive secondary metabolites in the intermediate steps of these pathways influencing insect growth and development are produced and these steps also release volatiles that attract insect's natural enemies (He et al. 2011).

The identification of MAP kinase and others in triggering herbivory has augured some interesting findings pertaining to the roles of signal transduction pathways on the expression of related defence genes, leading to activation of related biosynthesis pathways (Howe and Jander 2008; Maffei et al. 2012). Kinases are activated by plants through plant defence elicitors from herbivores and the defensive process is initiated which involves also phytohormones (Maffei et al. 2012; Pieterse et al. 2012). Some important plant signalling phytohormones are JA, SA and ethylene (ET). The signalling pathways against chewing insects and phloem-feeding insects are mediated by JA and SA (Howe and Jander 2008; Pieterse et al. 2012), while in rice plants, resistance to leaf folder *Cnaphalocrocis medinalis* is mediated by SA and ET signalling pathways (Wang et al. 2011). Plant infestations by sucking insect pests get isochorismate and phenylalanine ammonium lyase pathways activated, resulting in synthesis of SA (Dempsey et al. 2011). The SA accumulation in plant tissues activates translocation of the non-expressor protein of the pathogenesis-related genes 1 (*NPR1*) to the nucleus, downstream of which is located the SA-responsive genes.

The WRKY transcription factor genes are targeted by NPR1 while interacting with TGA-type transcription factors and leading to upregulation of pathogenesis-related (PR) proteins (Durrant and Dong 2004). Insect eggs are also known to activate SA signalling pathway (Reymond 2013; Erb and Reymond 2019). Thines et al. 2007 have discussed the mechanism of transcriptional activators MYC2 action leading to repression of JA signalling pathway. This pathway operates through two mechanisms, viz., one, MYC2, that regulates defence against insect herbivores and two, the ethylene response factor (ERF), that regulates plant defence against necrotrophic pathogens (Pieterse et al. 2012). In general, it is found that multiple herbivore attack triggers crosstalk between signalling pathways to induce specific responses against insect herbivores (Pieterse et al. 2012). The crosstalk between JA and SA occurs antagonistically; and the process is mediated by the MAPKs, WRKY, NPR1 and ET (Pieterse et al. 2012). These crosstalks between signalling pathways benefit herbivory.

Herbivores rescind plant defences by deactivating their biochemical defence, suppressing effector-mediated defence signalling, and behavioural changes controlled chemically (Erb and Reymond 2019). The molecular mechanisms of plant–herbivore interactions are currently well known for some model systems, and hence this approach needs to be expanded to the unexplored areas of plant–insect interactions in future researches (Erb and Reymond 2019).

In spite of so many useful implications of the studies on insect–plant interactions described above, many challenges still remain to be overcome. And, one of the key challenges is to develop durable pest control strategies in relation to the local ecological context of the pest and crop species. For this, one must understand the dynamics of insect responses in relation to numerous selection pressures an insect species is subjected to. These pressures mainly include the physical and chemical defences of the plant, environmental factors, presence of predators and natural enemies, etc.

Investigations pertaining to insect–plant interactions so far have been mainly carried out in the controlled conditions in many state-of-the-art laboratories across the globe. Great efforts have been made to simulate the near-field conditions in the labs, yet there are many lacunae that need to be addressed in future studies, especially those pertaining to recording of insect behaviour in response to the factors governing it. It needs to be emphasized here that the insect responses to cues (from its surroundings) are not fixed. They rather depend on the context in which such cues are perceived by the insect, its prior exposure to them, as well as the state of its physiology. Bruce (2015) has also indicated that space and time play crucial roles in influencing the outcome of interactions between insects and plants. For example, prior exposure of the plant to the attack by an insect species would alert and augment the plant defence mechanisms to thwart the subsequent insect attack (Ton et al. 2007; Jinwon et al. 2011). This presence of other entities associated with the plant, viz. other insects, plant pathogens or mycorrhizal fungi, can also alter the responses of insects and plants towards each other. Since the plant phenotype is plastic and can be changed by epigenetic factors, it enables the plant to adapt itself and cope up with the number of challenges during periods of biotic stress.

Unfortunately, in most of the studies, efforts have been made or are being made to solve the complex puzzle of insect–plant relationships in isolation, giving extreme importance to a single factor or parameter, while negating the significance of many other factors which may be working in conjunction with each other for eliciting an entirely different response from the experimental insect in the field as against that observed in the fine-tuned experimental conditions in the lab. This eventually does more harm than good and seriously affects the outcome of pest management programs.

Exhaustive studies by Bruce et al. (2005) and Bruce and Pickett (2011) have revealed the way in which insects use the volatiles, which usually emanate from their host plants as *blends* of commonly occurring volatile chemicals in specific combinations or ratios. That the time dimension is of major significance in eliciting insect responses has been emphasized by Bruce (2015). This is because the type of behavioural response elicited in the insect can very much change depending on whether or not odours impinge simultaneously on the chemoreceptors located on its antennae. The specific combinations of the chemical constituents in the 'blend' play a crucial role in eliciting the orientational responses of the black bean aphid, *Aphis fabae*, as demonstrated by Webster et al. (2010). These odours, when presented individually in an olfactometer, were repellent but when put together as a blend proved to be attractive to the aphid. Han et al. (2012) extended such studies and have established that a combination of olfactory and visual cues can further enhance attraction.

Many studies including that of Maffei et al. (2007) have established that the plant defences are synchronized in both time and space by highly complex regulatory networks that can be themselves modulated by interactions with other signalling pathways. Plant defences can be *constitutive* or *induced*. Time plays a crucial role in determining whether defences are induced or primed (Conrath et al. 2006; Ton et al. 2007; Bruce et al. 2007). Those plants that are subjected to insect attack for the first time (or are primed) respond more quickly and strongly during the subsequent attack (Ton et al. 2007; Jinwon et al. 2011). All these strategies work towards the efficient allocation of metabolites and energy to defensive activities by the plant, especially when it is able to recognize the herbivore challenge, and adaptively modulate its metabolism at the most appropriate time (Mithoefer and Boland 2012). According to the findings of Agrawal et al. (2012), there is rapid evolution of traits that confer resistance to herbivores when herbivores are present. On the other hand, it leads to the evolution of traits that confer increased competitive ability when herbivores are absent.

4 Conclusion

In light of the significant findings and our observations stated above, it needs to be taken into serious consideration that the major factors governing insect responses in the field are in relation to the various factors that an insect population is subjected to, such as the day-night regime, environmental factors prevailing at that particular space and time, physiological state of the insect, morphological and chemical factors of the host plant (which is itself surrounded by a complex maze of other plants), and above all the competition (for food and space) from other members of its own species and the continuous threat from its natural enemies/predators in the wild ecosystem. It is quite clear that the phytophagous insects face not only a range of plant defences that are heterogeneously distributed through time and space, but also severe selection pressures from their predators, parasitoids and competitors in their natural environment. As a result, the degree and nature of their responses to plant factors vary depending on the geographical distribution of plant phenotypes, the specificity of the insect–plant association and the biotic and abiotic factors in the ecosystem of the insect. In other words, the selection pressures imposed by the dynamic nature of the ecosystem to which a particular insect species has evolved its survival strategies during the course of its long evolutionary lifespan can in no way be replicated in the laboratory conditions; and as such, it is extremely difficult to unravel the mechanisms governing insect–plant interactions if we continue to depend on research findings from lab studies alone. It thus becomes imperative to extend the studies carried out under controlled conditions in the lab to the field which would yield more realistic and vital information for devising successful pest management programs in future. Since field investigations demand immense hard work, patience, continuous observations, a lot of commitment and dedication on part of the researchers to carry out exhaustive research with a holistic approach to better elucidate the complex insect–plant interactions, field studies have not found much-needed initiative from the young researchers. The important review by Harris et al. (2015) has emphasized upon the crucial role of **curiosity-driven fundamental research** in providing novel insights to drive methods for crop protection and how breeding for insect-resistant crops can lead to the development of a greater mechanistic understanding of the plant–insect interaction. One of the latest studies by Giron et al. (2018) has also emphasized that the future studies should work towards addressing the major challenge of working beyond the model systems and to transfer laboratory-based knowledge into natural ecosystems, including in multispecies interactions. Another important challenge is to develop a reliable and feasible framework to integrate the diversity of factors that modulate insect–plant interactions, and these range from multiple biotic and abiotic factors, in relation to the other players in the field, i.e. the beneficial and detrimental micro- and macroorganisms, along with the defence mechanism/s of the plant itself.

Another very important aspect to be noted here is that the various research teams involved in such investigations on insect–plant relationships are themselves separated in time, space, and more importantly, whether the researchers in this field have access to the necessary technical and financial support and the conditions they need to be working in. In a major way, all these factors greatly affect the outcome of research studies besides directing the future course of action for budding researchers. So, it calls for taking a pause and to ponder, and to reflect upon the lacunae in the previous studies. This, in turn, will tend to empower us to comprehend the complex insect–plant relationships, which in turn would help devise smarter strategies for the management of these insect pests (tiny yet mighty creatures), to our advantage.

Acknowledgements AVS gratefully acknowledges the guidance and mentorship of her Ph.D. guide, Prof. A.K. Singh (Retd.) at the Insect Behavior Laboratory, Department of Zoology, University of Delhi, India. Valuable inputs in respect of the molecular aspects of insect–plant interactions by Dr Anjana Singha Naorem, Assistant Professor, Cotton University, Assam, are gratefully acknowledged. She also expresses her thanks and gratitude to Dr. Anju Srivastava, Principal, Hindu College, University of Delhi, for providing all the necessary support and encouragement during the course of writing this chapter. VM is thankful to the colleagues who helped in providing literature resources for the article.

References

Agrawal AA, Hastings AP, Johnson MT, Maron JL, Salminen JP (2012) Insect herbivores drive real-time ecological and evolutionary change in plant populations. Science 338:113–116

Anderson JT, Mitchell-Olds T (2011) Ecological genetics and genomics of plant defences: evidence and approaches. Funct Ecol 25:312–324

Bell WJ (1990) Searching behavior patterns in insects. Annu Rev Entomol 35:447–467

Bernays EA, Chapman RF (1994) Host plant selection by phytophagous insects. Chapman and Hall, New York

Bernays EA, Jarzembowski EA, Malcolm SB (1991) Evolution of insect morphology in relation to plants. Philos Trans R Soc B Biol Sci 333:257–264

Bruce TJA (2010) Tackling the threat to food security caused by crop pests in the new millennium. Food Sec 2:133–141

Bruce TJA (2015) Interplay between insects and plants: dynamic and complex interactions that have coevolved over millions of years but act in milliseconds. J Exp Bot 66:455–465

Bruce TJA, Pickett JA (2011) Perception of plant volatile blends by herbivorous insects – finding the right mix. Phytochemistry 72:1605–1611

Bruce TJ, Wadhams LJ, Woodcock CM (2005) Insect host location: a volatile situation. Trends Plant Sci 10:269–274

Bruce TJ, Matthes MC, Napier JA, Pickett JA (2007) Stressful memories of plants. Plant Sci 173:603–608

Calatayud P, Sauvion N, Thiéry D (2018) Plant-insect interactions- ecology-Oxford bibliographies. https://doi.org/10.1093/OBO/9780199830060-0193

Carvalho MG, Bortolotto OC, Ventura MU (2017) Aromatic plants affect the selection of host tomato plants by *Bemisia tabaci* biotype B. Entomol Exp Appl 162:86–92

Conrath U, Beckers GJM, Flors V, Garcia-Augustin P, Jakab G, Mauch F, Newman M-A, Pieterse CMJ, Poinssot B, Pozo MJ, Pugin A, Schaffrath U, Ton J, Wendehenne D, Zimmerli L, Mauch-Mani B (2006) Priming: getting ready for battle. Mol Plant-Microbe Interact 19:1062–1071

Crane PR, Marie EF, Pedersen KR (1995) The origin and early diversification of the angiosperms. Nature 374:27–33

Darwin CR (1862) On the various contrivances by which British and foreign orchids are fertilized by insects, and on the good effects of intercrossing. John Murray, London

Darwin CR (1899) The various contrivances by which orchids are fertilized by insects, 2d edn. John Murray, London

Dempsey DA, Vlot AC, Wildermuth MC, Klessig DF (2011) Salicylic acid biosynthesis and metabolism. The Arabidopsies Book 9:e0156

Dethier VG (1941) Chemical factors determining the choice of food plants by *Papilio* larvae. Am Nat 75:61–73

Duffey SS (1980) Sequestration of plant natural products by insects. Annu Rev Entomol 25:447–477

Durrant WE, Dong X (2004) Systemic acquired resistance. Annu Rev Phytopathol 42:185–209

Ehrlich PR, Raven PH (1964) Butterflies and plants: a study in coevolution. Evolution 18:586–608

Erb M, Reymond P (2019) Molecular interactions between plants and insect herbivores. Annu Rev Plant Biol 70:527–557

Fabre J-H (1879) Etudes sur l'instinct et les moeurs des insectes. Souvenirs Entomologiques. Librairie Ch. Delagrave, Paris

Follett PA (2017) Insect-plant interactions: host selection, herbivory, and plant resistance – an introduction. Entomologia Experimentalis et Apeplicata Special Issue – Insect-Plant Interactions: Host Selection, Herbivory, And Plant Resistance, 162: 1-3

Fraenkel GS (1959) The raison d'être of secondary plant substances. Science 129:1466–1470

Fritz RS, Simms EL (1992) Plant resistance to herbivores and pathogens. University of Chicago Press, Chicago

Giron D, Dubreuil G, Bennett A, Dedeine F, Dicke M, Dyer LA, Erb M, Harris MO, Huguet E, Kaloshian I, Kawakita A, Lopez-Vaamonde C, Palmer TM, Petanidou T, Poulsen M, Salle A, Simon JC, Terblanche JS, Thiery D, Whiteman NK, Woods HA, Pincebourde S (2018) Promises and challenges in insect-plant interactions (16th International symposium on insect-plant relationships). Entomol Exp Appl 166:319–343

Gordon KHJ, Waterhouse PM (2007) RNAi for insect proof plants. Nat Biotechnol 25:1231–1232

Han B, Zhang Q-H, Byers JA (2012) Attraction of the tea aphid, Toxoptera aurantii, to combinations of volatiles and colors related to tea plants. Entomol Exp Appl 144:258–269

Hare JD (2012) How insect herbivores drive the evolution of plants. Science 338:50–51

Harris MO, Friesen TL, Xu SS, Chen MS, Giron D, Stuart J (2015) Pivoting from *Arabidopsis* to wheat to understand how agricultural plants integrate responses to biotic stress. J Exp Bot 66:513–531

Haverkamp A, Bing J, Badeke E, Hansson BS, Knaden M (2016) Innate olfactory preferences for flowers matching proboscis length ensure optimal energy gain in a hawkmoth. Nat Commun 7:11644

He J, Chen F, Chen S, Lv G, Deng Y, Fang W, Liu Z, Guan Z, He C (2011) Chrysanthemum leaf epidermal surface morphology and antioxidant and defense enzyme activity in response to aphid infestation. J Plant Physiol 168:687–693

Hogenhout SA, Jorunn IB (2011) Effector proteins that modulate insect-plant interactions. Curr Opin Plant Biol 14:422–428

Hopkins AD (1916) Economic investigations of the scolytid bark and timber beetles of North America. In: US Department of Agriculture, 353 (ed) US Department of Agriculture Program of Work for 2017. US Department of Agriculture, Washington, DC

Howe GA, Jander G (2008) Plant immunity to insect herbivores. Annu Rev Plant Biol 59:41–66

Jinwon K, Quaghebeur H, Felton GW (2011) Reiterative and interruptive signaling in induced plant resistance to chewing insects. Phytochemistry 72:1624–1634

Johnson MTJ (2011) Evolutionary ecology of plant defences against herbivores. Funct Ecol 25:305–311

Kerchev PI, Fenton B, Foyer CH, Hancock RD (2012) Plant responses to insect herbivory: interactions between photosynthesis, reactive oxygen species and hormonal signalling pathways. Plant Cell Environ 35:441–453

Kergoat GJ, Meseguer A, Jousselin E (2017) Evolution of insect-plant interactions: insights from macroevolutionary approaches in plants and herbivorous insects. In: Sauvion N, Calatayud P-a, Thiéry D (eds) Insect-plant interactions in a crop protection perspective, Advances in Botanical Research 81. Academic Press, London, pp 25–54

Kevan P, Shipp L (2011) Biological control and biotechnological amelioration in managed ecosystems. Compr Biotechnol 4:757–761

Labandeira CC (2006) The four phases of plant-arthropod associations in deep time. Geol Acta 4:409–438

Labandeira CC (2013) A paleobiologic perspective on plant-insect interactions. Curr Opin Plant Biol 16:414–421
Labandeira CC, Sepkoski JJ Jr (1993) Insect diversity in the fossil record. Science 261:310–315
Labandeira CC, Kvacek J, Mostovski MB (2007) Pollination drops, pollen, and insect pollination of Mesozoic gymnosperms. Taxon 56:663–695
Lieutier F, Bermudez-Torres K, Cook J (2017) From plant exploitation to mutualism. In: Sauvion N, Calatayud P-a, Thiéry D (eds) Insect-plant interactions in a crop protection perspective, Advances in Botanical Research 81. Academic Press, London, pp 55–109
Lipke H, Fraenkel GS (1956) Insect nutrition. Annu Rev Entomol 1:17–44
Maffei ME, Mithöfer A, Boland W (2007) Before gene expression: early events in plant–insect interaction. Trends Plant Sci 12:310–316
Maffei ME, Arimura G, Mithöfer A (2012) Natural elicitors, effectors and modulators of plant responses. Nat Prod Rep 29:1288–1303
Mithoefer A, Boland W (2012) Plant defense against herbivores: chemical aspects. Annu Rev Plant Biol 63:431–450
Nel A (1997) The probabilistic inference of unknown data in phylogenetic analysis. Mémoires du Muséum national d'histoire naturelle 173:305–327
Paine TD, Kenneth RR, Harrington TC (1997) Interactions among scolytid bark beetles, their associated fungi, and live host conifers. Annu Rev Entomol 42:179–206
Painter RH (1951) Insect resistance in crop plants. The Macmillan Company, New York
Pedigo LP (1999) Entomology and pest management, 3rd edn. Prentice Hall, Upper Saddle River, NJ, USA
Peñalver E, Labandeira CC, Barrón E (2012) Thrips pollination of Mesozoic gymnosperms. Proc Nat Acad Sci U S A 109:8623–8628
Pieterse CM, Van der Does D, Zamioudis C, Leon-Reyes A, Van Wees SC (2012) Hormonal modulation of plant immunity. Annu Rev Cell Dev Biol 28:489–521
Prado JR, Segers G, Voelker T, Carson D, Dobert R (2014) Genetically engineered crops: from idea to product. Annu Rev Plant Biol 65:769–790
Reichstein T, von Joseph E, Parsons JA, Rothschild M (1968) Heart poisons in the monarch butterfly. Science 161:861–866
Renner T, Specht CD (2013) Inside the trap: gland morphologies, digestive enzymes, and the evolution of plant carnivory in the Caryophyllales. Curr Opin Plant Biol 16:436–442
Reymond P (2013) Perception, signaling and molecular basis of oviposition-mediated plant responses. Planta 238:247–258
Saxena KN (1969) Patterns of insect-plant relationships determining susceptibility or resistance of different plants to an insect. Entomol Exp Appl 12:751–766
Schatz, B., Sauvion, N., Kjellberg, N. & Nel, A. 2017. Insect-plant interactions: a palaeontological and an evolutionary perspective. In Insect-plant interactions in a crop protection perspective. Sauvion, N., Calatayud, Paul-andré & Thiéry, D. (Eds). Advances in Botanical Research 81. Academic Press, London, pp. 1-24
Schoonhoven LM, Dethier VG (1966) Sensory aspects of host-plant discrimination by lepidopteraous larvae. Arch Néerl Zool 16:497–530
Schoonhoven LM, van Loon JJA, Dicke M (2005) Insect-plant biology, 2nd edn. Oxford University Press, Oxford
Scott MI, Thaler SJ, Scott GF (2010) Response of a generalist herbivore Trichoplusia ni to jasmonate-mediated induced defense in tomato. J Chem Ecol 36:490–499
Seger J, Brockmann JH (1987) What is bet-hedging? In: Harvey PH, Partridge L (eds) *Oxford surveys in evolutionary biology*, vol 4. Oxford University Press, Oxford, pp 182–211
Snelling RO (1941) Resistance of plants to insect attack. Bot Rev 7:543–586
Thines B, Katsir L, Melotto M, Niu Y, Mandaokar A, Liu G, Nomura K, He SY, Howe GA, Browse J (2007) JAZ repressor proteins are targets of the SCF(COI1) complex during jasmonate signalling. Nature 448:661–665

Thompson JN, Pellmyr O (1991) Evolution of oviposition behavior and host preference in Lepidoptera. Annu Rev Entomol 36:65–89

Ton J, D'Alessandro M, Jourdie V, Jakab G, Karlen D, Held M, Mauch-Mani B, Turlings TCJ (2007) Priming by airborne signals boosts direct and indirect resistance in maize. Plant J 49:16–26

Tumlinson JH, Lewis WJ, Louise EMV (1993) How parasitic wasps find their hosts. Sci Am

van Veen FFJ, Morris RJ, Godfray CJ (2006) Apparent competition, quantitative food webs, and the structure of phytophagous insect communities. Annu Rev Entomol 51:187–208

Varshney AK (1998) Certain aspects of the behavioural responses of maize stem borer, Chilo partellus (swinhoe) to certain maize cultivars in relation to their susceptibility and resistance, Ph. D. thesis, University of Delhi, India

Varshney AK, Singh AK (2000) Effect of maize leaf extracts on the feeding responses of *Chilo partellus* (Swinhoe) larvae. J Entomol Res 24(3):207–211

Varshney AK, Babu R, Singh AK, Agarwal HC, Jain SC (2003) Ovipositional responses of *Chilo partellus* (Swinhoe) (Lepidoptera: Pyralidae) to natural products from leaves of two maize (*Zea mays* L.) cultivars. J Agric Food Chem 51(14):4008–4012

Varshney AK, Singh AK, Agarwal HC (2007) Orientational responses of *Chilo partellus* (Swinhoe) (Lepidoptera: Pyralidae) to leaf volatiles of two maize (*Zea mays* L.) cultivars. Int J Trop Insect Sci 27(1):15–20

Viallanes H (1887) Etudes histologiques et organologiques sur les centres nerveux et les organes des sens des animaux articulés. Masson, Paris

von Frisch K (1953) The dancing bees: an account of the life and senses of the honey bee. Harcourt, Brack, New York

Wang X, Hu L, Zhou G, Cheng J, Lou Y (2011) Salicylic acid and ethylene signaling pathways are involved in production of rice trypsin proteinase inhibitors induced by the leaf folder Cnaphalocrocis medinalis (Guenée). Chin Sci Bull 56:2351–2358

War AR, Taggar, GK, Hussain, B, Taggar, MS, Nair, RM, Sharma, HC (2018) Plant defence against herbivory and insect adaptations. AoB Plants 10: ply037.

Webster B, Bruce T, Pickett J, Hardie J (2010) Volatiles functioning as host cues in a blend become nonhost cues when presented alone to the black bean aphid. Anim Behav 79:451–457

Yuan Y-W, Byers KJRP, Bradshaw HD Jr (2013) The genetic control of flower-pollinator specificity. Curr Opin Plant Biol 16:422–428

Zhang HM, Bernonville TD, Body M, Glévarec G, Reichelt M, Unsicker SB, Bruneau M, Renou J, Huguet ET, Dubreuil G, Giron D (2016) Leaf-mining by *Phyllonorycter blancardella* reprograms the host-leaf transcriptome to modulate phytohormones associated with nutrient mobilization and plant defense. J Insect Physiol 84:114–127

Zhao LY, Chen JL, Cheng DF, Sun JR, Liu Y, Tian Z (2009) Biochemical and molecular characterizations of Sitobion avenae-induced wheat defense responses. Crop Prot 28:435–442

Zhou W, Kügler A, McGale E, Haverkamp A, Knaden M, Guo H, Beran F, Yon F, Li R, Lackus N, Kollner TG, Bing J, Schuman MC, Hansson BS, Baldwin IT, Xu S (2017) Tissue-specific emission of (E)-α-bergamotene helps resolve the dilemma when pollinators are also herbivores. Curr Biol 27:1336–1341

Züst T, Heichinger C, Grossniklaus U, Harrington R, Kliebenstein DJ, Turnbull LA (2012) Natural enemies drive geographic variation in plant defenses. Science 338:116–119

Molecular and Biochemical Aspect of Insect-Plant Interaction: A Perspective for Pest Management

Pallee Shree, Mohit Kumar, and Dileep K. Singh

Abstract

The interactions of insect and their host plants are multifaceted and comprehensive. In an ecosystem to counter the effects of insect attack, plants respond to attack via several biochemical, morphological, and molecular mechanisms. The mechanisms of defense in plants against insect pest are extensive, dynamic, and mediated by direct as well as indirect defenses. Plants synthesize defensive compounds constitutively, or they get induced in response to the insect attack. Plants synthesize a plethora of chemicals which comprises elicitor and secondary metabolites which provide a foremost obstruction to herbivore. Plants also release volatile compounds which attract natural enemy's population against herbivores. The plant resistance is basically due to the sum of genetically inherited qualities which determine the eventual degree of damage of plants by the insect pest. Influencing and manipulating the expression of the genotype related to resistance in the crop is one of the imperative parts of pest management. Induced resistance in plants can be exploited as a significant tool for pest management as this will reduce the use of chemical pesticides for controlling pest population. Induced resistance in plants could be manipulated by employing secondary metabolites, elicitors, and VOCs which would further provide resistance to plants against herbivores. Understanding the mechanism of induced resistance could be helpful in predicting which kind of insect pests are likely to be get impacted by this plant's induced response. With the advancement in genetic engineering, induced responses can also be engineered genetically, so that the defense compounds could be constitutively produced after attacked by herbivores in plants. Plants' induced resistance trait can also be utilized to develop crop varieties, which can give defensive response upon mild infestation. These host plant resistance

P. Shree · M. Kumar · D. K. Singh (✉)
Department of Zoology, University of Delhi, New Delhi, India

I. K. Singh, A. Singh (eds.), *Plant-Pest Interactions: From Molecular Mechanisms to Chemical Ecology*, https://doi.org/10.1007/978-981-15-2467-7_17

strategies could be one of the significant parts of integrated pest management (IPM) for the sustainable production of crops.

Keywords

Elicitor · Insect-plant Interaction · Pest management

1 Introduction

Species interaction in an ecosystem is significant for the survival of organism. Plants are important for the survival of all organisms and interact with different species in different ways for their own survival. Plants interact both with beneficial and antagonistic organisms of the ecosystem. Beneficial relation improves nutrition of plants and also helps to conquer biotic as well as abiotic stresses. In natural and agricultural conditions, plants are not having only mutual relation but are also attacked by multitude of predator herbivores like mites, insects, and pathogens. These herbivores cause damage to plants, impede production, and cause crop losses. Plants have broad spectrum of defense strategies to counter with insect pests for survival (Zhu-Salzman and Zeng 2015). Defensive strategies include various forms of morphological, biochemical, and molecular defense strategies that restrict the attack of insect pests (Belete 2018; Mitchell et al. 2016). During the process of defense, both plant and insect pest evolve different strategies to avoid defense strategies of each other's, and this leads to co-evolution of them (Hougen-Eitzman and Rausher 1994). The evolutionary race between insect pest and plants leads to co-evolution of plants and insect pests which aids in an ingenious defense system development in plants. Therefore plants develop an ability to distinguish self and non-self molecules and signals from their damaged part which activate the immune response of plant against the pest in a similar manner occurring in animals (Hare 2011). Plant generally shows two kinds of defense strategies, that is, indirect and direct defense (Fig. 1) (Aljbory and Chen 2018; Chen 2008). Direct defense includes all characters of plants which enhance the host plant resistance either by providing mechanical protection or biochemical protection by the synthesis of toxic chemicals; these defense strategies affect the physiology as well as behavior of the insect attackers (Karban 2011; Ali and Agrawal 2012). Indirect defense strategies of plants involve the production of several types of volatiles, which particularly attract natural enemies of insect pest, or it may attract pest's natural enemies by providing food source like extra nectar as well as shelter which augment the efficiency of natural enemies (Hanley et al. 2007; Mithöfer and Maffei 2017). Defensive compound in plants is produced consecutively, or it might get induced by attack of herbivore.

In case of direct induced resistance after the insect attack, plants become phenotypically plastic to minimize the loss, and tissues also become less nutritious and appetizing to herbivore (Karban 2011; Lin and Kogan 1990). Induced resistance is not confined to damaged part of the plants but can also be there in undamaged portions of same plant as well as in the adjacent plants (Glinwood et al. 2011). Plant's induced resistance also shows trans-generational effect (Holeski et al. 2013; Holeski et al. 2012; Agrawal 2002).

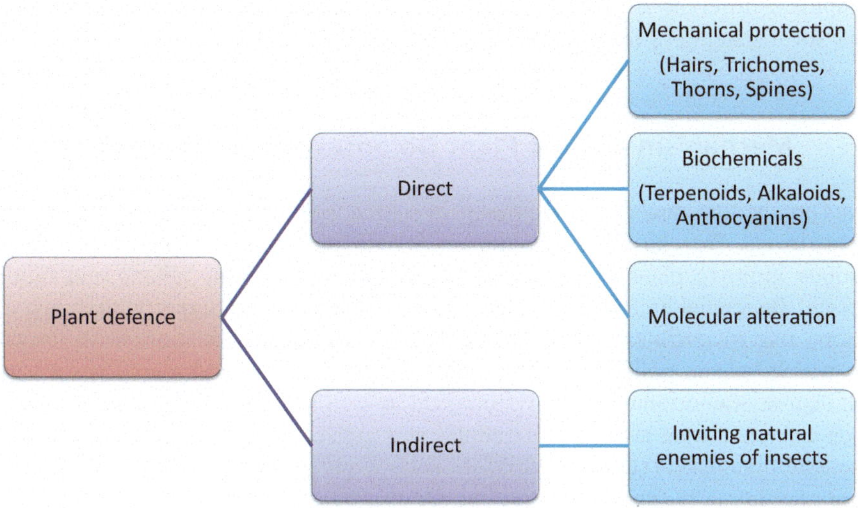

Fig. 1 An overview of Plant defense strategies

The human population is increasing, and it is difficult to meet the need of the growing population as we have limited land for agriculture, so we need a method to increase the production in a sustainable way. Nearly 40% of all the crop production is lost directly due to insect pest. We need a strategy of pest control which could be less detrimental for the ecosystem and thus could be a significant tool to lessen the use of chemical pesticides. Plant resistance to pest which is induced after the attack could be utilized as a strategy to control pest population below threshold level in insect pest management. Induced response is a significant part of pest management strategies, and it has been exploited for regulating population of insect pest below the economic injury level (War et al. 2011; Howe and Jander 2008). This induced resistance could be exploited further to minimize the pest population below threshold level.

Increase in pressure to minimize the application of pesticides in agricultural production and the development of technologies has augmented the role of plant pest resistance in the pest management strategies. Till now plant resistance against pest is an underutilized strategy for pest management. In the past plant breeding and biological control were generally considered an independent pest management practices (Thomas and Waage 1996; Onstad 1987). Nowadays scientists are more focused on selecting cultivars having higher natural defense against insect pests. In plants induced resistance could be manipulated by the use of elicitors which leads to the synthesis of plethora of secondary metabolites which ultimately enhance natural immunity. The understanding of plant immunity as well as potential of plant defense elicitors to augment the crop yield and quality is important. Induced compound can also be produced by genetic modification, and they can also be modified genetically so that they can be produced consecutively by minor attack. This review basically focuses on insect plant interaction and defense mechanism and different ways to manipulate different attributes like morphological and chemical composition of crops for managing of insect pests in agro-ecosystems in a sustainable way.

2 Mechanism of Host Plant Defenses Against Insects

Plants are the transducer; they are the main source for transfer of energy at different trophic levels. They have both beneficial and antagonist relation (Fig. 2). Beneficial relations improve plant nutrition and help them to overcome abiotic and biotic stresses. Beneficial relations of plants are symbiotic associations with fungi or bacteria like mycorrhiza and rhizobacteria. It has beneficial relations with insects also which help in pollination. The mutualistic relation generally involves a very high degree of coordination at physiological as well as morphological level, which needs an unremitting signaling at cellular and molecular level between plants and beneficial organisms (Harrison 2005). In natural and agricultural conditions, plants are not having only mutual relation but are also attacked by multitude of predator herbivores like mites, insects, and pathogens. These insect and pathogen cause damage to plants, impede production, and cause production loss. Plants show array of defensive strategies to avoid damage by insect pest for their survival (War et al. 2012a, b).

In order to combat insect attack, plant responds via direct as well as indirect defense mechanisms. Defense mechanism could be constitutive, or it may get induced after damage caused by insect pest attack. Plants integrate signals induced by pathogens, pest, and microbe in a very appropriate adaptive response. To counter the herbivore attack and for efficient discrimination, plants produce several chemicals which distinguish insect pest with a high degree of sophistication. Plants either produce a very specialized morphological defense structures like lignifications of cell wall and silica deposition or may synthesize secondary metabolites like terpenoids, alkaloids, phenolics, volatiles, reactive oxygen species (ROS), and

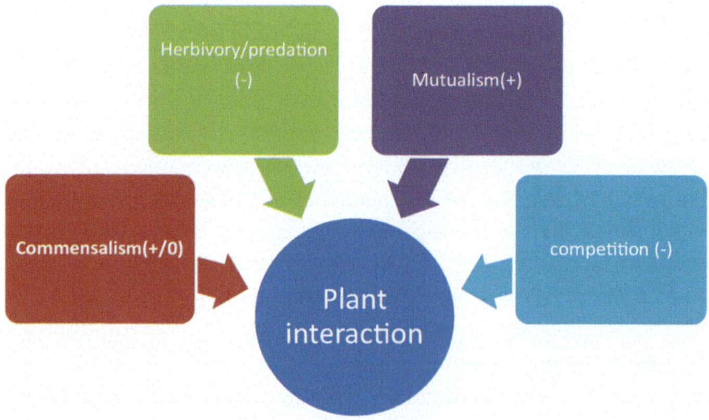

Fig. 2 Different types of beneficial and antagonistic relationships between plants and different organisms

Fig. 3 Different ways by which plant defense toxins impact insect pest

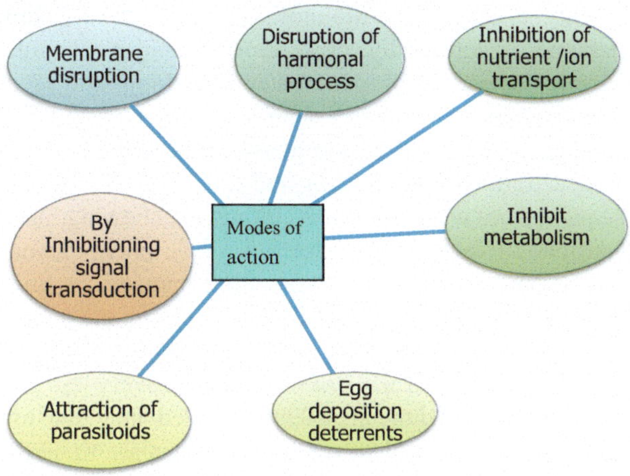

proteins which have anti-nutritional, toxic, or repellent impact on the pest population (Maffei et al. 2007; War et al. 2011). Therefore host plant resistance is a significant form of plant defense strategy for herbivore. During attack, insects release elicitors like salivary amylase and ovipositor fluid on the damaged part of the plants; these elicitors trigger JA/SA pathway which further brings biochemical and morphological changes as plant defense mechanism (Cheng et al. 2013; Feys and Parker 2000). Substantial progress has been observed in understanding the mechanism of plant's induced response against different abiotic and biotic stress over the past few decades, and it has also become a significant topic in ecology and evolutionary biology. Understanding the concept of defense, i.e., mode of release and initiation, is an important aspect of pest management and crop protection. Plant defense can be consecutive which is present without any external stimuli, but the induced defense is triggered by insect attack or elicitor. In order to cope up with attack, direct change in biochemical and morphological structure is important, and among these biochemical is most effective when compared with others (Kariyat et al. 2013). Synergistic effects of diverse defensive components like morphological, biochemical, and molecular enhance the defense mechanism of plants against the insect pest. As studied in tomato plant, phenolics, alkaloids, oxidative enzymes, and proteinase inhibitors comparatively give a better result when acting in a synergistic manner instead of separately working on pest (Duffey and Stout 1996).

Plant morphological traits such as lignifications, secondary metabolites and protein inhibitors act as deterrents or toxins which impact insects' development, digestibility, growth, etc. (Fig. 3) (Mitchell et al. 2016).

Plant morphological structures are the first line of defense which protect plant from insect attack, and thus they are the significant contributor in host plant resistance to insect pest. These morphological traits provide plants a fitness advantage over others as it deters the attack of herbivores from feeding. Thickening of cell wall, lignifications, suberization/corky tissue formation, silica deposition, surface wax, trichomes, thorns, etc. form the first physical obstruction for the herbivore

(Hanley et al. 2007; Malinovsky et al. 2014). Structures like spines, thorns, trichomes, sclerophylly, presence of granular minerals, tissue toughness in plant, epicuticular wax films, divaricated branching, and secretory structures etc. provide plants first line of defense against the insect pest (Hanley et al. 2007; Howe and Schaller 2008). Due to morphological modification, toughness in leaves interferes the penetration of the piercing and sucking mouthparts of the insects; it also protects the plant from wear and tear by biting and chewing insects and therefore protects the plant parts (Raupp 2008). Sclerophylly a phenomenon where hardening of leaves occurs, protects the plant tissue from damage, as this reduce the palatability and digestion of plant tissues, therefore decreasing the damage (Handley et al. 2005; War et al. 2012a, b). Chamarthi et al. (2011) studied that *Sorghum bicolor* damage by shoot fly *Atherigona soccata* was comparatively less due to leaf sheath pigmentation and glossiness. Agrawal (1999) reported that due to insect attack, induction in trichomes occurs in plant species. The increase in density of trichome after damage by *Phratora vulgatissima* (leaf beetles) adult in *Salix cinerea* was observed only in new developing leaves (Belete 2018).

There are diverse biochemical defense mechanisms present in plants against herbivore attack (Table 1). The biochemical defense mechanism is highly dynamic and can be both direct and indirect. Plants produce several chemicals for defense which generally belong to various classes of secondary metabolites which provide defense against pest attack. The biochemicals are produced in plant consecutively, or it may be induced after the attack of insects. Many of these chemicals act directly on herbivores, or they may act indirectly by attracting herbivore's natural enemy, thus protecting the plant. Biochemical can be toxic, anti-nutritive, or repellent in nature for the herbivores attacking the host plant. Induced defense is triggered by several endogenous signaling molecules which are produced after the attack of insect pest such as salicylic acid (SA) and jasmonic acid (JA) (Fig. 4). War et al. (2012a); Feys and Parker (2000).

Defense in plant is consecutive as well as induced. In case of induced defense, plants after insect attack produce secondary metabolites, volatile compounds and reactive oxygen species, and insecticidal gene products like polyphenol oxidase and proteinase inhibitors (PIs) which deter insect metabolism, physiology, as well as digestion (Fürstenberg-Hägg et al. 2013). Insecticidal genes are of different types; they all contribute in providing resistance in a different ways. The genes encode proteins after wounding due to herbivory are categorized into three groups: anti-nutritional protein product of defense gene, followed by signal pathway gene products, and the third gene synthesize proteinases (Ryan 2000). This cascade of gene activation provides an overview that how up regulation of these three classes of genes occur, which encode defensive proteins such as PI and enzymes for biosynthesis of secondary metabolites. It also gives an insight about genes involved in volatile compound production which are involved in signaling pathways along with the genes concerned with rerouting of plant metabolism synthesis of defensive compounds, such as proteinases.

Plants have the capability to differentiate between herbivore attack, ovipositioning, and mechanical damage which is a very significant trait needed to

Table 1 List of plant defensive compounds produced against insect attack

Defensive compounds	Plants	Insect attack	References
Phytoecdysteroids (secondary metabolite)	*Spinacia oleracea*	*Otiorhynchus sulcatus*	Wielkopolan and Obrępalska-Stęplowska (2016) and Schmelz et al. (1999)
Hydrogen peroxide (H_2O_2)	*Glycine max*	*Helicoverpa zea*	Alam (2018) and Bi and Felton (1995)
	Arabidopsis thaliana	*Heterodera glycines*	Waetzig et al. (1999)
Mitogen-activated protein kinase (MAPK)	*Arabidopsis thaliana*	*Heterodera glycines*	McNeece et al. (2019)
Phenols, hydrogen peroxide, malondialdehyde	Groundnut	*Spodoptera litura*	War et al. (2012a, b)
Proteinase inhibitor	Tomato and potato	*Leptinotarsa decemlineata*	Ryan (1978)
	Tobacco	*Manduca sexta*	Van Dam et al. (2001)
	Nicotiana attenuata	*Spodoptera exigua*	Steppuhn and Baldwin (2007)
	Arabidopsis/ oilseed rape	*Spodoptera exiguaPlutella xylostella*	De Leo et al. (2001)
Lectins	Rice	*Nilaparvata lugens*	Sun et al. (2002)
	Galanthus nivalis	Plant hoppers	Chandrasekhar et al. (2014)
Glucosinolate (secondary metabolite)	*Brassica napus*	*Psylliodes chrysocephala*	Bartlet et al. (1999)
	Brassica juncea	*S. eridania* and *P. xylostella*	Fürstenberg-Hägg et al. (2013)

avoid wastage of expensive defense resources as plant tradeoff between growth, development, and defense resources production. Therefore, the production of defense resource after herbivore attack can benefit only herbivore-challenged plants. After insect attack plants show a dynamic change in gene activity like transcriptomes, entire complement of proteins which is expressed, and metabolomes which leads to chemical and physical cues like compounds in oral secretions and oviposition fluids. Plants can also assess the quantity and quality of tissue damage; this was studied via an experiment. This characteristic of plant was studied by a designed experiment in which continuous mechanical wounding similar to caterpillar feeding was given repeatedly on lima bean leaves which leads to the discharge of volatile compounds similar to those produced after the attack of larvae (Mithöfer et al. 2005). Plants generally show responses against insect attack, which are mediated by elicitors present in oviposition fluids, regurgitates, and oral secretions of insect pest. The defense strategies initiated in plants by several elicitors differ

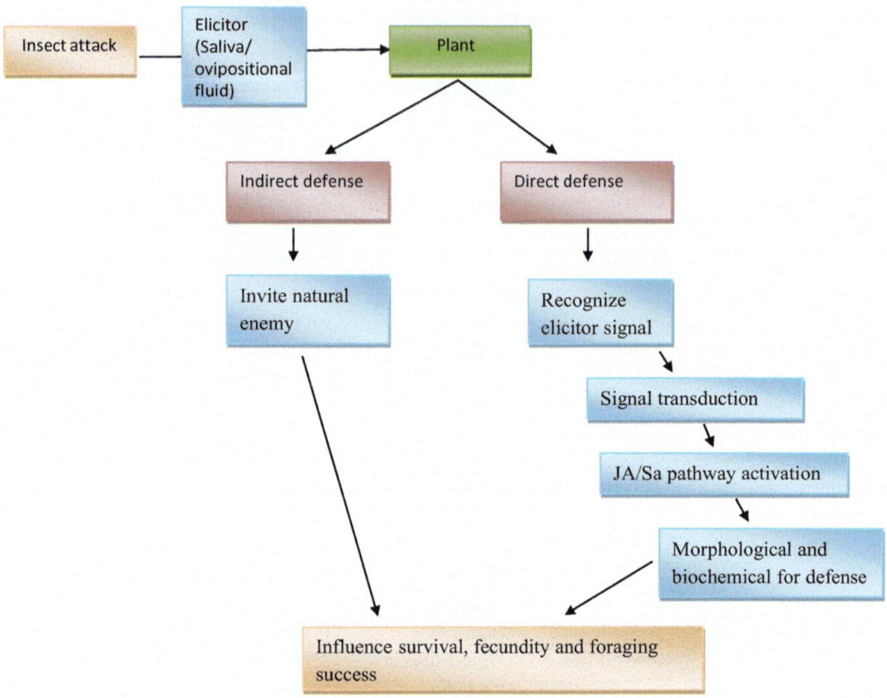

Fig. 4 An overview of induced resistance mechanism in plants

from each other on the basis of concerned biological processes and components elicitors (Garcia-Brugger et al. 2006; Pauwels et al. 2009). In a study, it has been reported that a β-glucosidase present in *Pieris brassicae* caterpillar regurgitate is a potential elicitor which can evoke emission of volatile mixture which attract natural enemy of the larva. An experiment was designed, in which, when cabbage leaves were mechanically wounded and subsequently treated with *P. brassicae* larvae, gut regurgitate attracted natural enemy of caterpillar *Cotesia glomerata* a parasitic wasp (Mattiacci et al. 1995). Other important components of insect secretion which act as elicitor are fatty acid-amino acid conjugates, volicitin, alkaline phosphatase, etc. When insect attacks plants, the plant senses the presence of insect by saliva or ovipositional fluid components. These fluids contain elicitors which provoke the plant immune system and increase plant defense (Wu et al. 2008).

Plants sense the presence of the response of insect by elicitors, and plant immune response gets aggravated. The use of model plant like Arabidopsis, tomato, etc. and application of plant genomics approaches helped in better conceptualization of the molecular mechanisms of plants for beneficial organisms and defense against insect pest. Plant hormones like salicylic acid, ethylene, and jasmonic acid play a significant role in plant defense by regulating signaling networks in plants (Bari and Jones 2009; Verma et al. 2016). Plant hormones like abscisic acid, auxins, and brassinosteroids are also having significant role in eliciting plant immune response against pest (Erb et al. 2012; Albrecht et al. 2012).

A study was conducted on tomato plants for understanding the molecular mechanism of induced resistance, which indicated that signaling molecule activates gene for the synthesis of proteinase inhibitor (PI) and polyphenol oxidase (PPO). The primary molecule of this cascade is prosystemin and photolytic cleavage of this release hormone systemin. Prosystemin is present at low level in leaf tissue which lacks signal peptidase or other targeting information. Photolytic cleavage forms primary signal molecule systemin. To prove that primary signaling molecule is systemin, an experiment was performed in which prosystemin expression was blocked with the help of antisense RNA strategy which showed several impairments in systemic response to damage caused by herbivore (McGurl et al. 1992; Orozco-Cardenas et al. 1993). Systemin is a 18-amino acid peptide in the vascular bundles of other Solanaceae family plants (Ryan and Pearce 1998). Systemin receptor signal transduction activates phospholipase via MAPK (mitogen-activated protein kinase), and then linolenic acid is released from membrane. Linolenic acid is significant for plant defense as it acts as the precursor of jasmonic acid, a signaling molecule which further activates gene for synthesis of secondary metabolites, PI and PPO. Jasmonic acid and oxylipid signaling molecule are involved in plant defense response via the octadecanoid pathway. Oxylipins are significant signaling molecules as they trigger cell division at wounding site and aid in the protective callus development (Somerville et al. 2000). After this the next step of plant defensive pathway leads to defense-response gene upregulation. This subsequently directs the synthesis of oligomeric polymers of galacturonic acid, that is, oligogalacturonides. Oligogalacturonides play various role in defense as they lead to the release of ROS via oxidative burst, G-protein activation, activation of phospholipase C, several kinases, and influx of Ca^{2+} (Apostol et al. 1989; Legendre et al. 1992; Legendre et al. 1993; Chandra and Low 1995). Signaling molecule activation activates gene for the synthesis of volatiles, secondary metabolites, reactive oxygen species, and insecticidal gene products like proteinase inhibitor (PI) and polyphenol oxidase (PPO) which deter digestion, reproduction, and metabolism (Fig. 5).

Secondary metabolites are the bioactive compounds which are concerned with plant protection against insect pest attack, pathogen infection, and also decrease in plant tissue palatability where they are produced (Bennett and Wallsgrove 1994; Howe and Jander 2008). These compounds generally target insect pest and attack their nervous system, endocrine organs, and digestive system (Rosenthal and Berenbaum 2012). Secondary metabolite production in plants could be constitutive or get induced in response to the herbivore attack. Several secondary metabolites are synthesized by plants like alkaloids, flavonoids, phenol, benzoxazinoides, glucosinolates, etc. Among secondary metabolites, phenol is one of the most common compound, which is produced by the plants, in response to insect attack, and it plays the most imperative role in imparting resistance to plants (Kubalt 2016). There are several instances where phenolic compounds are used by plants in defense response against insect pest, for example, cereal aphids (*Rhopalosiphum padi*) are less attracted to wheat varieties having comparatively higher phenolics (Leszczynski

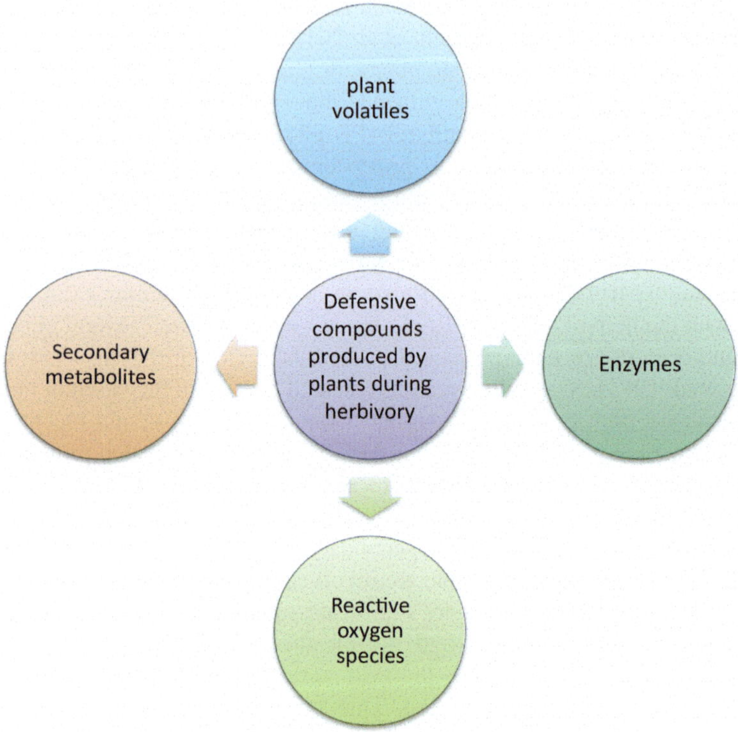

Fig. 5 Plant defensive compound productions during induced defense

1995). Lignin, flavonoids, and isoflavonoids are another significant phenolic compound; production of these compounds are induced by insect attack, which decreases insect pest behavior, fecundity, growth, and development (Johnson et al. 2009; War et al. 2012a, b). Another secondary metabolite induced by insect attack is tannins which have strong deleterious effect on insect pest, as these act as feeding deterrents to pest and thus influence the growth as well as development.

Thousands of volatile organic compounds (VOCs) are also synthesized from plants consisting volatile fatty acid derivatives, aldehydes, esters, alcohols, benzenoids, methyl salicylate, terpenoids, and amino acid-derived nitrogen- and sulfur-containing compounds (Dudareva et al. 2006; Dudareva et al. 2013; Riedlmeier et al. 2017). VOCs are induced by biotic as well as abiotic stresses in plants and perform significant functions as they enhance effectiveness by inducing direct and indirect resistance in plants (Cofer et al. 2018). Volatile compounds are also produced by plants in huge amount when attacked by insect pest (Engelberth et al. 2004). They attract natural predators of herbivore or drive back herbivores and also aids in communication between or within plants (McCormick et al. 2012). Volatile compounds play a decisive role in plant defense because they rapidly get to distant parts of plants because of their volatility (Mauch-Mani et al. 2017).

Plants maintain memory of any of the biotic and abiotic stress event they experienced, and the memory impacts the plant response for imminent stressful situations (Molinier et al. 2006; Crisp et al. 2016). Pre-exposed plants with stress conditions show a prior, faster, and stronger response against stress, which results in augmented tolerance and resistance against insect pest (Conrath et al. 2015).

It has been well documented that damaged plants release volatiles that attract parasitoids. Volatiles provide clues to the parasitoids in searching their prey and also provide adaptive benefit to the plants emitting volatiles. All these interactions and communications are very particular to a specific insect and plant. For instance, damage caused by larvae of corn rootworm in maize plant roots induces the release of (E)-β-caryophyllene a volatile compound that attracts a nematode *Heterorhabditis megidis* which in turn feeds on *D. virgifera* a corn rootworm larvae (Anbesse and Ehlers 2010). Plant defense responses to attract natural enemies of herbivore have a crucial job in defense mechanism of plants toward pest attack. Volatile compounds and extra floral nectar production arbitrate plant interactions with their natural enemies of herbivore, that is, predator or parasitoids, which decrease the population of pest. These kinds of induced indirect defense responses received attention recently, and their genetic, physiological, biochemical, and ecological levels have been studied for pest management.

Reactive oxygen species (ROS) is a significant contributor in defensive responses of plants toward infection. Insect or pathogen attacks plants, and the elicitors present in salivary amylase or ovipositional fluid cause an oxidative burst which is characterized by the synthesis of hydrogen peroxide (Lamb and Dixon 1997). After wounding of plant tissues due to insect pest attack, hydrogen peroxide is produced (Orozco-Cardenas and Ryan 1999). In response to insect attack, jasmonic acids and oligogalacturonic acids are synthesized in plants which leads to the production of hydrogen peroxide. Legendre et al. (1993) reported that in soybean cell cultures, oligogalacturonic acid induced oxidative burst. Oxidative burst in cultured tomato cells is induced by systemin (Stennis et al. 1998). In plant tissues hydrogen peroxide is synthesized upon wounding by pest and pathogen which leads to local as well as systemic response (Orozco-Cardenas and Ryan 1999). High concentration of hydrogen peroxide leads to hypersensitive response toward pest and pathogen which is concerned with cell death (Lamb and Dixon 1997).

Plant also produces defensive protein in response to defense. These defensive proteins cause imbalance in digestion as well as utilization of plant proteins by the insects which impact the insect physiology. These changes lead to alteration in expression of gene which results in quantitative and qualitative changes in proteins. These changes in protein expression play an imperative role in oxidative defense. Proteinase inhibitors (PIs) are one of the most abundant defensive proteins synthesized against insect attack in plants. PI concentration is highest in the storage organs like tubers and seeds. About 1–10% of the total proteins in the plants include PIs, which have a significant contribution in plant defense by inhibiting different types of enzymes (Dunse et al. 2010). PIs slow down the development of the insects as they bind with digestive enzymes of insect gut and restrain their activities and thus decrease the pace of protein digestion which further cause shortage of amino acids (Azzouz et al. 2005).

3 Plant Defense Molecule Perspective in Pest Management

Pest management is a serious issue. The target of integrated pest management (IPM) is to minimize the use of chemical pesticides and maximize the use of biological control, modification of cultural practices, and introduction of resistant cultivars. It is an ecosystem-based strategy which focuses on maintaining pest population below threshold level. All plants have different strategies for pest management. Some are having higher resistance as compared to another. So by understanding the induced and direct defense strategy of plants for pest resistance we can infer herbivore specific signals and molecular mechanism of pest resistance basically controls the pest population naturally. This defense mechanism is an unexplained field and can be used in pest management. The pathways are highly conserved among plants, identification of genes encoding proteins, secondary metabolites and volatile components can be helpful in conferring resistance in plants by genetic transformation. Understanding of elicitors of plant immune response can be helpful. Plants have several strategies to protect them from insect pests such as biochemical and morphological traits. All these are controlled ultimately by genes. Some plant varieties have higher tolerance as compared to their counterpart due to difference in their genetic makeup. Selective breeding of plants with desired characters, i.e., pest-resistant traits, could be a helpful tool in pest management strategies. Both biochemical and morphological defense are important to fight insect attack. This defense strategy of plants against insect pests is an important part of pest management (Maffei et al. 2012).

Proper understanding of direct, indirect, and induced resistance can help in exploiting plant defense in management of insect pest below economic injury level and also for the development of new strategies for crop protection. The induced resistance could be genetically engineered, in order to produce defensive compounds constitutively in plants when challenged by insect pest. Bioengineering and plant selective breeding could generate plants with desired characters. Those plants can produce repellents, protective metabolites, and toxins which can provide strength to the crop to withstand against pest attack. Knowledge of pathways will help in modification of existing pathways by utilizing metabolic engineering, for example, up- or downregulation of metabolic pathways could be helpful in attaining desired compound. Crop cultivar can also be developed by biotechnology in which induced resistance could be exploited to develop new crop varieties, where mild infestation can trigger inducible response. All these techniques are the significant components of IPM for the sustainable production of crops.

Generally elicitors are fatty acid conjugates. First elicitor identified in *Spodoptera exigua* oral secretion is volicitin whose application on maize leads to the secretion of volatile compound which attracts the natural enemy of insect pests (Alborn et al. 1997). Therefore we can say elicitors are extra significant compound involved in

Table 2 List of most common Elicitors

Elicitor	Plant	Insect	References
Glucose oxidase (GOX)	*Nicotiana tabacum*	*Helicoverpa zea*	Musser et al. (2002)
	Solanum lycopersicum	Lepidoptera and Hymenoptera	Eichenseer et al. (2010)
	Solanum lycopersicum	*Helicoverpa zea*	Wang et al. (2017) and Tian et al. (2012)
Volicitin	*Zea mays*	Lepidoptera	Frey et al. (2000)
	Corn	*Spodoptera exigua*	Alborn et al. (1997)
	Solanum melongena and *Nicotiana tabacum*	*Manduca sexta*	Yoshinaga et al. (2014)
Fatty acid-amino acid conjugates (FACs)	*Nicotiana attenuata*	*Manduca sexta*	Halitschke et al. (2001) and Hettenhausen et al. (2014)
β-Glucosidase	*Brassica oleracea*	*Pieris brassicae*	Mattiaccie et al. (1995)
	Phaseolus lunatus	*Pieris brassicae*	VanDoorn et al. (2010)
Alkaline phosphatase	Cotton and melon	*Bemisia tabaci*	Funk (2001)
Caeliferins	*Zea mays*	*Schistocerca americana*	Alborn et al. (2007)
Inceptins	*Vigna unguiculata*	*Fall armyworm*	Schmelz et al. (2007)
Bruchins, alpha, omegas, diols	*Pisum sativum*	*Callosobruchus maculatus*	Doss et al. (2000)

triggering plant defense response. They induce a whole array of response. Similarly another elicitor N-linolenoyl-glu present in the saliva of *Manduca sexta*, when applied on damaged leaves activate signaling pathway to fight insect attack (Halitschke et al. 2001). Glucose oxidase in *Helicoverpa zea* and *Ostrinia nubilalis* mediates defense signaling pathway in tomato (Louis et al. 2013; Tian et al. 2012). Elicitors of *Arabidopsis thaliana* are protein, oligosaccharides, glycol proteins, and fatty acids in nature which evoke in immune response. So, we can say recent development in understanding the chemical elicitors in induced response is important in manipulating plant defense. Therefore it is an important tool for crop protection. These elicitors have the potential to activate both direct and indirect response, thus augmenting host plant resistance. This is an ecological method to manipulate agricultural ecosystem. These elicitors induce response similar to pest attack. These are underutilized method for protection of plants. They are means of temporary activation of indirect and direct defense mechanism of crop plants at appropriate time during growing season of crop. It increases the immunity of crop plant. This is not an isolated strategy but can be an important component of pest management strategies. Similarly there are several elicitors which evoke plant immune response. Some of them are listed in Table 2.

Plant direct and indirect or induced resistance can be genetically engineered using biotechnology and bioengineering techniques to produce defensive compounds constitutively. Selective breeding also plays an important role in pest management as it generates plants with desired characters. Advancement in metabolic engineering can produce terpenoids or other secondary metabolites to reduce infestation by insects, and this technology is expected to be significant and more prominent in future pest management. Volatiles which play important role in plant defense strategies can be manipulated using genetic engineering techniques, so that it will increase plant potential to release target volatiles in elevated quantities. Tobacco plants when metabolically engineered synthesize higher amount of terpenoid and diterpene cembratriene-ol which alter the insect pest behavior (Lücker et al. 2004). Another study reported transgenic maize plant with overexpression of corn TPS10 gene products (E)-alpha-bergamotene, (E)-b-farnesene, and sesquiterpene, which are a kind of herbivore-induced hydrocarbon, attracts *C. marginiventris*, a parasitic wasp (Schnee et al. 2006). In *Arabidopsis* plant overexpression of a strawberry nerolidol synthase and terpene synthase produces (3S)-(E)-nerolidol, a sesquiterpene alcohol, which attracts *P. persimilis* a predatory mite (Kappers et al. 2005). Overexpression of nerolidol or linalool synthase from strawberry (*Fragaria x anannasa)* in *Arabidopsis* chloroplasts leads to a successful alteration plant defense mediated by volatiles which repel aphid *Myzus persicae* (Aharoni et al. 2003). Arabidopsis transgenic lines produce linalool a terpene alcohol that considerably repels aphids, *Myzus persicae* (Aharoni et al. 2003).

Expression of defensive proteins PIs in transgenic plants could enhance plant resistance for insects, and this property was studied against several hemipteran and lepidopteran pests (Dunse et al. 2010; Azzouz 2004). Transgenic plants which express PIs against insect attack have necessitated the understanding of mechanisms of PI release and their interactions among themselves and with other defensive compound. De Leo et al. (2001) studied effectiveness of PI release against herbivore is associated with the level of expression in the plants as well as their activities toward the proteinases of targeted insect and target insect adaptive capacity. Genes from plants which encode two kinds of PI proteins, that is, cowpea trypsin inhibitor and pea lectin, when expressed in transgenic tobacco plants showed enhanced resistance to *Heliothis virescens* (Boulter et al. 1990). In a study, oryzacystatin I, a cysteine protease inhibitor (PI) when administered in an artificial diet of aphids, showed considerable growth inhibition of aphid species. In transgenic *Brassica napus*, oryzacystatin I expression under CaMV-35S promoter control synthesizes oryzacystatin I which significantly decreased fecundity, growth, and mean adult weight of aphid *Myzus persicae* (Rahbé et al. 2003). Similarly oryzacystatin I affects as it inhibits extra-digestive proteolytic activities and also influence their reproduction (Rahbé et al. 2003). For pest management a better understanding of structure of protein and their post-translational modifications is needed which contributes in predicting toxicity, stability in pest gut, and mechanism of action of plant resistance proteins. Advancement in proteomics as well as microarray techniques have let to the discovery that a wide spectrum of resistance proteins are involved in plant defense against pest.

4 Conclusion

Crop protection for increasing the production for human benefit can not only be done by the use of pesticides but also by exploiting and employing traditional strategies of farming which can be based on plant chemistry and host plant resistance which would be sustainable and ecologically justified. Host plant resistance could play a significant role in future pest management programs. To comprehend the potential of plant resistance against insect, it is essential to develop a broad understanding of the mechanism of plant resistance. In order to make use of specialized compounds, the knowledge concerning efficiency and mode of action of defensive compounds is a prerequisite. These defensive compounds could be useful for developing new strategies insect pests' management. Selective breeding and bioengineering could be helpful in this as they can be helpful in generating plants which can produce repellents, toxins, or some other defensive compounds which can help plant to withstand pest attacks. Metabolic engineering could be helpful in modifying the existing pathways by up-/downregulation of biosynthetic steps to get desired compounds. Phytohormone jasmonate, salicylic acid regulation and production of secondary metabolites or other defense compounds could be a successful tool for pest control. Understanding of host plant interaction mechanism will lead to new avenues for further research on signaling cascades and ecological consequences in the ecosystems. Defensive compounds produced during host -plant interaction can be a better alternative for pest management, but its mass scale production suffers from certain limitations like the high costs associated with mass production, marketing of synthetic compounds, development of technologies, the process of formulation and complications. However for sustainable development, there is an urgent need to understand the mechanism, biochemistry, and ecological significance of host -plant resistance.

References

Agrawal AA (1999) Induced responses to herbivory in wild radish: effects on several herbivores and plant fitness. Ecology 80:1713–1723

Agrawal AA (2002) Herbivory and maternal effects: mechanisms and consequences of transgenerational induced plant resistance. Ecology 83(12):3408–3415

Aharoni A, Giri AP, Deuerlein S, Griepink F, de Kogel WJ, Verstappen FW et al (2003) Terpenoid metabolism in wild-type and transgenic Arabidopsis plants. Plant Cell 15(12):2866–2884

Alam MZ (2018) Comparative analysis of Transcriptomic expression of the salivary glands, midgut, and body tissues of Helicoverpa zea caterpillars after herbivory on soybean leaves. Western Illinois University, Macomb

Alborn HT, Hansen TV, Jones TH, Bennett DC, Tumlinson JH, Schmelz EA, Teal PE (2007) Disulfooxy fatty acids from the American bird grasshopper Schistocerca americana, elicitors of plant volatiles. Proc Natl Acad Sci 104(32):12976–12981

Alborn HT, Turlings TCJ, Jones TH, Stenhagen G, Loughrin JH, Tumlinson JH (1997) An elicitor of plant volatiles from beet armyworm oral secretion. Science 276(5314):945–949

Albrecht C, Boutrot F, Segonzac C, Schwessinger B, Gimenez-Ibanez S, Chinchilla D et al (2012) Brassinosteroids inhibit pathogen-associated molecular pattern–triggered immune signaling independent of the receptor kinase BAK1. Proc Natl Acad Sci 109(1):303–308

Ali JG, Agrawal AA (2012) Specialist versus generalist insect herbivores and plant defense. Trends Plant Sci 17(5):293–302

Aljbory Z, Chen MS (2018) Indirect plant defense against insect herbivores: a review. Insect Sci 25 (1):2–23

Anbesse S, Ehlers RU (2010) Attraction of Heterorhabditis sp. toward synthetic (E)-beta-cariophyllene, a plant SOS signal emitted by maize on feeding by larvae of Diabrotica virgifera virgifera. Commun Agric Appl Biol Sci 75(3):455–458

Apostol I, Heinstein PF, Low PS (1989) Rapid stimulation of an oxidative burst during elicitation of cultured plant cells: role in defense and signal transduction. Plant Physiol 90(1):109–116

Azzouz H, Cherqui A, Campan EDM, Rahbe Y, Duport G, Jouanin L et al (2005) Effects of plant protease inhibitors, oryzacystatin I and soybean Bowman–Birk inhibitor, on the aphid Macrosiphum euphorbiae (Homoptera, Aphididae) and its parasitoid Aphelinus abdominalis (Hymenoptera, Aphelinidae). J Insect Physiol 51(1):75–86

Azzouz H (2004) Risques d'exposition et effets potentiels d'inhibiteurs de protéases sur des parasitoïdes de pucerons (Doctoral dissertation, Paris 13)

Bari R, Jones JD (2009) Role of plant hormones in plant defence responses. Plant Mol Biol 69 (4):473–488

Bartlet E, Kiddle G, Williams I, Wallsgrove R (1999) Wound-induced increases in the glucosinolate content of oilseed rape and their effect on subsequent herbivory by a crucifer specialist. In: Proceedings of the 10th international symposium on insect-plant relationships. Springer, Dordrecht, pp 163–167

Belete T (2018) Defense mechanisms of plants to insect pests: from morphological to biochemical approach. Trends Tech Sci Res 2:555584

Bennett RN, Wallsgrove RM (1994) Secondary metabolites in plant defence mechanisms. New Phytol 127(4):617–633

Bi JL, Felton GW (1995) Foliar oxidative stress and insect herbivory: primary compounds, secondary metabolites, and reactive oxygen species as components of induced resistance. J Chem Ecol 21(10):1511–1530

Boulter D, Edwards GA, Gatehouse AM, Gatehouse JA, Hilder VA (1990) Additive protective effects of different plant-derived insect resistance genes in transgenic tobacco plants. Crop Prot 9(5):351–354

Chamarthi SK, Sharma HC, Sahrawat KL, Narasu LM, Dhillon MK (2011) Physico-chemical mechanisms of resistance to shoot fly, Atherigona soccata in sorghum, Sorghum bicolor. J Appl Entomol 135(6):446–455

Chandra S, Low PS (1995) Role of phosphorylation in elicitation of the oxidative burst in cultured soybean cells. Proc Natl Acad Sci 92(10):4120–4123

Chandrasekhar K, Vijayalakshmi M, Vani K, Kaul T, Reddy MK (2014) Phloem-specific expression of the lectin gene from Allium sativum confers resistance to the sap-sucker Nilaparvata lugens. Biotechnol Lett 36(5):1059–1067

Chen MS (2008) Inducible direct plant defense against insect herbivores: a review. Insect Sci 15 (2):101–114

Cheng X, Zhu L, He G (2013) Towards understanding of molecular interactions between rice and the brown planthopper. Mol Plant 6(3):621–634

Cofer TM, Seidl-Adams I, Tumlinson JH (2018) From acetoin to (Z)-3-hexen-1-ol: the diversity of volatile organic compounds that induce plant responses. J Agric Food Chem 66 (43):11197–11208

Conrath U, Beckers GJ, Langenbach CJ, Jaskiewicz MR (2015) Priming for enhanced defense. Ann Rev Phytopathol 53:97–119

Crisp PA, Ganguly D, Eichten SR, Borevitz JO, Pogson BJ (2016) Reconsidering plant memory: intersections between stress recovery, RNA turnover, and epigenetics. Sci Adv 2(2):e1501340

De Leo F, Bonadé-Bottino M, Ceci LR, Gallerani R, Jouanin L (2001) Effects of a mustard trypsin inhibitor expressed in different plants on three lepidopteran pests. Insect Biochem Mol Biol 31 (6–7):593–602

Doss RP, Oliver JE, Proebsting WM, Potter SW, Kuy S, Clement SL et al (2000) Bruchins: insect-derived plant regulators that stimulate neoplasm formation. Proc Natl Acad Sci 97 (11):6218–6223

Dudareva N, Negre F, Nagegowda DA, Orlova I (2006) Plant volatiles: recent advances and future perspectives. Crit Rev Plant Sci 25(5):417–440

Dudareva N, Klempien A, Muhlemann JK, Kaplan I (2013) Biosynthesis, function and metabolic engineering of plant volatile organic compounds. New Phytol 198(1):16–32

Duffey SS, Stout MJ (1996) Antinutritive and toxic components of plant defense against insects. Arch Insect Biochem Physiol: Published in Collaboration with the Entomological Society of America 32(1):3–37

Dunse KM, Stevens JA, Lay FT, Gaspar YM, Heath RL, Anderson MA (2010) Coexpression of potato type I and II proteinase inhibitors gives cotton plants protection against insect damage in the field. Proc Natl Acad Sci 107(34):15011–15015

Eichenseer H, Mathews MC, Powell JS, Felton GW (2010) Survey of a salivary effector in caterpillars: glucose oxidase variation and correlation with host range. J Chem Ecol 36 (8):885–897

Engelberth J, Alborn HT, Schmelz EA, Tumlinson JH (2004) Airborne signals prime plants against insect herbivore attack. Proc Natl Acad Sci 101(6):1781–1785

Erb M, Meldau S, Howe GA (2012) Role of phytohormones in insect-specific plant reactions. Trends Plant Sci 17(5):250–259

Feys BJ, Parker JE (2000) Interplay of signaling pathways in plant disease resistance. Trends Genet 16(10):449–455

Frey M, Stettner C, Paré PW, Schmelz EA, Tumlinson JH, Gierl A (2000) An herbivore elicitor activates the gene for indole emission in maize. Proc Natl Acad Sci 97(26):14801–14806

Funk CJ (2001) Alkaline phosphatase activity in whitefly salivary glands and saliva. Arch Insect Biochem Physiol: Published in Collaboration with the Entomological Society of America 46 (4):165–174

Fürstenberg-Hägg J, Zagrobelny M, Bak S (2013) Plant defense against insect herbivores. Int J Mol Sci 14(5):10242–10297

Garcia-Brugger A, Lamotte O, Vandelle E, Bourque S, Lecourieux D, Poinssot B et al (2006) Early signaling events induced by elicitors of plant defenses. Mol Plant-Microbe Interact 19 (7):711–724

Glinwood R, Ninkovic V, Pettersson J (2011) Chemical interaction between undamaged plants–effects on herbivores and natural enemies. Phytochemistry 72(13):1683–1689

Halitschke R, Schittko U, Pohnert G, Boland W, Baldwin IT (2001) Molecular interactions between the specialist herbivoremanduca sexta (Lepidoptera, Sphingidae) and its natural host nicotiana attenuata. III. Fatty acid-amino acid conjugates in herbivore oral secretions are necessary and sufficient for herbivore-specific plant responses. Plant Physiol 125(2):711–717

Handley R, Ekbom B, Ågren J (2005) Variation in trichome density and resistance against a specialist insect herbivore in natural populations of Arabidopsis thaliana. Ecol Entomol 30 (3):284–292

Hanley ME, Lamont BB, Fairbanks MM, Rafferty CM (2007) Plant structural traits and their role in anti-herbivore defence. Perspect Plant Ecol Evol Syst 8(4):157–178

Hare JD (2011) Ecological role of volatiles produced by plants in response to damage by herbivorous insects. Annu Rev Entomol 56:161–180

Harrison MJ (2005) Signaling in the arbuscular mycorrhizal symbiosis. Annu Rev Microbiol 59:19–42

Hettenhausen C, Heinrich M, Baldwin IT, Wu J (2014) Fatty acid-amino acid conjugates are essential for systemic activation of salicylic acid-induced protein kinase and accumulation of jasmonic acid in Nicotiana attenuata. BMC Plant Biol 14(1):326

Holeski LM, Jander G, Agrawal AA (2012) Transgenerational defense induction and epigenetic inheritance in plants. Trends Ecol Evol 27(11):618–626

Holeski LM, Zinkgraf MS, Couture JJ, Whitham TG, Lindroth RL (2013) Transgenerational effects of herbivory in a group of long-lived tree species: maternal damage reduces offspring allocation to resistance traits, but not growth. J Ecol 101(4):1062–1073

Hougen-Eitzman D, Rausher MD (1994) Interactions between herbivorous insects and plant-insect coevolution. Am Nat 143(4):677–697

Howe GA, Jander G (2008) Plant immunity to insect herbivores. Annu Rev Plant Biol 59:41–66

Howe GA, Schaller A (2008) Direct Defenses in plants and their induction by wounding and insect herbivores. University of Hohenheim, Institute of Plant Physiology and Biotechnology, Stuttgart, pp 7–29

Johnson MT, Smith SD, Rausher MD (2009) Plant sex and the evolution of plant defenses against herbivores. Proc Natl Acad Sci 106(43):18079–18084

Kappers IF, Aharoni A, Van Herpen TW, Luckerhoff LL, Dicke M, Bouwmeester HJ (2005) Genetic engineering of terpenoid metabolism attracts bodyguards to Arabidopsis. Science 309 (5743):2070–2072

Karban R (2011) The ecology and evolution of induced resistance against herbivores. Funct Ecol 25 (2):339–347

Kariyat RR, Balogh CM, Moraski RP, De Moraes CM, Mescher MC, Stephenson AG (2013) Constitutive and herbivore-induced structural defenses are compromised by inbreeding in Solanum carolinense (Solanaceae). Am J Bot 100(6):1014–1021

Kubalt K (2016) The role of phenolic compounds in plant resistance. Biotechnol Food Sci 80 (2):97–108

Lamb C, Dixon RA (1997) The oxidative burst in plant disease resistance. Annu Rev Plant Biol 48 (1):251–275

Legendre L, Heinstein PF, Low PS (1992) Evidence for participation of GTP-binding proteins in elicitation of the rapid oxidative burst in cultured soybean cells. J Biol Chem 267 (28):20140–20147

Legendre L, Yueh YG, Crain R, Haddock N, Heinstein PF, Low PS (1993) Phospholipase C activation during elicitation of the oxidative burst in cultured plant cells. J Biol Chem 268 (33):24559–24563

Leszczynski B (1995) The influence of phenolic compounds on the preference of winter wheat cultivars by cereal aphids. Insect Sci Its Appl 6:157–158

Lin H, Kogan M (1990) Influence of induced resistance in soybean on the development and nutrition of the soybean looper and the Mexican bean beetle. Entomol Exp Appl 55(2):131–138

Louis J, Peiffer M, Ray S, Luthe DS, Felton GW (2013) Host-specific salivary elicitor (s) of E uropean corn borer induce defenses in tomato and maize. New Phytol 199(1):66–73

Lücker J, Schwab W, Franssen MC, Van Der Plas LH, Bouwmeester HJ, Verhoeven HA (2004) Metabolic engineering of monoterpene biosynthesis: two-step production of (+)-trans-isopiperitenol by tobacco. Plant J 39(1):135–145

Maffei ME, Mithöfer A, Boland W (2007) Before gene expression: early events in plant–insect interaction. Trends Plant Sci 12(7):310–316

Maffei ME, Arimura GI, Mithöfer A (2012) Natural elicitors, effectors and modulators of plant responses. Nat Prod Rep 29(11):1288–1303

Malinovsky FG, Fangel JU, Willats WG (2014) The role of the cell wall in plant immunity. Front Plant Sci 5:178

Mattiaccie L, Dicke M, Posthumus MA (1995) beta-Glucosidase: an elicitor of herbivore-induced plant odor that attracts host-searching parasitic wasps. Proc Nat Acad Sci U S A *92*(6):2036

Mauch-Mani B, Baccelli I, Luna E, Flors V (2017) Defense priming: an adaptive part of induced resistance. Annu Rev Plant Biol 68:485–512

McCormick AC, Unsicker SB, Gershenzon J (2012) The specificity of herbivore-induced plant volatiles in attracting herbivore enemies. Trends Plant Sci 17(5):303–310

McGurl B, Pearce G, Orozco-Cardenas M, Ryan CA (1992) Structure, expression, and antisense inhibition of the systemin precursor gene. Science 255:1570–1573

McNeece BT, Sharma K, Lawrence GW, Lawrence KS, Klink VP (2019) The mitogen activated protein kinase (MAPK) gene family functions as a cohort during the Glycine max defense response to Heterodera glycines. Plant Physiol Biochem 137:25–41

Mitchell C, Brennan RM, Graham J, Karley AJ (2016) Plant defense against herbivorous pests: exploiting resistance and tolerance traits for sustainable crop protection. Front Plant Sci 7:1132

Mithöfer A, Maffei ME (2017) General mechanisms of plant defense and plant toxins. In: Plant toxins. Springer, Dordrecht, pp 3–24

Mithöfer A, Wanner G, Boland W (2005) Effects of feeding Spodoptera littoralis on lima bean leaves. II. Continuous mechanical wounding resembling insect feeding is sufficient to elicit herbivory-related volatile emission. Plant Physiol 137(3):1160–1168

Molinier J, Ries G, Zipfel C, Hohn B (2006) Transgeneration memory of stress in plants. Nature 442(7106):1046

Musser RO, Hum-Musser SM, Eichenseer H, Peiffer M, Ervin G, Murphy JB, Felton GW (2002) Herbivory: caterpillar saliva beats plant defences. Nature 416(6881):599

Onstad DW (1987) Calculation of economic-injury levels and economic thresholds for pest management. J Econ Entomol 80(2):297–303

Orozco-Cardenas M, Ryan CA (1999) Hydrogen peroxide is generated systemically in plant leaves by wounding and systemin via the octadecanoid pathway. Proc Natl Acad Sci 96 (11):6553–6557

Orozco-Cardenas M, McGurl B, Ryan CA (1993) Expression of an antisense prosystemin gene in tomato plants reduces resistance toward Manduca sexta larvae. Proc Natl Acad Sci USA 90:8273–8276

Pauwels L, Inzé D, Goossens A (2009) Jasmonate-inducible gene: what does it mean? Trends Plant Sci 14(2):87–91

Rahbé Y, Deraison C, Bonadé-Bottino M, Girard C, Nardon C, Jouanin L (2003) Effects of the cysteine protease inhibitor oryzacystatin (OC-I) on different aphids and reduced performance of Myzus persicae on OC-I expressing transgenic oilseed rape. Plant Sci 164(4):441–450

Raupp MJ (2008) Effects of leaf toughness on mandibular wear of the leaf beetle, Plagiodera versicolora. Ecol Entomol 10:73–79

Riedlmeier M, Ghirardo A, Wenig M, Knappe C, Koch K, Georgii E et al (2017) Monoterpenes support systemic acquired resistance within and between plants. Plant Cell 29(6):1440–1459

Rosenthal GA, Berenbaum MR (2012) Herbivores: their interactions with secondary plant metabolites: ecological and evolutionary processes, vol 2. Academic Press, San Diego

Ryan CA (2000) The systemin signaling pathway: differential activation of plant defensive genes. Bioch Biophys Acta (BBA) Protein Struct Mol Enzymol 1477(1–2):112–121

Ryan CA, Pearce G (1998) Systemin: a polypeptide signal for plant defensive genes. Annu Rev Cell Dev Biol 14(1):1–17

Ryan CA (1978) Proteinase inhibitors in plant leaves: a biochemical model for pest-induced natural plant protection. Trends Biochem Sci 3(3):148–150

Schmelz EA, Grebenok RJ, Galbraith DW, Bowers WS (1999) Insect-induced synthesis of phytoecdysteroids in spinach, Spinacia oleracea. J Chem Ecol 25(8):1739–1757

Schmelz EA, LeClere S, Carroll MJ, Alborn HT, Teal PE (2007) Cowpea chloroplastic ATP synthase is the source of multiple plant defense elicitors during insect herbivory. Plant Physiol 144(2):793–805

Schnee C, Köllner TG, Held M, Turlings TC, Gershenzon J, Degenhardt J (2006) The products of a single maize sesquiterpene synthase form a volatile defense signal that attracts natural enemies of maize herbivores. Proc Natl Acad Sci 103(4):1129–1134

Somerville C, Jaworski JG, Ohlrogge JB, Buchanan BB, Gruissem W, Jones RL (2000) Biochemistry and molecular biology of plants. American Society of Plant Physiologists, Rockville, pp 456–527

Stennis MJ, Chandra S, Ryan CA, Low PS (1998) Systemin potentiates the oxidative burst in cultured tomato cells. Plant Physiol 117(3):1031–1036

Steppuhn A, Baldwin IT (2007) Resistance management in a native plant: nicotine prevents herbivores from compensating for plant protease inhibitors. Ecol Lett 10(6):499–511

Sun X, Wu A, Tang K (2002) Transgenic rice lines with enhanced resistance to the small brown planthopper. Crop Prot 21(6):511–514

Thomas M, Waage J (1996) Integration of biological control and host plant resistance breeding: a scientific and literature review. CTA, Wageningen

Tian D, Peiffer M, Shoemaker E, Tooker J, Haubruge E, Francis F et al (2012) Salivary glucose oxidase from caterpillars mediates the induction of rapid and delayed-induced defenses in the tomato plant. PLoS One 7(4):e36168

Van Dam NM, Hermenau U, Baldwin IT (2001) Instar-specific sensitivity of specialist Manduca sexta larvae to induced defences in their host plant Nicotiana attenuata. Ecol Entomol 26 (6):578–586

VanDoorn A, Baldwin IT, Bonaventure G (2010) Lipoxygenase-mediated modification of insect elicitors: generating chemical diversity on the leaf wound surface. Plant Signal Behav 5 (12):1674–1676

Verma V, Ravindran P, Kumar PP (2016) Plant hormone-mediated regulation of stress responses. BMC Plant Biol 16(1):86

Waetzig G, Sobczak M, Grundler F (1999) Localization of hydrogen peroxide during the defence response of Arabidopsis thaliana against the plant-parasitic nematode Heterodera glycines. Nematology 1(7):681–686

Wang J, Peiffer M, Hoover K, Rosa C, Zeng R, Felton GW (2017) Helicoverpa zea gut-associated bacteria indirectly induce defenses in tomato by triggering a salivary elicitor (s). New Phytol 214(3):1294–1306

War AR, Paulraj MG, War MY, Ignacimuthu S (2011) Herbivore-and elicitor-induced resistance in groundnut to Asian armyworm, Spodoptera litura (Fab.)(Lepidoptera: Noctuidae). Plant Signal Behav 6(11):1769–1777

War AR, Paulraj MG, Ahmad T, Buhroo AA, Hussain B, Ignacimuthu S, Sharma HC (2012a) Mechanisms of plant defense against insect herbivores. Plant Signal Behav 7(10):1306–1320

War AR, Paulraj MG, War MY, Ignacimuthu S (2012b) Herbivore-induced resistance in different groundnut germplasm lines to Asian armyworm, Spodoptera litura (Fab.)(Lepidoptera: Noctuidae). Acta Physiol Plant 34(1):343–352

Wielkopolan B, Obrępalska-Stęplowska A (2016) Three-way interaction among plants, bacteria, and coleopteran insects. Planta 244(2):313–332

Wu J, Hettenhausen C, Schuman MC, Baldwin IT (2008) A comparison of two Nicotiana attenuata accessions reveals large differences in signaling induced by oral secretions of the specialist herbivore Manduca sexta. Plant Physiol 146(3):927–939

Yoshinaga N, Ishikawa C, Seidl-Adams I, Bosak E, Aboshi T, Tumlinson JH, Mori N (2014) N-(18-hydroxylinolenoyl)-L-glutamine: a newly discovered analog of volicitin in Manduca sexta and its elicitor activity in plants. J Chem Ecol 40(5):484–490

Zhu-Salzman K, Zeng R (2015) Insect response to plant defensive protease inhibitors. Annu Rev Entomol 60:233–252

Ecology and Evolution of Insect-Plant Interactions

Anjana Singha Naorem and Sengodan Karthi

Abstract

Interaction between herbivores and plants is essential for ecosystem functioning. Phytochemical variation in plants is one of the most fascinating yet bewildering properties of the natural world and has important implications for both human health and the functioning of the ecosystem. One of the key aspects of plant phytochemical research is to study the insects that feed on plants which are one of the driving forces behind the development of chemical diversity in plants. Decoding their interaction from molecular to the ecological level is important for developing a comprehensive understanding of their interaction that has shaped their evolutionary history. Applications of advanced technologies and collaborative work between molecular biologists, geneticists, ecologists, evolutionary biologists, and biochemists will unravel their complex interactions for meeting future challenges. In this chapter, we have emphasized on the ecological perspective of the two interacting system and how it has led to evolution of certain traits in them.

Keywords

Insect herbivory · Constitutive defense · Induced defense · Secondary metabolites · Coevolution

A. S. Naorem (✉)
Department of Zoology, Cotton University, Guwahati, Assam, India
e-mail: anjanasingha@cottonuniversity.ac.in

S. Karthi
Division of Biopesticides and Environmental Toxicology, Manonmaniam Sundaranar University, Alwarkurichi, Tamil Nadu, India

I. K. Singh, A. Singh (eds.), *Plant-Pest Interactions: From Molecular Mechanisms to Chemical Ecology*, https://doi.org/10.1007/978-981-15-2467-7_18

1 Introduction

Plant-insect interactions are one of the oldest kinds of interactions that nature has been witnessing since the time the two came into existence. The evolution of plants on Earth gave birth to the diverse form of life; most important and the oldest relationship that it has shared is with insects. The association between them probably dates back to 390 million years which is indicated by the molecular dating of the records of a holometabola (Rehm et al. 2011). Due to the preservational artiefact and fragmentary remains, the actual earliest coexistence cannot be ascertained so far. However, with the findings of the first complete insect fossil of Late Devonian, at least their existence during that period is established (Garrouste et al. 2012). Since then, these two organisms are constantly interacting with each other for their own benefit and to develop strategies to overpower the other.

The large diversity of insects, which is estimated to be more than 500,000 known species, is often attributed to their close association with land plants (Kerogoat et al. 2017). Insects derive benefits for their food, shelter, mate selection, and oviposition from plants, and plants, in turn, try to evade the herbivore attack by developing physical and chemical mechanisms (Ehrlich and Raven 1964). It is not just the damage that plants have to face from insects, but they do get benefit from insects in their defense and pollination (Panda and Khush 1995). In doing so, each has to pay the cost or, in other words, be subjected to the pressure of developing strategies to overcome the attack by the insects. Thus the interaction between the two systems exists in dynamicity, which is subjected to variation depending on the kind of pressure exerted by one on the other. Insect feeding impacts both the phenotype of the plant (Kessler and Baldwin 2002; Mithofer and Boland 2012) and the well-defined plant chemistry (Schoonhoven et al. 2005), which in turn impacts plant interactions with its other linked community (Stam et al. 2014). A diverse array of chemicals that are synthesized by plants are usually allocated into primary and secondary chemicals, though the two remain intertwined, and the division is arbitrary and for convenience. The primary metabolites such as carbohydrates, lipids, and proteins are the basic need for plant biological processes and they also serve vital nutrients for herbivorous insects (Scriber and Slansky 1981; Schoonhoven et al. 2005). Plant-derived secondary chemicals, which have recently been termed "specialized metabolites," are the compounds like glucosinolates, alkaloids, terpenoids, tannins, etc. which often not openly intricate in basic plant metabolism but have evolved to cope with heavy herbivory, and they have grown in different plant heredities to help them adjust to specific ecological situations and are essential to improve plant's fitness (Schilmiller et al. 2008; Pichersky and Lewinsohn 2011; de Brito and Martinoia 2018). Diversity of specialized chemicals derived from plants across different species is tremendous and likely exceeds 200,000 (Tissier et al. 2014). Insects also evolved to survive against these metabolites by developing counter mechanisms to detoxify or circumvent plant defense systems. Studies have shown that coevolution is the outcome of community interactions across time and space. Therefore mere understanding of the mechanistic process between the two would provide the base of this relationship, but research in entomological, ecological, and evolutionary context and genetic and molecular approach would probably give us a panoramic view of this interaction. In the last few years, researchers in both basic and applied sciences have been continually working to decipher the complex interaction existing between the two. This has been revolutionized with the advancement of genomic tools including next-generation sequencing (NGS), gene-editing technology, etc. (Hermsmeier et al. 2001; Hui et al. 2003). In this review, the ecological as well as evolutionary aspects of insect-plant interaction will be dealt with more recent findings.

2 Ecological Traits of Insect-Plant Interaction

There are two types of interactions that exists between insects and plants – the plant-pollinator interactions and the herbivore-plant interaction both of which are considered to be specialized interactions (Futuyma and Moreno 1988; Cook and Rasplus 2003). However, plant-pollinator is more generalized in terms of its niche width than herbivore (Fontaine et al. 2009). These two interactions though different as one is greatly mutualistic and the other one is antagonistic but they are important in shaping the way the plants are and for their diversity, which in turn impacts the various adaptive modifications in insects. However, we will deal with the ecological concepts of insect herbivory only as insect-pollinator concept is beyond the scope of this book.

Recognition of the host plant by insect is a complex process involving multiple sensory modalities – olfactory, visual, tactile, and gustatory cues. With the successful input received from these cues, the insect is capable to choose its host. The selection of plants is determined at two levels – oviposition and larval-feeding stages (Schoonhoven et al. 2005). In both the stages, the selection of a plant is dependent on two factors-response of the insects to plant and characteristics of the plants which shapes the insect responses (Saxena 1969). In both situations, chemistry of plants plays a significant role in host plant selection (Schulz 1988) that acts both from a distance and in contact. Once in the vicinity of the plant or in contact with the plant, the physical characteristics of both plant and insects come into play. However, for the first phase of interaction to occur, it is the plant volatiles that are of prime importance. Insect tries to adopt the best of its strategies to overcome the noise of volatile bouquet emanating from different host as well as non-host species in the field (Randlkofer et al. 2010) to be able to focus its senses efficiently to land on the appropriate plant for feeding or oviposition. As a result specialized and sensitive neurons have developed in the insects as an adaptation to the information processing problem (Bernays 1996). This may be one of the factors that largely favors the specialist insect herbivory with reduced diet breadth. Strong specialization is not only seen on specific plant but on specific parts and tissue of that plant (Gutbrodt et al. 2011; Kimmerer and Potter 1987; Gall 1987). This specificity depends on several parameters like the texture of the plant part, its age, morphology, kind of mouthpart insect possesses, etc. (Pecetti et al. 2010; Schoonhoven et al. 2005). Insect tries its best to utilize the resources of the host plant to maximize its fitness. After landing on the plant, the structural characteristics like trichomes, wax crystals, surface texture, etc. and the chemical profile of the plants play a vital role in final selection of the host by the insects.

2.1 Plant Actions to Insect Herbivory

Responses of plants to herbivores are complex. For the plants that are rooted and immobilized, nature has bestowed upon them the plasticity in their morphology, growth, development, etc. to respond to environmental variability. These adaptabilities are the result of their ability to anticipate or detect change in ecological conditions and respond accordingly (Casal et al. 2004). They have developed sensory mechanisms to recognize the danger of herbivory and activate their defense mechanism to respond accordingly. Both physical and chemical defense have developed in plants to counter the attack of insects and pathogens. It is the interplay of these chemical and physical defense mechanisms that defends the plant from herbivory, and these defense systems have complex interactions from molecular or genetic to the level of ecology (Dicke et al. 2009). For understanding such a complex process, researchers adopt a reductionist approach by concentrating on only one level of plant defense either molecular or genetic or ecological and then try to extrapolate their findings to get a comprehensive understanding of this interaction. In short, it can be stated that the interaction of the plant and insects at the ecological level is the outcome of the interactions of the cascade of genes at the molecular level. Herbivore attack results in a cascade of gene activation. With the advancement in molecular approaches, such as whole-genome sequences, microarrays, differential cloning techniques, etc., changes in gene expression profiles in herbivore-damaged plants have been investigated extensively in recent years (Korth 2003; Hermsmeier et al. 2001; Hui et al. 2003). However, the discussion in this chapter is limited to the ecological perspective.

2.1.1 Defensive Types

Plants have adapted themselves to insect and pathogen attack by evolving constitutive and induced defense mechanisms. Defenses through constitutive way are inherently present in the plants, and induced defenses are the ones which are activated by the attack of herbivore or pathogens. These defenses can be provided by the mechanical structures like thorns, leaf surface wax, trichomes, etc. which form the primary physical barriers to the ovipositing and feeding herbivore and/or can be provided by the chemical barriers such as nonvolatile chemicals (e.g., antinutritive enzymes, toxins, and deterrents) that form the next barrier defending the plants from their succeeding attack (Kessler and Baldwin 2002; Hanley et al. 2007; Schilmiller et al. 2008; Runyon et al. 2010) or volatile organic chemicals (VOCs) that are released against the attack of herbivore. The physical traits and nonvolatile chemicals are considered as direct defense system as they act directly against the aggressor, whereas the VOCs are considered as the indirect defense system of the plant as they deploy natural predators of herbivore like parasitoids to manage the populations of the herbivores, and they also serve to pass signals for plant-plant communication (Karban and Baldwin 1997; Dudareva and Pichersky 2008).

Apart from the volatile chemicals, some physical structures of the plants also contribute towards the indirect defense system. Plants, by possessing domatia, provide nesting or refuge sites to predators to get indirect defense from herbivores (Heil 2008). Recently, glandular trichome structures have been found to contribute in the indirect defensive mechanisms (Weinhold and Baldwin 2011). Glandular trichomes emit diversified secondary molecules including terpenoids, flavonoids, and alkaloids that can act as a chemical and structural defense (Hanley et al. 2007; Sharma et al. 2009). Diverse defensive chemicals are plant toxins, so the chemicals for constitutive defense are synthesized and stored separately as inactive precursors, like glycosides, to avoid poisoning themselves (Jones and Vogt 2001), which ensures separation from the activating enzymes.

Both the direct and indirect defenses can be constitutive and induced defense system. In response to herbivory on lima beans (*Phaseolus vulgaris*) and *Acacia* tree, the plants were found to discharge nectar that attracted ants which protected the host from herbivores (Heil et al. 2004; Heil 2004). While volatiles are usually secreted in a programmed manner but the quality and quantity of the VOCs may change under stress from herbivore or pathogen attack (Heil 2008). The herbivore-induced plant volatile (HIPV) contains more than 200 chemicals (Dicke and van Loon 2000) (Fig. 1) which function in an orchestrated manner to defend the plant from the insect herbivory. During quantitative changes, the volatiles existing in the non-induced plants are secreted in higher amounts or the comparative abundance is changed (e.g., Mumm et al. 2003; Bukovinszky et al. 2005). Qualitative changes indicate production of de novo many chemicals in plants resulting in the qualitative modifications of the volatile emissions (Turlings et al. 1998; Dicke 1999; Leitner et al. 2005). Plant genotypes are also associated with the kind of the volatile mixture released after such attack (Halitschke et al. 2000; Fritzsche-Hoballah et al. 2002). Other important factors include light or nutrients (Gouinguene and Turlings 2002; Fritzsche-Hoballah et al. 2002), the mode of damage (Mithöfer et al. 2005), developmental stage of the aggressive herbivores and the species (Ozawa et al. 2000; Heil 2008), kind of the herbivore (Paré and Tumlinson 1997), and egg deposition (Hilker and Meiners 2006). The plant volatile emissions are triggered by the elicitors present in the saliva of the insects but cannot be induced by the mechanical damage of the plant (Truitt and Pare 2004; Schmelz et al. 2006). For example, volicitin (N-(17-hydroxylinolenoyl)-L-glutamine) compound isolated from the oral emissions of caterpillars (beet armyworm) was found to attract parasitic wasps and the natural predators of the caterpillars, but the mechanical injury of the leaves did not attract the natural enemies (Alborn et al. 1997). The HIPVs are also involved in decreasing the rates of oviposition by the herbivores and thus are a part of both direct and indirect defense systems (Dicke and van Loon 2000; Kessler and Baldwin 2001).

Previous research gives an insight into the plant acute sensory capacity highlighting the mechanism of chemoreception in plants (Mescher and de Moraes

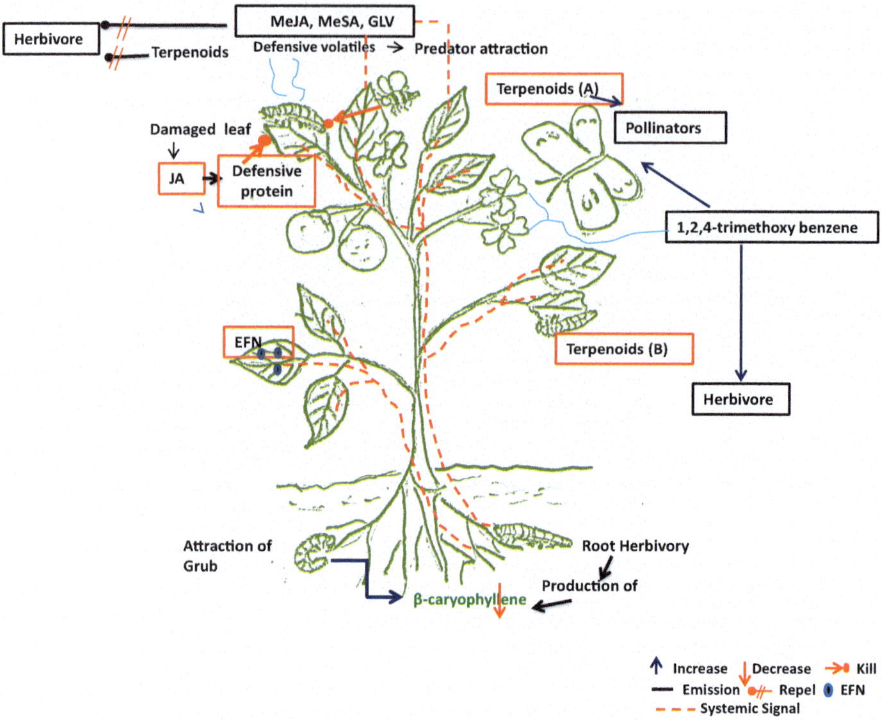

Fig. 1 Communication between plants and insects. Root herbivory results in production and increase of β-caryophyllene, which attracts beetle larvae. Herbivory-damaged leaves work in two ways – firstly by release of defensive volatiles [terpenoids (A), MeJA (methyl jasmonic acid), MeSA (methyl salicylic acid), and GLV (green leaf volatiles)] that attract the predator and secondly by the release of jasmonic acid (JA), defensive proteins, and terpenoids (B) which will repel the insects from further attack. Other volatiles like 1,2,4-trimethoxy benzene from the plants serve to attract both the pollinators and herbivores, but the terpenoids (B) stored in glandular trichomes will serve to repel the herbivore. Root herbivory also increases EFN (extrafloral nectar) (Das et al. 2012)

2015). Mechanical response like sudden closure of leaves of *Mimosa pudica* on touch is commonly alleged to function as a herbivory defense, possibly by startling or physically displacing herbivores, decreasing the leaf apparency, or promoting the defensive thorn display (Braam 2005), which is again a type of direct defense of plants. Similarly, vibrations linked with herbivore mastication result in the release of defensive chemicals in *Arabidopsis*, even over less distance from the wounding site (Appel and Cocroft 2014). Such vibrations along with other phytohormonal signalling pathways like jasmonic acid or salicylic acid pathway cause a systemic response in the plants. The vibrations also travel between plants through stems or roots, sending the distress signals to the nearby plants, signalling them for the herbivore presence and triggering the defense processes in them (Cokl and Virant-Doberlet 2003).

2.1.2 Plant Specialized Metabolites Against Insect Herbivores

Chemical defense in plants is provided by the specialized metabolites that work against insects and pathogens (Fraenkel 1959) besides their role to mediate interactions with other biotic components and abiotic stresses (Sharkey et al. 2001; Munné-Bosch 2005; Karban et al. 2006; Heil and Karban 2010; Lucas-Barbosa et al. 2011). So far, >1700 chemicals have been isolated from different families of plants (Pichersky and Gershenzon 2002). Their diversity in terms of number, functionality, structure, etc. makes it difficult to draw a clear classification system. However, a simple classification system widely followed divides these metabolites into three main groups: the terpenoids (composed almost entirely of carbon and hydrogen), nitrogen-containing alkaloids, and phenolics (Schoonhoven et al. 2005). The specialized plant chemicals which are found active against insect herbivores are listed in Table 1.

These metabolites have variable impacts on the fitness of the insects as well as their behavior and physiology. Only the effects of a few representative chemicals are discussed in this chapter being dealt with in detail. Terpenoids are the chief group of all known secondary derived compounds and exhibit extreme structural diversity and as a result diverse chemical properties and functionality (Nagegowda 2010). Monoterpenes among them are the most abundant volatile terpenoids that play a crucial role in herbivore defense. They act as feeding deterrents largely but are also reported to affect the oviposition behavior of the insect (Mumm et al. 2008). Another group of secondary metabolites, glucosinolates, present in plants belonging to Brassicaceae family are important component in plants' constitutive defense system (Hopkins et al. 2009).

Alkaloids, another important secondary metabolites, are natural organic compounds containing a heterocyclic ring with nitrogen, which serve as plant growth inhibitors as well as feeding deterrents (Saunders et al. 1992). Studies have shown that alkaloids exhibit high feeding deterrency to *Lymantria dispar*, *Spodoptera litura,* and *Lipaphis erysimi* (Shields et al. 2008; Ge et al. 2015). Plant phenols are another common group of plant defense playing a chief role in host resistance and are effective on both herbivore insects and microorganisms (War et al. 2012). Synthesis of lignin, a phenolic heteropolymer, is induced by herbivory, and it results in increasing the toughness of the leaf, thereby reducing the herbivore feeding (Johnson et al. 2009). Flavonoids and isoflavonoids also provide protection against the pests as they are potent antifeedants and have effects on the development and growth of the insects. Approximately, more than 5000 flavonoids have been characterized in plants. For example, flavonoids extracted from Tephrosia..T. vogelii were found to be a strong feeding deterrent against Spodoptera sp. (Simmonds et al. 1990). Overexpression of transcription factor, which manages major flavonoid synthesis in *Arabidopsis* was found to confer resistance against *Spodoptera frugiperda* (Johnson and Dowd 2004). Tannins, another secondary metabolite, reduced the growth rate of pests (Schoonhoven et al. 2005), act as feeding deterrent (Feeny 1968; Bernays 1981), and affect the composition of insect communities (Whitham et al. 2006).

Table 1 List of some of the common secondary plant metabolites

Source	Compounds	Plant family	Chemical structure	References
Steroids	Phytoecdyson	Ranunculaceae	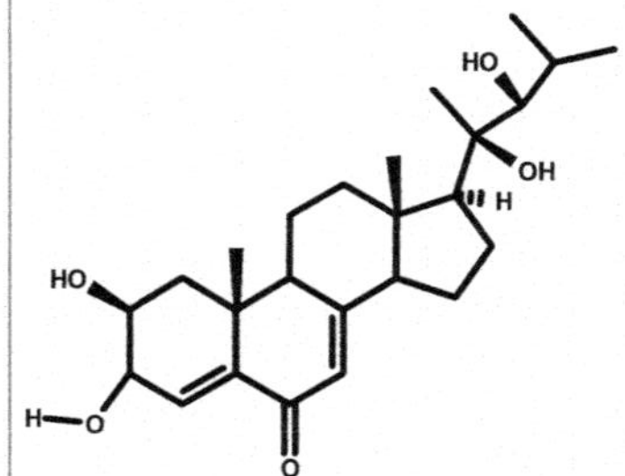	Bakkali et al. (2008)
Terpenoids	(E)-Beta-farnesene	Ubiquitous	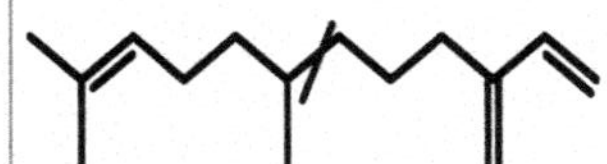	Aharoni et al. (2003)
Alkaloids	Nicotine	Solanaceae	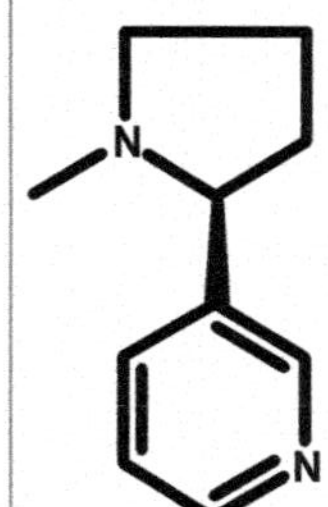	Mithofer and Boland (2012)

Glucosinolates	Sinigrin	Capparales	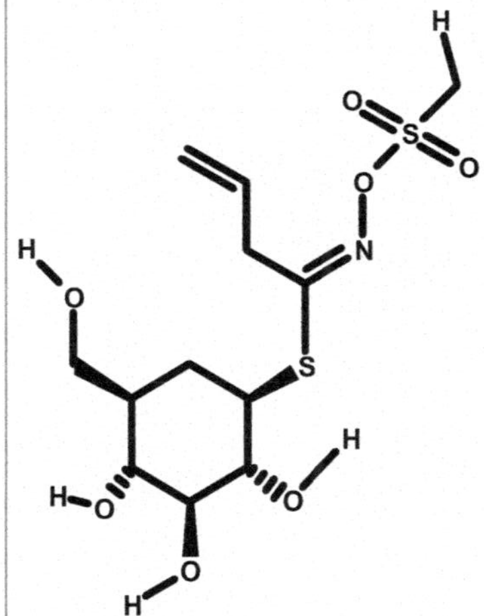	Hopkins et al. (2009)
Phenolics	Tannin	Ubiquitous	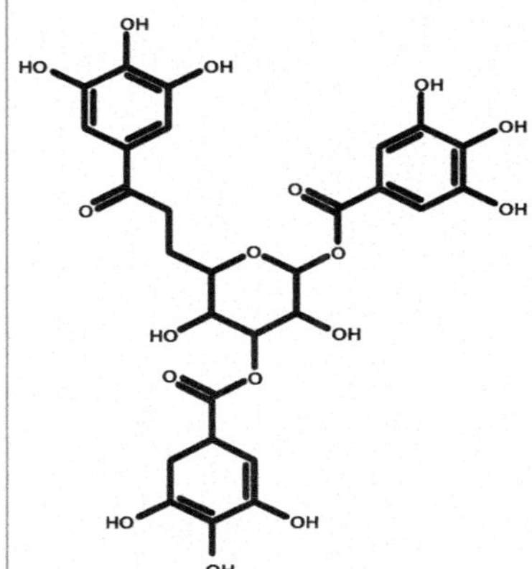	Coley (1986)

(continued)

Table 1 (continued)

Source	Compounds	Plant family	Chemical structure	References
Amino acids	γ-Aminobutyric acid	Fabaceae	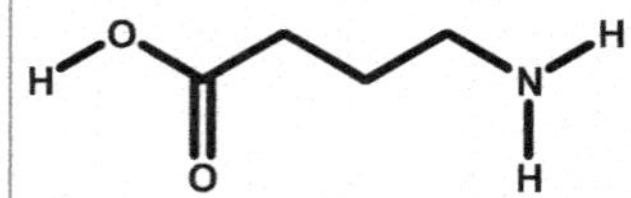	Huang et al. (2011)
Glucosides	Dhurrin	Rosaceae	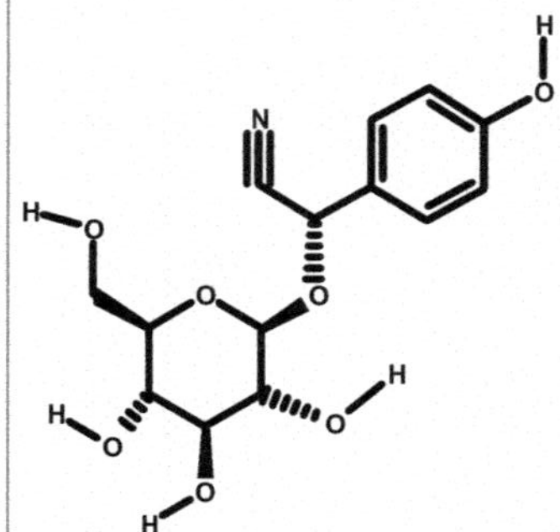	Vetter (2000)
Silica	Silicon dioxide	Poaceae		Reynolds et al. (2009)
Cardenolides	Digoxigenin	Plantaginaceae	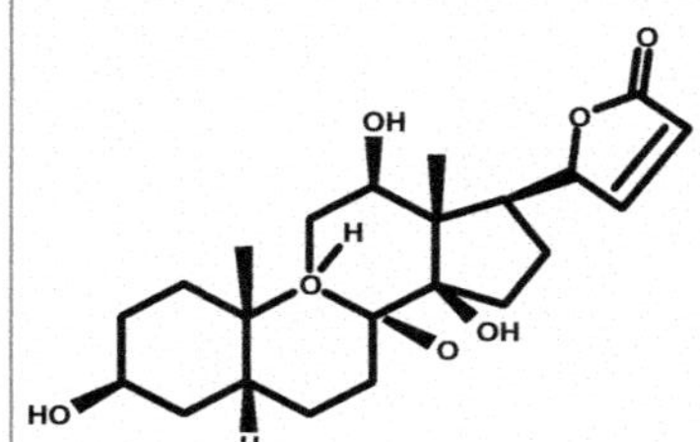	Mithofer and Boland (2012)

2.2 Insect Responses to Plant

Insects have developed various ways to enable them to exploit their hosts (Panda and Khush 1995). Insects are dependent on plants, and therefore their community composition is largely influenced by the plant morphology, physiology, chemistry, and genetics (Bukovinszky et al. 2010; Ohgushi 2008). These attributes of plants influence the behavior of insects on their settlement, feeding, growth and development, oviposition, etc. However, it is evident that whatsoever strategies plants evolve to evade insect attack, insects also developed ways to overcome them and are one of the most successful organisms in terms of variety of species and their geographical distribution (Price 1997).

Generalist insects are mostly highly sensitive to defensive compounds of the plants, while specialist insects will utilize these compounds as cues of recognition (Ali and Agrawal 2012; Gols et al. 2008; Schoonhoven et al. 2005). The difference in the responses of the two groups is due to the fact that insects have adapted to plant chemicals through specific sequestration or detoxification mechanism, whereas generalist insects are either incapable to grow or show reduced survival on these plants (Ali and Agrawal 2012; Hopkins et al. 2009). Insects have developed a wide array of enzymes as defense against the various compounds produced by the plants. One of the strategies adopted by insects is to detoxify the plant chemicals by reduction, oxidation, conjugation, or hydrolysis of the compounds (Scott and Wen 2001). Largely the detoxification of plant compounds by insects is done by cytochrome P450 (CYP450)-dependent monooxygenases (Mello and Silva-Filho 2002). In some cases, insects sequester the toxic compound in a manner that enable them to use it for their own defense against parasitoids (Nishida 2002) or feed on those part of the plants that lack this compound (Zangerl 1990). In the tobacco hornworm, insects are able to store the nicotine of the plant which is otherwise toxic to other pests, and they utilize the sequestered nicotine as parasitoid deterrents (de Bruxelles and Roberts 2001).

3 Evolution of Insect-Plant Interaction

Unraveling the evolutionary relationship between insects and plants has been a matter of extensive research in the biologist community. Different workers have given different theories to underline the evolutionary relationship between them. The most common perspective is that of coevolution (Ehrlich and Raven 1964), which explains the diversity of insects as a filler of niches resulted due to diversification of the plants, which was, in turn, the result of the development of the defensive chemicals in plants in response to herbivory. However, the term "coevolution" being very broad is used in different perspectives like *co-speciation and reciprocal adaptation* and as a concept of "*escape and radiation*" to explain the evolutionary processes that have modelled the insect and plant diversity.

According to Janzen (1980) and Futuyma and Slatkin (1983), coevolution can be explained by the process of reciprocal adaptation where evolutionary changes in a species bring about changes in evolution in another species through the natural selection. The account of co-speciation of plants and insects can be well understood by studying the interaction that exists between *Ficus* and agaonine wasps (Herre et al. 1996). In plant-insect herbivore interaction, the insect is the only one that is dependent on its partner, unlike in plant-pollinator interactions, which can be better explained by phylogenetic tracking evolution (escape and radiate) rather than co-speciation (Kerogoat et al. 2017).

Secondary plant metabolites present in plants are important from defense point of view (Fraenkel 1959) and are the result of the coevolutionary war between insects and plants that resulted in subsequent increase in the variety of different defensive compounds in plants (Ehrlich and Raven 1964). In an insect attack, the plants suffer by undergoing tissue damage and also by investing energy for the development of resistance. These defense systems come with the cost of investing energy for the toxin synthesis and sequestration, developing resistance against the insect pests or self-toxicity (Gershenzon 1994; Baldwin and Preston 1999; Purrington 2000). These costs are a crucial component for the resistance evolution since an evolutionary equilibrium needs to be maintained for herbivory reduction and plant fitness (Paul et al. 2000). One way to reduce the investment by the plants is by developing inducible defenses whereby plant will invest in defense only (Agrawal 2000). Plants that are rarely attacked have therefore evolved the mechanism to reduce the cost involved by synthesizing defensive chemicals only upon a herbivore or pathogen damage (induced defense). In contrast, those plants which are likely to get frequent attack are better off in investing energy in constitutive defensive system (Wittstock and Gershenzon 2002). Review by Becerra (2015) highlights the macroevolutionary intensification of chemicals in plants as a result of selection pressure exerted by insect herbivore. Such pressures are responsible for escalation in varied chemical compounds and biosynthetic pathway through time in plants which in some cases have resulted in the addition of novel synthetic pathway, consequently delaying the counter adaptation by the herbivore indicating the *escape and radiate* kind of evolution (Ehrlich and Raven 1964; Kerogoat et al. 2017). Phylogenetic analysis based on character evolution and association is important to understand the underlying evolutionary trends of insect-plant interactions (Wanntorp et al. 1990; Harvey and Pagel 1991; Mitter et al. 1991). Studies comprising geographical variations in these interactions may provide an insight for such processes (Thompson 1999). The interaction between the two systems is very diverse and complex in true sense and needs more and more multidisciplinary research to understand the evolution of plant resistance and defense (Futuyma 2000).

4 Future Prospect of Understanding Insect-Plant Interactions

Understanding the complex interactions existing between plants and insects at the different multitrophic level has a significant implication in the field of agriculture and forestry. Communication between them which initiates at the molecular level and reflected ultimately at the ecological level gives us an insight about their diversification and speciation process. Both the systems have tremendous role in the ecosystem; in fact it would not be exaggerating to say that the life system on the Earth revolves around the interaction and the functionality of the two. Research in this direction has revealed a lot of information between the two interacting systems, but with advancements in the molecular technology, more insights need to be developed that can be utilized for the more pertinent issues like ecological balance, food security, and sustainable development.

References

Agrawal AA (2000) Benefits and costs of induced plant defense for *Lepidium virginicum* (brassicaceae). Ecology 81(7):1804–1813

Aharoni A, Giri AP, Deuerlein S, Griepink F, de Kogel WJ, Verstappen FWA, Verhoeven HA, Jongsma MA, Schwab W, Bouwmeester HJ (2003) Terpenoid metabolism in wild-type and transgenic Arabidopsis plants. Plant Cell 15:2866–2884

Alborn HT, Turlings TCJ, Jones TH, Stenhagen G, Loughri JH, Tumlinson JH (1997) An elicitor of plant volatiles from beet armyworm oral secretion. Science 276:945–949

Ali JG, Agrawal AA (2012) Specialist versus generalist insect herbivores and plant defense. Trends in Plant Sci 17:293–302

Appel HM, Cocroft RB (2014) Plants respond to leaf vibrations caused by insect herbivore chewing. Oecologia:1–10

Bakkali F, Averbeck S, Averbeck D, Waomar M (2008) Biological effects of essential oils: a review. Food Chem Toxicol 46:446–475

Baldwin IT, Preston CA (1999) The eco-physiological complexity of plant responses to insect herbivores. Planta 208:137–145

Becerra JX (2015) Macroevolutionary and geographical intensification of chemical defense in plants driven by insect herbivore selection pressure. Curr Opin Insect Sci 8:15–21

Bernays EA (1981) Plant tannins and insect herbivores: an appraisal. Ecol Entomol 6:353–360

Bernays EA (1996) Selective attention and host-plant specialization. Entomol Exp Appl 80 (1):125–131

Braam J (2005) In touch: plant responses to mechanical stimuli. New Phytol 165:373–389

Bukovinszky T, Gols R, Posthumus MA, van Lenteren JC, Vet LEM (2005) Variation in plant volatiles and the attraction of the parasitoid *Diadegma semiclausum* (Hellen). Journal of Chem Ecol 31:461–480

Bukovinszky T, Gols R, Kamp A, De Oliveira-Domingues F, Hamback PA, Jongema Y, Bezemer TM, Dicke M, Van Dam NM, Harvey JA (2010) Combined effects of patch size and plant nutritional quality on local densities of insect herbivores. Basic Appl Ecol 11:396–405

Casal JJ, Fankhauser C, Coupland G, Blázquez MA (2004) Signalling for developmental plasticity. Trends Plant Sci 9:309–314

Čokl A, Virant-Doberlet M (2003) Communication with substrate-borne signals in small plant-dwelling insects. Annu Rev Entomol 48:29–50

Coley PD (1986) Costs and benefits of defense by tannins in a neotropical plant. Oecologia 70:238–241

Cook JM, Rasplus J-Y (2003) Mutualists with attitude: coevolving fig wasps and figs. Trends Ecol Evol 18:241–248

Das A, Lee S, Hyun TK, Kim S, Kim J (2012) Plant volatiles as method of communication. Plant Biotechnol Rep 7:9–26

de Brito FR, Martinoia E (2018) The vacuolar transportome of plant specialized metabolites. Plant Cell Physiol *59*(*7*):1326–1336. https://doi.org/10.1093/pcp/pcy039

de Bruxelles GL, Roberts MR (2001) Signals regulating multiple responses to wounding and herbivores. Crit Rev in Plant Sci 20(5):487–521
Dicke M (1999) Evolution of induced indirect defense of plants. In: Tollrian R, Harvell CD (eds) The ecology and evolution of inducible defenses. Princeton University Press, Princeton, pp 62–88
Dicke M, van Loon JJA (2000) Multitrophic effects of herbivore-induced plant volatiles in an evolutionary context. Entomol Exp Appl 37:237–249
Dicke M, van Loon JJA, Soler R (2009) Chemical complexity of volatiles from plants induced by multiple attack. Nat Chem Biol 5:317–324
Dudareva N, Pichersky E (2008) Metabolic engineering of plant volatiles. Curr Opin Biotechnol 19:181–189
Ehrlich PR, Raven PH (1964) Butterflies and plants: a study in coevolution. Evolution 18:586–608
Feeny PP (1968) Effect of oak leaf tannins on larval growth of the winter moth *Operophtera brumata*. J Insect Physiol 14:805–817
Fontaine C, Thébault E, Dajoz I (2009) Are insect pollinators more generalist than insect herbivores? Proceed Royal Soc B Biol Sci 276:3027–3033
Fraenkel GS (1959) The Raison d'Être of secondary plant substances. Science 129:1466–1470
Fritzsche-Hoballah ME, Tamo C, Turlings TCJ (2002) Differential attractiveness of induced odors emitted by eight maize varieties for the parasitoid *Cotesia marginiventris*: is quality or quantity important? J Chem Ecol 28:951–968
Futuyma DJ (2000) Some current approaches to the evolution of plant-herbivore interactions. Plant Species Biol 15:1–9
Futuyma DJ, Moreno G (1988) The evolution of ecological specialization. Annu Rev Ecol Syst 19:207–233
Futuyma DJ, Slatkin M (1983) Introduction. In: Futuyma DJ, Slatkin M (eds) Coevolution Sinauer Associates, Sunderland, pp 1–13.
Gall LF (1987) Leaflet position influences caterpillar feeding and development. Oikos 49:172–176
Garrouste R, Clement G, Nel P, Engel MS, Grandcolas P, D'Haese C, Lagebro L, Denayer J, Gueriau P, Lafaite P, Olive S, Prestianni C, Nel A (2012) A complete insect from the Late Devonian period. Nature 488:82–86
Ge Y, Liu P, Yang R, Zhang L, Chen H, Camara I, Liu Y, Wangpeng S (2015) Insecticidal Constituents and Activity of Alkaloids from *Cynanchum mongolicum*. Molecules 20:17483–17492
Gershenzon J (1994) Metabolic costs of terpenoid accumulation in higher plants. J Chem Ecol 20:1281–1328
Gols R, Bukovinszky T, Van Dam NM, Dicke M, Bullock JM, Harvey JA (2008) Performance of generalist and specialist herbivores and their endoparasitoids differs on cultivated and wild Brassica populations. J Chem Ecol 34:132–143
Gouinguene SP, Turlings TCJ (2002) The effects of abiotic factors on induced volatile emissions in corn plants. Plant Physiol 129:1296–1307
Gutbrodt B, Modv K, Dorn S (2011) Drought changes plant chemistry and causes contrasting responses in lepidopteran herbivores. Oikos 120(11):1732–1740
Halitschke R, Keßler A, Kahl J, Lorenz A, Baldwin IT (2000) Ecophysiological comparison of direct and indirect defenses in *Nicotiana attenuata*. Oecologia 124:408–417
Hanley ME, Lamont BB, Fairbanks MM, Rafferty CM (2007) Plant structural traits and their role in antiherbivore defense. Perspec. Plant Ecol Evol Syst 8:157–178. https://doi.org/10.1016/j.ppees.2007.01.001
Harvey PH, Pagel M (1991) The Comparative Method In Evolutionary Biology. Oxford Ser Ecol Evol 1:248
Heil M (2004) Direct defense or ecological costs: Responses of herbivorous beetles to volatiles released by wild lima bean (*Phaseolus lunatus*). J Chem Ecol 30:1289–1295
Heil M (2008) Indirect defence via tritrophic interactions. New Phytol 178:41–61

Heil M, Karban R (2010) Explaining evolution of plant communication by airborne signals. Trends Ecol Evol 25:137–144

Heil M, Greiner S, Meimberg H, Kruger R, Noyer JL, Heubl G, Linsenmair KE, Boland W (2004) Evolutionary change from induced to constitutive expression of an indirect plant resistance. Nature 430:205–208

Hermsmeier D, Schittko U, Baldwin IT (2001) Molecular interactions between the specialist herbivore *Manduca sexta* (Lepidoptera, Sphingidae) and its natural host *Nicotiana attenuata*. I. Large-scale changes in the accumulation of growth- and defense-related plant mRNAs. Plant Physiol 125:683–700. https://doi.org/10.1104/pp.125.2.683

Herre EA, Machado CA, Bermingham E, Nason JD, Windsor DM, McCafferty SS, van Houten W, Bachmann K (1996) Molecular Phylogenies of Figs and Their Pollinator Wasps. J Biogeogr 23:521–530

Hilker M, Meiners T (2006) Early herbivore alert: insect eggs induce plant defense. J Chem Ecol 32:1379–1397

Hopkins RJ, van Dam NM, van Loon JJA (2009) Role of Glucosinolates in Insect-Plant Relationships and Multitrophic Interactions. Ann Rev Entomol 54:58–73

Huang T, Jander G, de Vos M (2011) Non-protein amino acids in plant defense against insect herbivores: representative cases and opportunities for further functional analysis. Phytochem 72:1531–1537

Hui D, Iqbal J, Lehmann K, Gase K, Saluz HP, Baldwin IT (2003) Molecular interactions between the specialist herbivore *Manduca sexta* (Lepidoptera, Sphingidae) and its natural host *Nicotiana attenuata*. V: Microarray analysis and further characterization of large-scale changes in herbivore-induced mRNAs. Plant Physiol 131:1877–1893. https://doi.org/10.1104/pp.102.018176

Janzen DH (1980) When is it coevolution? Evolution 34(3):611–612

Johnson ET, Dowd PF (2004) Differentially enhanced insect resistance, at a cost, in *Arabidopsis thaliana* constitutively expressing a transcription factor of defensive metabolites. J Agric Food Chem 52:5135–5138. PMID:15291486. https://doi.org/10.1021/jf0308049.

Johnson MTJ, Smith SD, Rausher MD (2009) Plant sex and the evolution of plant defenses against herbivores. Proc Natl Acad Sci U S A 106:18079–18084.; PMID:19617572. https://doi.org/10.1073/pnas.0904695106

Jones P, Vogt T (2001) Glycosyltransferases in secondary plant metabolism: tranquilizers and stimulant controllers. Planta 213:164–174

Karban R, Baldwin IT (1997) Induced responses to herbivory. Chicago University Press, Chicago, 319 pp

Karban R, Shiojiri K, Huntzinger M, McCall AC (2006) Damage-induced resistance in sagebrush: volatiles are key to intra- and interplant communication. *Ecology* 87:922–930

Kerogoat GJ, Meseguer AS, Jousselin E (2017) Evolution of plant-insect interaction: Insights from evolutionary approaches in plant and herbivorous insects. Adv Bot Res 81:25–53

Kessler A, Baldwin IT (2001) Defensive function of herbivore-induced plant volatile emissions in nature. Science 291(5511):2141–2144

Kessler A, Baldwin IT (2002) Plant responses to insect herbivory: the emerging molecular analysis. Annu Rev Plant Biol 53:299–328

Kimmerer TW, Potter DA (1987) Nutritional quality of specific leaf tissues and selective feeding by a specialist leafminer. Oecologia 71:548–551

Korth KL (2003) Profiling the response of plants to herbivorous insects. Genome Biol 4(7):221

Leitner M, Boland W, Mithöfer A (2005) Direct and indirect defences induced by piercing-sucking and chewing herbivores in *Medicago truncatula*. New Phytol 67(2):597–606

Lucas-Barbosa D, van Loon JJA, Dicke M (2011) The effects of herbivore-induced plant volatiles on interactions between plants and flower-visiting insects. Phytochem 72:1647–1654

Mello MO, Silva-Filho MC (2002) Plant-insect interactions: an evolutionary arms race between two distinct defense mechanisms. Braz J Plant Physiol 14(2):71–81

Mescher MC, de Moraes CM (2015) Role of plant sensory perception in plant-animal interactions. J Exp Bot 66(2):425–433. https://doi.org/10.1093/jxb/eru414. Epub 2014 Nov 4

Mithofer A, Boland W (2012) Plant defense against herbivores: chemical aspects. Annu Rev Plant Biol 63:431–450

Mithöfer A, Wanner G, Boland W (2005) Effects of feeding Spodoptera littoralis on Lima bean leaves. II. Continuous mechanical wounding resembling insect feeding is sufficient to elicit herbivory-related volatile emission. Plant Physiol 137:1160–1168

Mitter C, Farell BD, Futuyma DJ (1991) Phylogenetic Studies of Insect-Plant Interactions Insights into the Genesis of Diversity. Trends Ecol Evol 6(9):290–293

Mumm R, Schrank K, Wegener R, Schulz S, Hilker M (2003) Chemical analysis of volatiles emitted by *Pinus sylvestris* after induction by insect oviposition. J Chem Ecol 29:1235–1252

Mumm R, Posthumus MA, Dicke M (2008) Significance of terpenoids in induced indirect plant defence against herbivorous arthropods. Plant Cell Env 31:575–585

Munné-Bosch S (2005) The role of alpha-tocopherol in plant stress tolerance. J Plant Physiol 162 (7):743–748

Nagegowda DA (2010) Plant volatile terpenoid metabolism: biosynthetic genes, transcriptional regulation and subcellular compartmentation. FEBS Lett 584(14):2965–2973

Nishida R (2002) Sequestration of defensive substances from plants by Lepidoptera. Annu Rev Entomol 47:57–92

Ohgushi T (2008) Herbivore-induced indirect interaction webs on terrestrial plants: the importance of non-trophic, indirect, and facilitative interactions. Entomol Exp Appl 128:217–229

Ozawa R, Arimura G, Takabayashi J, Shimoda T, Nishioka T (2000) Involvement of jasmonate- and salicylate-related signaling pathways for the production of specific herbivore-induced volatiles in plants. Plant Cell Physiol 41:391–398

Panda N, Khush GS (1995) Host plant resistance to insects. CAB International, Wallingford

Pare PW, Tumlinson JH (1997) De novo biosynthesis of volatiles induced by insect herbivory in cotton plants. Plant Physiol 114:1161–1167

Paul ND, Hatcher PE, Taylor JE (2000) Coping with multiple enemies: an integration of molecular and ecological perspectives. Trends Plant Sci 5:220–225

Pecetti L, Biazzi E, Tava A (2010) Variation in saponin content during the growing season of spotted medic *Medicago arabica* (L.) Huds. J Sci Food Agric 90:2405–2410

Picersky E, Lewinsohn E (2011) Convergent evolution in plant specialized metabolism. Annu Rev Plant Biol 62:549–566

Pichersky E, Gershenzon J (2002) The formation and function of plant volatiles: perfumes for pollinator attraction and defense. Curr Opin Plant Biol 5:237–243

Price PW (1997) Insect ecology, 3rd edn. Wiley, New York

Purrington CB (2000) Costs of resistance. Curr Opin Plant Biol 3(4):305–308

Randlkofer B, Obermaier E, Hilker M, Meiners T (2010) Vegetation complexity – the influence of plant species diversity and plant structure on plant chemical complexity and arthropods. Basic Appl Ecol 11(5):383–395

Rehm P, Borner J, Meusemann K, von Reumont BM, Simon S, Hadrys H, Misof B, Burmester T (2011) Dating the arthropod tree based on large-scale transcriptome data. Mol Phylogenet Evol 61:880–887

Reynolds OL, Keeoing MG, Meyer JH (2009) Silicon-augmented resistance of plants to herbivorous insects: a review. Ann Appl Biol 155:171–186

Runyon JB, Mescher MC, Felton GW, de Moraes CM (2010) Parasitism by *Cuscuta pentagona* sequentially induces JA and SA defence pathways in tomato. Plant Cell Environ 33:290–303

Saunders JA, O'Neill NR, Romeo JT (1992) Alkaloid chemistry and feeding specificity of insect herbivores. In: Pelletier SW (ed) Alkaloids: chemical and biological perspective. Springer, New York, pp 151–196

Saxena KN (1969) Patterns of insect-plant relationships determining susceptibility or resistance of different plants to an insect. Entomol Exp Appl 12:751–766

Schilmiller AL, Last RL, Pichersky E (2008) Harnessing plant trichome biochemistry for the production of useful compounds. Plant J 54:702–711

Schmelz EA, Carroll MJ, LeClere S, Phipps SM, Meredith J, Chourey PS, Alborn HT, Teal PEA (2006) Fragments of ATP synthase mediate plant perception of insect attack. Proc Nat Acad Sci USA 103:8894–8899

Schoonhoven LM, van Loon JJA, Dicke M (2005) Insect-plant biology, 2nd edn. Oxford University Press, Oxford

Schulz JC (1988) Many factors influence the evolution of herbivore diets, but plant chemistry is central. Ecology 69(4):896–897

Scott JG, Wen ZM (2001) Cytochromes P450 of insects: he tip of the iceberg. Pest Manag Sci 57:958–967

Scriber JM, Slansky F (1981) The nutritional ecology of immature insects. Ann Rev Entomol 26:183–211

Sharkey TD, Chen X, Yeh S (2001) Isoprene increases thermotolerance of fosmidomycin-fed leaves. Plant Physiol 25(4):2001–2006

Sharma HC, Sujana G, Rao DM (2009) Morphological and chemical components of resistance to pod borer, *Helicoverpa armigera* in wild relatives of pigeonpea. Arthropod Plant Interact 3:151–161. https://doi.org/10.1007/s11829-009-9068-5

Shields VDC, Smith KP, Arnold NS, Gordon IM, Shaw TE, Warancjh D (2008) The effect of varying alkaloid concentrations on the feeding behavior of gypsy moth larvae, *Lymantria dispar* (L.) (Lepidoptera: Lymantriidae). Arthropod Plant Interact 2(2):101–107

Simmonds MSJ, Blaney WM, Fellows LE (1990) Behavioural and electrophysiological study of antifeedant mechanisms associated with polyhydroxyalkaloids. J Chem Ecol 16:3167–3196. https://doi.org/10.1007/BF00979618.

Stam JM, Kroes A, Li Y, Gols R, van Loon JJ, Poelman EH, Dicke M (2014) Plant interactions with multiple insect herbivores: from community to genes. Annu Rev Plant Biol 65:689–713

Thompson JN (1999) Specific hypothesis on the geographic mosaic of coevolution. Am Nat 153 (Suppl):S1–S14

Tissier A, Ziegler J, Vogt T (2014) Specialized plant metabolites: diversity and biosynthesis. In: Krauss, G-J, Nies DH (eds) Ecological biochemistry: environmental and interspecies interactions, 1. Wiley-VCH Verlag GmbH & Co. KGaA

Truitt CL, Paré PW (2004) In situ translocation of volicitin by beet armyworm larvae to maize and systemic immobility of the herbivore elicitor *in planta*. Planta 218:999. https://doi.org/10.1007/s00425-003-1173-6

Turlings TCJ, Bernasconi M, Bertossa R, Bigler F, Caloz G, Dorn S (1998) The induction of volatile emissions in maize by three herbivore species with different feeding habitats: possible consequences for their natural enemies. Biol Control 11:122–129

Vetter J (2000) Plant cyanogenic glycosides. Toxicon 38:11–36

Wanntorp HE, Brooks DR, Nilsson R, Nylin E, Ronfquist F, Steams SC, Wedell N (1990) Phylogenetic approaches in ecology. Oikos 57:119–132

War AR, Paulraj MG, Ahmad T, Buhroo AA, Hussain B, Ignacimuthu S, Sharma HC (2012) Mechanisms of plant defense against insect herbivores. Plant Signal Behav 7:1306–1320. https://doi.org/10.4161/psb.21663

Weinhold A, Baldwin IT (2011) Trichome-derived *O*-acyl sugars are a first meal for caterpillars that tags them for predation. Proc Nat Acad Sci USA 108:7855–7859

Whitham TG, Bailey JK, Schweitzer JA, Shuster SM, Bangert RK, LeRoy CJ, Lonsdorf EV, Allan GJ, DiFazio SP, Potts BM, Fischer DG, Gehring CA, Lindroth RL, Marks JC, Hart SC, Wimp GM, Wooley SC (2006) A framework for community and ecosystem genetics: from genes to ecosystems. Nature Rev Genet 7:510–523

Wittstock U, Gershenzon J (2002) Constitutive plant toxins and their role in defense against herbivores and pathogens. Curr Opin Plant Biol 5:1–8

Zangerl AR (1990) Furanocoumarin induction in wild parsnip: evidence for an induced defense against herbivores. Ecology 71(5):1926

Glossary

Abiotic Factors Non-living or physical conditions which impact the distribution of population in an ecosystem.

Abiotic Stresses Abiotic stress is the negative impact of non-living factors on living organisms in a specific environment.

Agglutination Process causing clumping of red blood cells or bacteria in the liquid as a serological response to specific antibody.

Alarming Signal An antipredator adaptation in the form of signals emitted by social animals in response to danger.

Allelochemicals Chemicals produced by living organisms that exerts an injurious physiological effect on individuals of another species when released into the environment.

Allergens Antigen that produces abnormally strong immune response.

Antagonism *Antagonism* is hostility that results in active resistance, opposition, or contentiousness.

Antennal Lobe The antennal lobe is the deutocerebral neuropil of insects which receives the input from the olfactory sensory neurons on the antenna.

Antibiosis Antibiosis is a biological interaction between two or more organisms that is damaging to at least one of the interacting partner; it can also be an antagonistic association between an organism and the metabolic substances produced by another.

Antioxidant Defense Antioxidants are compounds that inhibit oxidation. Oxidation is chemical reactions that produce free radicals leading to chain reactions that damage the cells of organisms. Enzymes that can efficiently sequester the free radicals build up the defense system.

Antixenoxis Refers to a plant that has characteristics that lead pests away from a particular host plant.

Attractant A chemical that causes an insect to orient towards the source of odor.

Behavior Behavior is the action made by organisms in conjunction with themselves or their environment.

Biennial Plants A biennial plant is a *flowering plant* that takes 2 years to complete its biological lifecycle.

Biogenic Volatile Organic Compounds (BVOCs) Volatile organic compounds (VOC) of biological origin and can be emitted from different plant parts including

I. K. Singh, A. Singh (eds.), *Plant-Pest Interactions: From Molecular Mechanisms to Chemical Ecology*, https://doi.org/10.1007/978-981-15-2467-7

leaves, stem, roots, and flowers. They serve as pollinator attractants, herbivore, and pathogen repellents and protect plants from abiotic stress and therefore play many ecological functions. They include alkanes, alkenes, aldehydes, ketones, esters, and alcohols.

Bioinformatics A multidisciplinary science that uses computational tools to extract knowledge from biological data.

Biosafety Prevention of comprehensive loss of biological integrity, focusing on both ecology and human health.

Biosynthetic Pathways The process of combining the smaller molecules to form larger and more complex ones in living organisms.

Biotic Stress Biotic stress is stress that occurs as a result of damage done to an organism by other living organisms, such as bacteria, viruses, fungi, parasites, beneficial and harmful insects, weeds, and cultivated or native plants.

Biotrophic Pathogen Biotrophic pathogen feed on living host tissue.

Brassinosteroids A class of polyhydroxylated steroidal phytohormones in plants with similar structures to animals' steroid hormones. Brassinosteroids regulate a wide range of physiological processes including plant growth, development, and immunity.

Chemical Ecology A study that integrates chemistry and biology to examine the chemical interactions of organisms with their environment.

Chemical Warfare Chemical warfare involves using the toxic properties of chemical substances as weapons. Examples are mustard gas, nerve gas, phosgene, cyanogen chloride, hydrogen, cyanide etc.

Chromatin Immune Precipitation Sequencing Chromatin immunoprecipitation (ChIP) is a type of immunoprecipitation experimental technique used to investigate the interaction between proteins and DNA in the cell.

Co-evolutionary Arms Race a struggle between competing sets of co-evolving genes, traits, or species that develop adaptations and counter-adaptations against each other, resembling an arms race.

Conspecific Species Two or more individual organisms, populations, or taxa are conspecific if they belong to the same species.

CRISPR-CAS9 Technology A unique genome-editing tool that enables to manipulate the genome by editing the DNA sequence.

Cytoplasmic Incompatibility The phenomenon of mating incompatibility where males and/or females infected with certain bacteria (such as *Wolbachia*) on mating results in offspring that die in early embryogenesis.

Damage Associated Molecular Patterns (DAMPs) These are molecules released by a stressed cell to initiate immunity in the cell.

Deep Sequencing Next-generation sequencing approach of sequencing of a genomic region multiple times.

DELLA Protein DELLA proteins are a family of putative transcriptional regulators that inhibit cell proliferation and expansion that drives the growth of plant organs.

Direct Defense Means characters of plants which enhance the host plant resistance by providing mechanical or biochemical protection and affects the physiology as well as behavior of the insect pest.

Disease Vectors Any agent which carries and transmits an infectious pathogen into another living organism.

DNA Barcoding It is an advanced molecular taxonomic method that uses a short genetic marker in an organism's **DNA** to ascertain its belonging to a particular species.

DNA Methylation Addition of a methyl (-CH_3) group to DNA leading to epigenetic modification of gene function.

Drumming Striking of a body part, usually the abdomen, directly against the substrate.

Ecological Niche The position of a species within an ecosystem, describing both the range of conditions necessary for persistence of the species and its ecological role in the ecosystem.

Ecology The branch of biology that deals with the relations of organisms to one another and to their physical surroundings

Economic Injury Level The smallest number of insects which can cause yield losses or injury to crops equal to the insect management costs.

Ecosystem A biological community of interacting organisms and their physical environment.

Effector Triggered Immunity (ETI) A second line of defense response triggered by the effector produced by pathogen.

Effectors Effectors are proteins secreted by microbial pathogens which can either trigger or compromise immunity depending on the ability of perception and response of the plant.

Elicitor Pathogen-derived molecule that triggers defense response in the host cells.

Elicitors Molecules which trigger chemical defense in plants.

Endocytosis A process of up taking substances in the cell.

Endosymbionts Any organism living inside another organism that may or may not have mutualistic relationship with the host.

Entomopathogens Microorganisms (bacteria, fungi, oomycetes. or viruses) that are pathogenic to insect and are often used as a biocontrol strategy to control insect pest.

Entomotoxic Substances causing mortality to insects.

Extra Floral Nectar Extrafloral nectaries are nectar-secreting plant glands that develop outside of flowers and are not involved in pollination.

Fecundity The physiological maximum potential reproductive output of an individual (usually female) over its lifetime

Feeding Deterrent A chemical that inhibits feeding.

Floral Scent *Floral scent* or flower *scent* is composed of all the volatile organic compounds (VOCs) or aroma compounds emitted by *floral* tissue.

Fluorescent Two-Dimensional "Difference Gel Electrophoresis" Two-dimensional difference gel electrophoresis (2D DIGE) is a modified form of 2D electrophoresis (2DE) that allows one to compare two or three protein samples simultaneously on the same gel.

Gene Editing Genetic engineering technique in which DNA is modified in the genome of a living organism.

Gene Expression The process in which the genetic information encoded in a gene is translated into RNA or protein molecules.

Generalist Insects *Generalist* species is able to thrive in a wide variety of environmental conditions and can make use of a variety of different resources.

Genome An organism's complete set of DNA.

Glomeruli The antennal lobe is composed of densely packed neuropils, termed glomeruli, where the sensory neurons synapse with the two other kinds of neurons, the projection neurons and the local neurons.

Glutathionylation It is the posttranslational modification of protein cysteine residues by the addition of glutathione, the most abundant and important low-molecular-mass thiol found in cells.

GMO Organisms having unnatural alteration in genetic material (DNA).

Gravid The term applied to a female insect containing fully developed eggs ready to be laid.

Gustation The act of tasting or the faculty of taste.

Habitat A type of natural environment in which a particular species of organism lives.

Herbivore-Associated Molecular Pattern (HAMPs) Chemical signatures that initiate plant response upon insect attack, for example, fatty acid-amino acid conjugates (FACs), sulfur-containing fatty acids like inceptins, peptides like Glycolate oxidase, and inceptins.

Herbivore-Induced Plant Volatiles Herbivore-induced plant volatiles (HIPVs) are volatile compounds involved in plant communication with natural enemies of the insect herbivores, neighboring plants, and different parts of the damaged plant.

Herbivores A herbivore is an animal anatomically and physiologically adapted to eating plants for the main component of its diet.

Herbivorous Insects Insects feeding only on plants.

Herbivory Feeding of plant material by insects. There can be chewing, sap-sucking, wood boring, leaf-mining, feeding on fruits, spores, etc.

Holobiont The assembly of the host organism together with all other species living within or around it as an ecological unit. It usually refers to a comprehensive unit that includes a eukaryotic host together with its associated microbial communities.

Horizontal Gene Transfer Transmission of genetic material between different genomes across species boundaries except transmission from parent to its offspring.

Host Selection Acceptance or rejection of a host.

Hydrophilic A hydrophile is a molecule or other molecular entity that is attracted to water molecules and tends to be dissolved by water.

Hydrophobic The physical property of a molecule (known as a hydrophobe) that is seemingly repelled from a mass of water.

Indirect Defenses Involves the production of volatiles or it may provide shelter or food source like extra nectar that attracts natural enemies of insect pest.

Induced Resistance Temporarily increased resistance in plants due to changed conditions of plants or abiotic.

Insect Morphology The study and description of the physical form of insects.

Insect Physiology The study of physiology and biochemistry of insect organ systems.

IPM Sustainable strategy for long-term prevention of pests and their damage using combinations of eco-friendly approaches.

Isobarric Tagged for Relative and Absolute Quantitation An isobaric labelling method used in quantitative proteomics by tandem mass spectrometry to determine the amount of proteins from different sources in a single experiment.

Isotope-Coded Affinity Tags An in vitro labelling procedure that involves tagging of protein or peptide samples with the ICAT reagent specifically at their Cysteine residues. The ICAT reagent consists of a biotin tag, a light or heavy linker chain, and a Cys-reactive group.

Jasmonic Acid A phytohormone involved in the regulation of many physiological processes in plant growth and development, and especially the mediation of plant responses to biotic and abiotic stresses.

Kenyon Cells *Kenyon cells* are the intrinsic neurons of the mushroom body, a neuropil found in the brains of most arthropods and some annelids.

Lectin Proteins with non-immune origin, non-antibiotic in nature with binding specificity to carbohydrate moieties.

Mandibles Insect mandibles are a pair of appendages near the insect's mouth, and the most anterior of the three pairs of oral appendages.

Mass Spectrometry An analytical technique that measures the mass-to-charge ratio of ions.

Membrane Depolarization Loss of the difference in charge between the inside and outside of the plasma membrane of a cell due to a change in permeability and migration of sodium ions to the interior.

Metamorphosis A biological process by which an animal physically develops after birth or hatching, involving a conspicuous and relatively abrupt change in the animal's body structure through cell growth and differentiation.

Microarray A laboratory technique for studying expression of thousands of genes at the same time.

Microbial Associated Molecular Patterns Signature patterns conserved in microbes detected by pattern recognition receptors on plant cell membrane.

Microbiome The dynamic community of microorganisms associated with plants and soil. This community includes bacteria, archaea, and fungi and has the potential for both beneficial and harmful effects on plant growth and crop yield.

microRNAs Small, non-coding RNAs of ~22 nucleotide in length that are involved in regulation of gene expression.

miRBase A biological database of microRNA sequences.

Multidimensional Protein Identification Technology Gel-free technique for separating and identifying individual components of complex protein and peptide mixtures.

Mycorrhiza It is a symbiotic association between a plant and fungus.

Natural Selection The process whereby organisms better adapted to their environment tend to survive and produce more offspring. The theory of its action was first fully expounded by Charles Darwin, and it is now regarded as the main process that brings about evolution.

Necrotrophic Pathogens Necrotrophic pathogens actively kill host tissue as they colonize and thrive on the contents of dead or dying cells.

Neonates Larvae less than 4 hours old.

Next Generation Sequencing Modern sequencing techniques that enable deep, high through-put and parallel sequencing of multiple DNA samples.

Non-Host Volatiles Volatiles released from a non-host source, e.g., non-host plants.

Northern Blotting A laboratory technique for studying gene expression by detecting a specific RNA in a pool of RNAs.

Nucleic Acid Hybridization A molecular biology technique wherein single stranded DNA or RNA molecule base pairs with complementary sequence.

Odor Plume Plumes form when turbulent wind or water currents disperse odor molecules from their source.

Olfaction Sense of smell for detection of volatile chemicals.

Olfactory Binding Proteins Low-molecular-weight soluble proteins highly concentrated in the nasal mucus of vertebrates and in the sensillum lymph of insects.

Olfactory System The *olfactory system*, or sense of smell, is the sensory *system* used for *olfaction*.

ORYSATA A mannose-specific jacalin-related lectin, was the first inducible lectin from the salt-treated rice seedling.

Osmotic Stress Osmotic stress is a sudden change in the solute concentration around a cell, causing a rapid change in the movement of water across its cell membrane.

Oviposition Deterrents A chemical that inhibits oviposition.

Oviposition Process of egg laying.

Oxylipins Oxylipins constitute a family of oxygenated natural products which are formed from fatty acids by pathways involving at least one step of dioxygen-dependent oxidation.

PAMP-Triggered Immunity (PTI) The preliminary immune response of a plant involving recognition of pathogen/microbe associated molecular patterns (PAMPs/ MAMP) by the pattern recognition receptors (PRRs) to prevent further pathogen establishment.

Parasitoids A parasitoid is an organism that lives in close association with its host and eventually kills the host.

Pattern Recognition Receptors They are present on the surface of a plant cell to analyze for molecules containing signature patterns conserved in microbes known as pathogen-/microbe-associated molecular patterns (PAMPs/MAMPs).

Peritrophic Membrane Tubular chitinous covering inside the midgut of many insects.

Pheromones These are chemical compounds released in response to communication between members of the same species. There are several categories of such chemical compounds such as sex pheromone, alarm pheromone, etc., that affect the behavior or physiology of the organism.

Phloem Feeders They are phloem-feeding insects that obtain phloem sap by inserting their stylets into the phloem. Phloem turgor pressure drives sap into the insect's gut.

Phosphorylation A biochemical process that involves the addition of phosphate to an organic compound.

Phosphorylation Cascades A sequence of events where one enzyme phosphorylates another, causing a chain reaction leading to the phosphorylation of thousands of proteins.

Phytoalexins Phytoalexins are antimicrobial and often antioxidative substances synthesized de novo by plants that accumulate rapidly at areas of pathogen infection. They are broad-spectrum inhibitors and are chemically diverse with different types characteristic of particular plant species.

Phytoanticipins Low molecular weight, antimicrobial compounds that are present in plants before challenge by microorganisms or are produced after infection solely from preexisting constituents.

Phytohormones Plant hormones are signal molecules produced within plants, which occur in extremely low concentrations. Plant hormones control all aspects of growth and development, from embryogenesis, the regulation of organ size, pathogen defense, and stress tolerance to reproductive development.

Plant Signaling Conveying information within and between plant cells from receptor systems to effectors. Signals can take many forms, including chemical and electrical, and signaling can occur locally within a single plant or between different plants, including plants of different species.

Plant Volatiles Plant volatiles are the metabolites that plants release into the air.

Pollination The act of transferring pollen grains from the male anther of a flower to the female stigma.

Polyphagous Insects Insects that feed on a wide range of plants under different plant families.

Polyphenols A compound containing more than one phenolic hydroxyl group.

Predators Predator is an organism that primarily obtains food by the killing and consuming of other organisms.

Priming Defenses Sensing environmental cues which are indicative of pathogens or herbivores, plants can "prime" appropriate defenses and deploy faster, stronger responses to subsequent attack.

Projection Neurons *Neurons* whose axons extend from the *neuronal* cell body within the central nervous system (CNS) to one or more distant regions of the CNS.

Protein Microarray A high-throughput method used to track the interactions and activities of proteins, and to determine their function, and determining function on a large scale. Its main advantage lies in the fact that large numbers of proteins can be tracked in parallel.

Protein–Protein Interaction It is the interaction between two or more proteins. They are the physical contacts of high specificity established between protein molecules as a result of biochemical events steered by electrostatic forces including the hydrophobic effect.

Proteome Set of protein sequences that can be derived by translation of all protein-coding genes of a completely sequenced genome, including alternative products such as splice variants for those species in which these may occur.

Push-Pull Strategy A strategy for controlling agricultural pests by using repellent "push" plants and trap "pull" plants.

qRT-PCR A laboratory technique that allows absolute and relative quantification of expression of a gene.

R Proteins R protein may detect a Pathogen-Associated Molecular Pattern or PAMP (alternatively called MAMP for microbe-associated molecular pattern). The R protein encodes enzyme that degrades a toxin produced by a pathogen.

Reactive Oxygen Species Partially reduced or excited forms of atmospheric oxygen, such as singlet oxygen ($^1O_2$), superoxide radical ($O_2^{\cdot-}$), hydrogen peroxide (H_2O_2), and hydroxyl radical ($OH^{\cdot}$), having different levels of reactivity, sites of production, and potential to cross biological membranes.

Receptor An organ or cell able to respond to light, heat, or other external stimulus and transmit a signal to a sensory nerve.

Receptor Potential A receptor potential, a type of graded potential, is the transmembrane potential difference produced by activation of a sensory receptor.

RNA Interference The process in which the gene expression is inhibited by a small RNA molecule that signals enzyme complexes to degrade mRNA, thereby preventing translation and in turn resulting in gene silencing.

RNAseq A technique that detects and quantifies the RNA molecules in a biological sample at a given point of time using next generation sequencing.

ROS Known as reactive oxygen species. ROS belongs to the group of many reactive species, free radicals, etc., derived from oxygen. Singlet oxygen, hydrogen peroxide, superoxide radical, and hydroxyl radical are the most common ROS.

Salicylic Acid A phenolic phytohormone that regulates many aspects of plant growth and development as well as thermogenesis and resistance to biotic and abiotic stress.

Sclerophylly Development of sclerenchyma in several desert plant leaves, resulting in thickened and hardened foliage that provide resistant to water loss.

Secondary Metabolites Distinct type of natural compound produced by specific groups of bacteria, fungi, plants, and animals and not directly involved in the normal growth, development, and reproduction of the organism. They are molecules which are primarily involved in the overall maintenance/homeostasis of the organism.

Secretome The set of proteins expressed by an organism and secreted into the extracellular space.

Senescence *Senescence* or biological aging is the gradual deterioration of functional characteristics.

Sensillum Lymph The fluid secreted by the trichogen and tormogen cells.

Sexual Dimorphism A condition in which two sexes of the same species exhibit different characteristics beyond the differences in their sexual organs.

Signal Transduction The process of transferring a *signal* throughout an organism, especially across or through a cell.

Single Sensillum Recording The process of recording the responses of odorant receptor neurons, which is an effective method for screening active ligands on specific odorant receptors.

Small RNA Short, non-coding RNA molecules that are regulators of gene expression.

Specialist Insects Specialist insect species can thrive only in a narrow range of environmental conditions or has a limited diet.

Stylet A stylet is a hard, sharp, anatomical structure found in some invertebrates

Suberin A polymer which forms a water-impermeable layer in the casparian strips of roots.

Symbiosis Island These are mobile, integrative, conjugative elements that carry genes that enable them to expunge from the chromosome to form closed circular molecule that eventually conjugate and recombine into recipient chromosomes through horizontal gene transfer.

Thiol-Disulfide Exchange It is a chemical reaction in which a thiolate group $-S^-$ attacks a sulfur atom of a disulfide bond -S-S-. The original disulfide bond is broken and another sulfur atom gets released as a new thiolate, carrying away the negative charge.

Thioredoxins TRX are small proteins characterized by active site sequence-Trp-Cys-Gly-Pro-Cys-with a redox active disulfide bridge.

Transcription Factors Transcription factors are proteins involved in the process of transcribing DNA into RNA.

Transcriptional Reprogramming Refers to the phenomenon of global changes in gene expression that are typically initiated by transcription factors. During transcriptional reprogramming, the expression of specific genes is elevated, whereas other genes are repressed, compared to the previous state.

Transcriptome Complete set of transcripts in a cell, and their quantity, for a specific developmental stage or physiological condition.

Transducer A device which converts one form of energy to another, for example, plants converts solar energy to chemical energy.

Transgenerational Effects It can bring modification in phenotypes of offspring and therefore plays a significant role in the ecology and evolution of various plant species.

Transgenerational Immunity Refers to the transfer of the parental immunological experience to its progeny. This may result in offspring protection from repeated encounters with pathogens that persist across generations.

Transgenic Plant Plant whose DNA is modified using genetic engineering technique.

Transmembrane Potential Membrane potential is the difference in electric potential between the interior and the exterior of a biological cell. With respect to the exterior of the cell, typical values of membrane potential, normally given in units of millivolts and denoted as mV, ranges from –40 mV to –80 mV.

Trichome A small unicellular and glandular outgrowth from the epidermis or appendages on plants, algae, lichens, and certain protists.

Tritrophic Interaction It is the interaction of three trophic levels, as in plant–herbivore–predator.

Two Dimensional Polyacrylamide Gel Electrophoresis A form of gel electropho resis commonly used to analyze proteins. Mixtures of proteins are separated on the basis of their individual isoelectric point and molecular mass.

Ubiquitination A process where a protein, ubiquitin, is inactivated by attaching ubiquitin to it. Ubiquitin is a small molecule. It acts as a tag that signals the protein-transport machinery to ferry the protein to the proteasome for degradation.

Unwanted Plant An undesirable *plant or a plant at a wrong place*. For example, a weed is undesirable plant in a particular situation.

Volatile Organic Compounds Organic gases, from artificial or natural sources, which can propagate a greater or lesser distance from their place of emission in the atmosphere and can result in direct or indirect effects on organisms.

Xenobiotic Detoxification The body removes xenobiotics by xenobiotic metabolism. This consists of the deactivation and the excretion of xenobiotics, and happens mostly in the liver. Excretion routes are urine, feces, breath, and sweat.

Yeast Two-Hybrid System *Two-hybrid screening* (originally known as ***yeast two-hybrid system*** or Y2H) is a molecular biology technique used to discover protein–protein interactions and protein–DNA interactions by testing for physical interactions between ***two*** proteins or a single protein and a DNA molecule, respectively.

The manufacturer's authorised representative in the EU is Springer Nature Customer Service Centre GmbH, Europaplatz 3, 69115 Heidelberg, Germany. If you have any concerns regarding our products, please contact ProductSafety@springernature.com

Printed and bound by CPI Group (UK) Ltd, Croydon, CR0 4YY
30/09/2026
02225248-0016